A Profile of Economic Plants

By

John C. Roecklein

PingSun Leung

Transaction Books
New Brunswick (U.S.A.) and Oxford (U.K.)

Copyright © 1987 by Transaction, Inc.
New Brunswick, New Jersey 08903

Library of Congress Catalog Number: 87-6040

ISBN: 0-88738-167-7

Printed in the United States of America

Library of Congress Cataloging in Publication Data

A Profile of Economic Plants

  Bibliography: p.
  Includes index.
  1. Crops. I. Roecklein, John C. II. Leung, PingSun, 1952-
SB91.P76   1987   631   87-6040
ISBN: 0-88738-167-7

v

CONTENTS

Acknowledgments

The authors are grateful to faculty members in the College of Tropical Agriculture and Human Resources who provided encouragement and helpful suggestions as the profile developed; to both Rebecca Powell and Lila Gardner, who collected the bulk of the data, verified sources, and entered all of the information into the computer; to Connie Tsai and Helene T. Chun, who developed the computer programs necessary to format the document; to Beth Miller, who diligently proof-read all material and provided overall coordination; and to Dr. Frank S. Scott, Jr., whose review and constructive comments have greatly enhanced the profile.

The Authors

John C. Roecklein is a junior researcher, Department of Agricultural and Resource Economics, College of Tropical Agriculture and Human Resources, University of Hawaii.

PingSun Leung is an associate professor in the same department.

Introduction

Here in Hawaii, as in many states throughout the nation, tremendous energy and resources are devoted to the identification, testing, and development of alternative crops. The recognition that agricultural diversification is necessary goes back at least five decades[1] and has been formally enunciated as a State governmental objective in the State Plan.[2] Alternative crop research has been ongoing in both Hawaii's private and public sectors with successes and failures along the way. Among the successes are the development of the papaya, macadamia nut, and nursery stock industries. However, a recent report by the Council for Agricultural Science and Technology (CAST) has characterized the new crop development efforts in the United States as costly, time consuming, and haphazard.[3] An internal review, by the authors, of the new crop research conducted in Hawaii suggests that local experiences do not differ from the CAST characterization.

The U.S. Department of Agriculture has funded a project with the goal of expediting the process of screening and identifying promising crops for Hawaiian agriculture. Work was divided among two teams, with the authors making up the team working on the socioeconomic aspects of new crop development. Two major socioeconomic objectives identified for this phase of the project are to develop concurrently a crop-screening methodology and a crop-economic database. A precursor to the development of the screening methodology is to identify relevant techniques, issues, and problems via a literature search. This task is completed and documented in a Research Extension Series report.[4]

A second required step was to define the scope of the inquiry into new crops. This publication identifies the crops included, at this time, as a possible solution set. In other words, the crops found in this set of profiles constitute the project's research universe before any economic or physical screening. It is felt that publication of the raw, unscreened list of what is thought to be "possible" crops will be valuable because agriculturalists will be able to use this information, which is mainly economic in nature, to suit their own particular sets of economic and physical conditions. The rationale for this is twofold. By necessity a screening model for the State of Hawaii will be aggregated and possibly overlook options for individuals with unique environments, skills, or resources. Second, individuals interested in a particular crop may develop specific knowledge concerning particular crop species, processing methods, and markets, allowing a more accurate assessment than can be done by the team. A final objective in publication is to illustrate the complexity of new crop research by tabulating the multiple uses among the various crops.

Research Objectives

One of the principal findings of the literature review is that candidate alternative crops are often selected on the basis of researcher preference, government mandate, historical importance, and a host of other unsupportable factors. The scope of inquiry is often limited, and there is no reason for one to assume that crops considered have a better chance of evolving into agricultural industries than those not considered.

A major element incorporated in the design criteria of the screening methodology is that the process provides a systematic, all-inclusive approach to the alternative crop search. The screening process will go from "many" crops to "few," employing economic and physical characteristics of the crops and their products to "de-select." But which crops should be considered? The question of scope may be answered only in light of the screening objective. Hawaii's varied set of "agro-climatic" environments means that a tremendous number of crops can physically be grown in the State. Having the ability to grow a crop is not, however, sufficient for an industry to evolve. Many additional conditions must be met. Paramount among them is the condition wherein a crop can be transformed into a product(s) people want to buy at a price they are willing to pay, leaving an acceptable profit for the individuals involved, from the retailer/wholesaler back through the marketing chain to the farmer. Thus the objective of a crop-screening methodology is to identify those crops that can be processed into products with positive profit potential in the Hawaii context. If the screening process is to be systematic and procedural, all crops that can possibly be grown in Hawaii should be placed on the original list so that they may be analyzed for profit potential.

Selection of Crops

This profile of crops includes 1,163 crops that have been or are presently of use to man somewhere on earth. Given the necessity not to prejudge or exclude crops without good reason, it is important at the outset to identify all crops that fit in the "possible" category. Developing a comprehensive list is no easy matter after moving beyond the 50 to 100 common crops used throughout the world. Fortunately, a list of 1,000 crops of economic importance was developed and published by J. D. Duke and E. E. Terrell[5] in 1974. Their list forms the basis of the crop list presented here. Tropical and subtropical timber species were not included on the Duke list. A publication by D. B. Webb et al.[6] provides information on 121 timber species included in the list. The 42 remaining crops were in neither of the above lists, but should be included.

One class of crops has been specifically excluded from the crop list-- nursery and cut flowers. The reason for this is logistical, not methodological. *Exotica 4*, by A. B. Graf, a well-recognized ornamental reference, has 13,900 entries.[7] Collection of data concerning uses, prices, and markets for such a large number of plants was deemed too resource demanding, especially from a time standpoint. The methodology, in its present form, will handle ornamentals. They could be included at a later date, resources permitting.

Errors and Omissions

Information published in this crop profile is primarily economic in nature. It is presented in good faith. All information is verified against the source. However, for many of the crops economic information is remarkably difficult to find and some data may prove to be dated or inaccurate.

Crop Profile Contents

Information contained in this crop profile is a subset of the data captured in an on-line database being developed to analyze these same crops. The following information is presented for each crop: (1) the common name or names of the crop; (2) the scientific name or names of the crop; (3) the geographic region or regions of origin or importance; (4) a listing of the general categories into which products derived from the crop might be placed; (5) an indication--yes or no--of whether the crop has been or is now cultivated; (6) number(s) identifying the relevant references found; and (7) a summary of the plant, its important parts, its end-uses or by-products, its areas of use or marketability, and limited data concerning general growth requirements.

A more detailed explanation of each of the seven fields follows.

Common Name(s).

Many names for crops were uncovered in searching through reference books. Only the most widely known of the common names are included in this field; i.e., all of the local names for widely dispersed crops have not been included here. If the crop is limited to an area that prescribes a local name, this name appears. The common names used by Duke and Terrell are included for the first 1,000 crops. Common names for timber species are taken from Webb et al.

Scientific Name(s).

Considerable uncertainty exists with respect to scientific names in the agricultural/botanical community. The handbook entitled *A Checklist of 3,000 Vascular Plants of Economic Importance* by E. E. Terrell, USDA (Washington, D.C., Agriculture Handbook No. 505, 1977), was used as the final arbitrator whenever discrepancies occurred.

Geographic Regions.

The following general geographic regions have been used:

1. Africa
2. Africa-Central
3. Africa-East
4. Africa-North
5. Africa-South
6. Africa-West
7. America
8. America-Central
9. America-North (U.S. and Canada)
10. America-South
11. Asia
12. Asia-Central
13. Asia-China
14. Asia-India (subcontinent)
15. Asia-Southeast
16. Australia
17. Eurasia
18. Europe
19. Hawaii
20. Mediterranean
21. Pacific Islands
22. Subtropical
23. Temperate
24. Tropical
25. West Indies
26. World-wide

These regions are developed as general locations to indicate either where the crop is being grown or its location of origin. By referring to the summary, the reader will know which type of geographical location is represented.

End-Use Categories.

The following end-use categories have been identified:

1. Beverage	14. Medicinal
2. Cereal	15. Natural Resin
3. Dye & Tannin	16. Nut
4. Energy	17. Oil
5. Erosion Control	18. Ornamental & Lawn Grass
6. Essential Oil	19. Spice & Flavoring
7. Fat & Wax	20. Sweetener
8. Fiber	21. Timber
9. Forage, Pasture & Feed Grain	22. Tuber
10. Fruit	23. Vegetable
11. Gum & Starch	24. Weed
12. Insecticide	25. Miscellaneous
13. Isoprenoid Resin & Rubber	

These categories provide a logical grouping of the products derived from the various crops. It is possible to subdivide several of these categories into separate smaller groups, but this was thought to be of marginal value. Please note the inclusion of ornamental as an end-use category. In spite of the decision to exclude them, 190 crops, primarily from the Duke list, do appear as ornamentals. They are included because 189 of those crops have other uses, and this allows verification of the ability to handle the ornamental plants from a methodological standpoint.

Cultivation.

The question of cultivation is akin to domestication and often hard to determine. A "yes" or "no" response is provided based on information from the sources listed in Index C. In cases of uncertainty, a "no" is used. Cultivation implies that the plant is controlled by man. Plants not planted by man, yet utilized in some manner, are not considered cultivated.

Citation Numbers.

The citation numbers found in this field are pointers that identify complete citations for the crop. A complete bibliographic listing of all the sources used is found in Index C of this publication.

Summary.

The crop summary is intended to give the reader a quick explanation of the crop. An effort was made to include the following economic information: important part(s) of the plant; product(s) derived from the plant; geographic region(s) where the products are used; regions where the plant is grown; general ranges of temperature, rainfall, and soil conditions necessary to produce the crop; and, for the more common crops, recent information on total production, yield, and value of production for the world, the United States, and Hawaii, as appropriate.

Statistical information is taken from the following:

1. Worldwide data: *FAO Production Yearbook*, Food and Agriculture Organization of the United Nations, Rome, FAO Statistics Series, No. 55, Vol. 37, 1984.
                               FAO Trade Yearbook, Food and Agriculture Organization of the United Nations, Rome, FAO Statistics Series, No. 57, Vol. 37, 1984.

2. U.S. data: *Agricultural Statistics 1984*, USDA, U.S. Government Printing Office, Washington, D.C. 1984.

3. Hawaii data: *Statistics of Hawaiian Agriculture*, Hawaii Agricultural Reporting Service, Honolulu, 1983.

Abbreviations Used

Within each summary, a standard set of abbreviations was developed for expediency and clarity. For unit of weight and measurement, the metric system was used. Below are common units and their abbreviations:

metric ton	= MT	meter	= m
hectare	= ha	centimeter	= cm
kilogram	= kg	decimeter	= dm
cubic meter	= cu.m	millimeter	= mm

Other abbreviations commonly used include percent (%), dollars ($), per (/) and numbers (1, 2, 3,. . .). Yield information found in most summaries typifies the use of abbreviations, e.g., 175 kg/ha/annum, or 175 kilograms per hectare per annum.

Crop Profile Organization

The crop profile consists of one major listing and a set of three related indexes. Crops by Principal End-Use, comprises the bulk of the information. For each crop, the prominent or major use was identified. A particular profile is found in the end-use category that is its principal use. The seven data elements or fields identified in an earlier section are given for each crop. All crops with a product or products belonging in that category are given in alphabetical order based on the common name. If a crop has multiple uses, it is listed in each appropriate category with a pointer to the category containing the profile. For example, consider the hypothetical case in which a crop may be used in the manufacture of both a cereal and a beverage. If the crop is considered primarily a beverage, the crop profile will appear, alphabetized by common name, in the beverage category and a "See Beverage" message will appear under the crop name in the cereal section.

To facilitate the looking up of crop information, three indexes are provided. Index A is an alphabetical listing of all common names used for all crops in the profile. A person with only the common name of a crop may quickly scan this list to identify the principal end-use category, the page in that section on which to look, and the common name under which that profile is found.

If only the scientific name is known, the same process may be followed using Index B, which is an alphabetized listing of crops by their scientific names.

A number of excellent reference sources were uncovered through the extensive research involved in the compilation of this crop profile. A third index, Index C, includes the bibliographic references. It contains all of the sources that have been used to generate the information contained in the major listing. It is ordered by reference number to allow quick access. The reference numbers correspond to those numbers found in field 6 of the listing by principal end-use. If, for example, a crop in that list contains numbers 45 and 179 as sources, by referring to this index, one can easily identify the books and journal articles that are relevant.

The Multiplicity of Uses for Crops

Figure 1 summarizes crop relationships among end-use categories. Perhaps it provides a

	Beverage	Cereal	Dye	Energy	Erosion	Essential Oil	Fat & Wax	Fiber	Forage	Fruit
Beverage	138									
Cereal	21	94								
Dye & Tannin	18	4	92							
Energy	8	5	13	84						
Erosion Control	4	10	17	25	132					
Essential Oil	25	0	7	6	6	137				
Fat & Wax	2	1	5	3	0	0	20			
Fiber	14	8	9	3	11	3	2	86		
For., Past., & Feed	36	58	23	18	82	5	5	30	331	
Fruit	62	3	23	6	1	8	3	14	22	202
Gum & Starch	14	11	11	8	6	7	2	10	27	12
Insecticide	4	0	3	1	2	10	0	0	3	3
Isop. Resin & Rubber	0	0	1	0	0	0	1	2	0	1
Medicinal	39	6	35	8	9	74	2	17	29	40
Natural Resin	1	0	2	5	2	9	1	1	1	1
Nut	6	4	8	3	2	4	3	2	7	4
Oil	19	13	18	11	3	7	8	23	41	28
Orna. & Lawn Grass	20	7	20	17	35	27	8	15	57	34
Spice & Flavoring	29	13	4	4	6	86	0	6	17	17
Sweetener	9	3	3	1	1	1	1	8	9	7
Timber	19	4	30	69	29	20	3	10	24	27
Tuber	5	4	0	1	3	0	0	1	8	3
Vegetable	24	27	10	2	13	12	2	15	51	21
Weed	5	5	2	0	8	1	0	8	23	2
Miscellaneous	22	13	15	12	4	16	3	16	34	23
(Row)										
# of single use crops (1)	4	11	4	0	3	3	6	8	77	47
# of mult. use crops (2)	134	83	88	84	129	134	14	78	254	155
No. crops in category (3)	138	94	92	84	132	137	20	86	331	202
Total No. of uses (4)	544	314	373	313	411	471	75	314	941	567
Ratio of No. of Uses/ (5)	3.9	3.3	4.1	3.7	3.1	3.4	3.8	3.7	2.8	2.8
No. of Crops										

Figure 1. Multiple Use Among Selected Crops.

partial explanation of why new crop development efforts are costly and time consuming as reported in the introduction. Information is found in two sections--top and bottom. The upper section illustrates the interrelationships among end-uses and the lower section summarizes the data.

In the upper section, the number in the diagonal element(top number in each column, in this case) represents the NUMBER OF CROPS that have a product or products falling into the category found on the far left. Any number below, but in the same column, indicates how many of those same crops also have a product or products in the end-use category to its left. For example, there are 94 crops identified as having product(s) related to Cereals. Of those 94, 4 have products in the Dyes and Tannins group, 5 are in the Energy category, 10 may be found in

Gum	Insecticide	Rubber	Medicinal	Natural Resin	Nut	Oil	Ornamental	Spice	Sweetener	Timber	Tuber	Vegetable	Weed	Miscellaneous
63														
2	28													
1	0	20												
15	11	2	262											
2	1	1	12	45										
4	0	0	9	1	57									
11	1	2	25	4	22	123								
4	5	2	51	5	8	16	190							
7	6	0	84	5	3	19	30	176						
5	0	0	6	0	3	6	3	5	27					
10	3	1	27	26	15	20	32	9	4	209				
14	2	0	6	0	0	1	4	4	2	0	43			
13	2	1	42	2	5	27	22	39	5	6	13	181		
3	0	0	9	1	0	0	10	5	2	1	5	12	48	
11	6	2	36	2	14	19	26	15	13	30	4	17	7	117
3	5	11	31	1	8	8	1	8	3	37	12	30	5	1
60	23	9	231	44	49	115	189	168	24	172	31	151	43	116
63	28	20	262	45	57	123	190	176	27	209	43	181	48	117
273	93	37	856	130	184	467	648	589	124	628	123	564	157	477
4.3	3.3	1.9	3.3	2.9	3.2	3.8	3.4	3.3	4.6	3.0	2.9	3.1	3.3	4.1

Erosion Control, and so on.

The lower section provides a summary. For each column, the value in Row 3 is the sum of the values in Rows 1 and 2. The value in Row 4 is the sum of the off-diagonal elements from the upper section. The value in Row 5 is the Row 4 value divided by the Row 3 value.

The figure illustrates two factors of significance. First, the number of crops with single uses are few. Economic analysis becomes complex, and therefore more time consuming, when dealing with multiple products, and joint costs.

Second, grouping by end-use category is not likely to provide a means of identifying end-use groups on which to focus. The ratios on the last line range from 1.9 to 4.6 but there is no way to identify the relative importance of any one category or the degree of difficulty implicit in dealing with a particular group.

References

[1] Krauss, F. G. "Balancing Hawaii's Agriculture," *Paradise of the Pacific*, December 1933.

[2] State of Hawaii. *Hawaii Revised Statutes*, Chapter 226-7.

[3] Council for Agricultural Science and Technology (CAST). "Development of New Crops: Needs, Procedures, Strategies, and Options," Report No. 102, Ames, Iowa, 1984.

[4] Roecklein, J. C., P. S. Leung, and John W. Malone, Jr. "Alternative Crops For Hawaii: A Bibliography of Methodologies for Screening," *Research Extension Series 062*, University of Hawaii, College of Tropical Agriculture and Human Resources, 1985.

[5] Duke, J. A. and E. E. Terrell. "Crop Diversification Matrix: Introduction," *Taxon* 23(5/6):759-799, 1974.

[6] Webb, D. B., Peter J. Wood, and Julie Smith. "A Guide to Species Selection for Tropical and Sub-Tropical Plantations." Tropical Forestry Paper No. 15, Department of Forestry, Commonwealth Forestry Institute, University of Oxford, 1980.

[7] Graf, A. B. *Exotica International Pictoral Cyclopedia of Exotic Plants from the Tropical and Near Tropic Regions*, Volume 1 & 2, Series 4, East Rutherford, New Jersey: Reohrs Company, 1982.

CROPS BY

PRINCIPAL END-USE

Beverage

Abata Kola
(See Nut)

African Locust Bean
(See Oil)

African Oil Palm, Oil Palm
(See Oil)

Apple
(See Fruit)

Arabica Coffee, Coffee

Sci. Name(s)	: *Coffea arabica*
Geog. Reg(s)	: Africa, America-North (U.S. and Canada), America-South, Asia-India (subcontinent), Asia-Southeast, Pacific Islands
End-Use(s)	: Beverage, Forage, Pasture & Feed Grain, Medicinal, Miscellaneous
Domesticated	: Y
Ref. Code(s)	: 170
Summary	: The arabica coffee tree (*Coffea arabica*) provides economically important coffee beans. The beans are produced mainly in Africa, Brazil, Colombia, and other Latin American countries. The United States, West Germany, and France are the main coffee importers. Coffee production in the United States is limited primarily to Hawaii. In 1983, the United States produced 1,000 MT of green coffee beans. In the same year, the United States imported 998,108 MT of green and roasted coffee beans at a value of $2.6 billion. Statistics are for all species of coffee, combined.

Other parts of the tree are used; the bark is made into pulp, parchment, manure, and mulch. In India, the pulp is occasionally fed to cattle. In Indonesia and Malaysia, the leaves are dried and made into infusions, serving various unspecified medicinal purposes.

Because coffee trees are an upland species, ideal growth conditions are along the equator at elevations of 1,500-2,100 m. In the subtropics, coffee is grown successfully at sea level. Trees begin bearing 3-4 years after planting. Arabica coffee is native to Africa.

Asparagus, Garden Asparagus
(See Vegetable)

Australian Desert Lime
(See Fruit)

Banana Passion Fruit, Curuba, Banana Fruit
(See Fruit)

Baobab, Monkey-Bread
(See Vegetable)

Barbados Cherry, Acerola, West Indian-Cherry, Barbados-Cherry, West Indian Cherry
(See Fruit)

Barley
(See Cereal)

Barley, Bulbous Barley
(See Cereal)

Bay, Bay-Rum-Tree
(See Essential Oil)

Bengal Coffee, Coffee
Sci. Name(s) : *Coffea bengalensis*
Geog. Reg(s) : Asia-India (subcontinent)
End-Use(s) : Beverage
Domesticated : Y
Ref. Code(s) : 79, 170
Summary : The Bengal coffee tree (*Coffea bengalensis*) is only occasionally cultivated for its coffee beans. The beans are small and therefore are considered economically inferior to larger coffee beans.

Bengal coffee grows wild in Bengal, Burma, and Sumatra. It is a smooth branched shrub with large, white, fragrant flowers. The fruit blackens when ripe. There is some cultivation in parts of Central and southern India and in the tropical areas of the Himalayas.

Bignay, Chinese Laurel, Salamander Tree
(See Fruit)

Billion Dollar Grass, Japanese Barnyard Millet, Japanese Millet
(See Cereal)

Black Currant, European Black Currant
(See Fruit)

Black Mulberry
(See Fruit)

Borage
(See Ornamental & Lawn Grass)

Camel's-Foot, Gemsbok Bean
Sci. Name(s) : *Bauhinia esculenta*

Geog. Reg(s) : Africa
End-Use(s) : Beverage, Fruit
Domesticated : N
Ref. Code(s) : 158
Summary : Camel's-foot is a woody shrub found in tropical Africa. It bears fruit pods that are sometimes eaten, but are most often pounded and boiled to make a beverage popular in South Africa.

Carob, St. John's-Bread
(See Gum & Starch)

Carrot, Queen Anne's Lace
(See Vegetable)

Cashew
(See Nut)

Cassava, Manioc, Tapioca-Plant, Yuca, Mandioca, Guacomole
(See Energy)

Cherimoya
(See Fruit)

Chia, Ghia
(See Essential Oil)

Chicory
Sci. Name(s) : *Cichorium intybus*
Geog. Reg(s) : Africa-North, America-North (U.S. and Canada), Eurasia, Europe
End-Use(s) : Beverage, Forage, Pasture & Feed Grain, Medicinal, Sweetener, Vegetable
Domesticated : Y
Ref. Code(s) : 170, 194
Summary : Chicory is a perennial herb native to Europe, North Africa, and western Asia. Certain types of the plant are grown as salad plants and for greens. These plants have large leaves that are blanched to reduce their bitterness. Chicory also has a stout tap-root that contains inulin, vitamins A and C, chicoric acid, esculitin, esculin, and other bitter compounds. Because of these characteristics, the root is roasted and ground and used as a coffee bean substitute. Cultivars with larger roots have been developed for this purpose.

The plant is also a potential source of fructose for the flavor industry and is a source of the natural sweetener, maltol. Some of its extracts are used in alcoholic and nonalcoholic beverages. The root has been used medicinally as a digestive aid, diuretic, tonic, laxative, and mild sedative. Chicory plants are also grown for forage.

Chicory reaches about 1-2 m in height and has bright blue flowers. It is grown commercially in Belgium, Holland, France, and the United States. Chicory will not produce well in hot tropical climates.

Chufa, Ground Almond, Tigernut, Yellow Nutsedge
(See Forage, Pasture & Feed Grain)

Cimarrona, Mountain Soursop
(See Fruit)

Clary, Clary Wort, Clary Sage
(See Essential Oil)

Cocoa, Cacao

Sci. Name(s)	: *Theobroma cacao*
Geog. Reg(s)	: America-South, Tropical
End-Use(s)	: Beverage, Forage, Pasture & Feed Grain, Spice & Flavoring
Domesticated	: Y
Ref. Code(s)	: 132, 170
Summary	: Cocoa (*Theobroma cacao*) is a tree growing to 10 m in height, that produces the commercially important cocoa beans. These beans are processed to make chocolate, cocoa butter, and milk chocolate. By-products from processed beans include cocoa powder and an expressed fat that can be made into chocolate or cocoa butter. Milk chocolate is made by adding powdered milk. Chocolate has many culinary uses and is a very popular snack food. Cocoa butter is used in cosmetics and for certain pharmaceutical preparations. Cocoa shells are used as stock feed or manure. Shells are a source of theobromine, shell fat, and vitamin D.

Trees are native to the lower eastern equatorial slopes of the Andes. They are cultivated in other parts of South America and the tropics for their cocoa beans. Trees are tropical in nature and will continue to produce in dense shade or under considerable exposure. They thrive on well-drained, nutrient-rich soil from sea level to altitudes approaching 500 m. First harvests may be made after 3-4 years. Yields have been recorded at 275-1,350 kg/ha/year of dry cocoa. Trees are sometimes interplanted with bananas, *Hevea* rubber, oil palms, and coconuts.

In 1984, world production of cocoa beans was 1.5 million MT. The major exporting countries were the Ivory Coast, with 3.9% of the world production, Ghana with 10.9% and Brazil with 14.4%. The United States imported 193,942 MT cocoa beans, valued at $412 million. Cocoa products are processed and consumed mainly in temperate countries.

Coconut
(See Oil)

Coltsfoot
(See Medicinal)

Common Fig, Adriatic Fig
(See Fruit)

Common Guava, Lemon Guava
(See Fruit)

Congo Coffee, Coffee

Sci. Name(s)	: *Coffea congensis*
Geog. Reg(s)	: Africa, Asia-Southeast
End-Use(s)	: Beverage
Domesticated	: Y
Ref. Code(s)	: 79, 170
Summary	: The Congo coffee tree (*Coffea congensis*) is native to Zaire. This coffee is similar to arabica coffee (*Coffea arabica*) and has been crossbred with robusta coffee (*Coffea canephora*) to produce a local beverage, 'congusta' coffee, which is popular in Java. It is a stunted shrub that thrives in areas with a high water table.

Cow-Tree
(See Gum & Starch)

Cowpea, Southern Pea, Black-Eyed Pea, Crowder Pea
(See Forage, Pasture & Feed Grain)

Cranberry, Large Cranberry, American Cranberry
(See Fruit)

Cupuacu
(See Fruit)

Dandelion, Common Dandelion

Sci. Name(s)	: *Taraxacum officinale*
Geog. Reg(s)	: Temperate
End-Use(s)	: Beverage, Medicinal, Natural Resin, Spice & Flavoring, Vegetable, Weed
Domesticated	: Y
Ref. Code(s)	: 194, 205, 220
Summary	: The dandelion is an herb whose thick taproot, leaves, and flowers are nutritious and high in vitamins A and C, and niacin. The root yields a bitter resin that contains taraxin. Taraxin is being researched as a possible stimulus for gastric secretions. Dried and ground roots are used in decaffeinated coffee type beverages, as a flavoring in coffee and cocoa, and as an added flavoring to salads. Tender leaves are often made into salads or used in soups. Dandelion flowers are used to make dandelion wine.

There are several horticultural varieties of dandelions available for commercial use. Plants can be grown in most soils and can easily become established as weeds. They thrive in areas with temperatures from 5-26 degrees Celsius, and an annual rainfall from 0.3-2.7 m. Dandelions are most common in temperate regions.

Date Palm
(See Fruit)

Dittany, Candle Plant, Gas-Plant
(See Medicinal)

Douglas-Fir
(See Timber)

Dyer's-Greenwood, Dyer's-Greenweed
(See Dye & Tannin)

Einkorn, One-Grained Wheat
(See Forage, Pasture & Feed Grain)

English Gooseberry, European Gooseberry
(See Fruit)

Eucalyptus
(See Timber)

European Strawberry, Strawberry
(See Fruit)

Field Horsetail
(See Medicinal)

Figleaf Gourd, Malabar Gourd
(See Oil)

Finger Millet, African Millet, Wimbi, Ragi
(See Cereal)

Fox Grape
(See Fruit)

Garden Rhubarb
(See Vegetable)

Gbanja Kola
(See Nut)

Giant Granadilla, Barbardine
(See Fruit)

Ginger
(See Spice & Flavoring)

Ginseng, American Ginseng
(See Medicinal)

Grapefruit
(See Fruit)

Guarana

Sci. Name(s)	: *Paullinia cupana*
Geog. Reg(s)	: America-South
End-Use(s)	: Beverage, Medicinal, Spice & Flavoring
Domesticated	: Y
Ref. Code(s)	: 31, 119, 170, 220
Summary	: Guarana is a woody, climbing vine that is cultivated in the Matta Grosso region of Brazil for its seeds, which have 3 times as much caffeine as coffee. The seeds are made into a paste used to produce a stimulating beverage. An alcoholic drink can also be made when mixed with cassava. Guarana seeds are also roasted, pounded, and used as a flavoring in bread. The seeds have been used to treat diarrhea.

Hog-Plum, Yellow Mombin, Jobo
(See Fruit)

Hops

Sci. Name(s)	: *Humulus lupulus*
Geog. Reg(s)	: America-North (U.S. and Canada), America-South, Asia, Europe, Temperate
End-Use(s)	: Beverage, Essential Oil, Forage, Pasture & Feed Grain, Medicinal, Vegetable
Domesticated	: Y
Ref. Code(s)	: 205, 220
Summary	: The hops vine produces cones that are dried and used by brewers in the production of beer. Leading producers are Germany, the United Kingdom, and the United States, which account for about 75% of beer production. There is also some commercial production in Chile.

In 1983 world production of hops was 123,000 MT from a total area of 88,000 ha. The United States produced 31,000 MT from an area of 15,000 ha, accounting for 25% of the world's production. The Federal Republic of Germany accounted for 30% of the world production.

The Federal Republic of Germany is also the principal hops exporter, supplying 48,104 MT of exports. The United States exported 8,961 MT of hops at a value of $66.2 million.

The cones serve other purposes. Remnants of the dried, boiled cones are used as fertilizer. An essential oil from the cones and flowers is used in perfumery. Parts of the plant have various medicinal qualities. Young leaves and shoots may be eaten as vegetables.

Hops vines are dioecious perennials reaching up to 6 m in height. They produce flowers that bloom from late summer to mid-autumn. Vines are native to Europe, Asia, and North America and can be grown in most temperate zones. They are found on damp, rich soils.

Hungry-Rice, Fonio, Fundi, Acha
(See Cereal)

Imbe

Sci. Name(s)	: *Garcinia livingstonei*
Geog. Reg(s)	: Africa-East

End-Use(s) : Beverage, Fruit, Gum & Starch, Medicinal, Miscellaneous, Timber
Domesticated : N
Ref. Code(s) : 36, 170, 220, 226
Summary : Imbe is a slow-growing tree that produces bright orange, edible fruits with slightly acidic, sweet pulp. The pulp is used in some countries to prepare a fermented beverage. A gum can be obtained from this tree. Extracts of the leaves and flowers have given positive antibiotic test results. Tree wood is yellowish-white in color and can be employed in various types of construction. It is susceptible to damage from borer pests. The slow growth rate of the tree renders it virtually uneconomical for any commercial exploitation. Trees are found in Africa, primarily in Tanzania and Zambia.

Indian Bael, Bael Fruit, Bengal Quince, Bilva, Siniphal, Bael Tree
(See Fruit)

Indian Jujube, Beri, Inu-Natsume, Ber Tree
(See Fruit)

Jerusalem-Artichoke, Girasole
(See Tuber)

Job's Tears, Adlay, Adlay Millet
(See Miscellaneous)

Kava, 'Awa
Sci. Name(s) : *Piper methysticum*
Geog. Reg(s) : Hawaii, Pacific Islands
End-Use(s) : Beverage, Medicinal
Domesticated : N
Ref. Code(s) : 154, 170
Summary : The kava plant is grown for its knotty, gray-green roots. The plant is native to the Pacific islands and has spread throughout Oceania. Kava roots have provided one of the most popular beverages in the Polynesian islands. To prepare the beverage, roots are pulped and fermented in water, and then are strained through a fiber. The resulting concoction has narcotic properties, and can act as a sedative, soporific, and hypnotic. The beverage is said to produce pleasant dreams and sensations, and when taken in large quantities, to induce sleep. Kava plants are closely connected with the social, political, and religious life of the Polynesians.

The active principle of the kawa root is a viscous resin. It contains methysticin and a glucoside, yangonin.

Leadtree, Lead Tree
(See Forage, Pasture & Feed Grain)

Lemon
(See Fruit)

Liberica Coffee, Liberian Coffee

Sci. Name(s)   : *Coffea liberica*
Geog. Reg(s)   : Africa-West, Asia, Asia-Southeast
End-Use(s)   : Beverage, Ornamental & Lawn Grass
Domesticated   : Y
Ref. Code(s)   : 79, 170
Summary   : The liberica coffee tree (*Coffea liberica*) has the least economic value of the cultivated Coffea species, as it provides less than 1% of the world's coffee beans. Trees are produced in Liberia and various Southeast Asian countries. There is a small amount of trade with some Scandinavian countries.

Because liberica coffee beans have poor liquoring quality, they are used as fillers in other coffees. The bitter flavored beans are popular in Eastern European countries.

The trees thrive in hot, wet, lowland forests and are usually raised from seed in nurseries. Current bean yield statistics are difficult to obtain but Malaysia has reportedly produced 672-896 kg/ha. Liberica coffee makes an attractive ornamental in tropical areas.

Lime
(See Fruit)

Mahua
(See Oil)

Mammy-Apple, Mammee-Apple
(See Fruit)

Mango
(See Fruit)

Morula
(See Fruit)

Muriti
(See Fiber)

Muscadine Grape, Southern Fox Grape
(See Fruit)

Naranjilla, Lulo
(See Fruit)

Negro-Coffee, Coffee Senna, Senna Coffee

Sci. Name(s)   : *Cassia occidentalis*
Geog. Reg(s)   : Africa-West, America-South, Tropical
End-Use(s)   : Beverage, Ornamental & Lawn Grass
Domesticated   : Y

Ref. Code(s) : 170, 225
Summary   : Negro-coffee, or coffee senna, is a tropical American shrub. Its seeds are used as a substitute for coffee in tropical America and West Africa. Some types are grown in the tropics as ornamental plants.

The shrub reaches from 0.6-1.5 m in height. It produces yellow flowers throughout most of the year.

Passion Fruit, Purple Granadilla

Sci. Name(s)   : *Passiflora edulis*
Geog. Reg(s)   : Africa-East, Africa-South, America-South, Australia, Hawaii, Pacific Islands, Tropical
End-Use(s)   : Beverage, Fruit, Spice & Flavoring
Domesticated   : Y
Ref. Code(s)   : 132, 170
Summary   : Passion fruit (*Passiflora edulis*) is a woody perennial vine that grows throughout the tropics and is cultivated for its sweet fruit. There are 2 main cultivated forms: *Passiflora edulis* forma *edulis* grows at higher-altitudes and produces purple fruit; *Passiflora edulis* forma *flavicarpa* is best suited to tropical lowlands and produces yellow fruit.

Ripe passion fruits are eaten, pulp and seeds, from the shell. The pulp is used for making jams and jellies and is processed into juice, nectars, and flavoring for use in some processed food products. Passion fruit pulp is a good source of vitamin A.

Passion fruit vines are cultivated commercially in Australia, South Africa, New Zealand, and Hawaii. For fruit production in Hawaii, a nitrogen, phosphorous, potassium mixture is often used at the rate of 0.9 kg/plant. Fruit harvest takes place after the first year and can continue for at least 3-5 years. In Kenya, average fruit yields have been recorded as 16,815 kg/ha/annum. The purple-fruited variety yields are much less than does the yellow-fruited variety. Vines grow in most temperate or warm tropical climates that are frost-free. Rainfall should be medium to heavy and well-distributed. Passion fruit vines grow in a variety of soils with good drainage. They are usually grown from seed or rooted cuttings.

Pataua, Seje Ungurahuay
(See Oil)

Peach Palm, Pejibaye
(See Fruit)

Pear, Common Pear
(See Fruit)

Pearl Millet, Bulrush Millet, Spiked Millet, Cat-tail Millet
(See Cereal)

Pepper Tree, Brazilian Pepper Tree, California Peppertree
(See Gum & Starch)

Peppermint
(See Essential Oil)

Phalsa
Sci. Name(s)   : *Grewia asiatica*
Geog. Reg(s)   : Asia-India (subcontinent)
End-Use(s)   : Beverage, Fiber, Fruit, Timber
Domesticated   : Y
Ref. Code(s)   : 154
Summary   : Phalsa is a medium-sized tree whose edible fruit pulp is used in drinks or sher-
bets. Phalsa wood is used in India for construction, and the bark is used for fiber.
When grown for fruit, phalsa trees must be cut down to the ground each year to pro-
vide for new shoots.

Pimento, Allspice
(See Spice & Flavoring)

Pineapple
(See Fruit)

Plum, Common Plum, European Plum, Garden Plum, Prune Plum
(See Fruit)

Pomegranate
(See Fruit)

**Proso Millet, Hog Millet, Common Millet, Brown Corn Millet, Broomcorn
Millet**
(See Cereal)

Pulasan
(See Fruit)

Quinoa
(See Cereal)

Rambai
(See Fruit)

Ramon, Breadnut
(See Fruit)

Red Currant, Northern Red Currant, White Currant, Garden Currant
(See Fruit)

Robusta Coffee
Sci. Name(s)   : *Coffea canephora*

Geog. Reg(s) : Africa
End-Use(s) : Beverage
Domesticated : Y
Ref. Code(s) : 79, 170
Summary : The robusta coffee tree (*Coffea canephora*) is the second most important member of the Coffea genus. The beans are increasingly used in the manufacture of popular instant coffee and are threatening the arabica coffee bean market. Although robusta beans lack the distinct coffee taste and aroma of arabica coffee, the beans are economical to produce, resulting in lower market prices that undercut the more expensive arabica beans. Separate coffee production figures are not available.

Most of the robusta coffee beans are produced in Africa. Plants are started from seed. Flowering occurs 9-10 months later and bean harvest immediately follows. Robusta trees require little care. The average yield is about 1,000-1,252 kg/ha of coffee beans.

It is a smooth branched shrub or tree with large broad leaves and white flowers. Robusta is most successful in areas receiving adequate rainfall with high-humidity.

Roman Chamomile, Chamomile
(See Essential Oil)

Rose Geranium, Geranium
(See Essential Oil)

Roselle
(See Fiber)

Rye
(See Cereal)

Sage, Garden Sage, Common Sage
(See Spice & Flavoring)

Seagrape
(See Fruit)

Smooth Crotalaria, Striped Crotalaria
(See Forage, Pasture & Feed Grain)

Sorghum, Great Millet, Guinea Corn, Kaffir Corn, Milo, Sorgo, Kaoling, Durra, Mtama, Jola, Jawa, Cholam Grains, Sweets, Broomcorn, Shattercane, Grain Sorghums, Sweet Sorghums
(See Cereal)

Sour Cherry, Pie Cherry
(See Fruit)

Soursop
(See Fruit)

Southernwood, Lad's Love, Southern Wormwood
(See Ornamental & Lawn Grass)

Soybean
(See Oil)

Spearmint
(See Essential Oil)

Starfruit, Carambola
(See Fruit)

Strawberry Guava, Cattley Guava, Waiawi-'ula'ula, Wild Guava, Purple Guava
(See Fruit)

Sugar-Apple, Sweetsop
(See Fruit)

Sugarcane
(See Sweetener)

Surinam-Cherry, Pitanga Cherry
(See Fruit)

Sweet Balm, Lemon Balm
(See Essential Oil)

Sweet Birch, Black Birch, Cherry Birch, Mountain Mahogany
(See Essential Oil)

Sweet Granadilla, Granadilla

Sci. Name(s)	: *Passiflora ligularis*
Geog. Reg(s)	: America-Central, Hawaii, Tropical
End-Use(s)	: Beverage, Fruit
Domesticated	: Y
Ref. Code(s)	: 17, 119, 167, 170, 192
Summary	: Sweet granadilla is a woody, climbing vine cultivated in the tropics for its sweet fruit which has a juicy, acid pulp used mainly in beverages. Vines grow at elevations of 1,800-2,200 m. They produce 1 fruit crop/year. Sweet granadilla has become naturalized in Hawaii and is grown in the mountainous areas of Mexico and Central America.

Sweet Orange, Orange
(See Fruit)

Tamarind
(See Fruit)

Tea

Sci. Name(s)	: *Camellia sinensis*
Geog. Reg(s)	: Africa, Asia, Asia-China, Asia-India (subcontinent), Europe
End-Use(s)	: Beverage, Oil
Domesticated	: Y
Ref. Code(s)	: 148, 170
Summary	: World production of tea is over 2 million MT. Asia is the largest producer of the plant accounting for 78% of the total world production in 1983. India is the largest tea producing country with nearly 30% of the world's tea being produced there. Other tea producing regions are Japan, Indonesia, Pakistan, Kenya, Mozambique, Uganda, Tanzania, and the USSR.

The largest consumer with 20% of the world imports, is the United Kingdom. The United States follows, importing 77,317 MT of tea valued at $131.4 million in 1983.

There are basically 2 types of tea. Green tea is made from leaves that are steamed and dried. For black tea, the leaves are withered, fermented, and dried providing the bulk of the world tea supply.

Tea plantations are confined mainly to the subtropical mountainous regions of the tropics. Tea seed contains 20% of a nondrying oil that is not extracted commercially.

Plants are usually kept cut to 2 m in height while untrimmed shrubs can reach 15 m. Because most tea plants are grown on hillsides, erosion control precautions are necessary. Among the most effective methods are terracing and planting cover crops.

Ti Palm, Ti
(See Miscellaneous)

Velvetbean
(See Forage, Pasture & Feed Grain)

Virginia Strawberry
(See Fruit)

Water Lemon, Jamaica Honeysuckle, Belle Apple, Pomme De Liane, Yellow Granadilla
(See Fruit)

Wheat, Bread Wheat, Common Wheat
(See Cereal)

Wild Chamomile, Sweet False Chamomile, German Chamomile
(See Essential Oil)

Wild Thyme, Creeping Thyme
(See Ornamental & Lawn Grass)

Wine Grape, Grape, European Grape, California Grape

Sci. Name(s) : *Vitis vinifera*
Geog. Reg(s) : America-North (U.S. and Canada), Europe, Mediterranean, Temperate, Worldwide
End-Use(s) : Beverage, Dye & Tannin, Forage, Pasture & Feed Grain, Fruit, Oil
Domesticated : Y
Ref. Code(s) : 148, 154, 220
Summary : The wine grape is a major economic crop grown throughout the temperate areas of the world. Its fruits are eaten raw or dried, pressed for juice, fermented as wine, or made into jams and jellies.

World production of grapes is focused mainly on France, Italy, and Spain with the North American grape industry centered in California. Ninety percent of the crop is used for making wine. The rest is sold fresh, or is dried as raisins and currants. Greece is the main producer of dried grapes.

Wine is the fermented grape juice. Red wines are the product of red grapes, and white wines are the product of green or white grapes. Currants, raisins, and sultanas are products of the sun-dried, whole fruits. The residue from pressed grapes is fed to livestock or used to make tannin and cream of tartar. Grape seeds yield a drying oil (grape seed oil) which is used for lighting, paints, and for cooking.

Grape cultivation is time-consuming and expensive because it requires a lot of hand care and pruning. Plants are commonly propagated by cuttings. Grapes, except in a few cases, have not proven successful in the tropics.

Wine Palm, Coco de Chile, Coquito Palm, Honey Palm, Wine Palm of Chile, Chile Coco Palm

Sci. Name(s) : *Jubaea chilensis, Jubaea spectabilis*
Geog. Reg(s) : America-South
End-Use(s) : Beverage, Fiber, Fruit, Miscellaneous, Oil, Sweetener
Domesticated : N
Ref. Code(s) : 220
Summary : The wine palm yields a sweet sap from its trunk that is concentrated to produce palm honey and fermented to make palm wine. Its seeds contain an edible oil sold locally in Chile. The fruits are candied and eaten as sweet meat and the leaves are used to make baskets. Wine palms are grown primarily in Chile.

Wintergreen
(See Essential Oil)

Yard-Long Bean, Asparagus Bean
(See Vegetable)

Yellow Gentian

Sci. Name(s) : *Gentiana lutea*
Geog. Reg(s) : Asia, Europe
End-Use(s) : Beverage, Medicinal, Spice & Flavoring
Domesticated : Y
Ref. Code(s) : 205, 220
Summary : Yellow gentian is an herbaceous perennial about 110 cm high. Its thick taproot is dried, fermented, and used as a bitter preparation in alcoholic beverages which are mixed with orange peel, cardamon seed, glycerine, alcohol, and water to flavor liqueurs. The roots are used as a tonic for the gastrointestinal system. The tonic is especially good for treatment of anorexia associated with dyspepsia; small doses are used to stimulate the appetite. The constituents in the roots combine to stimulate the secretion of bile from the gall bladder.

Roots are collected from both wild and cultivated plants. There is some commercial cultivation in parts of eastern Europe and North America. Plants grow wild in mountain pastures and thinly wooded mountain forests. They are most successful on calcareous soils that are fairly porous and water-retentive. Yellow gentian is native to Central and southern Europe and parts of Asia Minor.

Yerba Mate, Paraguay Tea, Mate

Sci. Name(s) : *Ilex paraguariensis*
Geog. Reg(s) : America-South
End-Use(s) : Beverage, Medicinal
Domesticated : Y
Ref. Code(s) : 154, 205
Summary : Yerba mate is a small evergreen reaching 6 m in height. It is cultivated for its leaves, which contain from 0.2-2% caffeine. In South America, the leaves are used to produce a beverage ranking next to coffee, tea, and cocoa in popularity. Dried leaves have been used as a tonic, nervine, diuretic, and as a stimulant. Leaves are picked when the plants are about a year old. Yerba mate trees grow mainly in the mountainous areas of Brazil, Argentina, Chile, Peru, and Paraguay.

Yoco

Sci. Name(s) : *Paullinia yoco*
Geog. Reg(s) : America-South
End-Use(s) : Beverage
Domesticated : N
Ref. Code(s) : 170, 220
Summary : Yoco is a woody, climbing vine whose seeds contain 3% caffeine. The Indians of southern Colombia, Peru, and Equador use the seeds to make a stimulant beverage.

Cereal

Abyssinian Oat

Sci. Name(s)	: *Avena abyssinica*
Geog. Reg(s)	: Africa, Temperate, Tropical
End-Use(s)	: Cereal, Forage, Pasture & Feed Grain
Domesticated	: Y
Ref. Code(s)	: 171
Summary	: The Abyssinian oat (*Avena abyssinica*) is a temperate cereal crop whose oats are used as feed for livestock. The oats are sometimes used for human consumption. The cereal is self-pollinated and natural crosses occur only about 0.5-1% of the time.

Although usually restricted to the cooler, wetter temperate climates, Abyssinian oat crops have proved successful in the mountainous regions of the tropics. The oat is native to Africa.

African Rice

Sci. Name(s)	: *Oryza glaberrima*
Geog. Reg(s)	: Africa-West
End-Use(s)	: Cereal
Domesticated	: Y
Ref. Code(s)	: 171
Summary	: African rice (*Oryza glaberrima*) originated in Africa and is now produced mainly in the flood plains of the Sahel and Sudan zones of West Africa. The rice is broadcast in the flood plain 4-5 weeks before flooding. It is harvested after the floods have receded and can give yields from 700-1,000 kg/ha.

Although African rice is considered to be more hardy than the common rice (*Oryza sativa*), common rice tends to replace African rice when the two are grown in the same field.

Bambarra Groundnut, Voandzou, Ground Pea
(See Vegetable)

Bard Vetch, Monantha Vetch
(See Forage, Pasture & Feed Grain)

Barley

Sci. Name(s)	: *Hordeum vulgare*
Geog. Reg(s)	: America-North (U.S. and Canada), Asia-China, Europe
End-Use(s)	: Beverage, Cereal, Forage, Pasture & Feed Grain
Domesticated	: Y
Ref. Code(s)	: 126, 171
Summary	: Barley is a cereal crop whose grain meets different requirements in different areas of the world. In the temperate areas of North America and Europe, it is used primarily in the brewing of beer and for livestock feed. When grown in countries like Ethiopia, Libya, Morocco, Tunisia, Algeria, and some Near Eastern countries, barley

is an important food crop and as such it is considered palatable and fairly nutritious. The grain contains from 65-68% starch and is used as an energy food. It contains 12-14% protein but is deficient in the amino acids, lysine and threonine. Because of this, foods prepared from the grain should be supplemented with other protein foods from animal products, food grain legumes, and oilseed meals.

This cereal is prepared for eating by boiling or parching the whole grain. It can then be ground for gruel or made into flour for baking. Grain husks are ground to make pearl barley, used in soups, or used as feed for livestock.

The cereal is considered a temperate crop but has been successfully grown in the higher-elevations of the tropics. Crops have also been produced in irrigated areas of the desert where the alkali of the soil is too high for other cereal crops to survive. Barley crops are most successful in deep, well-drained, fertile loams.

Barley is a member of the grass family. Plants begin flowering approximately 2-4 months after planting. The maturing and ripening of the kernels requires one additional month. There are 2 basic types of barley: winter annuals, which need very cool temperatures in order to stimulate heading, and spring annuals, which can form heads without cool temperatures.

The world's major barley producers are the USSR, the United Kingdom, Germany, China, and the United States with the highest yields being from the United Kingdom. In 1983, the United States produced 11.3 million MT of barley and imported an additional 141,141 MT valued at $20.1 million. In the same year, the United States exported 1,521,443 MT of barley valued at $195.7 million.

Barley, Bulbous Barley

Sci. Name(s)	:	*Hordeum bulbosum*
Geog. Reg(s)	:	Temperate, Worldwide
End-Use(s)	:	Beverage, Cereal
Domesticated	:	Y
Ref. Code(s)	:	17
Summary	:	Bulbous barley (*Hordeum bulbosum*), one of the hardiest cereal crops, is grown in temperate areas throughout the world. There is very little specific information on this particular cereal. The grain is used mainly for malting and as a food source.

Barnyardgrass, Barnyard Millet
(See Forage, Pasture & Feed Grain)

Benefing
(See Oil)

Billion Dollar Grass, Japanese Barnyard Millet, Japanese Millet

Sci. Name(s)	:	*Echinochloa crusgalli* var. *frumentacea*, *Echinochloa frumentacea*, *Panicum frumentacea*
Geog. Reg(s)	:	America-North (U.S. and Canada), Asia, Asia-India (subcontinent), Tropical
End-Use(s)	:	Beverage, Cereal, Forage, Pasture & Feed Grain
Domesticated	:	Y
Ref. Code(s)	:	58, 123, 171

Summary : Billion dollar grass is one of the quickest growing annual cereals. It produces a crop in about 6 weeks. It is grown to some extent in the Orient and India but is not considered an important crop there. The grass produces tall, erect culms that reach 1-2 m in height.

In Asia, billion dollar grass is grown as a substitute for rice. In northern India, it is a rainy season crop that can be produced at low elevations on hillsides. It is sometimes mixed with rice and fermented to make beer. In the United States and Japan, the grass is a possible fodder. Experimental plantings have produced as many as 8 fodder crops/year. The grass seed grain has been used as an experimental ingredient in bird seed mixes. It is now recommended for planting in wetland areas for wildfowl.

Billion dollar grass is harvested after 6 weeks and yields from 700-800 kg/ha of grain and 1,000-1,500 kg/ha of straw.

Bitter Vetch, Ervil
(See Forage, Pasture & Feed Grain)

Black Fonio, Hungry-Rice
Sci. Name(s) : *Digitaria iburua*
Geog. Reg(s) : Africa-West
End-Use(s) : Cereal
Domesticated : Y
Ref. Code(s) : 171, 183
Summary : Black fonio (*Digitaria iburua*) is a white-seeded form of hungry rice (*Digitaria exilis*) that can reach 50 cm in height. In Nigeria, it is cultivated as a cereal crop and is either grown alone or mixed with millet. There is very little information as to the economic viability of this cereal.

Black Gram, Urd, Wooly Pyrol
(See Vegetable)

Black Oat, Bristle Oat, Sand Oat, Small Oat
(See Weed)

Blue Panicgrass, Blue Panicum, Giant Panicum
(See Forage, Pasture & Feed Grain)

Browntop Millet
(See Forage, Pasture & Feed Grain)

Buckwheat
Sci. Name(s) : *Fagopyrum esculentum*
Geog. Reg(s) : Asia-Central, Eurasia, Europe, Temperate
End-Use(s) : Cereal, Forage, Pasture & Feed Grain, Medicinal, Miscellaneous
Domesticated : Y
Ref. Code(s) : 154, 170, 214, 220

Summary   : Buckwheat is an important cereal crop cultivated in Europe and the USSR. Crops are grown in cool, moist, temperate regions but will tolerate tropical climates at higher-altitudes.

     The cereal grain is ground into flour and provides stock and poultry feed. Plants are grown for green manure. A medicinal substance, rutin, is obtained from the leaves and flowers and is used in the treatment of hypertension. The flowers are a source of honey. Buckwheat is native to parts of Central Asia.

Cassava, Manioc, Tapioca-Plant, Yuca, Mandioca, Guacomole
(See Energy)

Catjang, Jerusalem Pea, Marble Pea, Catjan
(See Vegetable)

Chufa, Ground Almond, Tigernut, Yellow Nutsedge
(See Forage, Pasture & Feed Grain)

Club Wheat
Sci. Name(s)   : *Triticum compactum*
Geog. Reg(s)   : America-North (U.S. and Canada), Europe
End-Use(s)   : Cereal
Domesticated   : Y
Ref. Code(s)   : 161, 220
Summary   : Club wheat is a cereal crop whose kernels are valued as a source of low-gluten flour. Plants produce short, dense ears that form white, yellowish, or red grains. The straw is usually 93-140 cm in height. Plants are fairly resistant to frosts, drought, and fungi. They will continue to thrive on poor soils.

     Plants are native to Europe and are also grown there and in the western parts of the United States.

Common Bean, French Bean, Kidney Bean, Runner Bean, Snap Bean, String Bean, Garden Bean, Green Bean, Haricot Bean
(See Vegetable)

Common Oat, Oat, Oats
Sci. Name(s)   : *Avena sativa*
Geog. Reg(s)   : America-North (U.S. and Canada), Europe, Temperate, Tropical
End-Use(s)   : Cereal, Forage, Pasture & Feed Grain
Domesticated   : Y
Ref. Code(s)   : 171
Summary   : The common oat (*Avena sativa*) is a temperate country or tropical mountain cereal. The oats are made into porridges and oatcakes and are often fed to livestock. The oat is an allohexaploid developed from the wild oat (*Avena fatua*). It is self-pollinating, only crossing naturally about 0.5-1% of the time.

The common oat is native to Europe where the major producers are found. In 1983, European countries accounted for 29% of the world's production of 43 million MT. In the same year, the United States produced 7 million MT of oats and imported 277,069 MT of oats valued at $27.6 million.

Crowfoot Grass
(See Weed)

Danicha
(See Fiber)

Dasheen, Elephant's Ear, Taro, Eddoe
(See Tuber)

Durum Wheat

Sci. Name(s) : *Triticum durum*
Geog. Reg(s) : Africa-South, America-Central, America-North (U.S. and Canada), America-South, Australia, Europe
End-Use(s) : Cereal
Domesticated : Y
Ref. Code(s) : 161, 171
Summary : Durum wheat is the most important tetraploid wheat. Plants produce hard high gluten grain that is made into flour and used mainly for the manufacture of paste products like macaroni and spaghetti. It is also used in breads.

Plants thrive on deep, rich soils, in areas with hot, dry climates and no threat of frost. They are fairly resistant to drought, rust, and smut fungi. Durum wheat was introduced to the United States from Russia, and is now grown in the Great Plains. Plants are grown in Europe, Canada, Central and South America, South Africa, and Australia.

Einkorn, One-Grained Wheat
(See Forage, Pasture & Feed Grain)

Emmer

Sci. Name(s) : *Triticum diococcon, Triticum dicoccum*
Geog. Reg(s) : Asia, Europe, Temperate
End-Use(s) : Cereal, Forage, Pasture & Feed Grain
Domesticated : Y
Ref. Code(s) : 161, 220
Summary : Emmer is a bearded wheat that is cultivated mainly as a source of flour and livestock feed. The grains can be ground into a starchy white flour that is favored for making fine pastries and cakes. The grains may also be added to soups and are considered better than pearl barley for this purpose. A quality starch can be prepared from the flour.

There are 2 main groups of emmer, Indo-Abyssinian and European. Emmer wheat is native to Europe and temperate Asia. Plants thrive on light soils in areas with warm, dry climates. They do not tolerate frost.

European Chestnut, Spanish Chestnut
(See Nut)

Fat Hen, Lamb's-Quarters, White Goosefoot, Lambsquarter

Sci. Name(s)	: *Chenopodium album*
Geog. Reg(s)	: Europe
End-Use(s)	: Cereal, Dye & Tannin, Forage, Pasture & Feed Grain, Medicinal, Vegetable
Domesticated	: Y
Ref. Code(s)	: 170, 205
Summary	: Fat hen, or lamb's-quarters, is a wild annual European herb cultivated for its leaves and seeds. The seeds are ground into flour and the leaves are eaten as green vegetables. Both the leaves and seeds are rich in iron, calcium, and vitamins B and C. They are thought to be more nutritious than spinach or cabbage. Parts of the herb produce a reddish-golden dye. The plant is used as fodder. Fat hen has no important medicinal uses except as a mild laxative. It grows best in nitrogen rich soils.

Finger Millet, African Millet, Wimbi, Ragi

Sci. Name(s)	: *Eleusine coracana*
Geog. Reg(s)	: Africa-Central, Africa-East, Asia-India (subcontinent)
End-Use(s)	: Beverage, Cereal
Domesticated	: Y
Ref. Code(s)	: 126, 171
Summary	: Finger millet is the most important annual cereal crop in Uganda. It is also an important staple food in other parts of East and Central Africa, and India. Finger millet is an especially good cereal crop because it can be stored for up to 10 years, which makes it an important resource in times of famine.

The grain is made into porridge and eaten with other foods. The protein content is fairly low. The grain is used for malting and brewing beer.

Finger millet thrives in moist climates at altitudes of 2,000 to 2,500 m. It grows in rocky, shallow soils and can reach heights of 1 m or more. Crops are most successful in weed-free seedbeds.

World hectarage and production of finger millet cannot be accurately estimated as statistics are aggregated with other millets. Crops mature in 3-5 months. A good grain yield grown under natural rainfall ranges from 2,000 to 4,000 kg/ha. Under irrigation, yields may be higher. Several high-yielding and disease resistant varieties of finger millet have been developed which may increase its economic importance.

Fishtail Palm, Toddy Palm
(See Sweetener)

Foxtail Millet, Italian Millet, German Millet, Hungarian Millet, Siberian Millet

Sci. Name(s)	: *Setaria italica*
Geog. Reg(s)	: Africa-North, Asia, Europe
End-Use(s)	: Cereal, Forage, Pasture & Feed Grain
Domesticated	: Y

Ref. Code(s) : 31, 126, 163, 171
Summary : The foxtail millet is a fast-growing, erect annual 90-150 cm high that was formerly grown as a catch crop, but is now cultivated as fodder. For human consumption the grain can be cooked and eaten as rice or ground up for porridge or pudding. It is used for birdseed. Plants will produce a substantial grain crop 60-90 days from planting.

Plants grow well in areas receiving limited rainfall from sea level to 2,000 m in elevation. They are able to withstand a wide range of soil conditions from light sands to heavy clays. Rain-fed crops have yielded approximately 800-900 kg/ha of grain. Higher yields are possible through irrigation. Crops are grown throughout Asia, southeastern Europe, and North Africa.

Garbanzo, Chickpea, Gram
(See Vegetable)

Geocarpa, Kersting's Groundnut, Geocarpa Groundnut
Sci. Name(s) : *Kerstingiella geocarpa*
Geog. Reg(s) : Africa-West
End-Use(s) : Cereal, Nut
Domesticated : Y
Ref. Code(s) : 126, 140
Summary : Geocarpa groundnut is a small annual herb whose immature and mature seeds are used as a protein-rich pulse crop. Plants are grown in the drier parts of West Africa as a subsistence crop. The seeds are highly nutritious and similar to the bambara groundnut. Geocarpa groundnut yields are generally poor.

Plants thrive on dry, poor, sandy soils and in direct sunlight with high temperatures. Crops mature in 90-150 days. Dry seed yields are about 500 kg/ha.

Giant Wildrye
(See Forage, Pasture & Feed Grain)

Grass Pea, Chickling Vetch, Chickling Pea
Sci. Name(s) : *Lathyrus sativus*
Geog. Reg(s) : Africa-North, Asia, Asia-India (subcontinent), Eurasia, Europe, Subtropical, Temperate
End-Use(s) : Cereal, Forage, Pasture & Feed Grain, Miscellaneous, Vegetable, Weed
Domesticated : Y
Ref. Code(s) : 126, 170
Summary : The grass pea is a winter annual or cool-climate legume that provides small but reliable crops even in times of drought. Grass peas are one of the cheapest pulse crops because plants grow as weeds in cultivated barley fields. Not all forms of grass pea can be consumed and care must be taken to harvest only the nontoxic types that are rich in protein, can be boiled, ground into flour, and fed to livestock. The leaves are cooked as potherbs. Grass peas when eaten with the common vetch (*Vicia sativa*) can have a paralytic effect upon the lower limbs. It is advised that all seeds be parched and carefully boiled.

Plants take 4 months to reach maturity. They grow up to 75 cm in height. About 1,000-1,121 kg/ha of seeds can be collected and 1,345-1,569 kg/ha of hay.

Grass peas are found in southern Europe, western Asia, India, the Near East, and North Africa. They grow in most subtropical and temperate areas but not in tropical areas. Crops are most successful at higher-altitudes in semiarid regions receiving less than 500-600 mm/annum of rain.

Horsegram

Sci. Name(s) : *Dolichos uniflorus, Macrotyloma uniflorum*
Geog. Reg(s) : Africa, Africa-West, America-North (U.S. and Canada), Asia, Asia-India (subcontinent), Asia-Southeast, Australia, West Indies
End-Use(s) : Cereal, Erosion Control, Forage, Pasture & Feed Grain, Spice & Flavoring
Domesticated : Y
Ref. Code(s) : 24, 128, 170
Summary : The horsegram is a twining annual or perennial herb with dense growth which provides a cover, fodder and pulse crop in India, Asia, Africa, the West Indies, and parts of the southern United States. In southern India, seeds are parched and eaten boiled or fried. The stems, leaves, and husks of the horsegram are used as fodder and the seeds are fed to cattle and horses. In Burma, dry seeds are processed into a sauce similar to soy sauce. The grass is palatable at all stages of development.

The pulse crop is harvested at 4-6 months and yields 68-136 kg/ha. Fodder crops are harvested at 6 weeks and yield 1.8-4.5 MT of green fodder. In Mali, yields vary from 16-27 MT/ha. In Australia, mature stands yield over 6 MT/ha of dry matter.

Horsegrams are fairly drought-resistant and grow in areas receiving seasonal rainfall of about 380 mm. Plants grow on most soils but do not tolerate waterlogging.

Hungry-Rice, Fonio, Fundi, Acha

Sci. Name(s) : *Digitaria exilis*
Geog. Reg(s) : Africa-South, Africa-West
End-Use(s) : Beverage, Cereal, Forage, Pasture & Feed Grain
Domesticated : Y
Ref. Code(s) : 171, 183, 222
Summary : Hungry-rice is grown as a cereal crop in West Africa. It is widely grown as a complimentary cereal with millet, rather than a staple crop in areas receiving a rainfall above 400 m. Crops can be harvested about 4 months after planting. It is not cultivated to any extent outside of South Africa.

This small grain contains about 81% carbohydrates and 9% protein. It is made into porridge, ground and mixed with other cereals, or used in brewing beer. The straw can be used as fodder.

Inca-Wheat, Quihuicha, Quinoa, Love-Lies-Bleeding

Sci. Name(s) : *Amaranthus caudatus*
Geog. Reg(s) : Africa-West, America-South, Europe
End-Use(s) : Cereal, Medicinal, Ornamental & Lawn Grass, Spice & Flavoring, Vegetable
Domesticated : Y
Ref. Code(s) : 50, 170, 220

Summary   : Inca-wheat or grain amaranth is a grain crop grown in the higher elevations of South America. In West Africa, amaranth leaves are used as potherbs and parts of the fresh plant are used as a diuretic. In certain areas of Europe, amaranth leaves are eaten as a substitute for spinach. Other varieties of the plant are grown as ornamentals (Love-lies-bleeding) for their attractive red or purple leaves and long, drooping flower stalks.

Indian Rice, Silkgrass, Indian Millet, Indian Ricegrass

Sci. Name(s)   : *Oryzopsis hymenoides, Oryzopsis cuspidata*
Geog. Reg(s)   : America-North (U.S. and Canada)
End-Use(s)   : Cereal, Forage, Pasture & Feed Grain
Domesticated   : Y
Ref. Code(s)   : 17, 83, 114, 119, 220
Summary   : Indian rice is a slender, tufted perennial grass grown either as a cereal or as forage. The seeds are eaten raw but are preferred dried and ground into flour. Indian rice is favored as forage because of its durable, densely tufted growth. It ranges from Manitoba to British Colombia, south to Texas, New Mexico, and California at elevations up to 2,900 m.

Job's Tears, Adlay, Adlay Millet
(See Miscellaneous)

Kodo Millet, Kodo, Kodra

Sci. Name(s)   : *Paspalum scrobiculatum* var. *scrobiculatum*
Geog. Reg(s)   : Asia-India (subcontinent)
End-Use(s)   : Cereal, Forage, Pasture & Feed Grain
Domesticated   : Y
Ref. Code(s)   : 41, 58, 119, 171
Summary   : Kodo millet is a perennial cereal that is considered a minor grain of southern Central India. The grain is used like rice, boiled or parched and ground into flour. Cultivated kodo millet is grown as a rainy season crop throughout the lower elevations of the Himalayan mountains. It will grow on poor soils. Kodo millet is fairly drought-resistant and matures very rapidly in about 4-6 months. It has a 60% milling ratio. Grain yields are 250-1,000 kg/ha. A wild form of the millet, variety commerson, grows well in damp places and provides good pasturage because it is palatable at all stages.

Lablab, Lablab Bean
(See Forage, Pasture & Feed Grain)

Lentil

Sci. Name(s)   : *Lens culinaris, Lens esculenta*
Geog. Reg(s)   : Africa-East, Africa-North, America-North (U.S. and Canada), Eurasia, Europe, Mediterranean, Subtropical, Temperate, Tropical
End-Use(s)   : Cereal, Vegetable
Domesticated   : Y
Ref. Code(s)   : 126, 170

36

Summary : Lentils are short-season, bushy annuals that can provide a nutritious pulse crop in 70-110 days. The seeds are thought to be an excellent supplement to other cereal grains because they are high in carbohydrates, with fair amounts of protein. In the tropics, lentils are often eaten in place of animal protein. As a commercial produce item, lentils can be sold almost as easily as other better known food grain legumes. Improvements in crop yields and harvesting techniques must be made if commercial production is to expand. Seeds are split in soups or ground as flour. The flour is often used in invalid and infant foods.

Crops can be grown in cooler climates as winter annuals or in the higher-altitudes of the tropics during the cool-season. Plants will not produce well in hot climates or during extremely dry weather. They must be grown in deep, fertile, well-drained soil that is kept relatively weed-free. Lentils are often grown as cool-season, short-season annuals together with a major, warm-season crop. Plants originated in Egypt, southern Europe, and western Asia.

Lentil crops mature in 3-3.5 months and yield 450-670 kg/ha. Yields of up to 1,682 kg/ha can be produced with irrigation. India is the major producer, with decreasing amounts of production from the Middle East, East and North Africa, Russia, and the United States.

Ma-ha-wa-soo
(See Oil)

Maize, Corn, Indian Corn

Sci. Name(s) : *Zea mays* subsp. *mays*
Geog. Reg(s) : Subtropical, Tropical, Worldwide
End-Use(s) : Cereal, Forage, Pasture & Feed Grain, Gum & Starch, Oil, Vegetable
Domesticated : Y
Ref. Code(s) : 17, 187, 220
Summary : Corn or maize, is one of the world's three most important cereal crops. It is grown primarily for food and domestic animals. Crops are grown throughout the world in most tropical areas.

The grain is ground to flour and used in starchy foods and breads. It is used in breakfast foods, i.e. corn-flake cereals. Fermented grain is made into whiskey and industrial alcohol. Corn starch is used in cosmetics, adhesives, glucose, and syrup. An oil extracted from the seed embryo (corn oil) is used as a salad oil, and to make linoleum, paints, varnishes, soaps, glycerine, and high fructose corn syrup.

The most prominent cultivars include, 'dent corn,' the number one United States market corn; 'sweet corn,' which is eaten as a frozen or canned vegetable; 'flint corn,' which has a low starch content and can grow in the northern United States and tropical lowlands; 'popcorn,' which is used in confectionery; 'flour corn,' which has a high starch content and is grown in the Andes; 'pod corn,' which has little or no economic importance; 'deratina,' which is grown in the Far East; and 'starchy sweet corn,' which has little or no economic importance.

In 1983-84, world corn production was 344.1 million MT from an area of 120.6 million ha. The United States produced 106.8 million MT of corn from an area of 20.9 million ha. The United States exported 47.6 million MT of corn, valued at $6.5

billion. In the same year, the United States also exported 110,953 MT of oil of maize, valued at $89.6 million.

Mango
(See Fruit)
Moth Bean, Mat Bean
(See Vegetable)

Mozinda, African Breadfruit
(See Nut)

Mung Bean, Green Gram, Golden Gram
(See Vegetable)

Oil-Bean Tree, Owala Oil
(See Oil)

Onion, Common Onion
(See Vegetable)

Oriental Wheat
Sci. Name(s)   : *Triticum turanicum*
Geog. Reg(s)   : Eurasia, Mediterranean
End-Use(s)   : Cereal
Domesticated   : Y
Ref. Code(s)   : 161, 219
Summary   : Oriental wheat is a cereal that is occasionally grown as a grain crop in some Mediterranean countries and the Near East. The grain is long, narrow, and flinty. It is ground into flour and used as a substitute for other more common grains. There are two main varieties of oriental wheat: the first, variety insigne or camel's tooth, is an early crop with thin, reedy straw 95-110 cm in height; the second, variety notabile or black hand, is distinguished from the first variety only by its color.

Pea, Garden Pea, Field Pea
(See Vegetable)

Pearl Millet, Bulrush Millet, Spiked Millet, Cat-tail Millet
Sci. Name(s)   : *Pennisetum americanum, Pennisetum typhoideum, Pennisetum glaucum, Pennisetum spicatum*
Geog. Reg(s)   : Africa, Asia-India (subcontinent)
End-Use(s)   : Beverage, Cereal, Energy, Forage, Pasture & Feed Grain
Domesticated   : Y
Ref. Code(s)   : 171
Summary   : Pearl millet is an annual cereal crop that provides a staple food in the drier parts of Africa and India. In these countries, nearly the entire plant is used in some way. The grains are cooked as rice, made into flour, or used to produce malt for beer. Whole grains can be fed to poultry and livestock. Green plants provide adequate fod-

der. Plant straw is coarse and pithy and is most often used for bedding, thatching, fencing, and fuel.

The millet is an ideal crop for most dry areas. It can grow and produce on poor sandy soil in low rainfall areas. Plants are resistant to most diseases but are very prone to bird damage. This millet has been introduced to the United States as fodder but is considered unimportant in the world market as it is replaced by sudangrass and sorghum hybrids. Pearl millet reportedly has more protein and fat in the grain than either corn or sorghum.

The grains develop about 8 days after the spikes emerge; it is completely developed 20-30 days later, and then, after fertilization, will ripen in about 40 days. The threshing percentage is about 55%. Yields in Africa and India vary from about 250-3,000 kg/ha, depending on conditions. This crop does not enter international trade.

Pecan
(See Nut)

Peppergrass, Virginia Pepperweed, Pepper Grass
Sci. Name(s) : *Lepidium virginicum*
Geog. Reg(s) : America-North (U.S. and Canada)
End-Use(s) : Cereal, Spice & Flavoring
Domesticated : N
Ref. Code(s) : 43
Summary : Peppergrass grows wild in waste places and along roads in the United States. The small seeds can be parched and ground to flour. Peppergrass leaves are roasted or used as potherbs. They have a very strong taste.

Persian Wheat
Sci. Name(s) : *Triticum carthlicum, Triticum persicum*
Geog. Reg(s) : Asia
End-Use(s) : Cereal
Domesticated : Y
Ref. Code(s) : 220
Summary : Persian wheat is a cereal crop that has been cultivated in Asia Minor. There is very little information available concerning the uses of or economics for this grain crop.

Polish Wheat, Astra Wheat, Giant Rye, Jerusalem Rye
Sci. Name(s) : *Triticum polonicum*
Geog. Reg(s) : Africa-North, Europe
End-Use(s) : Cereal
Domesticated : Y
Ref. Code(s) : 161, 220
Summary : Polish wheat is a bearded cereal with hard grains that are ground into flour and used in certain breads. The flour generally is not used in pastes because of the hardness of the grains.

Plants are fairly delicate, spring crops that require hot climates and fertile soils for adequate production. They produce large ears and are among the tallest wheats in the Triticum genus, with culms 140-160 cm in height. Polish wheat is found in Algeria, Spain, and Italy. The wheat is native to Central Europe.

Potato, European Potato, Irish Potato, White Potato
(See Tuber)

Poulard Wheat, Cone Wheat, Rivet Wheat

Sci. Name(s) : *Triticum turgidum*
Geog. Reg(s) : Africa-North, Africa-South, America-Central, America-North (U.S. and Canada), America-South, Asia, Australia, Europe
End-Use(s) : Cereal, Forage, Pasture & Feed Grain
Domesticated : Y
Ref. Code(s) : 161, 220
Summary : Poulard wheat or rivet wheat, produces grains that are ground into a flour and used for making paste products. The grain can also be fed to livestock. Plants are often grown in mixtures with the more durable Canadian and Russian wheats.

Plants are fairly delicate and are easily damaged by frost or steady rain. They are highly resistant to rust fungi. Poulard wheat thrives on warm, well-drained soils in Europe, Central, South and North America, North and South Africa, and Australia. Plants are native to Asia Minor.

Proso Millet, Hog Millet, Common Millet, Brown Corn Millet, Broomcorn Millet

Sci. Name(s) : *Panicum miliaceum*
Geog. Reg(s) : Africa-North, America-North (U.S. and Canada), Asia-Central, Asia-India (subcontinent), Europe, Mediterranean
End-Use(s) : Beverage, Cereal, Forage, Pasture & Feed Grain
Domesticated : Y
Ref. Code(s) : 76, 94, 171, 220
Summary : Proso millet is an annual cereal crop reaching 60 cm in height with a low water requirement. It can be grown in hot climates in areas with sparse rainfall and poor soil. It will adapt to plateau conditions and high-altitudes. The grain has a nutty flavor and is highly nutritious, containing 10% protein, 4% fat, and starch. The grain is fed to livestock, is cooked like rice, or ground to flour. Proso millet straw is poor, but green plants provide good fodder. In parts of Central and eastern Europe and Asia Minor, the grain is used to make an alcoholic beverage. It is often used as a substitute for maize or sorghum.

Crops mature in 60-90 days. They are not bothered by serious diseases or pests and can produce substantial yields. In India, crop yields of 450-650 kg/ha have been reached. With irrigation, 1,000-2,000 kg/ha are possible. Proso millet is domesticated in Central and East Asia, India, Russia, the Middle East, and the United States.

Quinoa

Sci. Name(s) : *Chenopodium quinoa*
Geog. Reg(s) : America-South
End-Use(s) : Beverage, Cereal
Domesticated : Y
Ref. Code(s) : 170

Summary : Quinoa is a South American herb which has replaced maize as a staple crop, especially in the higher-altitudes of the Andes mountains. The seeds are used in soups, made into an alcoholic beverage, or ground into flour. Crops will mature in 5-6 months.

This herb is an annual that reaches from 1-2 m in height.

Red Oat, Mediterranean Oat, Algerian Oat

Sci. Name(s) : *Avena byzantina*
Geog. Reg(s) : Asia, Temperate, Tropical
End-Use(s) : Cereal, Forage, Pasture & Feed Grain
Domesticated : Y
Ref. Code(s) : 171
Summary : The red oat (*Avena byzantica*) is grown in temperate countries and tropical highlands as a cereal crop. It is fed to livestock or eaten in porridge and oatcakes. The cereal was probably developed from the sterile oat (*Avena sterilis*). The red oat is native to Asia.

Rice

Sci. Name(s) : *Oryza sativa*
Geog. Reg(s) : America-North (U.S. and Canada), Asia, Asia-China, Asia-Southeast
End-Use(s) : Cereal, Energy, Fat & Wax, Forage, Pasture & Feed Grain, Gum & Starch, Miscellaneous
Domesticated : Y
Ref. Code(s) : 132, 139, 154, 171
Summary : Rice (*Oryza sativa*) is one of the major cereal crops of the world. The commercially important parts of the plant are the fibrous roots, culms, leaves, and panicles. Rice is mainly grown for human consumption but is also used in cosmetics, laundering starch, foods, and textiles. A wax is obtained from crude rice-bran oil and used to make plastic packaging materials for food products. Rice husks are used as fuel for rice mills, as an aggregate for light-weight concrete, for making hardboard, and as an abrasive. Rice straw is an inferior cereal straw for livestock, but does provide bulkfeed. In China and Thailand, rice straw is used for mushroom culture.

In 1983, Asia produced 93% of the world's rice. China was the largest Asian producer with 38% of the world total. Only limited amounts of rice reach international trade because most of what is grown is consumed locally. World production figures for rice showed that 450 million MT were produced from a total area of 144.5 million ha. In the United States, 4.5 million MT of rice were produced from 878,000 ha.

The United States and Thailand are major rice exporters. In 1983, the United States accounted for 21% of the total world exports. Thailand, in the same year, accounted for 31%.

Rice is an annual crop and transplanting seedlings generally gives better results than broadcasting. Harvesting is generally done by hand which requires a lot of labor. Rice needs warm water for best growth; generally water 20-30 degrees Celsius is favorable. Plants should be kept weed-free. The growing season lasts 4-6 months.

Roselle
(See Fiber)

Rye

Sci. Name(s)	: *Secale cereale*
Geog. Reg(s)	: Africa-South, America-North (U.S. and Canada), Asia-Central, Europe, Temperate
End-Use(s)	: Beverage, Cereal, Fiber, Forage, Pasture & Feed Grain
Domesticated	: Y
Ref. Code(s)	: 116, 171, 220
Summary	: Rye is a tufted annual or perennial cereal crop whose gluten-rich grain contains 13.4% protein, 1.8% fat, 80.2% carbohydrates, and 2-3% fiber. It is used for making black bread, whiskey, gin, and beer, but is most often used as livestock feed. Mature plant stalks are too fibrous for fodder but are used for animal bedding, paper pulp, cardboard, thatching, and hats. Rye is an important cereal because of its winter hardiness, resistance to drought, and its ability to grow on light, sandy, and acidic soils.

Crops are produced mainly in the USSR, Poland, Germany, Czechoslovakia, Hungary, and the United States. Plants are successful in most cool, temperate climates. Cultivation extends from sea level to 4,250 m in elevation. There are both spring and winter forms of rye.

In 1984, world imports for rye were 866,333 MT valued at $115 million. The United States imported 10,382 MT of rye valued at $874,000 and exported 12,894 MT valued at $1.4 million. Japan and Korea were major rye importers with Japan importing 341,961 MT valued at $42 million, while Korea brought in 263,896 MT valued at $35 million.

Sanduri, Tschalta Sanduri, Timopheri

Sci. Name(s)	: *Triticum timopheevii*
Geog. Reg(s)	: Eurasia
End-Use(s)	: Cereal
Domesticated	: Y
Ref. Code(s)	: 220
Summary	: Sanduri is a bearded cereal that is grown as a domestic grain crop in the southern parts of the USSR. There is very little information concerning the physical description and economic potential of this crop.

Sorghum, Great Millet, Guinea Corn, Kaffir Corn, Milo, Sorgo, Kaoling, Durra, Mtama, Jola, Jawa, Cholam Grains, Sweets, Broomcorn, Shattercane, Grain Sorghums, Sweet Sorghums

Sci. Name(s)	: *Sorghum bicolor, Holchus sorghum, Andropogon sorghum, Sorghum vulgare, Sorghum caffrorum, Sorghum dochna, Sorghum drummondii, Sorghum durra, Sorghum nervosum, Sorghum guineense*
Geog. Reg(s)	: Africa, Worldwide
End-Use(s)	: Beverage, Cereal, Dye & Tannin, Fiber, Forage, Pasture & Feed Grain, Gum & Starch, Miscellaneous, Oil, Sweetener
Domesticated	: Y
Ref. Code(s)	: 171

42

Summary   : Sorghum is a true perennial grass that is most often grown as an annual. It is an extremely important cereal crop with world production exceeded only by wheat, rice, and maize. This hardy, drought-resistant crop can be grown throughout the world. Developed countries use sorghum as stockfeed while developing countries value the grain as a staple food.

In the drier regions of tropical Africa, India, and China, the grain is ground into flour. In India, the most common cultivar grown is pop sorghum, which can be used like popcorn or sweet corn. Dark-grained varieties are used for the manufacture of certain beers, and are important nutritionally because of their high vitamin B content. Plant bases and stems have been used for thatching, fencing, and basketry. In West Africa, a red dye obtained from the plant has been used for dyeing leather.

In the United States, sorghum is used primarily for the manufacture of sucrose which is obtained from the sweet, juicy plant stems. Sucrose is used in syrups. There is a good demand for sorghum grain as stockfeed and a moderate export industry has developed for the United States. Sorghum straw is very palatable to cattle, and stock graze the ratoon stems. Sorghum also provides good forage and silage. An important variety of Sorghum, variety dochna, or broomcorn, is also produced in the United States. Variety dochna is used for starch production, adhesives, for sizing paper and textiles, for gum used on stamps and envelopes, and for thickening pies and gravies. Embryos yield an oil used in cooking and salad oils.

Maize has been the only real competition for sorghum and has replaced sorghum in parts of Africa as the staple cereal because of its potential for higher yield and its wide acceptance as a nutritious food grain. Maize does not have the wide range of environmental adaptation that sorghum has, and therefore cannot completely replace it.

Sorghum world production in 1983 was 62.4 million MT, from a total area of 46.5 million ha. The United States was the leading producer with 12.3 million MT of sorghum, representing 20% of world production. In the United States, 4 million ha of sorghum was planted for grain, producing 12.1 million MT of grain, valued at $1.4 billion. An area of 254,745 ha was planted for silage, producing 5,856,499 MT of silage, and an area of 302,535 ha was planted for forage.

Soybean
(See Oil)

Spanish Greens, Huauhtli

Sci. Name(s)   : *Amaranthus cruentus*
Geog. Reg(s)   : America-Central
End-Use(s)   : Cereal
Domesticated   : Y
Ref. Code(s)   : 170, 220
Summary   : Spanish greens (*Amaranthus cruentus*), a branching herb resembling Inca wheat (*Amaranthus caudatus*), was widely cultivated as a grain crop in Central America before the Spanish conquests. Today, this grain crop is cultivated to a limited extent in Guatemala and other parts of Central America.

Spelt

Sci. Name(s)   : *Triticum spelta*
Geog. Reg(s)   : Europe
End-Use(s)   : Cereal, Forage, Pasture & Feed Grain
Domesticated : Y
Ref. Code(s)   : 17, 161, 171, 220
Summary   : Spelt is a widely grown cereal crop. Its grain is ground into a flour that has less sugar and dextrin but more starch than other bread flours and is valued in confectionery and pastries. Husk grains and straw provide livestock feed.

Plants are cultivated mainly in Germany, with less production in northern Spain, Switzerland, France, and Italy. They are most successful on light, dry soils. Plants are usually resistant to fungal diseases, mainly rust. Spelt wheat produces straw from 100-120 cm in height.

Sterile Oat, Animated Oat

Sci. Name(s)   : *Avena sterilis*
Geog. Reg(s)   : Asia, Temperate, Tropical
End-Use(s)   : Cereal, Forage, Pasture & Feed Grain
Domesticated : Y
Ref. Code(s)   : 171
Summary   : The sterile oat (*Avena sterilis*) is a cereal crop whose oats are made into porridges and oatmeal. The oats are also used as livestock feed. Crops are most successful in temperate climates and in the mountainous regions of the tropics. The sterile oat is native to Asia.

Summer-Cypress, Kochia

Sci. Name(s)   : *Kochia scoparia*
Geog. Reg(s)   : America-North (U.S. and Canada), Asia, Europe, Temperate
End-Use(s)   : Cereal, Vegetable
Domesticated : N
Ref. Code(s)   : 220
Summary   : Summer-cypress is a plant whose young shoots are eaten as vegetables and seeds ground to flour. There is little information concerning other uses or economic potential for this plant. Summer-cypress grows throughout Central Europe and temperate Asia and has naturalized in North America.

Tacoutta, Wing Sesame
(See Oil)

Tamarind
(See Fruit)

Tartary Buckwheat, Indian Wheat, Siberian Buckwheat, Tartarian Buckwheat

Sci. Name(s)   : *Fagopyrum tataricum*
Geog. Reg(s)   : Asia
End-Use(s)   : Cereal

Domesticated : Y
Ref. Code(s) : 214, 220
Summary : Tartary buckwheat is considered more hardy than common buckwheat (*Fagopyrum esculentum*). It is ground into flour for breads and other flour products. Tartary buckwheat is native to parts of Asia.

Teff
Sci. Name(s) : *Eragrostis tef*
Geog. Reg(s) : Africa
End-Use(s) : Cereal, Fiber, Forage, Pasture & Feed Grain, Ornamental & Lawn Grass
Domesticated : Y
Ref. Code(s) : 25, 144, 171, 183
Summary : Teff is an iron-rich cereal crop grown primarily in Ethiopia. The grain is ground to flour. Straw provides highly palatable fodder. Its main attributes are quick growth, complete in 6 weeks, and its quick regeneration after a first mowing. It has a capacity to grow in all kinds of soils, especially sandy soil. The straw is used for weaving fine double mats, baskets, and hats. Teff is sometimes grown as an ornamental.

Crops are planted during summer months and reach maturity in 4 months. Grain yields are 3-30 quintals/ha depending on management practices.

Teosinte
(See Forage, Pasture & Feed Grain)

Tepary Bean
(See Vegetable)

Triticale
Sci. Name(s) : *Triticosecale* X
Geog. Reg(s) : America-North (U.S. and Canada)
End-Use(s) : Cereal, Miscellaneous
Domesticated : Y
Ref. Code(s) : 78, 160, 168
Summary : Triticale is a member of a genus that was created by scientists from crosses between tetraploid or hexaploid wheat with rye. The resulting species, triticale, is an experimental cereal that shows high-yield potential. The kernels are higher in protein content than wheat. Plants demonstrate resistance to diseases that can hinder wheat production. Despite these advantages, triticale plants do not form the fully developed kernels like wheat and are of a lower test-weight than wheat. These limitations are being improved. Through research in Canada, several hybrids of triticale have been developed, among them Carman triticale--a spring cereal, OAC Wintri--a winter hexaploid cereal, and OAC Trivell--another spring cereal.

Two-Rowed Barley
Sci. Name(s) : *Hordeum distichon, Hordeum vulgare*
Geog. Reg(s) : Asia-India (subcontinent), Temperate
End-Use(s) : Cereal
Domesticated : Y

Ref. Code(s) : 25, 225
Summary : Two-rowed barley (*Hordeum distichon*), a variety of barley (*Hordeum vulgare*), is a cereal crop with little or no commercial value. In India, crops have been cultivated in Khorasan without irrigation below 914 m in altitude. Crops ripen in about 3 months and are therefore sometimes planted as a second crop.

The cereal is either an annual or perennial. It has erect culms with broad, flat leaves and is usually taller than the variety hexastichon, or six-rowed barley. The barley spikelets are sterile resulting in only 2 rows of fruiting. Two-rowed barley grows mainly in temperate areas.

Velvetbean
(See Forage, Pasture & Feed Grain)

Vine Mesquite, Vine Mesquitegrass
(See Forage, Pasture & Feed Grain)

Wheat, Bread Wheat, Common Wheat
Sci. Name(s) : *Triticum aestivum*
Geog. Reg(s) : Asia, Subtropical, Temperate, Worldwide
End-Use(s) : Beverage, Cereal, Fiber, Forage, Pasture & Feed Grain, Gum & Starch
Domesticated : Y
Ref. Code(s) : 161, 171, 220
Summary : Wheat (*Triticum aestivum*) is a member of the species vulgare, many of which are important cereal crops throughout the world. *Triticum aestivum*, usually referred to as bearded spring wheat, varies in individual culm height, but is generally about 75 cm tall. The high gluten grain is ground to flour and used in breads, biscuits, and confectionery. Fermented grain is made into various alcoholic drinks. Industrial alcohol and wheat starch are used as cloth-stiffeners. Wheat straw feeds and beds livestock and is used in basketry and other woven products.

Plants are native to Afghanistan and are now grown in many temperate and subtropical regions throughout the world as highly mechanized, annual crops. In 1983, world wheat production was close to 500 million MT, with yields of about 2.6 MT/ha. Russia is a leading producer with 16% of the world's total. The United States produced 65.9 million MT of wheat, valued at $8.6 billion. Of this, 17 million MT of wheat was used for human consumption, 2.9 million MT was used for seed, and 12.2 million MT was used for livestock feed. Wheat exports from the United States were 32.1 million MT.

White Lupine, Egyptian Lupine
(See Forage, Pasture & Feed Grain)

Wild Oat
Sci. Name(s) : *Avena fatua*
Geog. Reg(s) : Mediterranean
End-Use(s) : Cereal, Forage, Pasture & Feed Grain

Domesticated : Y
Ref. Code(s) : 171
Summary : The wild oat (*Avena fatua*) is now a domesticated cereal crop. The oats are made into oatmeal or rolled oats and are eaten in porridges and oatcakes. Wild oats are used as feed for livestock. The wild oat is native to the Mediterranean.

Wild Rice, Southern Wildrice, Indian Rice, Tuscarora Rice

Sci. Name(s) : *Zizania aquatica*
Geog. Reg(s) : America-North (U.S. and Canada)
End-Use(s) : Cereal, Ornamental & Lawn Grass
Domesticated : Y
Ref. Code(s) : 17, 135, 171, 220
Summary : Wild rice is a cereal found in the marshy areas of the eastern United States. It was formerly an important part of the Native American diet. Wild rice is often planted as food and shelter for wild ducks and other waterfowl.

Grain is usually collected for domestic use or sale. Once it is harvested, wild rice is sun-dried, boiled, or steamed for use in a variety of dishes. The grain has a higher protein content than other cereals but is comparable in the vitamin content. Wild rice is difficult to harvest. Plants are occasionally grown as ornamentals.

Wild Sesame
(See Oil)

Winged Bean, Goa Bean, Asparagus Pea, Four-Angled Bean, Manilla Bean, Princess Pea
(See Vegetable)

Yague

Sci. Name(s) : *Brachiaria deflexa*
Geog. Reg(s) : Africa-West
End-Use(s) : Cereal, Forage, Pasture & Feed Grain
Domesticated : Y
Ref. Code(s) : 16, 171
Summary : Yague (*Brachiaria deflexa*) is a cereal similar in appearance to hungry rice (*Digitaria exilis*) which is grown in Africa. Yague is a wild grain reaching heights of up to 1 m. The tufted annuals are fairly drought-resistant and provide good fodder in drier areas. Plants mature in 70-75 days. The crop is reportedly grown in the mountainous regions of western Guinea.

Yard-Long Bean, Asparagus Bean
(See Vegetable)

Yellow Lupine, European Yellow Lupine
(See Forage, Pasture & Feed Grain)

Dye & Tannin

African Locust Bean
(See Oil)

Alfalfa, Lucerne, Sativa
(See Forage, Pasture & Feed Grain)

American Chestnut
(See Timber)

Annatto

Sci. Name(s)	: *Bixa orellana*
Geog. Reg(s)	: America, Tropical, West Indies
End-Use(s)	: Dye & Tannin, Ornamental & Lawn Grass, Timber
Domesticated	: Y
Ref. Code(s)	: 26, 170
Summary	: Annatto is a small shrub or tree that is often grown as a hedge plant. Its seeds produce a red dye called annatto. This dye can be used to produce various shades of red and orange although aniline dyes have largely replaced it. Annatto dyes are still used as food coloring in butter, margarine, cheese, and chocolate.

Trees are native to tropical America and the West Indies. They have been widely introduced in the tropics. Trees have soft, even-grained timber. They are often grown as ornamentals.

Avaram

Sci. Name(s)	: *Cassia auriculata*
Geog. Reg(s)	: Africa-East, Africa-South, Asia-India (subcontinent)
End-Use(s)	: Dye & Tannin, Erosion Control
Domesticated	: Y
Ref. Code(s)	: 170
Summary	: Avaram is a sturdy shrub with yellow flowers that is used as a green manure and for replenishing barren soils. Avaram bark contains tannin. At one time, India produced about 45,350 MT of tannin annually but competition from wattle bark tannin imported from eastern and southern Africa reduced the avaram market.

Tannin crops are harvested when shrubs are 2-3 years old. The branches and twigs are cut off and the bark is stripped to yield 18% tannin, approximately 20% by weight. Green bark yields are 1,566 kg/ha.

Babul
(See Gum & Starch)

Baniti
(See Fruit)

Barbatimao

Sci. Name(s)	: *Stryphnodendron adstringens, Stryphnodendron barbatimam*

Geog. Reg(s) : America-South
End-Use(s) : Dye & Tannin, Medicinal
Domesticated : N
Ref. Code(s) : 5
Summary : Barbatimao is a small tree whose bark yields about 40% tannins. The bark has also been used medicinally in tonics, antidiarrhetics and hemostatic astringents. Tree wood is usually twisted and considered unattractive for refined furniture and carpentry work. Trees are native to tropical South America, primarily Brazil, Costa Rica, and Guiana.

Barwood

Sci. Name(s) : *Pterocarpus soyauxii*
Geog. Reg(s) : Africa-West
End-Use(s) : Dye & Tannin, Timber
Domesticated : N
Ref. Code(s) : 36, 56, 170
Summary : Barwood is a West African tree 25-35 m in height whose roots, stems, and heartwood provide useful dyes. The dye from the roots and stems can be used on fabrics and fibers such as raffia. When mixed with palm oil, the dye can be used in cosmetics. Trees also produce a fairly heavy timber valued in West African construction.

Trees are common in mixed deciduous forests in parts of tropical West Africa and can be found in the dense, evergreen forests of Cameroun. Barwood trees need a deep, rich tropical soil with an annual rainfall from 15-17 dm. This timber has been an important export from Cameroun.

Barwood, Camwood

Sci. Name(s) : *Baphia nitida*
Geog. Reg(s) : Africa-West
End-Use(s) : Dye & Tannin, Medicinal, Ornamental & Lawn Grass, Timber
Domesticated : Y
Ref. Code(s) : 50, 170
Summary : Barwood is an attractive West African tree commonly planted as an ornamental. A dye is extracted from the dark red heart of the tree. The dye is insoluble in water but is easily dissolved in alkalis. In West Africa, the yellowish white timber is used for house posts and rafters. A paste made from the heartwood is mixed with shea-butter and used in home remedies to heal wounds. The barwood tree is often started from cuttings or seeds and grown as a natural fence.

Black Mulberry
(See Fruit)

Black Wattle

Sci. Name(s) : *Acacia mearnsii*
Geog. Reg(s) : Africa-South, Australia
End-Use(s) : Dye & Tannin, Energy, Timber
Domesticated : Y
Ref. Code(s) : 170, 229

Summary : The black wattle tree produces 30-45% commercial tannin from the bark. The tannin is used in the production of tannin glue and for tanning hides and skins for various classes of leather. Minor uses are for treating boiler water, for dye production, and for mining work. Black wattle tannin is considered of a better quality than that of the quebracho tannins produced from 2 South American trees (*Schinopsis balansae* and *Schinopsis lorentzii*). The lightweight timber is used for fuel, charcoal, poles, and in wood and paper manufacture. The tree is popular as an agricultural shade source.

Black wattle production centers are mainly in South Africa where trees often spread out of control due to their fast growth rate. Wattle tannin is exported as an extract. In South Africa, crops grow best at higher-altitudes on deep, well-drained soils. Trees will not tolerate extremely hot or cold climates. Black wattles are native to Australia.

Blue Clitoria, Butterfly Pea, Asian Pigeon-Wings, Butterfly Bean, Kordofan Pea
(See Forage, Pasture & Feed Grain)

Brazilwood
Sci. Name(s) : *Caesalpinia echinata*
Geog. Reg(s) : America-South
End-Use(s) : Dye & Tannin
Domesticated : N
Ref. Code(s) : 16, 170
Summary : Brazilwood (*Caesalpinia echinata*) is a South American tree whose heartwood yields a red dye similar to that of the sappanwood tree (*Caesalpinia sappan*). The tree has prickly branches and yellow flowers.

Butternut
(See Nut)

Cashew
(See Nut)

Castorbean
(See Oil)

Chinese Chestnut
(See Nut)

Chinese Tallow-Tree, Tallow-Tree
(See Oil)

Cochineal Cactus
Sci. Name(s) : *Nopalea cochenillifera*
Geog. Reg(s) : America, Subtropical, Tropical, West Indies
End-Use(s) : Dye & Tannin, Fruit

Domesticated : Y
Ref. Code(s) : 100, 154
Summary : Cochineal cactus is a tropical American plant that produces edible red berries. The cochineal insect eats the berry and can then be made to produce a red dye. The cactus is grown commercially for this red dye.

Plants are most successful in full sunlight on moist lime soils. They thrive in the subtropical climates of Haiti, Cuba, Jamaica, and Puerto Rico.

Colchicum, Meadow Saffron, Autumn Crocus
(See Medicinal)

Coltsfoot
(See Medicinal)

Common Guava, Lemon Guava
(See Fruit)

Common Indigo, Indigo
Sci. Name(s) : *Indigofera tinctoria, Indigofera sumatrana*
Geog. Reg(s) : Africa, America-South, Asia-China, Asia-India (subcontinent), Asia-Southeast
End-Use(s) : Dye & Tannin, Erosion Control, Forage, Pasture & Feed Grain
Domesticated : Y
Ref. Code(s) : 56
Summary : Common indigo is a shrublike, herbaceous plant from 1-2 m high whose leaves are collected in India, China, Java, Africa, Malagasay, and tropical America and processed to make a blue dye. India is the major producer and exporter of the dye. Plants are also grown as cover crops and green manure. The leaves are rich in potash and are palatable to cattle.

Plants can be grown from sea level to 300 m in altitude. They are sensitive to waterlogged soils and high winds. Leaves are harvested when plants first begin to flower. Yields vary depending on the area, season, and cultivation. A yield of 75-130 kg bundles/ha of plant material is produced and 1.6-5.4 kg/ha of dye. After 5 months, 2.5 MT/ha of wet matter is obtained and increases to 7 MT after a year.

Divi-Divi
Sci. Name(s) : *Caesalpinia coriaria*
Geog. Reg(s) : America-Central, America-South, Tropical
End-Use(s) : Dye & Tannin, Timber
Domesticated : Y
Ref. Code(s) : 170
Summary : Divi-divi is a small tree or shrub whose pods provide a commercial source of tannin. It has hard, dark colored wood used in general carpentry. The pods also are a source of a black dye used for ink in Mexico. The tree is native to tropical America.

Douglas-Fir
(See Timber)

Dundas Mahogany
(See Timber)

Dyer's-Greenwood, Dyer's-Greenweed

Sci. Name(s) : *Genista tinctoria*
Geog. Reg(s) : America-North (U.S. and Canada), Asia, Europe, Temperate
End-Use(s) : Beverage, Dye & Tannin, Medicinal, Ornamental & Lawn Grass, Spice & Flavoring
Domesticated : N
Ref. Code(s) : 56, 205
Summary : Dyer's-greenwood is a perennial shrub from 10-200 cm in height. It bears yellow flowers that bloom from midsummer to early autumn which can produce a yellow or green dye. When grown for dye production, the flowering tops should be gathered during the summer months. The flower buds are used to flavor sauces, often replacing capers. The seeds have been used as a coffee substitute. The leaves have a bitter principle that can cause cows to produce bitter milk.

The flowers and seeds of this plant have a wide variety of medicinal uses. They have been used as a purgative, diuretic, and have been shown to be slightly cardio-active. The plant is of minor importance in traditional medicine. It is grown most often as an ornamental for borders and rockeries in gardens.

Plants grow in temperate areas on calcareous or sandy soils. They are hardy shrubs that can adapt to drier conditions and are sometimes planted to protect dry sandy banks or rocky slopes. Plants are native to Europe, western Asia and Siberia. Dyer's-greenwood has been introduced to North America where it grows best in dry, woody areas.

Elephant Bush, Spekboom
(See Ornamental & Lawn Grass)

Emblic, Myrobalan, Emblica
(See Fruit)

English Walnut, Persian Walnut
(See Nut)

Eucalyptus
(See Timber)

European Strawberry, Strawberry
(See Fruit)

Fat Hen, Lamb's-Quarters, White Goosefoot, Lambsquarter
(See Cereal)

Flat-Topped Yate
(See Timber)

Fox Grape
(See Fruit)

Fustic-Mulberry

Sci. Name(s)	: *Chlorophora tinctoria*
Geog. Reg(s)	: America-North (U.S. and Canada), West Indies
End-Use(s)	: Dye & Tannin
Domesticated	: N
Ref. Code(s)	: 170
Summary	: The fustic-mulberry tree yields a khaki dye (fustic) obtained from the heart wood. The tree is native to the West Indies and Central and South America.

Gambier, White Cutch

Sci. Name(s)	: *Uncaria gambir*
Geog. Reg(s)	: Asia-Southeast
End-Use(s)	: Dye & Tannin, Medicinal
Domesticated	: Y
Ref. Code(s)	: 36, 170, 220
Summary	: Gambier is a woody, climbing plant whose leaves and young branches produce a tannin called gambier or white cutch. The tannin and dye are extracted from the leaves and twigs. It is used for tanning soft leather and for dyeing and printing. It dyes cotton, wool, and silk a fast brown color. This dye is especially useful in helping to weight silk. The tannin is sometimes used to clear beer. Plant leaves are chewed with the betel nut. The plant's astringency is due to large amounts of catechin and catechu-tannin. The leaves are used in a concoction to treat diarrhea and dystentery, and as a gargle for sore throats.

Plants are cultivated in the wetter parts of western Malaysia and the western half of Java.

Golden Wattle

Sci. Name(s)	: *Acacia pycnantha*
Geog. Reg(s)	: Australia
End-Use(s)	: Dye & Tannin, Erosion Control, Essential Oil, Gum & Starch
Domesticated	: Y
Ref. Code(s)	: 170, 193, 229
Summary	: The bark of golden wattle yields over 40% tannin but the small size of the tree renders it uneconomical for large-scale production. The tree matures in approximately 10 years. Golden wattle flowers yield an essential oil used in perfumery. The tree produces a good gum and has been used for erosion control. The highly adaptable Australian tree will grow in most soils. Its habitat ranges from coastal areas to inland semiarid regions.

Green Wattle
(See Timber)

Hairy Indigo

Sci. Name(s)	: *Indigofera hirsuta*

Geog. Reg(s)   : Africa-West, America-North (U.S. and Canada), Asia, Asia-India
                   (subcontinent), Subtropical, Tropical
End-Use(s)   : Dye & Tannin, Erosion Control, Forage, Pasture & Feed Grain
Domesticated   : Y
Ref. Code(s)   : 56
Summary   : Hairy indigo is a valuable summer annual or biennial herb from 0.6-2.3 m high.
Plants are most successful on fertile, sandy loams. They are grown as cover crops and
for soil improvement. Hairy indigo has been mixed with permanent grass for grazing.
Its leaves are a source of indigo dye in West Africa. Hairy indigo is native to tropical
Asia and the India plains. It has been introduced and grown in other tropical and
subtropical areas.

In the United States, yields of green matter reach 22 MT/ha/year. In Florida,
hairy indigo is mixed with pangola or bahiagrass to yield 5.5 MT/ha of dry matter.
Average seed yields are from 100-300 kg/ha.

Harmala Shrub, African-Rue, Harmel Piganum
(See Medicinal)

Heartnut, Cordate Walnut, Siebold Walnut
(See Nut)

Henna
(See Ornamental & Lawn Grass)

Horehound, White Horehound
(See Medicinal)

Huang T'eng
(See Medicinal)

Indian Bael, Bael Fruit, Bengal Quince, Bilva, Siniphal, Bael Tree
(See Fruit)

Indian Jujube, Beri, Inu-Natsume, Ber Tree
(See Fruit)

Indian Mulberry

Sci. Name(s)   : *Morinda citrifolia, Morinda trifolia*
Geog. Reg(s)   : Asia-Southeast
End-Use(s)   : Dye & Tannin, Insecticide, Medicinal, Vegetable
Domesticated   : Y
Ref. Code(s)   : 17, 220
Summary   : The leaves of the Indian mulberry tree yield a red dye used for batik work in
Java. The roots provide a yellow dye. Mulberry leaves are eaten as vegetables, but
the fruit is believed to be poisonous. An infusion of the bark, roots, and fruit has
been used to dress wounds. Tree wood is burned as an insect repellent. The Indian
mulberry tree is native to Malaysia and Indonesia.

Indian Wood-Apple
(See Fruit)

Indian-Almond, Country Almond, Tropical-Almond, Myrobalan, Almendro
(See Oil)

Inula, Elecampane
(See Essential Oil)

Jackfruit, Jack
(See Fruit)

Kadjatoa
(See Gum & Starch)

Lac-tree, Ceylon Oak, Kussum Tree, Malay Lac-tree
(See Fat & Wax)

Malabar Nut
(See Medicinal)

Marking-Nut Tree

Sci. Name(s)	: *Semecarpus anacardium*
Geog. Reg(s)	: Asia-India (subcontinent), Asia-Southeast, Australia
End-Use(s)	: Dye & Tannin, Energy, Fruit, Medicinal, Miscellaneous, Oil
Domesticated	: N
Ref. Code(s)	: 16, 36, 98, 154
Summary	: The marking-nut tree is a moderate-sized, deciduous tree whose nuts produce

a juice used as an indelible, black, marking ink for linen. The fruit provides a dark gray dye when mixed with alum and lime. The fleshy fruit base is eaten roasted. In India, immature fruit is pickled. The nut kernels contain 9.2% sweet oil. The pericarp also yields an oil that acts as an irritant and has been used in certain external medicines. Ripe fruit has been used as an internal medicine. Tree wood has been used as charcoal.

Trees will continue to thrive at altitudes approaching 1,066 m and are found throughout the Malay Archipelago and in northern Australia.

Menteng, Kapundung
(See Fruit)

Natal Indigo

Sci. Name(s)	: *Indigofera arrecta*
Geog. Reg(s)	: Africa, Africa-North, Africa-South, Asia, Asia-India (subcontinent), Asia-Southeast, Mediterranean
End-Use(s)	: Dye & Tannin, Erosion Control, Forage, Pasture & Feed Grain
Domesticated	: Y
Ref. Code(s)	: 56

Summary   : Natal indigo is a stout, perennial herb whose leaves are the source of the blue dye, indigo. The herb reaches heights from 1-2 m and is occasionally grown as a cover crop or an intercrop with cocoa, coffee, oil palm, tea, or rubber. Plants help control weeds and improve the soil. They can tolerate dry weather better than other common cover crops such as the pigeon pea. Natal indigo is also grown as a green manure crop.

Plants are most successful in hot, moist climates in areas receiving 175 cm/annum of rain. In Africa, crops thrive from 300-2,700 m in altitude. They should be planted in well-drained, light, friable soils. India, Assam, and Java are major producing regions. Plants are native to tropical and South Africa, southern Arabia, and Madagascar.

Approximately 22-100 MT of green matter has reportedly been produced in India. About 1,200 kg/ha of seeds are produced in Assam and 135-325 kg/ha of indigo cake. Natal indigo is supposedly one of the highest leaf yielding members of the Indigofera species.

Oiticica Oil, Oiticica
(See Oil)

Pamque

Sci. Name(s) : *Gunnera tinctoria, Gunnera chilensis*
Geog. Reg(s) : America-South
End-Use(s) : Dye & Tannin, Vegetable
Domesticated : N
Ref. Code(s) : 220
Summary : Pamque is an herb whose roots have been used for tanning. Young leaf stalks can be eaten as vegetables. There is little other information available about this plant. The herb is native to Chile.

Papaya
(See Fruit)

Pecan
(See Nut)

Pokeweed, Poke, Skoke, Pigeon Berry
(See Weed)

Pomegranate
(See Fruit)

Quebracho

Sci. Name(s) : *Schinopsis balansae*
Geog. Reg(s) : America-South
End-Use(s) : Dye & Tannin
Domesticated : N
Ref. Code(s) : 170, 220

Summary : Quebracho is a South American tree whose heartwood contains about 30% tannin. The tannin is extracted by a pressurized steam process and is produced mainly in Argentina and Paraguay. The United States has been a major importer of the tannin.

Quinine, Quinine Tree, Peruvian Bark
(See Medicinal)

Rambai
(See Fruit)

Red Clover
(See Forage, Pasture & Feed Grain)

Red Ironbark
(See Timber)

Red Quebracho
Sci. Name(s) : *Schinopsis lorentzii, Schinopsis quebracho-colorado*
Geog. Reg(s) : America-South
End-Use(s) : Dye & Tannin, Miscellaneous, Timber
Domesticated : N
Ref. Code(s) : 116, 170
Summary : Red quebracho is a medium-sized deciduous tree whose heartwood contains about 30% tannins by weight. The bark and sapwood contain about 3-8% tannins. Red quebracho tannin is fairly inexpensive and is quick-acting, producing strong, firm, and tough leather. A quebracho extract in the form of a dark red powder is used in oil-drilling mud, and in the dyeing and printing of certain fabrics. The hard, heavy wood has been used in cabinetry and fence post-making.

Trees have relatively slow growth and do not develop significant amounts of tannin until they reach 40-50 years. They mature in 100 years and are cut at 100-120 years. Red quebracho thrives in lowland areas and swamps and will tolerate some soil salinity. They are native to southern South America. Tannin is produced in Argentina and Paraguay. The United States has been a major importer.

Red Sanderswood, Red Sandalwood, Calialur Wood
Sci. Name(s) : *Pterocarpus santalinus*
Geog. Reg(s) : Asia-India (subcontinent)
End-Use(s) : Dye & Tannin, Medicinal, Timber
Domesticated : Y
Ref. Code(s) : 36, 56, 170
Summary : Red sanderswood is an East Indian tree from 5-6 m in height from which a rose-tinted dye is obtained. The dye is valued for dyeing leather red, for staining wood, and for calico printing. It is used for making caste marks in India. Tree timber is hard and impenetrable to white ants and termites and can be polished to a high-luster. In India, the timber has been used for various types of construction. Preparations of the wood have astringent properties and are used in soothing external

applications for inflammation, headaches, fevers, and boils, and as a diaphoretic and treatment for scorpion stings. There is little or no use of red sanderswood in traditional western medicines.

Trees grow well in dry, hot areas on rocky soils but they thrive on well-drained alluvial soils. They require considerable sunlight and an annual rainfall from 13.6-27.8 cm. Southern India has been the major timber and dye producing region. A regular export market has evolved for this timber.

Rose Apple, Jambos
(See Fruit)

Roselle
(See Fiber)

Safflower
(See Oil)

Saffron, Saffron Crocus

Sci. Name(s)	: *Crocus sativus*
Geog. Reg(s)	: Asia-China, Asia-India (subcontinent), Eurasia, Europe
End-Use(s)	: Dye & Tannin, Medicinal, Spice & Flavoring
Domesticated	: Y
Ref. Code(s)	: 194, 220
Summary	: Saffron is a perennial, low-growing herb with upright leaves from 0.15-0.3 m in

height. It is cultivated for its fragrant flowers that yield the dye, saffron. The dye contains the yellow glucoside, crocin, and is used as a coloring agent in foods. Saffron is marketed as individual stigmas, ground, or crushed. It is used in these forms as a cooking spice, and flavor in apertif beverages. Medicinally, saffron has been used in traditional preparations as an anodyne, antispasmodic, aphrodisiac, diaphoretic, emmenagogue, expectorant, and sedative. The high cost of saffron production makes other competitive dye crops such as tumeric, and synthetically produced colorants such as tartrazine, attractive.

Plants are cultivated primarily in Spain, with lesser production in Greece, Turkey, India, France, Italy, and China. Saffron is most successful when grown in well-drained soils with medium fertility, in areas receiving an annual rainfall from 0.1-1.1 m.

Salt River Mallett
(See Energy)

Sappanwood, Japan Wood

Sci. Name(s)	: *Caesalpinia sappan*
Geog. Reg(s)	: Asia-India (subcontinent), Asia-Southeast, Tropical
End-Use(s)	: Dye & Tannin, Timber
Domesticated	: N
Ref. Code(s)	: 29, 170, 220

Summary : Sappanwood is a native tree of Southeast Asia. Its red heartwood produces a red dye, called bresil. When mixed with iron salts, the bark produces a black dye which is used in India. The wood itself is used for cabinet work in the tropics. It is a small or middle-sized thorny tree that is common throughout the Malay Peninsula and Archipelago.

Senegal Rosewood, Barwood, West African Kino, Red Barwood

Sci. Name(s) : *Pterocarpus erinaceus*
Geog. Reg(s) : Africa-West
End-Use(s) : Dye & Tannin, Forage, Pasture & Feed Grain, Gum & Starch, Timber
Domesticated : Y
Ref. Code(s) : 36, 56, 170
Summary : The Senegal rosewood is a West African tree growing to 17 m high whose heartwood yields a dye and kino. Kino is a West African term referring to the gum from which the first European medicines were derived. The term now refers to any ruby red or purple black astringent gum. The cost of producing barwood kino is too high for it to be an economical tanning material. Trees have hard, fine-grained, rose-red timber that has been used for fence posts, light construction, and carpentry. The foliage provides forage for domestic animals.

Trees thrive in dry savannah forests, along riverbanks, and on shallow soils in areas receiving 17.3-40.3 dm/year of rainfall. Senegal rosewood trees require little cultivation once established. Timber is cut as needed, and foliage is either grazed or cut. Trees have been a valuable resource in tropical West Africa.

Shittim-Wood
(See Forage, Pasture & Feed Grain)

Sorghum, Great Millet, Guinea Corn, Kaffir Corn, Milo, Sorgo, Kaoling, Durra, Mtama, Jola, Jawa, Cholam Grains, Sweets, Broomcorn, Shattercane, Grain Sorghums, Sweet Sorghums
(See Cereal)

Spicata Indigo

Sci. Name(s) : *Indigofera spicata, Indigofera endecaphylla, Indigofera hendecaphylla*
Geog. Reg(s) : Africa-South, America-South, Asia-India (subcontinent), Asia-Southeast, Hawaii
End-Use(s) : Dye & Tannin, Erosion Control
Domesticated : Y
Ref. Code(s) : 56
Summary : Spicata indigo is a perennial herb from 1-2 m in height grown as a dye source in tropical South Africa, Southeast Asia, Sri Lanka, India, tropical America, and the Philippines. It has been grown successfully in Hawaii. The dye is obtained from the leaves. Plants are grown as cover crops for coffee, tea, and rubber. Spicata indigo is not recommended as a forage crop because of its toxicity.

Crops are short-day plants that are most successful in areas receiving 600-1,500 mm of rain/year. Plants can be grown in partial shade from sea level to 2,700 m in altitude. They do well on clay soils and provide good cover on sandy soils. A

2 month old plant yields 5 MT/ha of green matter, while a 6 month plant yields 25 MT/ha. Average seed yields are approximately 500 kg/ha.

Spotted Burclover
(See Erosion Control)

Sunflower
(See Oil)

Surinam-Cherry, Pitanga Cherry
(See Fruit)

Tan Wattle
(See Timber)

Tara

Sci. Name(s)	: *Caesalpinia spinosa*
Geog. Reg(s)	: America-South
End-Use(s)	: Dye & Tannin
Domesticated	: N
Ref. Code(s)	: 170
Summary	: Tara is a tree native to tropical America. Its pods provide a commercial source of tannin.

Tuart
(See Timber)

Turmeric
(See Spice & Flavoring)

Vogel Fig
(See Isoprenoid Resin & Rubber)

Wine Grape, Grape, European Grape, California Grape
(See Beverage)

Yellow Flame, Yellow Poinciana, Soga
(See Ornamental & Lawn Grass)

Energy

Almond
(See Nut)

Apple
(See Fruit)

Arizona Cypress
(See Timber)

Australian Blackwood
(See Timber)

Babassu, Orbignya speciosa
(See Oil)

Babul
(See Gum & Starch)

Balsam Amyris, Amyris
(See Essential Oil)

Beach, She Oak
(See Timber)

Black Box
(See Timber)

Black Locust, False Acacia
(See Timber)

Black Wattle
(See Dye & Tannin)

Blue Gum
(See Timber)

Cassava, Manioc, Tapioca-Plant, Yuca, Mandioca, Guacomole

Sci. Name(s)	: *Manihot esculenta, Manihot dulcis, Manihot melanobasis, Manihot tilissima*
Geog. Reg(s)	: Africa, America-Central, America-South, Asia-Southeast, Tropical
End-Use(s)	: Beverage, Cereal, Energy, Forage, Pasture & Feed Grain, Gum & Starch, Spice & Flavoring, Tuber, Vegetable
Domesticated	: Y
Ref. Code(s)	: 44, 126, 132, 154, 170
Summary	: Cassava is a tropical tuber crop which is not known in a wild state. Cassava can provide ethanol for fuel if there is an efficient source of energy for distillation.

The tubers are a valuable source of food, carbohydrates, phosphorous, iron, and calcium. A starch is made from the tubers that is highly prized in cooking. Tapioca, a popular component in puddings, bisquits and other confectionery, is obtained by heating the cassava flour. Parts of the cassava plant are used in the manufacture of adhesives, cosmetics, textiles and paper making. In many countries it is a major food staple. It has a potential market as a substitute for cereal flours and as an energy source for animals. Sweet cassavas are grown much more than bitter cassavas.

Young plant leaves are edible and can be cooked much like spinach. Alcoholic beverages are made from the tubers and the plant leaves are used as pot herbs. A thick paste, fufu, is made by peeling, cutting, boiling and pounding the roots. A disadvantage to growing the crop is that it is scavenged by wild animals.

Cassava is believed to have originated in western and southern Mexico and parts of Guatemala. It is grown as a reserve crop in some areas. Maturity for short-season cassava cultivars are 6-10 months, while the long-season cultivar takes approximately 2 or more years. Yields are from 2.25-27 MT/ha.

In 1984, total world production of cassava was 115 million MT on 14 million ha. The United States imported 136,050 MT of cassava flour and starch in 1984. The main exporting countries are Thailand with 12 million MT, Indonesia with 408,150 MT and China with 362,800 MT.

Cassia
(See Timber)

Cooba
(See Timber)

Coolabah
(See Timber)

Damas
(See Timber)

Desert Date
(See Fruit)

Douglas-Fir
(See Timber)

Dundas Mahogany
(See Timber)

Eucalyptus
(See Timber)

Eucalyptus urophylla
(See Timber)

Eurabbie
(See Timber)

European Beech
(See Ornamental & Lawn Grass)

Fenugreek
(See Spice & Flavoring)

Flat-Topped Yate
(See Timber)

Flooded Gum, Rose Gum
(See Timber)

Forest Red Gum, Mysore "Hybrid", Izabl, Eucalyptus "C"
(See Timber)

Green Wattle
(See Timber)

Grey Ironbark
(See Timber)

Gum-Barked Coolibah
(See Timber)

Horsebean
(See Timber)

Jambolan, Java Plum
(See Fruit)

Khasya Pine, Khasi Pine, Benguet Pine
(See Timber)

Leucaena-Hawaiian Type
(See Timber)

Leucaena-Salvadorian Type
(See Timber)

Longleaf Pine
(See Essential Oil)

Lukrabao, Chaulmogra Tree
(See Medicinal)

Maidens Gum
(See Timber)

Marking-Nut Tree
(See Dye & Tannin)

Mesquite, Algorrobo
(See Timber)

Messmate
(See Timber)

Mindanao Gum
(See Timber)

Monterey Cypress
(See Timber)

Montezuma Pine
(See Timber)

Narrow-Leaved Ironbark
(See Timber)

Neem
(See Timber)

Opepe, Bilinga
(See Timber)

Orange Wattle

Sci. Name(s) : *Acacia cyanophylla, Acacia saligna*
Geog. Reg(s) : Australia
End-Use(s) : Energy, Erosion Control
Domesticated : Y
Ref. Code(s) : 228
Summary : Orange wattle is a small evergreen tree that reaches 4-8 m in height. It is grown on the coastal strip of West Australia as a source of charcoal and fuel. The trees provide a good windbreak and hinder erosion. Two to 4 cu.m/ha/annum of trees are produced.

The orange wattle grows at altitudes up to 2,500 m. Necessary rainfall varies from 300-700 mm annually, with temperatures ranging from 16-26 degrees Celsius. These trees need considerable sunlight for optimum growth. They favor light- to medium-textured soils but tolerate shallow and moderately saline soils.

Paperbark
(See Timber)

Pearl Millet, Bulrush Millet, Spiked Millet, Cat-tail Millet
(See Cereal)

Pecan
(See Nut)

Pino Blanco
(See Timber)

Ponderosa Pine
(See Timber)

Red Ironbark
(See Timber)

Red Mahogany
(See Timber)

Red River Gum
(See Timber)

Rice
(See Cereal)

Russian Olive
(See Timber)

Salmon Gum
(See Timber)

Salt River Mallett

Sci. Name(s)	: *Eucalyptus sargentii*
Geog. Reg(s)	: Australia
End-Use(s)	: Dye & Tannin, Energy, Erosion Control
Domesticated	: Y
Ref. Code(s)	: 228
Summary	: The salt river mallett of Southwest Western Australia is an evergreen whose roundwood is used for fuel and charcoal. Its timber is never sawn. It is the source of a tannin. The tree itself serves as a shade, shelter, and windbreak.

This evergreen tree stands 8-10 m in height. It demands considerable light and is most often located at altitudes up to 1,500 m. Temperatures of 18-25 degrees Celsius and rainfall of 330-450 mm is required. This tree prefers light- to medium-textured soils and will tolerate saline soils.

Sesame, Simsim, Beniseed, Gingelly, Til
(See Oil)

Shingle Tree
(See Timber)

Silky Oak
(See Timber)

Siris, East Indian Walnut, Koko
(See Timber)

Southern Mahogany
(See Timber)

Spotted Gum
(See Timber)

Sugar Gum
(See Timber)

Sugarcane
(See Sweetener)

Swamp Cypress
(See Timber)

Swamp Mahogany
(See Timber)

Sweet Birch, Black Birch, Cherry Birch, Mountain Mahogany
(See Essential Oil)

Tamarind
(See Fruit)

Tamarisk
(See Timber)

Tamarugo, Tamarugal
(See Timber)

Tan Wattle
(See Timber)

Teak, Tec, Teca
(See Timber)

Tuart
(See Timber)

Umbrella Pine, Stone Pine
(See Timber)

Yellow Box
(See Timber)

Yemane, Gmelina
(See Timber)

Erosion Control

African Bermudagrass, Masindi Grass
(See Ornamental & Lawn Grass)

Aleppo Pine
(See Timber)

Alfalfa, Lucerne, Sativa
(See Forage, Pasture & Feed Grain)

Alyceclover, Oneleaf-Clover
(See Forage, Pasture & Feed Grain)

American Beachgrass

Sci. Name(s)	:	*Ammophila breviligulata*
Geog. Reg(s)	:	America-North (U.S. and Canada)
End-Use(s)	:	Erosion Control
Domesticated	:	Y
Ref. Code(s)	:	164
Summary	:	American beachgrass is a coarse, tough perennial grass important in the United States as a sand binder along the Atlantic coast and Great Lakes.

American Licorice, Wild Licorice
(See Spice & Flavoring)

Arizona Cypress
(See Timber)

Australian Blackwood
(See Timber)

Australian Bluestem
(See Forage, Pasture & Feed Grain)

Australian Tea-Tree

Sci. Name(s)	:	*Leptospermum laevigatum*
Geog. Reg(s)	:	America-North (U.S. and Canada), Australia
End-Use(s)	:	Erosion Control, Ornamental & Lawn Grass
Domesticated	:	Y
Ref. Code(s)	:	127
Summary	:	The Australian tea-tree is a small evergreen from 2-5 m in height which is often planted as a large ornamental hedge or sand binder in Australia and California. Australian tea-trees are common beach trees in Tasmania, Victoria, New South Wales, and Queensland. They are hardy and reproduce quickly.

70

Avaram
(See Dye & Tannin)

Babul
(See Gum & Starch)

Bahia Grass, Bahiagrass
(See Forage, Pasture & Feed Grain)

Ball Clover
(See Forage, Pasture & Feed Grain)

Bamboo, Phyllostachys
(See Timber)

Bard Vetch, Monantha Vetch
(See Forage, Pasture & Feed Grain)

Barnyardgrass, Barnyard Millet
(See Forage, Pasture & Feed Grain)

Batai
(See Timber)

Beach, She Oak
(See Timber)

Beach Plum, Shore Plum

Sci. Name(s)	: *Prunus maritima, Prunus acuminata*
Geog. Reg(s)	: America-North (U.S. and Canada)
End-Use(s)	: Erosion Control, Fruit, Ornamental & Lawn Grass
Domesticated	: Y
Ref. Code(s)	: 15, 16, 17, 89, 212
Summary	: The beach plum is a shrubby tree from 0.6-1.8 m in height that produces tart, edible fruit. Trees make attractive ornamentals and good sand binders. They are hardy and require little care. Trees are common along the coastal areas of the eastern United States.

Bermudagrass, Star Grass, Bahama Grass, Devilgrass

Sci. Name(s)	: *Cynodon dactylon*
Geog. Reg(s)	: Europe, Tropical
End-Use(s)	: Erosion Control, Forage, Pasture & Feed Grain, Ornamental & Lawn Grass, Weed
Domesticated	: Y
Ref. Code(s)	: 171
Summary	: Bermudagrass is a spreading perennial from 6-90 cm in height and is one of the most drought-resistant pasture and lawn grasses of the tropical and warm

countries of the world. It remains green longer than other grasses during times of drought, is very palatable when young, and is able to withstand heavy grazing. The grass can become a serious weed of cultivated land and once established, is difficult to eradicate. It is a good soil binder.

The grass is most common on sandy grounds and is native to the sandy shores of southwestern England. It is very hardy and can adapt to most environments but will not tolerate shade.

Bhutan Cypress
(See Timber)

Birdsfoot Trefoil

Sci. Name(s)	: *Lotus corniculatus*
Geog. Reg(s)	: America-North (U.S. and Canada), America-South, Asia, Asia-Central, Asia-India (subcontinent), Australia, Eurasia, Europe
End-Use(s)	: Erosion Control, Forage, Pasture & Feed Grain
Domesticated	: Y
Ref. Code(s)	: 56
Summary	: Birdsfoot trefoil is a major pasture legume that thrives alone or mixed with various grasses. It is used extensively for soil improvement and erosion control. This legume grows on infertile, poorly drained soils, and makes a good, drought-resistant ground cover. Its seeds are reported to contain trypsin inhibitors, cytisine, malonic acid, and saponin. Trefoil yields vary from 55-160 kg/ha. Plants are found in Eurasia, Europe, Australia, North and South America, Asia and India.

Black Gram, Urd, Wooly Pyrol
(See Vegetable)

Black Locust, False Acacia
(See Timber)

Blackpod Vetch, Narrow Leaf Vetch
(See Forage, Pasture & Feed Grain)

Blue Wildrye
(See Ornamental & Lawn Grass)

Boer Lovegrass

Sci. Name(s)	: *Eragrostis chloromelas*
Geog. Reg(s)	: America-North (U.S. and Canada), Asia-India (subcontinent)
End-Use(s)	: Erosion Control, Forage, Pasture & Feed Grain
Domesticated	: Y
Ref. Code(s)	: 17, 226
Summary	: Boer lovegrass is a strongly tufted, long-lived perennial grass, used for fodder and erosion control. It has been used for reclaiming semidesert grazing lands in the United States and India. Its forage is considered quite palatable.

This lovegrass grows to a height from 40-90 cm and forms dense clumps with deep, wide-spreading, well-distributed roots. Boer lovegrass is well-adapted to semidesert conditions but is not frost-resistant. It grows on a wide range of soils, can be established by seeding, and should be fertilized if high-quality forage is desired.

Brazilian Lucerne
(See Forage, Pasture & Feed Grain)

Buffalograss
(See Forage, Pasture & Feed Grain)

Butterfly Pea

Sci. Name(s)	: *Clitoria laurifolia*
Geog. Reg(s)	: America-South, Asia-Southeast, Tropical
End-Use(s)	: Erosion Control, Forage, Pasture & Feed Grain, Vegetable
Domesticated	: Y
Ref. Code(s)	: 36, 56, 170
Summary	: The butterfly pea is a pulse crop of tropical America. It was one of the first experimental green manure crops but because it does not spread, is used primarily as a cover crop. Butterfly peas are cultivated in the Malay Pennisula and southeast Asia for fodder and green manure and occasionally as a cover crop. Because the plant has stiff, erect stems and throws out suckers, it is sturdy enough to be used in contour hedges that hold up soil. The plant is known to increase the nitrogen content in soils.

It is a perennial shrub reaching up to 2-9 dm in height. It can tolerate heavy soil, sand, and waterlogging and is most successful in areas with an annual rainfall between 5.3-27.8 dm and temperatures between 21-27.3 degrees Celsius.

California Burclover, Toothed Burclover
(See Forage, Pasture & Feed Grain)

Canada Wildrye
(See Forage, Pasture & Feed Grain)

Cassia
(See Timber)

Caucasian Bluestem
(See Forage, Pasture & Feed Grain)

Centipede Grass
(See Ornamental & Lawn Grass)

Cherry Laurel, English Laurel

Sci. Name(s)	: *Prunus laurocerasus, Laurocerasus officinalis*
Geog. Reg(s)	: Asia, Europe
End-Use(s)	: Erosion Control, Medicinal, Ornamental & Lawn Grass, Spice & Flavoring
Domesticated	: Y

Ref. Code(s) : 16, 17, 21, 212, 220
Summary : The cherry laurel is a shrub or small evergreen tree often grown as an ornamental. It tolerates light pruning and is popular as hedges and covers for soil banks. An extract of the leaves (cherry laurel water) is used as a pain reliever, sedative, and treatment for coughs. This extract has been substituted for bitter almond extract. Shrubs are native to eastern Europe and southwestern Asia.

Common Indigo, Indigo
(See Dye & Tannin)

Common Lavender, English Lavender
(See Essential Oil)

Common Vetch
(See Forage, Pasture & Feed Grain)

Crimson Clover
(See Forage, Pasture & Feed Grain)

Crowfoot Grass
(See Weed)

Crownvetch, Trailing Crownvetch
(See Forage, Pasture & Feed Grain)

Dallisgrass
(See Forage, Pasture & Feed Grain)

Dichondra
(See Ornamental & Lawn Grass)

Dundas Mahogany
(See Timber)

Elephant Grass, Napier Grass
(See Forage, Pasture & Feed Grain)

Eucalyptus
(See Timber)

European Beachgrass, Marram Grass
Sci. Name(s) : *Ammophila arenaria*
Geog. Reg(s) : Europe
End-Use(s) : Erosion Control, Fiber
Domesticated : Y
Ref. Code(s) : 42

Summary   : European beachgrass is a stout perennial grass grown extensively as a sand binder. Its leaves are used in basketry, mats, and sandals. European beachgrass grows along the coasts of the British Isles and western Europe.

Fairway Crested Wheatgrass
Sci. Name(s)   : *Agropyron cristatum*
Geog. Reg(s)   : America-North (U.S. and Canada)
End-Use(s)   : Erosion Control, Forage, Pasture & Feed Grain
Domesticated   : Y
Ref. Code(s)   : 164
Summary   : Fairway crested wheatgrass has proved very useful for regrassing abandoned croplands and depleted ranges in the northern Great Plains area of the United States. The grass is a valuable forage species and is sometimes found in mixed plantings of *Agropyron desertorum*.

Flat-Topped Yate
(See Timber)

Gamba Grass
(See Forage, Pasture & Feed Grain)

Golden Wattle
(See Dye & Tannin)

Gow-Kee, Chinese Matrimony Vine, Chinese Wolfberry
(See Vegetable)

Green Wattle
(See Timber)

Hairy Indigo
(See Dye & Tannin)

Hairy Vetch, Winter Vetch, Russian Vetch
(See Forage, Pasture & Feed Grain)

Hirta Grass
Sci. Name(s)   : *Hyparrhenia hirta*
Geog. Reg(s)   : Africa-North, America-North (U.S. and Canada), Asia-Southeast, Australia, Mediterranean, Tropical
End-Use(s)   : Erosion Control, Forage, Pasture & Feed Grain
Domesticated   : Y
Ref. Code(s)   : 17, 25, 96, 220
Summary   : Hirta grass is a perennial not more than 1 m in height cultivated for erosion control. It has little forage value but has potential as a fodder crop because it adapts to adverse conditions. This grass grows in the drier parts of the Mediterranean, the Middle East, and tropical Africa. It is cultivated at the Florida State Experimental

Station. Hirta grass is native to Africa, Sri Lanka, the Philippines, and eastern Australia.

Horsebean
(See Timber)

Horsegram
(See Cereal)

Hungarian Vetch
(See Forage, Pasture & Feed Grain)

Indian Melilot, Sourclover, Indian Sweetclover, Annual Yellow Sweetclover
(See Forage, Pasture & Feed Grain)

Intermediate Wheatgrass, Grenar Intermediate Wheatgrass
(See Forage, Pasture & Feed Grain)

Itabo, Izote, Palmita, Ozote, Spanish Bayonnette
(See Vegetable)

Japanese Lespedeza, Japanese Clover, Common Lespedeza, Striate Lespedeza
(See Forage, Pasture & Feed Grain)

Kikuyu Grass
(See Forage, Pasture & Feed Grain)

Korean Lespedeza, Korean Clover

Sci. Name(s)	: *Lespedeza stipulacea*
Geog. Reg(s)	: America-North (U.S. and Canada), Asia, Asia-China, Temperate
End-Use(s)	: Erosion Control, Forage, Pasture & Feed Grain
Domesticated	: Y
Ref. Code(s)	: 56
Summary	: Korean lespedeza is an annual herb grown for forage, hay, green manure, soil

improvement, erosion control, and as a cover crop. This herb is more tolerant of acidic soils than alfalfa, sweet clover, or red clover and is fairly drought-hardy.

Plants grow in warm temperate climates in most rich soils. They are native to eastern Asia, northern China, Korea, and Manchuria and have been introduced to the United States where they are grown extensively as soil improvement and erosion control agents.

Yields vary greatly, depending on the soil and length of the growing season. Average yields are 2.5 MT/ha of hay and 300-400 kg/ha of unhulled seeds. Crops can be grazed when 5 cm in height. Highest quality hay is cut just prior to first bloom.

Kudzu Vine, Kudzu
(See Tuber)

Lanceleaf Crotalaria

Sci. Name(s)	: *Crotalaria lanceolata, Crotalaria longirostrata*
Geog. Reg(s)	: Africa, Tropical
End-Use(s)	: Erosion Control, Forage, Pasture & Feed Grain
Domesticated	: Y
Ref. Code(s)	: 56, 170
Summary	: Lanceleaf crotalaria is an annual herb reaching 1.7 m in height used for soil erosion control because of its rapid and persistent growth. It is sometimes cultivated for green pasture and forage and can produce a substantial pasture quickly. Plants are native to Africa and are common in most tropical areas. They thrive in damp soils anywhere from sea level to 2,150 m in elevation.

Leadtree, Lead Tree
(See Forage, Pasture & Feed Grain)

Lehmann Lovegrass

Sci. Name(s)	: *Eragrostis lehmanniana*
Geog. Reg(s)	: Africa-South, America-North (U.S. and Canada), Asia-India (subcontinent), Subtropical, Temperate, Tropical
End-Use(s)	: Erosion Control, Medicinal
Domesticated	: Y
Ref. Code(s)	: 17, 226
Summary	: Lehmann lovegrass is a hardy South African grass naturalized in Arizona but originally introduced to the southwestern United States as an erosion control agent. It should be surface sown for best germination results. It occurs naturally in tropical and subtropical South Africa and is cultivated mostly in the subtropics or warm temperate areas. In Africa, it is used as a remedy for colic, diarrhea, and typhoid fever.

Leucaena-Hawaiian Type
(See Timber)

Leucaena-Salvadorian Type
(See Timber)

Lily Turf, Dwarf Lily Turf, Mondo Grass, Snake's Beard
(See Ornamental & Lawn Grass)

Manila Grass, Zoysia, Japanese Carpet Grass
(See Ornamental & Lawn Grass)

Monterey Cypress
(See Timber)

Moth Bean, Mat Bean
(See Vegetable)

Natal Indigo
(See Dye & Tannin)

Orange Wattle
(See Energy)

Paperbark
(See Timber)

Pinus pumilio, Swiss Mountain Pine, Dwarf Pine
(See Essential Oil)

Pinyon Pine, Silver Pine, Nut Pine, Pinyon
(See Nut)

Ponderosa Pine
(See Timber)

Prisolilia

Sci. Name(s)	: *Calopogonium mucunoides*
Geog. Reg(s)	: America-South, Tropical
End-Use(s)	: Erosion Control
Domesticated	: Y
Ref. Code(s)	: 170
Summary	: Prisolilia is a climbing perennial valued as a cover crop for its rapid growth and dense, matted foliage which is hairy and unpalatable to livestock. It is native to South America and widely introduced throughout the tropics.

Purple Vetch

Sci. Name(s)	: *Vicia benghalensis, Vicia atropurpurea, Vicia broteriana*
Geog. Reg(s)	: Africa-North, Europe, Subtropical, Temperate, Tropical
End-Use(s)	: Erosion Control, Forage, Pasture & Feed Grain
Domesticated	: Y
Ref. Code(s)	: 24, 42, 141, 220
Summary	: The purple vetch is a climbing annual, biennial, or short-lived perennial cultivated in mild, temperate areas as a cover crop and for fodder, hay, and silage. Plants are grown in mixtures with winter cereals and in the tropics and subtropics at high-altitudes. They are also found wild. Purple vetch is native to southern Europe and North Africa.

Quack Grass, Torpedograss
(See Weed)

Red Clover
(See Forage, Pasture & Feed Grain)

Reed Canarygrass
(See Forage, Pasture & Feed Grain)

Russian Olive
(See Timber)

Russian Wildrye
(See Forage, Pasture & Feed Grain)

Salt River Mallett
(See Energy)

Sand Bluestem, Turkey Foot

Sci. Name(s) : *Andropogon hallii*
Geog. Reg(s) : America-North (U.S. and Canada)
End-Use(s) : Erosion Control
Domesticated : N
Ref. Code(s) : 96
Summary : Sand bluestem is a North American grass with potential as an erosion control
agent. It grows on the sand hills and sandy soils of North Dakota, eastern Montana,
Texas, Wyoming, Arizona, and Iowa.

Sand Dropseed

Sci. Name(s) : *Sporobolus cryptandrus*
Geog. Reg(s) : America-Central, America-North (U.S. and Canada)
End-Use(s) : Erosion Control, Nut
Domesticated : N
Ref. Code(s) : 220
Summary : The sand dropseed is an erect perennial grass from 60-90 cm in height often
used for soil reclamation. In the United States, it grows on sandy soils from New
England to Michigan and from Oregon to Mexico. Its seeds are edible.

Sea Rush, Sparto
(See Ornamental & Lawn Grass)

Sericea Lespedeza
(See Forage, Pasture & Feed Grain)

Serradella
(See Forage, Pasture & Feed Grain)

Sesbania
(See Forage, Pasture & Feed Grain)

Sheep Fescue
(See Ornamental & Lawn Grass)

Showy Crotalaria

Sci. Name(s) : *Crotalaria spectabilis, Crotalaria retzii, Crotalaria cuneifolia, Crotalaria leschenaultii, Crotalaria macrophylla, Crotalaria sericea*
Geog. Reg(s) : America, America-North (U.S. and Canada), Asia, Asia-India (subcontinent), Tropical
End-Use(s) : Erosion Control, Fiber, Forage, Pasture & Feed Grain, Ornamental & Lawn Grass
Domesticated : Y
Ref. Code(s) : 39, 170, 220
Summary : Showy crotalaria is an annual herb cultivated for erosion control and green manure in India, the southeastern United States, and tropical America. Plants contain a toxic alkaloid, monocrotaline, which is poisonous to livestock. Leaves yield a strong, durable fiber. Plants are often grown as ornamentals. Crops produce about 50 MT/ha of organic material. They are native to parts of tropical Asia.

Sickle Medic, Yellow-Flowered Alfalfa
(See Forage, Pasture & Feed Grain)

Single-Flowered Vetch, Monantha Vetch, One-Flowered Vetch

Sci. Name(s) : *Vicia articulata*
Geog. Reg(s) : America-North (U.S. and Canada), Asia, Mediterranean
End-Use(s) : Erosion Control, Forage, Pasture & Feed Grain
Domesticated : Y
Ref. Code(s) : 17, 138, 141, 220
Summary : The single-flowered vetch is cultivated as a cover crop particularly in orchards. It is also grown for forage. This plant is more winter-hardy than common vetch. It does best on sandy loam soil and is well-adapted to the Cotton Belt area of the United States. It does not do well in the Pacific Coast states. This herb is native to the Mediterranean region and Asia Minor.

Siris, East Indian Walnut, Koko
(See Timber)

Slenderleaf Crotalaria
(See Forage, Pasture & Feed Grain)

Smilograss
(See Forage, Pasture & Feed Grain)

Smooth Crotalaria, Striped Crotalaria
(See Forage, Pasture & Feed Grain)

Southern Mahogany
(See Timber)

Spicata Indigo
(See Dye & Tannin)

80

Spotted Burclover

Sci. Name(s) : *Medicago arabica*
Geog. Reg(s) : America-North (U.S. and Canada), America-South, Australia, Europe
End-Use(s) : Dye & Tannin, Erosion Control, Forage, Pasture & Feed Grain
Domesticated : Y
Ref. Code(s) : 56
Summary : Spotted burclover is used as pasture, erosion control, and soil-improvement. This crop makes a good winter pasture. It is native to southern Europe and has been introduced to the United States. It is cultivated in Argentina and Australia. Among the ingredients contained in this clover are citric acid, pantothenic acid, pectin, and tannin. Its seeds contain trypsin inhibitors.

Burclover thrives in slightly acidic, limestone soils. About 5-7.5 MT/ha of hay is produced. Seed yields are approximately 400 kg/ha.

Springer Asparagus, Jessop
(See Ornamental & Lawn Grass)

Standard Crested Wheatgrass, Crested Wheatgrass

Sci. Name(s) : *Agropyron desertorum*
Geog. Reg(s) : America-North (U.S. and Canada)
End-Use(s) : Erosion Control, Forage, Pasture & Feed Grain
Domesticated : Y
Ref. Code(s) : 164
Summary : Standard crested wheatgrass is useful for regrassing abandoned croplands and depleted ranges in the northern Great Plains states of the United States. The grass is a valuable forage species and is sometimes found in mixed plantings.

Sub Clover, Subterranean Clover
(See Forage, Pasture & Feed Grain)

Sugar Gum
(See Timber)

Sugi
(See Timber)

Switchgrass
(See Forage, Pasture & Feed Grain)

Swordbean
(See Forage, Pasture & Feed Grain)

Tamarisk
(See Timber)

Tan Wattle
(See Timber)

Tepary Bean
(See Vegetable)

Townsville Stylo, Townsville Lucerne
(See Forage, Pasture & Feed Grain)

Tropical Kudzu
(See Forage, Pasture & Feed Grain)

Tuart
(See Timber)

Veldtgrass, Perennial Veldtgrass
(See Forage, Pasture & Feed Grain)

Velvetbean
(See Forage, Pasture & Feed Grain)

Vetiver, Khus-Khus
(See Essential Oil)

Vogel Tephrosia

Sci. Name(s) : *Tephrosia vogelii*
Geog. Reg(s) : Africa, Africa-East, Asia-India (subcontinent), Tropical
End-Use(s) : Erosion Control, Forage, Pasture & Feed Grain, Insecticide, Ornamental & Lawn Grass
Domesticated : Y
Ref. Code(s) : 17, 36, 56, 170
Summary : Vogel tephrosia (*Tephrosia vogelii*) is a woody herb or short-lived shrub from 1-4 m in height grown as a cover crop. Its leaves and seeds are high in rotenone. Leaves contain the toxic substances tephrosin and deguelin which make them unfit for livestock consumption. They are cultivated as an insecticide, and in East Africa, as a fish poison and mollucide. Vogel tephrosia is an important grass in the African grasslands because it is not completely destroyed by range fires. These herbs are sometimes grown as ornamentals. In Sri Lanka, plants are used as green manure.

Plants are not affected by certain pests that attack other members of the Tephrosia genus. Vogel tephrosia thrives on poor soils from sea level to altitudes approaching 2,100 m. Leaves and branches are collected on demand. Leaves are collected for rotenone production 6 months from planting. Plants are successful in the tropics.

Weeping Lovegrass
(See Forage, Pasture & Feed Grain)

West Indian Lemongrass, Sere Grass, Lemon Grass
(See Essential Oil)

White Clover, Ladino Clover
(See Forage, Pasture & Feed Grain)

White Lupine, Egyptian Lupine
(See Forage, Pasture & Feed Grain)

White Tephrosia

Sci. Name(s)	: *Tephrosia candida*
Geog. Reg(s)	: Asia-Southeast, Hawaii, Tropical
End-Use(s)	: Erosion Control, Forage, Pasture & Feed Grain, Ornamental & Lawn Grass
Domesticated	: Y
Ref. Code(s)	: 17, 36, 56, 170
Summary	: White tephrosia is a shrub from 1-3 m in height that is grown as a pasture legume. Plants are tolerant of drought, heavy grazing, and low soil fertility, and will thrive in shade and on slopes and waterlogged soils. Plants are not frost-tolerant. They are also used for soil-improvement, cover crops, and, in Java, Sumatra, and Hawaii, for green manure. Plants are sometimes grown as ornamentals.

In the tropics, plants are grown up to 1,650 m in elevation. Green manure yields are 15 MT/ha with about 6% nitrogen.

Winged Bean, Goa Bean, Asparagus Pea, Four-Angled Bean, Manilla Bean, Princess Pea
(See Vegetable)

Yellow Melilot, Yellow Sweetclover
(See Forage, Pasture & Feed Grain)

Essential Oil

Anise

Sci. Name(s)	: *Pimpinella anisum*
Geog. Reg(s)	: America-Central, Asia, Asia-India (subcontinent), Europe, Mediterranean
End-Use(s)	: Essential Oil, Medicinal, Spice & Flavoring
Domesticated	: Y
Ref. Code(s)	: 170, 194
Summary	: Anise is an annual herb whose seeds are a source of an important oil and spice.

Through distillation, anise seeds yield an essential oil used in perfumery, soaps, beverages, and, to some extent, in cough medicines and lozenges. Anise seeds are used to flavor curries, sweets, and certain liqueurs, such as anisette. Anise seed oil is reported to have diuretic and diaphoretic properties.

The herb is native to the Mediterranean region. It grows throughout most parts of the world and is especially widespread in Europe, Asia Minor, India, and Mexico. Plants are most successful in deep, friable soils, and respond favorably to nitrogen fertilizer. Individual plants reach heights of 0.5 m. They require a long, warm, frost-free growing period of 120 days. Anise was valued by the ancient Egyptians, Hebrews, Greeks, and Romans as a powerful medicinal source. In 1984, the United States imported 48.7 MT of anise as an essential oil, valued at $552,000.

Arbor-Vitae, Northern White-Cedar

Sci. Name(s)	: *Thuja occidentalis*
Geog. Reg(s)	: America-North (U.S. and Canada)
End-Use(s)	: Essential Oil, Medicinal, Timber
Domesticated	: Y
Ref. Code(s)	: 99, 180, 220
Summary	: Arbor-vitae is a tree whose leaves and twigs are a source of thuja oil. This oil

is composed of a nerve poison, thujaplicine and a poisonous skin irritant, thujone. In small quantities, this oil is harmless and is used in scenting technical preparations, shoe polishes, and perfumes. It is also used to relieve rheumatism, as an expectorant, to control menstruation, and to soothe skin diseases. Oil content is approximately 0.6-1.0%. The coarse wood is light and durable and used for fencing and roofs.

The states of New York, Michigan, and Vermont and Canada are major producers of thuja oil. In 1983, the United States imported 4,376 MT of oil, worth $2.3 million.

Balsam Amyris, Amyris

Sci. Name(s)	: *Amyris balsamifera*
Geog. Reg(s)	: America-Central, America-South, West Indies
End-Use(s)	: Energy, Essential Oil
Domesticated	: Y
Ref. Code(s)	: 77, 180, 220
Summary	: Balsam amyris is a tree whose leaves and twigs yield an essential oil that has a

higher specific gravity, optical rotation, and alcohol content than West Indies sandalwood oil. This oil is used as a perfume fixative in soaps, low-priced perfumes and cosmetics, and, in the West Indies, it is used as a substitute for sandalwood oil. Oil yield varies from 2-4% depending on the quality of the wood. The oil content makes the leaves and twigs burn and is used by fishermen to provide light for night fishing.

Amyris is grown in Central and South America. In 1983, sandalwood oil imports were 24 MT, worth $2 million.

Trees grow in small thickets in areas with low rainfall, calcerous soils and eroded mountain slopes. They grow wild in Haiti.

Balsam Fir, Canada Balsam, American Silver Fir

Sci. Name(s)	: *Abies balsamea*
Geog. Reg(s)	: America-North (U.S. and Canada)
End-Use(s)	: Essential Oil, Natural Resin, Timber
Domesticated	: Y
Ref. Code(s)	: 77, 227
Summary	: The balsam fir is an evergreen tree which reaches heights of 12 m. It yields an essential oil made into Canadian turpentine and resembles hemlock and spruce oil. The leaves and twigs contain many ingredients, one of which is 17.6% bornyl acetate. It produces an oleoresin rich in volatile terpene collected in the United States and Canada for use in mounting microscope specimens and cementing optical glass elements. The wood is used for wood pulp and paper manufacturing, crates and barrels.

Balsamo

Sci. Name(s)	: *Bulnesia sarmienti*
Geog. Reg(s)	: America-South
End-Use(s)	: Essential Oil
Domesticated	: N
Ref. Code(s)	: 220
Summary	: Balsamo is a South American tree of commercial value. An oil is extracted from its wood by steam distillation. This oil has a roselike scent and is used in perfumery and soap manufacture. It goes by several names: Essence de Bois Gaiac, Champaca Wood Oil, Oleum Ligni Guaiaci, and Oil of Guaiac Wood. Trees are grown in Argentina and Paraguay.

Basil, Sweet Basil

Sci. Name(s)	: *Ocimum basilicum*
Geog. Reg(s)	: Africa-North, America-Central, America-North (U.S. and Canada), Asia, Europe, Mediterranean, Worldwide
End-Use(s)	: Essential Oil, Medicinal, Spice & Flavoring
Domesticated	: Y
Ref. Code(s)	: 170, 194, 205
Summary	: Basil is an aromatic annual herb from 30-60 cm in height whose leaves yield an essential oil used as a flavoring for soups, salads, fish and meat dishes, and in perfumery as a substitute for a mignonette scent. The essential oil is composed mainly of estragol. Fresh basil leaves have been used as an antispasmodic, galactagogue, stomachic, carminative, and mild sedative.

Plants are primarily cultivated in the United States, Central and southern Europe, North Africa, Asia, and in subtropical America, and is native to southern Asia, Iran, and the Middle East. It is now widely distributed throughout the world. Plants thrive in moist, well-drained soils in full sunlight. Leaves can be harvested 2-5 times a year. Two major types of basil dominate the world market: European sweet basil and Reunion, or African, basil. European sweet basil is of higher-quality with better aroma while African basil smells like camphor and is of inferior quality.

Bay, Bay-Rum-Tree

Sci. Name(s)	: *Pimenta racemosa, Pimenta acris*
Geog. Reg(s)	: West Indies
End-Use(s)	: Beverage, Essential Oil, Spice & Flavoring
Domesticated	: Y
Ref. Code(s)	: 119, 170
Summary	: The bay tree grows to about 10 m in height and has aromatic fruits, leaves, and

bark which are processed for use as a spice. Distillation of the leaves provides bay oil that is used in perfumes and bay rum. Bay rum is valued for its soothing, antiseptic properties, which are important in toilet preparations and hair tonics. Bay leaf oil content on a fresh weight basis is from 1-3%. The rum may also be consumed as a beverage. The fruit and bark are minor sources of dried spice or essential oils.

Although trees are cultivated in parts of the Caribbean, they are usually considered to be wild trees. Leaves have been harvested from trees on the smaller islands of the West Indies and have been shipped to Trinidad for the manufacture of bay rum.

Bergamot

Sci. Name(s)	: *Citrus aurantium* subsp. *bergamia*
Geog. Reg(s)	: Europe
End-Use(s)	: Essential Oil, Medicinal, Miscellaneous, Spice & Flavoring
Domesticated	: Y
Ref. Code(s)	: 71, 77, 170, 180
Summary	: Bergamot is a small tree that reaches a height of 4 m. Its fruit rinds, flowers,

and leaves produce a valuable, pleasant smelling oil with a copper content and bitter taste. It is used in toilet waters, colognes, and soaps. Fruit rinds yield approximately 0.5% of an oil composed of about 30-40% linayl actate. As plants mature, their ester content increases. Bergamot leaf oil is often found mixed with other oils. Fruit is too bitter for fresh consumption but is valued for marmalades, as flavoring, and in liqueurs. The peel has medicinal uses. This plant makes a hardy rootstock for grafting with lemon, sweet orange, and grapefruit.

Light and frequent pruning promotes best growth. Plants are adversely affected by strong winds, too much sun, and excessive heat. They prefer mild climates with temperatures ranging from 2-37 degrees Celsius and well-drained, alluvial soils. They are found at altitudes ranging from 400-500 m. Plants bear fruit in 6-7 years. They are grown extensively in Calabria, Southern Italy. In 1984, the United States imported 29.8 MT of bergamot oil, valued at $880,000.

Bitter Almond

Sci. Name(s)	: *Prunus amygdalus* var. *amara*
Geog. Reg(s)	: Africa-North, Europe
End-Use(s)	: Essential Oil, Medicinal, Nut, Spice & Flavoring
Domesticated	: Y

Ref. Code(s) : 77, 180
Summary : The bitter almond tree produces nuts whose dried ripe kernels produce 0.5-0.7% of a volatile oil containing 2-4% hydrocyanic acid. This oil is used for flavoring baked goods and soft drinks and was used to scent cosmetics and soaps. Its main constituent is benzaldehyde, which is now produced synthetically. Apricot kernel oil is often substituted for bitter almond oil because of its higher oil yield. Powdered press cake from bitter almond oil extraction contains amygdalin. Bitter almond oil requires special care during storing to prevent exposure to air.

Trees are cultivated in Morocco, Algeria, Spain, and France. In 1984, the United States imported 104.9 MT of bitter almond oil as an essential oil, valued at $461,000.

Bitter Orange

Sci. Name(s) : *Citrus aurantium* subsp. *amara*
Geog. Reg(s) : Africa, America-South, Europe, Mediterranean, West Indies
End-Use(s) : Essential Oil, Spice & Flavoring
Domesticated : Y
Ref. Code(s) : 77, 180
Summary : The bitter orange tree is the source of an essential oil called petitgrain bigarade. This oil is obtained from the leaves by steam distillation with an average yield of 0.2%. Neroli oil, used in flavoring and perfumery, yields .07-.12%. Adulteration of these oils is practiced extensively.

France, Haiti, Paraguay, Italy, Maghreb, Spain, and Guinea are the main commercial oil producers. French and Haitian oil contains 70% linalyl acetate and Paraguayan oil about 50%. Linalyl acetate is the most important constituent of this oil. Other constituents are linelool, linonene, and beta-pinene. In 1984, the United States imported 0.64 MT of neroli oil as an essential oil, valued at $285,000 and 96.1 MT of petitgrain as an essential oil, valued at $1.4 million.

Black Mustard
(See Spice & Flavoring)

Broad-Leaved Lavender

Sci. Name(s) : *Lavandula latifolia*
Geog. Reg(s) : Europe, Mediterranean
End-Use(s) : Essential Oil, Medicinal, Miscellaneous, Ornamental & Lawn Grass
Domesticated : Y
Ref. Code(s) : 194, 220
Summary : Broad-leaved lavender is an aromatic shrub that is grown for an essential oil extracted from its leaves and flowers. The oil is used in perfumery and porcelain paints. Leaves are collected for perfumery in southern France and Spain. Lavender leaves and flowers are also ingredients of potpourri which is used to scent dresser drawers and clothes. At one time, the leaves were used to control menstruation or to induce abortions. Shrubs are often planted as ornamentals. Broad-leaved lavender is native to the Mediterranean. For trade statistic information on lavender as an essential oil, see common lavender.

Buchu, Bacco

Sci. Name(s) : *Barosma betulina, Agathosma betulina*
Geog. Reg(s) : Africa-South
End-Use(s)   : Essential Oil, Medicinal
Domesticated : N
Ref. Code(s) : 220
Summary   : Buchu is an evergreen tree whose leaves are dried for an essential oil (diosphenol). The oil is used by some South Africans as a carminative and diuretic. The tree is native to South Africa.

Burdock, Great Burdock, Edible Burdock
(See Medicinal)

Cabbage Rose

Sci. Name(s) : *Rosa centifolia*
Geog. Reg(s) : Africa-North, Europe, Mediterranean
End-Use(s)   : Essential Oil, Spice & Flavoring
Domesticated : Y
Ref. Code(s) : 77, 180, 220
Summary   : The cabbage rose bush is a source of rose oil. This oil provides a high-quality fragrance used in cosmetics and perfumes and as flavoring in tobacco and foods. It is also a source of rose water. The oil is extracted from the flowers by steam distillation. Oil yields are from 0.02-0.03%. This plant is fairly drought-resistant.

Bulgaria, Egypt, Turkey, Netherlands, France, and Switzerland are important producers of rose oil. In 1984, the United States imported 28.8 MT of rose oil as an essential oil, valued at $3.5 million.

Cajeput
(See Medicinal)

Camphor-Tree, Camphor
(See Medicinal)

Caraway
(See Spice & Flavoring)

Cardamom, Cardamon
(See Spice & Flavoring)

Carrot, Queen Anne's Lace
(See Vegetable)

Cassia

Sci. Name(s) : *Cinnamomum cassia*
Geog. Reg(s) : Asia-China
End-Use(s)   : Essential Oil, Spice & Flavoring
Domesticated : Y

Ref. Code(s) : 172
Summary : Cassia (*Cinnamomum cassia*) is an aromatic evergreen tree that reaches 18 m in height. It produces a cassia oil of commerce from its distilled leaves and twigs. This oil contains cinnamaldehyde and is used to flavor soft drinks, candies, and mouthwashes. In 1984, the United States imported 97.6 MT of cassia oil, valued at $6.9 million.

Dried, unripe fruit, or cassia buds, have a sweet pungent aroma and are used for flavoring sweet pickles. The buds are cut when trees are 5-7 years. Cassia bark is cut into strips and exported as spice or distilled for its oil and oleoresins.

Trees are cultivated in China, mainly in the Kwangsi and Kwangtung provinces. They are most often planted on terraced hillsides at 90-300 m in elevation.

Celery
(See Vegetable)

Chervil
(See Spice & Flavoring)

Chia, Ghia
Sci. Name(s) : *Salvia hispanica*
Geog. Reg(s) : America-Central, America-South, Asia-Southeast, West Indies
End-Use(s) : Beverage, Essential Oil, Spice & Flavoring
Domesticated : Y
Ref. Code(s) : 17, 36, 107, 119
Summary : Chia is a plant reaching 1.5 m in height that grows from Mexico to Peru and has naturalized in the West Indies. Its watery seeds are used to make a popular beverage and contain 34% of an essential oil, ghia seed oil. In Java, the plant is used as a substitute flavoring for basil. Plants are usually most successful wild, but some cultivated forms continue flowering in Singapore and other parts of the Malay Pennisula.

Cimaru, Tonka Bean
(See Spice & Flavoring)

Cinnamon
(See Spice & Flavoring)

Clary, Clary Wort, Clary Sage
Sci. Name(s) : *Salvia sclarea*
Geog. Reg(s) : Eurasia, Europe, Temperate
End-Use(s) : Beverage, Essential Oil, Miscellaneous, Ornamental & Lawn Grass, Spice & Flavoring
Domesticated : Y
Ref. Code(s) : 17, 194, 220
Summary : Clary is a biennial herb reaching 1 m in height whose leaves are an important source of an aromatic oil, oil of clary. This oil is marketed as concrete and absolute and is often blended with lavender or jasmine for use in soaps, detergents, creams, powders, perfumes, and lotions. Fresh and dried leaves are used as flavoring in wines

like Vermouth and Muscatel, and in certain medicines like digitalis. Flowers are used in herbal teas, sachets, potpourris, and beverages. Clary is grown as an ornamental.

Plants are cultivated in temperate areas. Major producing areas are in France, Hungary, and the USSR. Clary grows well on dry, well-drained, calcareous soils at high-elevations and requires from 0.7-2.6 m of rainfall annually. About 2 harvests/plant are made each year. Harvest occurs at the end of the blooming period. Clary is native to southern Europe.

Clove
(See Spice & Flavoring)

Common Lavender, English Lavender

Sci. Name(s)	: *Lavandula angustifolia, Lavandula officinalis, Lavandula vera*
Geog. Reg(s)	: Africa-East, Europe, Mediterranean
End-Use(s)	: Erosion Control, Essential Oil, Ornamental & Lawn Grass, Spice & Flavoring
Domesticated	: Y.
Ref. Code(s)	: 17, 180, 194
Summary	: Common or English lavender, is an aromatic shrub widely cultivated for an essential oil, flavoring, and as an ornamental. Distilled flowers yield an essential oil used in perfumes, toilet waters, cosmetics, shaving creams, lotions, and soaps. Leaves are used to flavor salads, dressings, fruit desserts, jellies, wines, and herbal teas and to scent linen, sachets, and tobacco. Lavender shrubs are grown as ornamental borders, in rock gardens, as potted outdoor plants and along highways for decoration and erosion control.

Plants are cultivated in southern France, Tanzania, and Bulgaria for their oil. They take several years to provide a substantial crop but are fairly long-lived with some plantings lasting 30 years. Plants thrive in areas with temperatures from 7-21 degrees Celsius with an annual rainfall of 0.3-1.3 m. They need well-drained, calcareous soils in sunny positions for optimum growth. Shrubs are native to the Mediterranean. In 1984, the United States imported 90.9 MT of lavender and spike lavender as an essential oil, valued at $1.3 million.

Common Mignonette, Sweet Reseda

Sci. Name(s)	: *Reseda odorata*
Geog. Reg(s)	: Africa, Mediterranean, Temperate
End-Use(s)	: Essential Oil, Ornamental & Lawn Grass
Domesticated	: Y
Ref. Code(s)	: 15, 17, 220, 226
Summary	: Common mignonette is an annual herb cultivated mainly as an ornamental. An essential oil (reseda oil) is obtained from its flowers and used in perfumery. Roots yield from 0.014-0.035% of a volatile oil (reseda root oil). Plants thrive on moderately rich soil in areas of partial shade. Plants are found in the temperate areas of Africa and the Mediterranean.

Common Rue, Garden Rue
(See Medicinal)

Common Thyme
(See Spice & Flavoring)

Coriander
(See Spice & Flavoring)

Corn Mint, Japanese Mint, Field Mint

Sci. Name(s) : *Mentha arvensis*
Geog. Reg(s) : America-North (U.S. and Canada), America-South, Asia, Asia-China, Europe, Temperate
End-Use(s) : Essential Oil, Medicinal, Spice & Flavoring
Domesticated : Y
Ref. Code(s) : 170, 220
Summary : Corn mint (*Mentha arvensis*) is a temperate herb to 1 m in height grown commercially in China, Japan, and Brazil for an oil extracted from its leaves. This oil contains 60% menthol and is a valuable flavoring in pharmaceutical products, toothpastes, food products, cigarettes, and alcoholic beverages. Although tests are still inconclusive, corn mint oil is thought to serve as an antifertility agent.

Leaves are harvested while plants are in bloom. Plants thrive in areas with temperatures of 6-27 degrees Celsius and an annual rainfall from 0.3-4.2 m. Plants are native to Europe, Asia, and North America. In 1984, the United States imported 287.2 MT of cornmint and peppermint as an essential oil, valued at $3.1 million.

Costmary, Mint Geranium, Alecost, Bible-Leaf Mace
(See Spice & Flavoring)

Crown Daisy, Tansy
(See Medicinal)

Cumin
(See Spice & Flavoring)

Curry-Leaf-Tree
(See Spice & Flavoring)

Damask Rose

Sci. Name(s) : *Rosa damascena* forma *trigintipelala*
Geog. Reg(s) : Africa-North, Europe, Mediterranean
End-Use(s) : Essential Oil, Spice & Flavoring
Domesticated : Y
Ref. Code(s) : 77, 180, 220
Summary : The damask rose bush has flowers that are a source of a rose-scented oil used as a scent and flavoring. About 0.5 g of oil is obtained from 1,000 g of flowers. A volatile oil containing citronello, geranol, nerol, and linalool is taken from the nearly opened flower buds.

Bulgaria, Egypt, the Netherlands, France, Switzerland, and Turkey are major producers of this oil. Plants grow wild in the Caucasus, Syria, Morocco, and

Andalusia. For 1984 United States import trade statistic for rose oil as an essential oil, see cabbage rose.

Dill
(See Spice & Flavoring)

East Indian Lemongrass, Malabar Grass

Sci. Name(s) : *Cymbopogon flexuosus*
Geog. Reg(s) : America-Central, Asia-China, Asia-India (subcontinent), Europe
End-Use(s) : Essential Oil, Spice & Flavoring
Domesticated : Y
Ref. Code(s) : 171, 194
Summary : East Indian lemongrass yields an essential oil from its leaves. This oil is considered commercially distinct from West Indian lemongrass oil. Because it contains 75-85% aldehydes, mainly citral, and is more soluble in 70% alcohol than West Indian lemongrass oil, it is preferred in perfume manufacture. This oil is also used in nonalcoholic beverages and herbal teas.

The grass is cultivated commercially in Guatemala, India, China, Paraguay, England, and Sri Lanka. It is most successful in areas with temperatures 18-29 degrees Celsius with annual rainfalls of 0.7-4.1 m. The grass thrives in full sun on sandy, well-drained soils. Leaf cuttings are taken after three months and afterwards every 6-8 weeks. In India, crops grown on sandy soils yield 2.5 MT/ha/annum of grass, giving an oil yield of 19-40 kg. In 1984, the United States imported 84.1 MT of lemon grass oil as an essential oil, valued at $757,000.

East Indian Sandalwood

Sci. Name(s) : *Santalum album*
Geog. Reg(s) : Asia-India (subcontinent), Asia-Southeast
End-Use(s) : Essential Oil, Medicinal, Miscellaneous, Timber
Domesticated : Y
Ref. Code(s) : 77, 180
Summary : East Indian sandalwood is an evergreen tree that produces an essential oil from its roots and heartwood. About 90% of the oil is used in perfumes, cosmetics, and soaps while its strong, sweet odor makes it an excellent fixative. Oil yields range from 4.5-6.3% with the roots yielding up to 10% of the total yield. It contains 60% sesquiterpene alcohol santalol. The oil is used in the treatment of gonorrhea and in USP preparations and pharmaceuticals. Tree chips and sawdust are made into incense and the wood is carved into art objects and specialized utensils. It is native to southern India and the Malay archipelago.

Trees grow at altitudes ranging from 609-914 m and reach heights of 18-20 m. They are fully mature at 60-80 years. Higher yielding trees grow in drier areas on red soils or stony ground. They thrive on well-drained, loamy soils on hillsides in direct sun. They require a minimum of 50.8-63.5 cm/year of rainfall. Trees are attacked by spike disease. In 1984, the United States imported 32.5 MT of sandalwood oil as an essential oil, worth 2.9 million.

Eucalyptus
(See Timber)

Eucalyptus

Sci. Name(s) : *Eucalyptus macarthurii*
Geog. Reg(s) : Africa-Central, America-South, Europe
End-Use(s) : Essential Oil, Medicinal
Domesticated : Y
Ref. Code(s) : 17, 77, 180, 220
Summary : The Eucalyptus tree is a spreading evergreen tree reaching 24.4 m in height. Its leaves yield an oil used industrially and in perfumes. The oil is composed of 70% geranyl acetate and 15% alpha-eudesmol, geraniol and eudesmol. It is used as a disinfectant and in medical and dental preparations. A rectified oil is used as a denaturant of alcohol and the residue is an excellent source of eudesmol (fixative).

Spain, Portugal and Zaire are the main commercial producers of eucalyptus oil. Look under eucalyptus, broad-leaved peppermint tree for United States trade statistics.

Eucalyptus, Broad-Leaved Peppermint Tree

Sci. Name(s) : *Eucalyptus dives*
Geog. Reg(s) : America-South, Australia, Europe, Temperate
End-Use(s) : Essential Oil, Medicinal, Timber
Domesticated : Y
Ref. Code(s) : 11, 17, 77, 180
Summary : Eucalyptus is a robust, wide-crowned evergreen tree of Australia, Spain, Portugal, and Zaire. It reaches 12.2-24.4 m in height and thrives at altitudes ranging from 150-1200 m in areas with a maximum warm temperature of 26.6 degrees Celsius and an annual rainfall of 9.8-19.7 cm.

Trees grow on most soils but are commonly found on poor, shallow soils in mixtures with other trees. The wood is hard, moderately strong and light, and used in construction. Leaves yield 2-3% of an essential oil comprised mainly of eucalyptol and piperitone. Leaf oil yields of 13.6-20.4 kg are obtained from 454 kg of leaves. This oil provides piperitone for synthetic thymol and menthol, and ketone and pinene used in blended turpentines. It is also used in dental products, inhalants, solvents, room sprays, and medicated soaps. Its lilac odor makes it useful in perfumes. In 1984, the United States imported 384.1 MT of eucalyptus oil as an essential oil, valued at $2 million.

European Hazel, European Filbert
(See Nut)

European Pennyroyal

Sci. Name(s) : *Mentha pulegium*
Geog. Reg(s) : Europe, Mediterranean
End-Use(s) : Essential Oil, Insecticide, Medicinal, Ornamental & Lawn Grass, Spice & Flavoring
Domesticated : Y

Ref. Code(s)   : 36, 205, 220
Summary   : European pennyroyal is a scented perennial herb growing to 30 cm in height. Its leaves yield an essential oil prepared on a small scale commercial basis, used to scent soaps and as an insect repellent in certain cosmetics. Pennyroyal oil may be used for flavoring only if the concentration of pulegone does not exceed 20 mg in 1 kg of the final product flavored. The oil and dried leaves are used to treat stomach problems.

Plants are native to Europe and the Mediterranean area. They are often cultivated for horticultural reasons. Prostrate forms are grown as lawn or aromatic ground cover. European pennyroyal will thrive in most fertile, sandy soils in full sunlight.

Fennel, Florence Fennel, Finocchio
(See Spice & Flavoring)

Fenugreek
(See Spice & Flavoring)

Fitweed, False Coriander
(See Spice & Flavoring)

French Lavender
Sci. Name(s)   : *Lavandula stoechas*
Geog. Reg(s)   : Europe, Mediterranean
End-Use(s)   : Essential Oil, Medicinal, Ornamental & Lawn Grass
Domesticated   : Y
Ref. Code(s)   : 17, 194, 205
Summary   : French lavender is an aromatic shrub grown as an ornamental or as a source of an essential oil. The leaves and flowers yield an essential oil used in perfumery and in certain medicinal preparations as a treatment for lung diseases, asthma, and cramps. Dried leaves and flowers are used in sachets and in home remedies as an antiseptic, mild sedative, and for treatment for nausea. French lavender is native to Spain and parts of the Mediterranean. For trade statistic information on lavender as an essential oil, see common lavender.

Galanga, Kent Joer
(See Ornamental & Lawn Grass)

Garden Angelica, Angelica
Sci. Name(s)   : *Angelica archangelica*
Geog. Reg(s)   : Asia-India (subcontinent), Europe, Temperate
End-Use(s)   : Essential Oil, Medicinal, Vegetable
Domesticated   : Y
Ref. Code(s)   : 194, 220
Summary   : Garden angelica is a perennial herb of temperate Europe whose roots, seeds, and hollow leaf stalks are used in a variety of ways. The roots and seeds yield essential oils (Oil of angelica root) used for flavoring liqueurs, gin, and vermouth, and are

components in perfumes, creams, soaps, salves, oils, shampoos, and cigarettes. Angelica fruit is sometimes used in herbal teas. Leaves can be eaten as salad vegetables, or used as garnish. Leaf stalks are candied and used in confectionery.

Medicinally, plants have been used as a carminative, expectorant, stomachic, tonic, and stimulant. An infusion of the roots has been used to treat indigestion and chest colds.

Garden angelica grows in areas with temperatures from 5-19 degrees Celsius, with an annual average rainfall of 0.5-1.3 m. Plants are most successful when grown on rich, well-drained, loamy soil. Roots are harvested during the plant's first year and leaves and stalks are not collected until the second year. Seeds are gathered when fully ripe. Plants die soon after flowering.

Garlic
(See Spice & Flavoring)

Gbanja Kola
(See Nut)

Ginger
(See Spice & Flavoring)

Golden Wattle
(See Dye & Tannin)

Grapefruit
(See Fruit)

Henna
(See Ornamental & Lawn Grass)

Hops
(See Beverage)

Horehound, White Horehound
(See Medicinal)

Huisache, Cassie Flower, Kolu, Sweet Acacia, Cassie

Sci. Name(s)	: *Acacia farnesiana*
Geog. Reg(s)	: Africa, America-Central, America-North (U.S. and Canada), Asia, Australia, Europe, Tropical
End-Use(s)	: Essential Oil, Timber, Weed
Domesticated	: Y
Ref. Code(s)	: 127, 170, 175, 193, 229
Summary	: Huisache is a shrub or small tree whose large orange-yellow flowers yield an essential oil extracted commercially in southern France for use in the perfume industry. The shrub also produces a heavy, close-grained timber.

Trees reach heights of 2.5-4 m and thrive in the dry, tropical areas of France, the United States, Asia, Africa and Australia. In the United States, shrubs are pasture weeds that hinder effective management of livestock.

Huon-Pine
(See Timber)

Hyssop

Sci. Name(s)	: *Hyssopus officinalis*
Geog. Reg(s)	: Asia, Europe, Mediterranean
End-Use(s)	: Essential Oil, Medicinal, Spice & Flavoring
Domesticated	: Y
Ref. Code(s)	: 17, 194, 220
Summary	: Hyssop is a subshrub from 0.5-1 m in height cultivated for its leaves, which yield an essential oil. The oil is used to flavor liqueurs, perfumes, soaps, creams, and various cosmetics. The flowering tops and leaves are used to flavor teas, tonics, and bitters, and are occasionally added to vegetable dishes, soups, salads, and confectionery. Hyssop leaves are made into poultices for bruises or are mixed with honey to combat coughs and colds.

Plants are grown commercially in France and other European countries. They are native to the Mediterranean, the Balkan States, Asia Minor, and Iran. Hyssop is easily grown and thrives on dry, limestone soils in areas with temperatures from 7-21 degrees Celsius and an annual rainfall of 0.6-1.5 m.

Indian Bael, Bael Fruit, Bengal Quince, Bilva, Siniphal, Bael Tree
(See Fruit)

Inula, Elecampane

Sci. Name(s)	: *Inula helenium*
Geog. Reg(s)	: America-North (U.S. and Canada), Asia-Central, Europe
End-Use(s)	: Dye & Tannin, Essential Oil, Medicinal, Natural Resin
Domesticated	: Y
Ref. Code(s)	: 205, 220
Summary	: Inula or elecampane is an attractive perennial to 2 m in height whose roots produce an essential oil used as a flavoring agent. The roots contain about 40% inulin, the essential oil, and a resin. Dried roots are used medicinally for the treatment of respiratory ailments, especially bronchitis, coughs, asthma, and catarrh. Preparations of the root have been used as an appetite-stimulant. Roots yield a blue dye.

Plants grow best in rich, damp soils. They are mostly wild, but there is some cultivation in parts of Central Europe. Inula is native to southern and Central Europe and Central Asia. It has naturalized in the United States.

Jasmine, Poet's Jessamine

Sci. Name(s)	: *Jasminum officinale*
Geog. Reg(s)	: Africa-North, Asia-Central, Asia-China, Asia-India (subcontinent), Europe, Mediterranean
End-Use(s)	: Essential Oil, Miscellaneous, Ornamental & Lawn Grass, Spice & Flavoring

Domesticated : Y
Ref. Code(s) : 180, 194, 220
Summary : Jasmine or poet's jessamine is a viny plant to 10 m in height that produces blossoms that yield an essential oil available as both jasmine concrete and jasmine absolute. Jasmine oil is a fragrance in quality perfumes, creams, oils, soaps, and shampoos. The flowers are used as flavoring in jasmine and other herbal or black teas. Plants are also used as ornamentals. Jasmine oil is produced in France, Italy, Spain, several Middle Eastern and North African countries, the Comoro Islands, India, China, Taiwan, and Japan.

Plants thrive on most soils in sunny positions in areas with temperatures from 11-17 degrees Celsius and an annual rainfall from 0.3-2.8 m. High-quality oil is obtained from plants grown at higher-altitudes. Plants are fairly cold-tolerant and are used as rootstocks for the more widely grown and more valuable royal jasmine (*Jasminum grandiflorum*). Jasmine is native to Central Asia.

Juniper, Common Juniper

Sci. Name(s) : *Juniperus communis*
Geog. Reg(s) : Europe
End-Use(s) : Essential Oil, Fiber, Medicinal, Ornamental & Lawn Grass, Spice & Flavoring
Domesticated : Y
Ref. Code(s) : 180, 220
Summary : Juniper is a plant whose dried ripe fruit is the source of a commercial essential oil. The oil is obtained through steam distillation and used as a flavoring agent in gins, liqueurs, and such cordials as sloe gin. The oil is also used as a diuretic and urogenital stimulant. An oil distilled from the wood has been used in veterinarian medicine. Juniper bark has been used as cordage. Juniper bushes are popular as ornamentals.

Italy, Czechoslovakia, Hungary, and Yugoslavia produce the best quality berries. A berry contains 0.5-1.5% essential oils, 10% resin, and 15-30% dextrose.

Labdanum

Sci. Name(s) : *Cistus ladaniferus*
Geog. Reg(s) : Mediterranean
End-Use(s) : Essential Oil, Natural Resin
Domesticated : N
Ref. Code(s) : 77, 180
Summary : Labdanum is a perennial shrub that contains a scented oil used in the perfume, cosmetic, and soap industries. The scent is similar to ambergris. Cistus oil is also extracted from labdanum concrete but the concrete and absolute forms are the most popular. Leaves and twigs are boiled in water to produce an alcohol-soluble, labdanum resin that has limited use.

Plants grow wild in the Mediterranean. They reach 2 m in height and grow in large stands in sunny, sheltered areas. Plants have a maximum resin content during the hot season.

Lavandin

Sci. Name(s) : *Lavandula hybrida*

Geog. Reg(s) : Africa-North, Europe
End-Use(s) : Essential Oil
Domesticated : Y
Ref. Code(s) : 77, 180, 194
Summary : Lavandin is a cross between true lavender and spike. Its flowers produce 1.0-1.8% of an essential oil, lavandin oil, which is a necessary ingredient in soaps and used to make imitation lavender oils and perfumes. It is composed of linalool and 18-25% linalyl acetate with a lower ester content than lavender oil but a harsher odor than lavender. Four thousand to 5,000 kg/ha are possible and 100 kg of lavandin plants yield from 1-1.8 kg oil. Plants are cross-pollinated by bees.

 Lavandin oil is produced mainly in France, Italy, Yugoslavia, and North Africa. It prefers sunny soils, does not tolerate prolonged drought, and grows wild at medium altitudes ranging from 500-700 m. Lavender and spike imported from France, Spain, and other countries to the United States, in 1983, were 105.9 MT, valued at $1.7 million.

Lemon
(See Fruit)

Lime
(See Fruit)

Longleaf Pine
Sci. Name(s) : *Pinus palustris, Pinus australis*
Geog. Reg(s) : America-North (U.S. and Canada)
End-Use(s) : Energy, Essential Oil, Insecticide, Natural Resin, Timber
Domesticated : Y
Ref. Code(s) : 77, 99, 180, 220, 228
Summary : The longleaf pine is a light-crowned evergreen tree found in the Atlantic and Gulf States of the United States. The pine tree is a major source of pine oil obtained from the bark by steam or fractional distillation. This oil is used as an herbicide and disinfectant, in soaps, insecticides, deodorants, polishes, and in perfumery-grade terpineol, anethole, fenchone and camphor, and in turpentine and rosin. Its timber is moderately durable but has a problem with resin checks. Its saw timber is used for heavy and light construction, boat building, and boxes. Its roundwood is used for transmission poles, fence posts, longfiber pulp, fuel, and charcoal. It produces a resin. Annual timber production ranges from 6-12 cu.m/ha.

 For optimum growth, this tree needs considerable light and soils of light- to medium-texture, although it is adaptable to most soils. It is a windfirm and moderately frost-resistant tree and is found at altitudes ranging from sea level to 1,500 m with a mean annual rainfall of 1,000-2,000 mm and a mean annual temperature of 16-22 degrees Celsius.

 This tree reaches heights of 40 m and survives on a wide variety of soils. It does best in deep soils with clay substructure. It is relatively fire-resistant and is planted for ornament, shelter and erosion control. The United States pine oil imports for 1983 were 291 MT, valued at $266,000.

Lovage, Common Lovage
(See Spice & Flavoring)

Mastic
(See Natural Resin)

Musk Okra, Musk Mallow, Ambrette
Sci. Name(s) : *Abelmoschus moschatus, Hibiscus abelmoschus*
Geog. Reg(s) : Asia-India (subcontinent), West Indies
End-Use(s) : Essential Oil, Medicinal, Vegetable
Domesticated : Y
Ref. Code(s) : 154, 170
Summary : Musk okra is an herb whose seeds produce a musk-scented oil used in perfumes and perfumed products. The seeds are exported from Martinique. The plant grows to 2 m in height and is used as a stomach and digestive tonic. Its young leaves are occasionally eaten in soups.

Myrtle, Indian Buchu
(See Spice & Flavoring)

Nardus Grass, Citronella Grass
Sci. Name(s) : *Cymbopogon nardus*
Geog. Reg(s) : Africa-East, America-Central, America-South, Asia-India (subcontinent), Asia-Southeast, West Indies
End-Use(s) : Essential Oil, Spice & Flavoring
Domesticated : Y
Ref. Code(s) : 171, 180
Summary : Nardus grass is a sturdy perennial to 1 m in height that yields 2 types of citronella oil: the Ceylon type, *Cymbopogon nardus* var. *lenabath*; and the Java type, *Cymbopogon nardus* var. *mahapengiri*. These oils are used for flavoring and scent.

The 1st type of citronella oil is used for perfuming low-priced technical preparations such as detergents, sprays, and polishes but rarely in insect repellents because of more efficient substances. Ceylon variety plants are cultivated in southern Sri Lanka and harvested 8 months after planting and for as long as 10-15 years.

The Java variety has a much higher oil yield than the Ceylon variety. Java citronella oil contains 40-50% citronellal and is more valuable for the synthesis of perfumed products and synthetic menthol. This oil is produced on a large commercial scale but has competition from synthetically produced citronella oils.

Commercial production is mainly in Central America and Taiwan. Smaller areas of production are in Brazil, Sri Lanka, Java, Guatemala, Honduras, Haiti, East Africa, Zaire, and the West Indies. The United States is a major importer of the oil.

The 2 varieties have different cultivation requirements. The Ceylon variety is more hardy and is grown on relatively poor soils. Its leaves have a lower citronella content than the Java type. The Java variety is grown in tropical areas and is not as hardy. It is most successful from sea level to 1,000 m in altitude and requires sunlight and sandy, well-drained loams. The first harvest is made 6-9 months from planting and then every 3-4 months for 4-5 years afterwards. In 1984, the United States imported 845 MT of citronella as an essential oil, valued at $3,539,000.

Nutmeg, Mace
(See Spice & Flavoring)

Ocotea, Carela sassafraz

Sci. Name(s)	: *Ocotea pretiosa*
Geog. Reg(s)	: America-South
End-Use(s)	: Essential Oil, Timber
Domesticated	: N
Ref. Code(s)	: 77, 180
Summary	: Ocotea is a tree 10-15 m in height. Its wood produces an essential oil that yields 90% percent safrole that is converted to heliotropine for use in technical preparations, soaps, and sprays. It is also used as a substitute for Japanese artificial sassafras oil. The wood is grainy and full of holes, splits easily, and is used for fence material.

In 1984, the United States imported 491 MT of sassofras as an essential oil, valued at $2.9 million. The tree grows wild in Brazil, Paraguay and Colombia.

Oregano, Pot Marjoram, Wild Marjoram
(See Spice & Flavoring)

Orris Root, Florentine Iris

Sci. Name(s)	: *Iris florentina*
Geog. Reg(s)	: Europe, Mediterranean
End-Use(s)	: Essential Oil, Spice & Flavoring
Domesticated	: N
Ref. Code(s)	: 180, 220
Summary	: Orris root is the source of orris root oil used as flavoring in a variety of foods and perfumed products, particularly those with a violet base. It has a solid consistency due to the presence of myristic acid and gets its odor from irone. In 1984, the United States imported 412 kg of oil, valued at $233,000.

Padang-Cassia
(See Spice & Flavoring)

Parsley
(See Spice & Flavoring)

Patchouli

Sci. Name(s)	: *Pogostemon cablin*
Geog. Reg(s)	: Africa-East, Africa-West, America-South, Asia-Southeast
End-Use(s)	: Essential Oil, Spice & Flavoring
Domesticated	: Y
Ref. Code(s)	: 119, 170

Summary   : Patchouli is a small shrub of the Philippines whose dry shoots yield an essential oil. Its fleshy leaves and young buds are used as culinary flavoring. For oil production, young shoots are harvested every 6 months for 2-3 years, dried, and the oil extracted. This oil is considered to be one of the best fixatives for perfumes and is widely used in soaps, hair tonics, and as flavoring for tobacco. Shrubs are grown in Sumatra, the Seychelles, Madagascar, and Brazil. In 1984 the United States imported 329 MT of patchouli oil, valued at $9.3 million.

Pepper, Black Pepper
(See Spice & Flavoring)

Pepper Tree, Brazilian Pepper Tree, California Peppertree
(See Gum & Starch)

Peppermint
Sci. Name(s) : *Mentha X piperita*
Geog. Reg(s) : America-North (U.S. and Canada), Europe, Temperate
End-Use(s)   : Beverage, Essential Oil, Medicinal, Spice & Flavoring
Domesticated : Y
Ref. Code(s) : 170, 194, 205
Summary   : Peppermint is a strongly scented perennial herb to 1 m in height whose distilled leaves yield an essential oil (peppermint oil) that contains approximately 50% menthol. This oil is a commercial item used in the food, flavor, and pharmaceutical industries. Dried leaves are used in herbal teas, fruit jams, and desserts. Peppermint oil is thought to be a natural carminative with some antispasmodic properties.

Most peppermint oil produced in the United States comes from English mint (*Mentha vulgaris*), a plant with high leaf oil yields. Peppermint is a hybrid of *Mentha spicata* and *Mentha aquatica*. It is native to Europe and thrives in most temperate areas with rich, damp soils in partially sunny situations. In 1984, the United States imported 6.1 MT of peppermint as an essential oil, valued at $230,000.

Peru Balsam, Balsam-of-Peru
(See Natural Resin)

Pili Nut, Elemi
(See Nut)

Pimento, Allspice
(See Spice & Flavoring)

Pinus pumilio, Swiss Mountain Pine, Dwarf Pine
Sci. Name(s) : *Pinus pumilio*
Geog. Reg(s) : Eurasia, Europe
End-Use(s)   : Erosion Control, Essential Oil, Medicinal, Timber
Domesticated : N
Ref. Code(s) : 77, 180, 220

Summary : Pinus pumilio is a wild, shrublike tree of Austria, Italy, and Yugosalvia. Its branches and leaves contain an important essential oil and its twigs and needles yield dwarf pine needle oil used in soaps, perfumes, air freshners, as expectorants, treatment for rheumatism, chest and bladder complaints, and skin diseases. Oil yields are 0.3-0.4%. The main constituents are terpenes which account for the oil's aroma. The wood is of good quality and the tree itself is used for erosion control.

In 1984, the United States imported 378 MT of pine needle oil as an essential oil, valued at $705,000.

Roman Chamomile, Chamomile

Sci. Name(s) : *Anthemis nobilis, Chamaemelum nobile*
Geog. Reg(s) : Mediterranean
End-Use(s) : Beverage, Essential Oil, Medicinal, Ornamental & Lawn Grass
Domesticated : Y
Ref. Code(s) : 170, 194, 220
Summary : Roman chamomile is a creeping, herbaceous perennial cultivated for an essential oil (Oil of Roman chamomile) extracted from the flowers. Essential oils obtained through a steam distillation process are used in alcoholic beverages, confectionery, desserts, perfumes, and cosmetics. Medicinally, the essential oil is considered to have antispasmodic, carminative, diaphoretic, sedative, and stomachic qualities. Flowers can also be dried and brewed in herbal teas. Flower heads have been used to manufacture herb beers. Because plants have attractive foliage and flowers, they are often grown as ground cover or ornamentals in flower gardens.

Plants grow to 0.3 m. Chamomile is cultivated in Europe, particularly Belgium, France, and England. Plants grow best in areas with temperatures of 7-26 degrees Celsius, and an annual rainfall of 0.4-1.4. When planted on a field scale, chamomiles require direct sunlight and well-drained soil for optimum growth.

Roquette, Garden Rocket, Rocket Salad, Roka
(See Oil)

Rose Geranium, Geranium

Sci. Name(s) : *Pelargonium graveolens*
Geog. Reg(s) : Africa-North, Africa-South, Subtropical, Temperate, Worldwide
End-Use(s) : Beverage, Essential Oil, Insecticide, Medicinal, Miscellaneous, Spice & Flavoring
Domesticated : Y
Ref. Code(s) : 180, 194, 220
Summary : The rose geranium is a perennial herb native to southern Africa, the Reunion Islands, Morocco, and Algeria. It is cultivated for its leaves and branches which yield an essential oil containing geraniol. This oil is used for soaps, perfume and for manufacturing commercial rhodinol. Dried leaves are used in sachets, potpourris, herbal teas, baked goods, fruit desserts and as an astringent. They also aid in the treatment of ulcers are thought to repel insects because of their citronella content.

Plants reach heights of 1 m. They flourish in full sun in temperate and subtropical climates throughout the world. Rose geraniums are also grown in the Mediterranean, Africa and generally worldwide in both tropical and subtropical areas. They prefer well-drained, fertile soils in humid climates but will tolerate drought and cold. In 1984, the United States imported 61 MT of the oil, valued at $3.1 million.

Rosemary
(See Spice & Flavoring)

Rosewood

Sci. Name(s) : *Aniba rosqedora* var. *amazonica*
Geog. Reg(s) : America-South
End-Use(s) : Essential Oil, Spice & Flavoring
Domesticated : N
Ref. Code(s) : 77, 180
Summary : Rosewood is a wild evergreen tree that grows in the forests of Brazil and Peru. Its wood is the source of Bois de rose oil and yields approximately 0.7-1.2 % oil whose main constituent is linalool, one of the most important aromatic isolates. This oil is used for artificial flavors, soaps, perfumes and lavender applications, and technical preparations.

This tree prefers clay soils. The trees should not be cut until 10 years old in order to have a large trunk. Rosewood trees are grown commercially for their essential oils. Bois de rose oil imports to the United States from Brazil and other countries, for 1984, were 108.8 MT, worth $1.8 million.

Rosha Grass, Palmarosa

Sci. Name(s) : *Cymbopogon martinii*
Geog. Reg(s) : Asia-India (subcontinent), Asia-Southeast
End-Use(s) : Essential Oil, Spice & Flavoring
Domesticated : Y
Ref. Code(s) : 171
Summary : Rosha grass is a perennial grass that reaches 2 m in height and grows wild in India. There are two varieties distinct in terms of habit, although they are similar in appearance. The flowering tops of both varieties yield essential oils.

Variety motia produces Palmarosa oil that is a natural source of, and contains about, 95% geraniol. Palmarosa oil is used as an adulterant of Turkish attar of roses, scents for soaps and cosmetics, and as a flavoring in tobacco. The oil has competition from ninde oil which is obtained from the shrub *Aeolanthus gamwelliae*. Variety motia grows in separate clumps on hillsides in dry, well-drained soil in areas receiving from 800-900 m/annum of rain.

Variety sofia yields gingergrass oil occasionally used in perfumery. This variety grows, in dense clumps at lower elevations, in moist, poorly-drained soils, and requires more rainfall than variety motia.

There has been some commercial exploitation of rosha grass in Java and the Seychelles. In 1984, the United States imported 17.2 MT of palmarosa oil, valued at $640,000.

Sage, Garden Sage, Common Sage
(See Spice & Flavoring)

Saigon Cinnamon, Cassia
(See Spice & Flavoring)

Scotch Spearmint, Red Mint

Sci. Name(s)	: *Mentha* X *gentilis, Mentha cardiaca*
Geog. Reg(s)	: America, Europe
End-Use(s)	: Essential Oil, Spice & Flavoring
Domesticated	: Y
Ref. Code(s)	: 220
Summary	: Scotch spearmint (*Mentha* X *gentilis*) is a hybrid of spearmint (*Mentha spicata*) and corn mint (*Mentha arvensis*). Its leaves yield an essential oil used most often as a flavoring agent. The herb grows wild in Europe and America. For United States import statistics for spearmint as an essentail oil, see spearmint.

Sesame, Simsim, Beniseed, Gingelly, Til
(See Oil)

Sour Orange, Seville Orange

Sci. Name(s)	: *Citrus aurantium*
Geog. Reg(s)	: America-North (U.S. and Canada), America-South, Asia-Southeast, Mediterranean, Subtropical, Tropical
End-Use(s)	: Essential Oil, Fruit, Medicinal, Miscellaneous, Spice & Flavoring
Domesticated	: Y
Ref. Code(s)	: 170, 180, 205
Summary	: The sour orange is a tree that reaches up to 10 m in height. The fruit, leaves, and flowers yield a volatile bigarade oil used in flavoring, liqueurs, and perfumery. The tree is often confused with the sweet orange (*Citrus sinensis*). Sour orange fruit is considered too bitter to be a fresh market fruit but is used in marmelades. Its fruit is thick-peeled, rough textured, and has an extremely sour and bitter pulp with numerous seeds. An oil is extracted from the fruit peel and produced commercially in the West Indies, Italy, Brazil, and Spain. It contains more than 90% d-limonene. The oxygenated portion of the oil gives its characteristic odor and flavor. The oil is used as a flavor in bakery products, soft drinks, candies, and also in cosmetic waters and perfumes.

The leaves and flowers have been used medicinally as an antispasmodic, sedative, cholagogue, tonic, and vermifluge. The flowers yield a volatile oil (Oil of Neroli) used in perfumery. In India, neroli oil is used in vaseline against leeches. Infusions of the leaves and flowers have sedative qualities. Orange-flower water is used to flavor medicines, foods, and perfumes. The trees are often used as root stock for lemon, sweet orange, and grapefruit.

The sour orange is native to southeast Asia and has been introduced into the tropics and subtropics. It has naturalized in southern Europe, Florida, and other parts of the United States. The tree is fairly resistant to a common Citrus disease, gummosis and is considered more hardy than the sweet orange. It is the fourth most important essence after roses, jasmine, and ylang-ylang and is produced commercially in the Mediterranean countries. In 1984, the United States imported 0.64 MT of neroli oil as an essential oil, valued at $285,000.

Southern Red Cedar, Eastern Red-Cedar

Sci. Name(s) : *Juniperus virginiana*
Geog. Reg(s) : America-North (U.S. and Canada)
End-Use(s) : Essential Oil, Insecticide, Miscellaneous, Ornamental & Lawn Grass, Timber
Domesticated : Y
Ref. Code(s) : 77, 180, 220
Summary : The southern red cedar is a slow-growing evergreen tree. This tree is a major source of cedarwood and cedarleaf oil. These oils are used as a fragrance, a fixative, in insecticides and moth repellants, and as an immersion oil for microscopic work. They are also used in shoe polish but medicinally, they are a dangerous abortive. The dull red, light-weight timber is not particularly strong but is used for making furniture, baskets and pencils. The trees are sold for Christmas trees and ornamentals in the southern United States and the leaves are used for incense.

These trees reach a height of over 18.2 m and grow throughout the southern United States. They prefer dry or moist limey soils and grow prolifically, seed themselves, and can be cut throughout the year.

The total United States import value for cedarwood oil in 1984 was 240 MT, worth $1.3 million. In 1984, the United States imported 9.7 MT of cedar leaf oil as an essential oil, valued at $367,000.

Spanish Broom, Weaver's Broom
(See Fiber)

Spanish Oregano, Origanum

Sci. Name(s) : *Coridothymus capitatus*
Geog. Reg(s) : Africa, Europe, Mediterranean
End-Use(s) : Essential Oil, Medicinal, Spice & Flavoring
Domesticated : Y
Ref. Code(s) : 77, 180, 220
Summary : Spanish oregano is an herb whose flowers produce about 0.9% origanum oil used as an antiseptic for pharmaceutical and dental products and as flavoring for soft drinks, desserts, and candies. This essential oil is used in soaps, cosmetics, lotions, and perfumes. This plant produces the prefered origanum oil and it also produces an inferior oregano spice. The main constituent is a noncrystallizable phenol, carvacrol. Origanum leaves are used to flavor olive oil.

The herb grows wild in Spain, Morocco, and the Near East. It prefers lower elevations and moist soils. In 1984, United States import figures for origanum oil as an essential oil were 4.65 MT, worth $150,000.

Spearmint

Sci. Name(s) : *Mentha spicata, Mentha viridis*
Geog. Reg(s) : America-North (U.S. and Canada), Asia-China, Asia-India (subcontinent), Europe, Worldwide
End-Use(s) : Beverage, Essential Oil, Spice & Flavoring
Domesticated : Y
Ref. Code(s) : 170, 194, 205

Summary   : Spearmint (*Mentha spicata*) is a scented perennial herb that reaches from 30-60 cm in height and is cultivated for its aomatic leaves and essential oil. Both the fresh and dried leaves are used in salads, sauces, soups, jellies, hot and cold beverages, and for garnishing.   They are also used to flavor ice creams, chewing gums, toothpastes, and mouthwashes. Spearmint oil is used to flavor toiletries and confectionery. It does not have extensive medicinal value.

Plants are grown mainly in the United States, the People's Republic of China, and India and is native to the temperate regions of Europe. It is cultivated throughout the world as a home garden plant. In 1984, the United States imported 75 MT of spearmint as an essential oil, valued at $933,000.

Star-Anise Tree, Star Anise

Sci. Name(s)	: *Illicium verum*
Geog. Reg(s)	: Asia-China, Asia-Southeast
End-Use(s)	: Essential Oil, Fruit, Medicinal, Spice & Flavoring
Domesticated	: Y
Ref. Code(s)	: 154, 205
Summary	: The star-anise tree is a Chinese tree with spicy, star-shaped fruit that yields an essential oil used as a substitute for the high priced European aniseed oil. This oil is used to flavor products requiring an aniseed flavor. The fruit is used as a spice in the East. Dried fruit has been used as a carminative, weak stimulant, and a mild expectorant.

The tree is a small evergreen reaching up to 5 m in height. It grows mainly in southern and southwestern China, northern Vietnam, and it has been introduced to other countries in well-drained soils at altitudes above 2,500 m.

Summer Savory
(See Spice & Flavoring)

Sweet Balm, Lemon Balm

Sci. Name(s)	: *Melissa officinalis*
Geog. Reg(s)	: Europe, Mediterranean
End-Use(s)	: Beverage, Essential Oil, Miscellaneous, Spice & Flavoring
Domesticated	: Y
Ref. Code(s)	: 194, 205, 220
Summary	: Sweet balm is an aromatic perennial herb with lemon-scented leaves that yield an essential oil used in perfumery, ointments, and furniture creams. The oil also has potential in culinary items requiring a lemon flavor. Sweet balm is a good bee plant. Plant leaves can be dried and used in potpourris, herb pillows, and in herb mixtures for scented baths. The leaves are used as flavoring in salads, soups, and liqueurs such as Benedictine and Chartreuse, or brewed as tea.

Plants are native to the Mediterranean and other parts of southern Europe and thrive in rich, moist soil in partial sun. In colder areas shelter is necessary to protect the plants from frost. Leaves are harvested only once during the flowering period of the first year. Much of the lemon scent is lost while the leaves are drying.

Sweet Bay, Bay, Laurel, Grecian Laurel
(See Spice & Flavoring)

Sweet Birch, Black Birch, Cherry Birch, Mountain Mahogany

Sci. Name(s)	: *Betula lenta*
Geog. Reg(s)	: America-North (U.S. and Canada)
End-Use(s)	: Beverage, Energy, Essential Oil, Miscellaneous, Ornamental & Lawn Grass, Timber
Domesticated	: N
Ref. Code(s)	: 17, 77, 180
Summary	: Sweet or black birch is a tree whose leaves and twigs are the source of birch bark oil obtained by way of a process of steeping and enzymatic action. The oil is composed of about 98% methyl-salicylate. The young twigs and bark are the main source, and the oil is much like oil of wintergreen and is lower in price. It is used in linaments, mouthwashes, and toothpastes. The dark brown wood is heavy and strong and used for furniture, utensils, agricultural tools as well as fuel. It is also cultivated as an ornamental. The sap has a high sugar content and is used to make birch beer.

These trees grow wild in the United States, particularly in Pennsylvania, and reach heights of 24 m. It thrives in mountainous areas, on sandy soils, and in moist woodlands. The trees are cut during the cold months of the year. The second growth timber is most often used for the production of the oil. The trees with dark bark produce more oil than those with reddish bark. Rough birch imports from Canada for 1983 were 7,491 MBF, worth $3.5 million and dressed birch imports were 5,078 MBF, valued at $2.5 million.

Sweet Flag, Calamus, Flagroot

Sci. Name(s)	: *Acorus calamus*
Geog. Reg(s)	: America-North (U.S. and Canada), Asia, Asia-India (subcontinent), Asia-Southeast, Europe
End-Use(s)	: Essential Oil, Medicinal, Spice & Flavoring
Domesticated	: Y
Ref. Code(s)	: 171
Summary	: Sweet flag is a perennial aquatic herb grown for its underground stem (calamus root), powdered, and used in sachets and toiletries. An oil distilled from the stem is used in perfumery and for flavoring beverages and other foodstuffs. In China, the powdered root has medicinal uses. This herb is native to Asia and North America. It also grows in India and Malaysia and has naturalized in Europe.

Sweet Marjoram, Knotted Marjoram
(See Spice & Flavoring)

Sweet Orange, Orange
(See Fruit)

Sweet Woodruff
(See Spice & Flavoring)

Tansy
(See Medicinal)

Tarragon, French Tarragon
(See Spice & Flavoring)

Texas Cedar

Sci. Name(s)	: *Juniperus mexicana*
Geog. Reg(s)	: America-Central, America-North (U.S. and Canada)
End-Use(s)	: Essential Oil, Insecticide
Domesticated	: Y
Ref. Code(s)	: 77, 180, 220
Summary	: Texas cedar is a small- to medium-sized tree whose heartwood produces an essential oil, cedarwood oil. The yield of crude oil varies from 1.8-2.3% and has a high specific gravity with a woody fragrance. It is used in room sprays, deodorants, insecticides, mothproofing, clothing bags, floor polishes, lubricating greases, janitor supplies, and as a fixative. The wood cracks easily and has irregularities and is rarely used in construction.

Trees reach 6 m in height and grow at altitudes of 200-760 m. It is usually found on rocky, limestone hills and prefers deep sand with a thin top soil. It grows slowly and develops considerable heartwood with a small amount of sapwood. Trees grow in Texas, Central Mexico, and Guatemala. For 1984 United States import statistics on cedarwood oil, see southern red cedar.

Thyme

Sci. Name(s)	: *Thymus zygis* var. *gracilis*
Geog. Reg(s)	: Europe, Mediterranean, Temperate
End-Use(s)	: Essential Oil, Medicinal, Spice & Flavoring
Domesticated	: Y
Ref. Code(s)	: 77, 180
Summary	: Thyme is a short, erect shrub whose leaves and flowers produce thyme oil. The plant has a strong aromatic odor. The oil contains thymol, which is a strong germicide and antiseptic. It is used in dental products, inhalants, sprays, lotions, and soap. It is reported to cause mental excitment and is helpful in stimulating collapse. The oil is also used as a condiment to flavor processed foods, sausages, meats, and sauces. Thyme oil is extracted through steam distillation and yields are approximately 0.7%. The oil contains a high phenol content and the phenols in thyme are crystallizable which increases the plants value.

This herb grows wild in Spain on arid mountain slopes. It is grown in countries around the Mediterrean, Europe, and in many temperate countries around the world. For United States import statistics on thyme as an essential oil and thyme as a spice, see common thyme.

Tolu Balsam, Balsam-of-Tolu
(See Natural Resin)

Tuberose

Sci. Name(s)	: *Polianthes tuberosa*
Geog. Reg(s)	: America-Central, America-South, Tropical
End-Use(s)	: Essential Oil, Ornamental & Lawn Grass, Vegetable
Domesticated	: N
Ref. Code(s)	: 119, 171
Summary	: The tuberose is a Mexican herb grown in the tropics for its white, fragrant flowers successful in some Central and South American markets. The flowers are usually eaten in soups. An essential oil may also be obtained that is the source of a very expensive perfume material, tuberose absolute. Plants make attractive ornamentals.

Turmeric

(See Spice & Flavoring)

Udjung Atup

Sci. Name(s)	: *Baeckea frutescens*
Geog. Reg(s)	: Asia-China, Asia-Southeast, Australia
End-Use(s)	: Essential Oil, Medicinal, Timber
Domesticated	: N
Ref. Code(s)	: 36, 220
Summary	: Udjung atup is a small tree of the Malay Peninsula and Sumatra whose leaves and flowers are used in medicinal and herbal teas. An essential oil (Essence de Bruere de Tonkin) is obtained from the leaves. This oil is exported from the Malay Peninsula and Sumatra to France where it is used in soap perfumery. Udjung atup timber is hard and durable but has not been commercially exploited. Trees grow along the coasts of southern China and Australia.

Vanilla

Sci. Name(s)	: *Vanilla planifolia, Vanilla fragrans*
Geog. Reg(s)	: Africa, America, America-Central
End-Use(s)	: Essential Oil, Medicinal, Spice & Flavoring
Domesticated	: Y
Ref. Code(s)	: 148, 154, 171
Summary	: Vanilla is an herb grown for its immature fruits processed to produce 'vanilla extract'. It is also used in perfumes and soaps and was formerly used in medicinal preparations as a mild stimulant and aphrodisiac.

Vanillin, the flavor and fragrance of vanilla, can be produced synthetically and inexpensively from the waste sulphite liquor of paper mills, from coal tar extracts, and from eugenol obtained from clove oil. Although synthetic vanillin is cheaper, natural vanillin is preferred. The USDA requires product labels to state whether natural or synthetic vanillin has been used.

The vanilla plant requires shade and a hot, damp climate. If there is too much rain while the fruit is ripening, the crop is damaged. About 112 kg/ha of cured pods are yielded. Vanilla requires a rainy season of 45 cm and a dry period of at least 3 months to ripen the fruits. The soil should be high in organic matter. The podlike fruits are picked green and cured after being left 3-5 weeks. The active principle of vanilla crystallizes outside the pod.

Vanilla is grown in Africa and Central America primarily, although it can be grown in the Pacific area and Hawaii. At one time, the largest importer of vanilla beans was the United States, with 450 to 900 MT/annum. The top exporter was once the Malagasay Republic.

Vetiver, Khus-Khus

Sci. Name(s)	: *Vetiveria zizanioides*
Geog. Reg(s)	: Tropical
End-Use(s)	: Erosion Control, Essential Oil, Fiber, Ornamental & Lawn Grass
Domesticated	: Y
Ref. Code(s)	: 17, 171
Summary	: Vetiver is grown as a hedge plant. Its roots yield an essential oil used as a fixative for more volatile ingredients, for scenting soaps, and in cosmetics and perfumes. In the tropics, the roots are used for basketry, mats, fans, screens, awnings, sachet bags, and other woven handicrafts. The grass is used for planting contours and as erosion control on borders and roads.

Plants require a hot, humid climate and moist soils. If the roots are left in the ground for several years, the oil yield diminishes. In 1984, the United States imported 76.1 MT of vetivert oil as an essential oil, valued at $3.4 million.

West Indian Lemongrass, Sere Grass, Lemon Grass

Sci. Name(s)	: *Cymbopogon citratus*
Geog. Reg(s)	: Africa-East, America-Central, America-South, Asia-India (subcontinent), Europe, West Indies
End-Use(s)	: Erosion Control, Essential Oil, Insecticide, Spice & Flavoring
Domesticated	: Y
Ref. Code(s)	: 171, 194
Summary	: West Indian lemongrass is a tufted, perennial herb that seldom flowers under cultivation. An essential oil is obtained from the leaves by steam distillation. Citral from the oil is used in insect repellents, for flavoring soft drinks, and in scenting soaps, various technical preparations, perfumery, and cosmetics. The oil is also used in the chemical isolation of citral, which is then converted into violet-scented ionones. Ionones are used in the commercial synthesis of vitamin A. Lemongrass oil is now produced synthetically. The grass is useful for bunds for soil improvement and erosion control.

The grass grows throughout the tropics and is cultivated for oil production in the West Indies, Guatemala, Brazil, Zaire, Tanzania, Paraguay, England, Sri Lanka, and the Malagasay Republic. Plants require warm climates, lots of sunlight, and adequate rainfall. Harvest begins when the grass is 4-8 months old, additional cuttings being taken every 3-4 months. Harvest may continue for 4 years. Fresh grass can yield from 0.2-0.4% oil, giving 40-112 kg/ha/annum of oil. For trade statistic information on lemon grass as an essential oil, see malabar grass.

White Cottage Rose, Rose

Sci. Name(s)	: *Rosa alba*

Geog. Reg(s) : Africa-North, Europe, Mediterranean
End-Use(s) : Essential Oil, Spice & Flavoring
Domesticated : Y
Ref. Code(s) : 77, 180
Summary : The white cottage rose is the source of a rose oil. Fresh flowers are steam-distilled to yield this high-quality oil. Yields range from 0.02-0.03%. The main constituents of rose oil are rhodinol, geraniol, and nerol. Rose oil can be used in high-grade cosmetics, toilet waters, and perfumes. It can also be used to flavor tobacco and certain foods. Bulgaria, Egypt, and Turkey are the main producers of rose oil. For United States import statistics for rose oil as an essential oil, see cabbage rose.

Wild Chamomile, Sweet False Chamomile, German Chamomile
Sci. Name(s) : *Matricaria chamomilla*
Geog. Reg(s) : Asia, Eurasia, Europe
End-Use(s) : Beverage, Essential Oil, Medicinal
Domesticated : **Y**
Ref. Code(s) : 154, 194, 220
Summary : Wild chamomile is a cultivated herb in both the Old and New World. Its leaves are used in teas, as home remedies for intestinal worms, or as mild sedatives. Flower heads provide a nerve tonic that induces sweating and treats upset stomachs. An essential oil is used in perfumes, shampoos, liqueurs, tobacco, and confections.

Wild Thyme, Creeping Thyme
(See Ornamental & Lawn Grass)

Winter Savory
(See Spice & Flavoring)

Winter's Grass, Citronella Grass
Sci. Name(s) : *Cymbopogon winteranus*
Geog. Reg(s) : Africa-East, America-Central, America-South, Asia, Asia-India (subcontinent), Asia-Southeast, West Indies
End-Use(s) : Essential Oil, Spice & Flavoring
Domesticated : Y
Ref. Code(s) : 171
Summary : Winter's grass is sometimes placed in a separate category in the Cymbopogon species. It is also considered a type of *Cymbopogon nardus* var. *mahapengiri*. The grass yields an essential oil that contains citronellal, making it useful both as a fragrance in perfumed products and as a flavoring. The oil is produced in Java, Taiwan, Guatemala, Honduras, Brazil, Sri Lanka, Zaire, East Africa, and the West Indies. It is not clear whether this particular grass is cultivated as a separate commercial product. For 1984 United States import statistics on citronella as an essential oil, see nardus grass.

Wintergreen
Sci. Name(s) : *Gaultheria procumbens*
Geog. Reg(s) : America-North (U.S. and Canada)

End-Use(s) : Beverage, Essential Oil, Fruit, Medicinal
Domesticated : Y
Ref. Code(s) : 194, 220
Summary : Wintergreen is a perennial, evergreen herb with white flowers that bloom in late summer. Scarlet-colored fruit appear in early autumn. About 0.5-0.8% of an essential oil can be distilled from the fresh leaves of the plant. The oil, containing 96-98% methyl salicylate is used to flavor confectionery, chewing gums, nonalcoholic beverages, herbal teas, toothpastes, and mouthwashes. The natural oil is being replaced by synthetic reproductions of methyl salicylate and an essential oil obtained from the twigs of *Betula lenta*. The berries are used in pies.

Shrubs are most successful in areas with temperatures 22-31 degrees Celsius with an average annual rainfall 0.6-1.3 m. Commercial cultivation takes place in some of the Appalachian and northeastern parts of the United States. Leaves should be harvested during June through September when oil content is the highest. Pickings can be made twice during a growing season. Most leaves are obtained from wild plants. Wintergreen is native to North America.

Wormseed, Mouse Food, Mexican Tea, Wormseed Goosefoot
(See Medicinal)

Wormwood, Absinthe, Absinthewood, Absinthium
Sci. Name(s) : *Artemisia absinthium*
Geog. Reg(s) : Africa-North, America-North (U.S. and Canada), Asia-Central, Asia-India (subcontinent), Europe, Mediterranean
End-Use(s) : Essential Oil, Medicinal
Domesticated : Y
Ref. Code(s) : 170, 194, 205, 220
Summary : Wormwood is an erect perennial herb whose leaves and flower tops yield an essential oil used in the preparation of wines (Vermouth and Muse Verte) and absinthe. The oil contains absinthin and thujon which, if taken in excess, will cause delirium, hallucinations, and possibly permanent mental illness. The extracted oil comprises from 0.5-1% of the fresh weight of the leaves. Preparations of the leaves and flowers have been used in tonics to treat stomachaches and intestinal worms.

The herb grows wild in the Mediterranean, southern Siberia, and Kashmir. It is cultivated commercially in Europe, North Africa, and the United States. Plantings last from 7-10 years, with top production in the second and third year. Harvesting is done twice a year, usually during late spring and full bloom periods.

Ylang-Ylang
Sci. Name(s) : *Cananga odorata*
Geog. Reg(s) : Africa, Asia-Southeast
End-Use(s) : Essential Oil
Domesticated : Y
Ref. Code(s) : 170
Summary : Ylang-ylang is an evergreen tree whose fragrant flowers provide two essential oils. The first is ylang-ylang oil used in high class perfumery. The second, cananga oil, is used in cheap perfume and for scenting soaps. Production of ylang-ylang was for-

merly confined to the Philippines but now includes Java and Reunion which have a corner on the market.

Cultivated trees begin to flower at 1-2 years of age. At 4 years they produce 4-5 kg/tree/annum of flowers and at 10 years approximately 9-11 kg/annum. The oil yield by fresh flower weight is 1.5-2.5% with both oils being in equal proportions. Trees stand from 3-30 m in height. In 1984, the United States imported 37.4 MT of ylang-ylang oil or cananga oil as an essential oil, valued at $1.7 million.

Zedoary, Shoti

Sci. Name(s)   : *Curcuma zedoaria*
Geog. Reg(s)   : Asia-China, Asia-India (subcontinent), Asia-Southeast
End-Use(s)   : Essential Oil, Gum & Starch, Medicinal
Domesticated   : Y
Ref. Code(s)   : 111, 171
Summary   : Zedoary, or shoti, is an herb about 1 m tall whose rhizomes provide an easily digested starch, similar to true arrowroot. The starch does not show promise as an industrial starch because of low rhizome yields. It is a valuable component in infant and invalid diets. Through steam distillation, about 1-2% of a light yellow essential oil can be obtained. In Asia, this oil is used in perfumery and medicinal preparations. Plant leaves are sometimes used for culinary purposes.

The herb is native to northeastern India and grows both wild and cultivated in India, Sri Lanka, China, Malaysia, Indonesia, and the Philippines. Zedoary requires hot, humid climates with average rainfall from 90-125 cm. Plants are found up to 900 m in elevation, usually in shady locations. Optimum rhizome yields can be obtained if plants are grown in loamy, well-drained soils. Rhizomes are harvested 10 months after planting.

Fat & Wax

Bayberry

Sci. Name(s) : *Myristica caroliniensis*
Geog. Reg(s) : America-South
End-Use(s) : Fat & Wax
Domesticated : Y
Ref. Code(s) : 139
Summary : Bayberry is a shrub grown in Colombia and the eastern and southern United States for the commercial production of bayberry wax. The green wax is from the processed berries of several species of myrtle. It has an aromatic odor and is used in candles and soaps. Shrubs reach 1-2 m in height.

Candelilla

Sci. Name(s) : *Euphorbia antisyphilitica*
Geog. Reg(s) : America-Central, America-North (U.S. and Canada)
End-Use(s) : Fat & Wax
Domesticated : N
Ref. Code(s) : 170, 220
Summary : Candelilla is a wild plant processed for a commercially important wax. This wax is used in leather polishes, waterproofing, insulation, dental molds, and other products. When mixed with paraffin, candelilla wax is made into candles. Other products in which candelilla wax is used are metal lacqueurs, paint removers, paper sizing, and lithographic colors. Candelilla plants are native to Mexico and the southwestern United States.

Carnauba

Sci. Name(s) : *Copernicia prunicia*
Geog. Reg(s) : America-South
End-Use(s) : Fat & Wax
Domesticated : N
Ref. Code(s) : 139
Summary : Carnauba is a palm tree of the semiarid, northeast section of Brazil. Carnauba wax is obtained from the cut and dried fronds. Depending on the frond age, 3 grades of wax can be obtained: yellow, light gray, and chalky.

Carnauba wax is used in paste polishes and carbon inks. The high-gloss polish is used in candies and pills. It slows down the rate of moisture loss on perishable products like citrus fruits. This wax is used on polystyrene records, leather finishes, and lipstick.

The wax is harvested twice a year. About 10,884 MT of wax are produced annually, 9.1 MT of which are exported. In 1983, the United States imported 3,755 MT, valued at $5.7 million. Because of unstable prices and availability, many industries are resorting to synthetic substitutes waxes.

Castorbean
(See Oil)

Chinese Tallow-Tree, Tallow-Tree
(See Oil)

Colchicum, Meadow Saffron, Autumn Crocus
(See Medicinal)

Douglas's Meadowfoam
(See Oil)

Douglas-Fir
(See Timber)

Gutta-Percha
(See Isoprenoid Resin & Rubber)

Japan Wax

Sci. Name(s)	: *Rhus succedanea*
Geog. Reg(s)	: Asia, Asia-China
End-Use(s)	: Fat & Wax, Natural Resin, Ornamental & Lawn Grass
Domesticated	: Y
Ref. Code(s)	: 17, 99, 139, 220
Summary	: The Japan wax tree is a shrub or tree with berries that are processed for a wax used in candles, polishes, ointments, plaster, and lubricants. They yield a natural lacquer. The tree is grown as an ornamental, mainly in China and Japan, for its shape, dense fruiting heads, and brilliant fall colors. Trees reach heights up to 9 m. The United States imports for 1983 were 50.5 MT, valued at $301,000.

Jojoba, Goat-Nut

Sci. Name(s)	: *Simmondsia chinensis*
Geog. Reg(s)	: America-North (U.S. and Canada), Subtropical, Tropical
End-Use(s)	: Fat & Wax, Medicinal, Nut, Oil, Ornamental & Lawn Grass
Domesticated	: Y
Ref. Code(s)	: 17, 101, 139, 182
Summary	: Jojoba is a bushy, evergreen shrub from 0.6-4.5 m in height that produces an important oil and wax from its fruit capsules and brown nuts. Jojoba oil is used in cosmetics, drugs, plastics, waxes, and leather. It has also been used as a substitute for sperm whale oil. Jojoba wax is used as a lubricant and has potential use in linoleum, detergents, pharmaceutical products, polishes, and carbon paper. It is also being considered as a substitute for carnauba and bee's wax. The filbert-flavored nuts are cured and eaten from the shell. Jojoba seed meal is a good source of concentrated vegetable proteins but has a high amount of toxic material, "simmondsin," which acts as an appetite-suppressent on laboratory test animals. Jojoba plants make attractive clipped hedges and slope covers.

Plants thrive in hot, dry environments from 609-1,218 m in elevation with an annual rainfall from 38.1-45.7 cm on well-drained soil. Plants live for about 200 years under the right conditions. In the United States, jojoba is grown primarily in southern

California. Plants produce well in other tropical and subtropical regions in exposed, sunny areas.

Lac-tree, Ceylon Oak, Kussum Tree, Malay Lac-tree

Sci. Name(s)   : *Schleichera oleosa, Schleichera trijuga*
Geog. Reg(s)   : Asia-India (subcontinent)
End-Use(s)   : Dye & Tannin, Fat & Wax, Fruit, Oil, Timber, Vegetable
Domesticated   : N
Ref. Code(s)   : 98, 154, 170, 220
Summary   : The lac-tree is grown in India for its fruit seeds, which contain an edible fat used in illumination and hair-oils. Lac-tree oil is probably the original source of Macassar oil. Fruit pulp is edible. Unripe fruits are eaten as pickles and young leaves as vegetables. The hard, red timber is used to make small boats and sugar mill rollers. Bark is used for tanning.

Ouricury Palm, Licuri, Nicuri

Sci. Name(s)   : *Syagrus coronata*
Geog. Reg(s)   : America, Tropical
End-Use(s)   : Fat & Wax, Forage, Pasture & Feed Grain, Fruit, Oil
Domesticated   : N
Ref. Code(s)   : 139, 220
Summary   : The ouricury palm is grown for its waxy leaves and edible fruit. The wax is collected and used as a substitute for carnauba wax in low grade items. The fruit is also fed to livestock. Fruit seeds contain a nondrying oil produced for the commercial manufacture of margarine. Palms grow in tropical America.

Rice
(See Cereal)

Shea-Butter, Butterseed

Sci. Name(s)   : *Butyrospermum paradoxum*
Geog. Reg(s)   : Africa-West
End-Use(s)   : Fat & Wax, Nut, Oil
Domesticated   : N
Ref. Code(s)   : 170
Summary   : The shea-butter tree produces nuts that yield a commercially important fat and oil. The seeds contain 45-60% fat called shea butter and 9% protein. Shea-oil is used in soap and candle making, cosmetics, and as an ingredient in the fillings used for chocolate cream.

Principal nut exporters are Nigeria, Ghana, Senegal, Mali, the Ivory Coast, Upper Volta, and Benin. Trees are an important part of society in the Savanna areas of West Africa. Shea-nut production in West Africa has been estimated at 450,000 MT/year. The small, deciduous trees bear fruit 12-15 years and mature in 30 years.

Sugarcane
(See Sweetener)

Virola

Sci. Name(s)	: *Virola sebifera*
Geog. Reg(s)	: America-South
End-Use(s)	: Fat & Wax
Domesticated	: N
Ref. Code(s)	: 220
Summary	: Virola is a tree whose seeds produce a fat that has economic importance when made into a wax. The wax is exported to the United States and Europe from northern South America to make soap and scented candles.

Wax Palm, South American Wax Palm

Sci. Name(s)	: *Ceroxylon alpinum, Ceroxylon andicola*
Geog. Reg(s)	: America-South
End-Use(s)	: Fat & Wax
Domesticated	: N
Ref. Code(s)	: 124, 171
Summary	: The wax palm, or South American wax palm, is a very tall palm reaching heights of 60 m. Its trunk and leaves are coated with wax (palm wax) often used as a substitute for carnauba wax because of similar chemical characteristics. There are only a few products marketed that are actually palm wax. Most products so labelled are more likely carnauba wax. The crude palm wax is a mixture of pure wax with a resinous substance removed through extensive crystallization processes from boiling alcohol.

Waxgourd, White Gourd
(See Vegetable)

White Meadowfoam, Meadowfoam

Sci. Name(s)	: *Limnanthes alba*
Geog. Reg(s)	: America-North (U.S. and Canada)
End-Use(s)	: Fat & Wax
Domesticated	: Y
Ref. Code(s)	: 70, 93, 106, 231
Summary	: White meadowfoam (*Limnanthes alba*) is an herb whose seeds provide a high-quality fatty oil. This oil has potential as a substitute for sperm whale oil and for use in high-quality waxes, lubricants, detergents, and plasticizers. Limnanthes oil produces a liquid wax similar in composition to jojoba oil. The expressed, defatted seed is considered a possible feedmeal for goats, chickens, and sheep.

In Alaska, Oregon, and California, experimental meadowfoam crops have shown new crop potential. Meadowfoam seed yields are 1,137-1,177 kg/ha. White meadowfoam is native to the Pacific Coast region of the United States, particularly in northern California and southern Oregon.

Fiber

Abaca, Manila Hemp

Sci. Name(s)	: *Musa textilis*
Geog. Reg(s)	: America-South, Asia-Southeast, Tropical
End-Use(s)	: Fiber
Domesticated	: Y
Ref. Code(s)	: 126, 154
Summary	: Abaca is a perennial herbaceous plant similar to the banana which is grown as a fiber crop in the Philippines. The fiber is collected from the overlapping leaf sheaths and was once considered an important source of marine cordage because of its durability in salt water. It no longer comprises a major portion of the marine cordage market because it cannot compete with synthetic fibers from a price stand-point. Demand for the fiber as a pulp product has been increasing. The Philippines is one of the leading abaca producers and Ecuador is another significant producer. Trees bear inedible fruit.

Abaca grows best in moist climates, especially tropical lowlands below 500 m in elevation. It should be planted in deep, well-drained, volcanic or alluvial soils. Leaves are cut 1.5-2 years after planting. Stalks are taken prior to flowering. Trees are harvested for 10-15 years.

African Locust Bean
(See Oil)

Alfalfa, Lucerne, Sativa
(See Forage, Pasture & Feed Grain)

Aramina Fiber, Congo Jute, Cadillo, Caesarweed, Aramina

Sci. Name(s)	: *Urena lobata*
Geog. Reg(s)	: Africa, America, America-South, Asia, Asia-Southeast, Mediterranean, Subtropical, Tropical, West Indies
End-Use(s)	: Fiber, Medicinal, Weed
Domesticated	: Y
Ref. Code(s)	: 154, 170, 220
Summary	: Aramina fiber is a fiber crop grown in the tropics and subtropics. Its fiber resembles jute and is used in hessian, ropes, and carpets. Most fiber production is centered in Zaire, Brazil, and Madagascar. It is used mainly for coffee bags. In Malaysia, the plant has certain undocumented medicinal purposes.

This crop does well in hot, humid climates with full sun and deep, fertile, well-drained soil. If neglected, it becomes a troublesome weed.

Babul
(See Gum & Starch)

Bambusa, Bamboo
(See Timber)

Baobab, Monkey-Bread
(See Vegetable)

Bolo-Bolo

Sci. Name(s)	: *Clappertonia ficifolia*
Geog. Reg(s)	: Africa-West
End-Use(s)	: Fiber, Ornamental & Lawn Grass
Domesticated	: N
Ref. Code(s)	: 50, 170
Summary	: Bolo-bolo is a fiber yielding shrub of tropical Africa. The fiber is comparable to jute in quality but has not yet been commercially exploited. In West Africa, the fiber is used for mat making. The bast is suitable for paper pulp. Bolo-bolo shrubs are grown as ornamentals.

Bowstring Hemp, Snake-Plant

Sci. Name(s)	: *Sansevieria trifasciata*
Geog. Reg(s)	: Africa-West, Tropical
End-Use(s)	: Fiber, Ornamental & Lawn Grass
Domesticated	: Y
Ref. Code(s)	: 154, 171
Summary	: Bowstring hemp is a tropical fiber crop. The leaves contain 2% fiber used for making mats, hammocks, bowstrings, and twine. Some forms of the plant are grown as ornamentals.

Plants are easily propagated by leaf tips. They are fairly hardy and resistant to cold temperatures. Fiber yields have been recorded at about 2,250 kg/ha.

Breadfruit
(See Vegetable)

Burweed

Sci. Name(s)	: *Triumfetta cordifolia*
Geog. Reg(s)	: Africa-West, Tropical
End-Use(s)	: Fiber, Gum & Starch, Medicinal, Spice & Flavoring
Domesticated	: Y
Ref. Code(s)	: 50, 170
Summary	: Burweed is a plant that has been experimented with as a small-scale commercial fiber crop. The fiber is obtained from the bark and has mucilaginous properties. It has been used to make strong cordage, bowstrings, and fishing line. Plant leaves can be used as a soup or sauce herb. Leaves have also served unspecified medicinal purposes. Burweed is found mainly in tropical Africa.

Castorbean
(See Oil)

China Jute, Indian Mallow, Velvetleaf, Butterprint

Sci. Name(s)	: *Abutilon theophrasti, Abutilon avicennae*

Geog. Reg(s)   : America-North (U.S. and Canada), Asia-China, Asia-India (subcontinent),
                   Eurasia
End-Use(s)   : Fiber, Weed
Domesticated   : Y
Ref. Code(s)   : 51, 112, 217
Summary   : China jute is a fiber crop of considerable economic importance in the Orient.
The soft fiber obtained from the stems is blended with other fibers or used alone to
manufacture sacks, ropes, and fishnets. Plants grown on moist alluvials or friable
loams produce the best quality fiber.

This jute is native to China and the USSR. In the United States, China jute
plants are most often found as weeds that compete with row-crop culture.

Clapper Polyandra
Sci. Name(s)   : *Clappertonia polyandra*
Geog. Reg(s)   : Africa-West, Tropical
End-Use(s)   : Fiber
Domesticated   : N
Ref. Code(s)   : 220
Summary   : Clapper polyandra is a tree whose bark yields fiber used for making mats and
paper pulp. The tree grows in tropical Africa. There is little information available as
to the economic possibilities of this tree.

Coconut
(See Oil)

Common Reed Grass, Carrizo
(See Miscellaneous)

Cow-Tree
(See Gum & Starch)

Danicha
Sci. Name(s)   : *Sesbania bispinosa, Sesbania aculeata*
Geog. Reg(s)   : Africa-South, Africa-West, America-North (U.S. and Canada), America-South,
                   Asia, Asia-India (subcontinent)
End-Use(s)   : Cereal, Fiber, Forage, Pasture & Feed Grain, Weed
Domesticated   : Y
Ref. Code(s)   : 56, 154
Summary   : Danicha is a low-growing, prickly shrub to 7 m in height that is grown for fiber
and fodder. Stems provide a strong, lasting fiber substitute for rope hemp, twine,
fishnet cordage, and sailcloth. In parts of South Africa, plant leaves are used as forage or poultry feed. Danicha seeds are edible and used as an alternate food supply in
times of famine.

Plants are most often found as weeds, especially in tropical Africa from Senegal
to Cameroun. They thrive in wet, heavy soils. Plants are most successful in areas receiving 5.7-22.1 dm/year of rainfall with temperatures of 19.9-27.3 degrees Celsius.
Seeds mature in about 5 months. In India, yields have been recorded at 600 kg/ha of

seed. Seed yields in Peru have been recorded at 900 kg/ha. In the United States, danicha grows mainly in California, where seed yields have been recorded at 1,000 kg/ha and fiber yields have been recorded at 100-1,000 kg/ha.

Date Palm
(See Fruit)

Dendrocalamus, Bamboo
(See Timber)

Dwarf Banana, Pisong Jacki
(See Fruit)

Elephant Grass, Napier Grass
(See Forage, Pasture & Feed Grain)

Ensete, Abyssinian Banana
(See Vegetable)

Esparto, Halfa

Sci. Name(s)	: *Stipa tenacissima*
Geog. Reg(s)	: America-North (U.S. and Canada), Mediterranean
End-Use(s)	: Fiber
Domesticated	: Y
Ref. Code(s)	: 20, 220
Summary	: Esparto is a grass with slender culms between sixty and eighty centimeters in height, and whose leaves and stems have been used to make fine paper, cordage, sails, and mats. The grass is found in the Mediterranean region, and in the United States, primarily in western Texas, New Mexico, and Arizona.

European Beachgrass, Marram Grass
(See Erosion Control)

Fishtail Palm, Toddy Palm
(See Sweetener)

Flax

Sci. Name(s)	: *Linum usitatissimum*
Geog. Reg(s)	: America, America-Central, America-North (U.S. and Canada), Asia, Asia-Southeast, Eurasia, Mediterranean, Temperate
End-Use(s)	: Fiber, Forage, Pasture & Feed Grain, Medicinal, Oil
Domesticated	: Y
Ref. Code(s)	: 154, 205, 220, 227
Summary	: Flax is an important fiber crop whose seeds provide linseed oil used in veterinary medicine and in industrial applications. The seeds contain 33-34% oil used in paints, varnishes, printer's inks, soap, painting and litho-varnishes, and cooking. Linseed oil is used in a mixture as an antiscaling compound to protect cement.

A small seeded variety of flax is the source for linen cloth. Flax fiber toughens when wet. Shorter fibers are used to make cigarette paper and other fine-textured paper. Seed residues from hot-pressing contain 33-43% protein and are used for cattle feed. The seed or powdered seed is applied as a poultice to burns and scalds, or used internally as a mild laxative. Wilted flax contains prussic acid dangerous to livestock. Flax thrives in cool, moist climates.

More than 0.5 million MT are produced annually in the USSR, France, Poland, and Czechoslovakia. Flaxseed production in the United States reached 187,000 MT in 1983. The United States exported 6,905 MT of linseed oil valued at $3.8 million. Linseed oilcake exports were 86,533 MT valued at $16.2 million. Imports were valued at $2.9 million dollars for 2,425 MT for the same year.

Giant Reed
(See Miscellaneous)

Goosegrass, Fowl Foot Grass, Yardgrass
(See Weed)

Gum Arabic, Senegal Gum, Kher
(See Gum & Starch)

Henequen

Sci. Name(s)	: *Agave fourcroydes*
Geog. Reg(s)	: America-Central, West Indies
End-Use(s)	: Fiber
Domesticated	: Y
Ref. Code(s)	: 126, 171
Summary	: Henequen is grown extensively as a fiber crop in the Yucatan peninsula and Campeche states in Mexico. The leaves contain 4% fiber and are decorticated and exported mainly to the United States. Cuba and El Salvador are minor producers. In 1984, the United States imported 272 MT of henequen and sisal combined valued at just over 1 million dollars. Henequen fiber is considered coarse and less attractive than sisal. Manufactured fiber is made into binder and baler twine and the lower grades into padding. Henequen fiber is being replaced by synthetics.

Plants are drought-resistant and most are successful on gravelly soils in areas with hot climates and low rainfall. In Mexico, about 1,000 plants/ha are planted. Crops are weeded when young and basal leaves harvested 4-6 years after planting. Leaves are harvested subsequently twice a year. Overcutting lowers crop yields.

Istle, Mexican Fiber, Lechuguilla, Tula Istle, Tampico Fiber

Sci. Name(s)	: *Agave lecheguilla*
Geog. Reg(s)	: America-Central, America-North (U.S. and Canada)
End-Use(s)	: Fiber, Weed
Domesticated	: N
Ref. Code(s)	: 171
Summary	: Istle plants produce a useful fiber from their immature leaves. Plants grow wild in Mexico at altitudes of 180-1,800 m in areas with low rainfall. In the United

States, plants are found in Texas as weeds and are not used commercially. The fiber is used in a variety of brushes and as a substitute for bristle. They are obtained by cutting and scraping the immature leaves of the central bud.

Flowering is prevented by cutting the central buds after 6 years. Plants continue producing central buds that are cut twice a year for another 6 years.

Itabo, Izote, Palmita, Ozote, Spanish Bayonnette
(See Vegetable)

Juniper, Common Juniper
(See Essential Oil)

Kamraj
(See Vegetable)

Kapok, Silk-Cotton-Tree, Ceiba

Sci. Name(s)	: *Ceiba pentandra, Bombax pentandra, Eriodendron anfractuosum* var. *indicum*
Geog. Reg(s)	: Africa-East, Africa-West, Asia, Asia-India (subcontinent), Asia-Southeast, Tropical
End-Use(s)	: Fiber, Forage, Pasture & Feed Grain, Oil, Timber
Domesticated	: Y
Ref. Code(s)	: 170
Summary	: Kapok is a rapid-growing deciduous tree 10-30 m tall. It bears fruit with seeds that yield buoyant and water-repellent single-celled fiber used in insulation and stuffing. The fiber (kapok) is a low thermal conductor and excellent for absorbing sound. These factors make it a popular thermal and acoustic insulator in airplanes. Kapok fiber is also spun into yarn and textiles. It contains 64% cellulose and 13% lignin. The seeds yield 20-25% edible oil, similar to cottonseed oil, used in cooking, lubricants, and soap. The expressed oil cake has 26% protein and is fed to livestock. Kapok wood is soft, light, and used to make canoes, stools, and furniture.

Trees are cultivated commercially in West Africa and Asia. They are highly adaptable but most successful at elevations below 46 m in the tropics. They bear at 3-4 years and produce about 100 pods. Trees mature in 7-10 years and yield 300-400 pods to provide 1.6-1.8 kg of kapok fiber. Trees bear for 60 years. Thailand is a major kapok producer. Other minor producing countries are Cambodia, Indonesia, East Africa, and India. The United States has been a major kapok importer.

Kenaf, Bimli, Bimlipatum Jute, Deccan Hemp

Sci. Name(s)	: *Hibiscus cannabinus*
Geog. Reg(s)	: Africa-North, America-South, Asia-China, Asia-India (subcontinent), Asia-Southeast, Eurasia, Subtropical, Tropical
End-Use(s)	: Fiber, Oil, Spice & Flavoring
Domesticated	: Y
Ref. Code(s)	: 126, 170
Summary	: Kenaf trees are grown in most tropical and subtropical areas as fiber crops. They are grown extensively in India, Thailand, and China, and to a lesser extent, in

the USSR, Brazil, Bangladesh, and Egypt. Thailand and Bangladesh export raw fiber, while other countries process the fiber for domestic use.

Fiber from the stems of wild plants is used in textile industries for rope, bag fabric, cordage, and carpet yarn. Kenaf shows good potential as a pulp fiber source. Kenaf seeds contain 20% oil used in lubricants, soaps, linoleum, and paint. Young plant leaves are used as potherbs.

India is the leading kenaf fiber producer with general yields of 1,121 kg/ha. Annual seed yields are 340-409 kg/ha. Kenaf plants reach 1-4 m in height. Crops mature earlier than roselle (*Hibiscus sabdariffa*).

Lagos Silk-Rubber, Funtumia Elastica, Lagos Silkrubber Tree, Silkrubber Tree
(See Isoprenoid Resin & Rubber)

Lapulla, Triumfetta
Sci. Name(s)   : *Triumfetta tomentosa*
Geog. Reg(s)   : Africa, Tropical
End-Use(s)   : Fiber
Domesticated : N
Ref. Code(s)   : 170, 220
Summary   : Lapulla is a plant that provides excellent fiber. The fiber is obtained from the bark and has mucilaginous properties. Lapulla fiber is widely used in tropical Africa for cordage and other binding materials and has been tried as a commercial source of fiber. Plants are most successful in tropical regions.

Levant Cotton, Arabian Cotton, Maltese Cotton, Syrian Cotton
Sci. Name(s)   : *Gossypium herbaceum, Gossypium obtusifolium*
Geog. Reg(s)   : Africa, Asia, Asia-India (subcontinent), Mediterranean
End-Use(s)   : Fiber, Oil
Domesticated : Y
Ref. Code(s)   : 117, 170
Summary   : Levant cotton (*Gossypium herbaceum*) is an annual or perennial tree with little economic value either as a cotton producer or as an oil crop. Within this species are several primitive cottons, most of which are wild plants. Two members are notable. Variety kuljanum is a hardy, annual subshrub that thrives in areas with short, hot summers and long, cold winters. It produces early cotton yields. A 2nd member, variety wightianum, is an annual shrub first developed in India. This variety produces some of the better Asiatic cottons. It is grown for domestic use in the drier parts of Africa and Asia.

Lukrabao, Chaulmogra Tree
(See Medicinal)

Marijuana, Hemp, Marihuana
Sci. Name(s)   : *Cannabis sativa*
Geog. Reg(s)   : Asia-Central, Asia-India (subcontinent), Eurasia, Europe, Temperate, Tropical
End-Use(s)   : Fiber, Medicinal, Natural Resin, Oil

124

Domesticated : Y
Ref. Code(s) : 170
Summary : Marijuana is a tall herb native to Central Asia. The plant provides 3 products--fiber from the stems, oil from the seeds, and narcotics from the leaves and flowers. Marijuana is grown mainly for fiber and narcotics. Private cultivation of marijuana as a narcotic source is generally restricted.

For fiber production, plants are grown in temperate countries. The USSR is the largest producer of hemp. The principal exporting countries are Italy and Yugoslavia. Hemp fiber contains 70% cellulose and is valued commercially for its long-length, strength, and durability. It is used as a substitute for flax in the manufacture of yarns and twines. When cultivated as an oil crop, female plants are left after male plants have been harvested for fiber. The seeds are then collected and pressed to yield 30-35% of a drying oil used as a substitute for linseed oil in paints and varnishes. The seeds also contain 22% protein.

Marijuana plants yield 3 types of narcotics. Bhang, or hashish, is the dried leaves and flowering shoots from male and female plants. Ganja is the dried unfertilized female inflorescences of special cultivars grown in India. Charas is the crude resin collected from the tops of the plants. This resin has both therapeutic and narcotic qualities, containing a mixture of cannabinol and allied compounds. These properties produce hallucinatory effects when used.

Although marijuana can be grown in most tropical and temperate countries, tropical conditions are best for resin production. Harvesting is done when the plants first flower, usually 4-5 months after sowing.

Mauritius Hemp

Sci. Name(s) : *Furcraea foetida*
Geog. Reg(s) : Africa, Tropical
End-Use(s) : Fiber, Ornamental & Lawn Grass
Domesticated : Y
Ref. Code(s) : 17, 61, 113, 136, 214
Summary : Mauritius hemp is a fiber crop cultivated throughout the tropics for its commercially important fiber. The long, soft fiber is used alone or with other fibers in twine, sacks, hammocks, and other products. It breaks down in salt water but withstands fresh water. It tolerates the use of natural or synthetic dyes and retains bright colors.

On Mauritius, 3,000-4,000 plants/ha are grown and require approximately 3 years till harvest and 3 years recovery. There were commercial plantations on the islands of Mauritius and St. Helena where a coarse cloth was produced and made into sugar bags. Shorter fibers are used for brush making and upholstery work and the plants for the manufacture of filter press cloth.

Compared to sisal, Mauritius hemp has many superior qualities but is not as strong. It is often used in mixtures with sisal. Plants are grown in desert gardens as ornamentals.

Menteng, Kapundung
(See Fruit)

Muriti

Sci. Name(s)	: *Mauritia vinifera, Mauritia flexuosa*
Geog. Reg(s)	: America-South
End-Use(s)	: Beverage, Fiber, Fruit
Domesticated	: N
Ref. Code(s)	: 220
Summary	: Muriti is a Brazilian fiber crop. The fiber is used primarily for cordage. Muriti fruits can be made into wine through a fermentation process. The pulp of the fruit is often eaten as a side dish.

New Zealand Flax, New Zealand Hemp

Sci. Name(s)	: *Phormium tenax*
Geog. Reg(s)	: Pacific Islands
End-Use(s)	: Fiber
Domesticated	: Y
Ref. Code(s)	: 16
Summary	: New Zealand flax is a large perennial herb considered an extremely useful fiber crop in New Zealand. It is second only to timber as an economic crop in New Zealand. Herbs are propagated by seed.

Okra, Lady's Finger, Gumbo
(See Vegetable)

Palmyra Palm, African Fan Palm
(See Sweetener)

Paroquet Bur, Pulut-Pulut

Sci. Name(s)	: *Triumfetta rhomboidea, Triumfetta bartramia*
Geog. Reg(s)	: Africa-South, Africa-West
End-Use(s)	: Fiber, Medicinal, Spice & Flavoring
Domesticated	: N
Ref. Code(s)	: 36, 50, 170
Summary	: Paroquet bur is a plant grown as a fiber crop in tropical Africa. The fiber is used for binding and is popular because it is easily prepared. Plant leaves are occasionally used as potherbs and are sometimes made into medicinal brans for horses. In South Africa, plant roots have been used as a Zulu women's medicine.

Peanut, Groundnut, Goober, Mani
(See Oil)

Phalsa
(See Beverage)

Pineapple
(See Fruit)

Ramie, Rhea, China Grass

Sci. Name(s)	: *Boehmeria nivea*
Geog. Reg(s)	: America-North (U.S. and Canada), Asia, Asia-China, Asia-Southeast
End-Use(s)	: Fiber, Forage, Pasture & Feed Grain
Domesticated	: Y
Ref. Code(s)	: 126, 170
Summary	: Ramie is an East Asian grass 1-3 m in height. Its stems provide a highly water-resistant vegetable fiber with a greater tensile strength than cotton. Its strength increases when wet. The fiber is made into twine, thread, and other flax and hemp products, and is spun as grasscloth or Chinese linen. Fiber production is difficult because stems are coated with a gummy pectin that makes fiber separation and decortication difficult. Ramie leaves and tops are palatable, high in protein, and used for livestock feed.

The major centers of large-scale fiber production are in Japan and China. Production costs are thought too high for ramie fiber to be economically feasible in other countries. Experimental crops have been grown with little success in the southern United States.

Ramie needs warm, moist climates, a rich, loamy soil, and frequent manuring for optimum growth. It does not tolerate waterlogging. The first harvest occurs in 10 months but is rarely used for fiber with subsequent harvests 2-3 times/year. Ramie is also found in Southeast Asia. In 1984, the United States imported approximately $30,000 worth of ramie fiber.

Rattan
(See Timber)

Red Silk-Cotton, Silk Cotton

Sci. Name(s)	: *Bombax ceiba*
Geog. Reg(s)	: Africa, America, Asia-Southeast, Tropical
End-Use(s)	: Fiber, Oil, Timber
Domesticated	: Y
Ref. Code(s)	: 170, 220
Summary	: Red silk-cotton is a tree from which a cotton fiber is obtained. The fiber is water-resistant and is a valuable component in life jackets. It is used in insulation, for stuffing, and its light wood is used in packing cases. The seeds produce an oil used in soap manufacture.

Indonesia produces over 90% of the world production of red silk-cotton seeds. This tree is native to tropical America and has been introduced to Africa and the Far East.

Roselle

Sci. Name(s)	: *Hibiscus sabdariffa*
Geog. Reg(s)	: Africa-West, America-Central, Asia, Asia-India (subcontinent), Asia-Southeast, Tropical
End-Use(s)	: Beverage, Cereal, Dye & Tannin, Fiber, Fruit, Oil, Spice & Flavoring, Vegetable
Domesticated	: Y

Ref. Code(s)   : 126, 132, 170
Summary   : Roselle is an annual or biennial plant, 1-4 m high, whose 2 main varieties are grown for fruit and fiber production. It is native to West Africa but can now be grown in most tropical areas.

The sabdariffe variety produces edible fruits that are boiled with sugar to make beverages, jellies, sauces, and preserves. Its leaves and stalks are eaten as salad vegetables, used as potherbs, and for seasoning. The seeds contain an edible oil and can be ground to make flour. Parts of the fruit can be used commercially as a coloring agent and beverage. Variety sabdariffe must be grown at low to medium altitudes on well-drained soil in areas receiving low to moderate rainfall. Fruits can be harvested in 3-4 months. A single plant produces about 1.4 kg of fruit.

The altissima variety is grown in India, Java, and the Philippines as a fiber crop. The fiber is used in textile industries, either alone or with jute, in the manufacture of bag fabric, twine, and carpet yarn. Pulp fiber has shown higher annual production yields than wood pulp, but crops are subject to severe damage from nematodes. Raw fiber is exported from Thailand and Bangladesh, while other producing countries manufacture the fiber for domestic consumption. Variety altissima is slower to mature than kenaf (*Hibiscus cannabinus*). It is usually harvested before flowering. Crops require tropical conditions for best development. They are fairly drought-resistant.

Rye
(See Cereal)

Sacred Lotus, Lotus, East Indian Lotus
(See Tuber)

Sea Island Cotton, American Egyptian Cotton, Extra Long Staple Cotton
Sci. Name(s)   : *Gossypium barbadense, Gossypium peruvianum, Gossypium vitifolium*
Geog. Reg(s)   : Africa-North, America-North (U.S. and Canada), America-South, Asia-China, Europe, West Indies
End-Use(s)   : Fiber, Forage, Pasture & Feed Grain, Miscellaneous, Oil, Ornamental & Lawn Grass
Domesticated   : Y
Ref. Code(s)   : 117, 170
Summary   : Sea Island cotton (*Gossypium barbadense*) is a perennial shrub or annual sub-shrub that provides the 2nd most important cotton entered in world trade. Its seeds are covered with long, strong lint made into creamy white, silky cotton valued in high-quality textiles, luxury fabrics, yarns, and sewing thread. This lint is of high-quality, but the trees give low yields, which raises the price of the cotton. Cotton seeds produce a valuable oil used on salads or as lard and butter substitutes. Expressed oil cakes are used as fertilizer, stock feed, in the making of soap, oil cloth, putty, and nitroglycerine. Sea Island cotton trees are sometimes grown for ornamental purposes.

In Egypt and the West Indies, Sea Island cotton is the main cotton producing species. Limbless cultivars are grown in the USSR. Other major producers are the Democratic Republic of the Sudan, Peru, and the United States. Countries with less production are Israel, Morocco, Columbia, China, and the Yemen. Attempts to grow

Sea Island cotton in Hawaii have failed because crops are destroyed by the pink boll-worm. Trees are native to the arid, mountainous regions of South America.

Sea Rush, Sparto
(See Ornamental & Lawn Grass)

Sericea Lespedeza
(See Forage, Pasture & Feed Grain)

Sesbania
(See Forage, Pasture & Feed Grain)

Showy Crotalaria
(See Erosion Control)

Sisal

Sci. Name(s)	: *Agave sisalana*
Geog. Reg(s)	: Africa-East, America-South, West Indies
End-Use(s)	: Fiber, Medicinal
Domesticated	: Y
Ref. Code(s)	: 126, 171
Summary	: Sisal is the most important of the fiber producing Agave species. It reaches

heights of 1-3 m. Tanzania, Brazil, Haiti, and Kenya are the leading producers, exporting primarily to the United States. See henequen for import information.

Sisal leaves yield a hard, coarse fiber used in the manufacture of twines and cordage with more than half used as agriculture and harvest twine. Sisal is often substituted for abaca and jute but sisal ropes have a lower tensile strength and tend to break suddenly. The fiber is made into sacks but cannot be spun as finely as jute. It is woven into open-mesh material for carpet-backing bags and industrial fabrics. Chopped fibers are used to reinforce plaster boards. Whole single fibers are used for strengthening bituminized paper. Sisal leaves also contain hecogenin used in the partial synthesis of the drug cortisone. Manufactured sisal fiber is being replaced by synthetic fibers.

All commercial sisal is propagated vegetatively by suckers or bulbils. About 4,000-6,000 plants/ha are planted for good yields. Crops are grown in areas with evenly distributed rainfall, moderate humidity, and short dry seasons. The leaves are cut by hand when plants are about 1.5 m tall. About 35-40 of the 100 leaves produced are of economic value. Yields of 678 kg/ha of fiber are obtained. Each green leaf contains 2-5% fiber and weighs about 0.7 kg.

Smooth Loofah, Sponge Gourd, Dishcloth Gourd, Vegetable Sponge, Luffa, Loofah
(See Miscellaneous)

Sorghum, Great Millet, Guinea Corn, Kaffir Corn, Milo, Sorgo, Kaoling, Durra, Mtama, Jola, Jawa, Cholam Grains, Sweets, Broomcorn, Shattercane, Grain Sorghums, Sweet Sorghums
(See Cereal)

Spanish Broom, Weaver's Broom

Sci. Name(s)	: *Spartium junceum*
Geog. Reg(s)	: America-North (U.S. and Canada), Europe, Mediterranean
End-Use(s)	: Essential Oil, Fiber, Ornamental & Lawn Grass
Domesticated	: Y
Ref. Code(s)	: 56
Summary	: Spanish broom is a shrub from 1-3 m in height grown as a fiber crop. This

fiber is used in the manufacture of ropes, canvas, coarse cloths, mats, pillows, and paper and is superior to flax and cotton because it does not rot or lose its strength in humid weather. Plant stems can be rubberized and used for conveyer belts. The large yellow flowers contain an essential oil extracted and used in the perfume industry. Shrubs are sometimes grown as hedges.

In France, Spanish broom has been cultivated on a small scale, commercial basis. Shrubs are native to southwestern Europe and the Mediterranean area and have been introduced to the southern and central areas of the United States. Fiber yields have been recorded at about 1.5 MT/ha. Approximately 1,200 kg of flowers have yielded 300-350 g of oil. First harvests are made when plants are about 3 years old. Subsequent cuts can be made at 18 month intervals.

Stinging Nettle, European Nettle
(See Weed)

Sugar Palm, Gomuti Palm
(See Gum & Starch)

Sugarcane
(See Sweetener)

Sunflower
(See Oil)

Sunn Hemp, Sann Hemp

Sci. Name(s)	: *Crotalaria juncea*
Geog. Reg(s)	: America-North (U.S. and Canada), Asia-India (subcontinent), Europe, Tropical
End-Use(s)	: Fiber, Forage, Pasture & Feed Grain
Domesticated	: Y
Ref. Code(s)	: 170, 220
Summary	: Sunn hemp (*Crotalaria juncea*) is an annual fiber plant that reaches from 1-3 m

in height. A whitish-gray or yellowish fiber with an average length of 1-1.5 m is obtained from the stems. The fiber is used in cordage products, mainly for the manufacture of twine, cord, sacks, nets, and other similar items. Compared to jute, sunn hemp is considered more durable. Plants are grown in the tropics as a green manure crop. Dried stalks are used as livestock feed.

India is a prominent producer of this fiber and much of it is exported to the United Kingdom, Belgium and the United States. In the United States, sunn hemp is being considered as a potential paper production crop.

Sunn hemp grows rapidly, shows great weed control, and is hardy and fairly drought-resistant. The highest quality fiber is from plants grown in light, loamy, well-drained soils. Plants should be sown close together to reduce branching. Little additional care is required. They are harvested when pods form at 3-3.5 months. Plants grown for green manure are plowed under 2 months after sowing.

Teff
(See Cereal)

Tikus, Tikug

Sci. Name(s)	: *Fimbristylis globulosa*
Geog. Reg(s)	: Asia, Asia-India (subcontinent), Asia-Southeast, Pacific Islands, Tropical
End-Use(s)	: Fiber, Forage, Pasture & Feed Grain, Weed
Domesticated	: Y
Ref. Code(s)	: 36, 220, 238
Summary	: Tikus is a reedlike plant used in India and the Pacific for basketry, matting, hats, and bags. Plant stems are used as string. Crops are cultivated year-round and last 3-9 years. They are used as green manure. Tikus is a weed of rice fields. Its cultivation is similar to rice.

Tree Cotton

Sci. Name(s)	: *Gossypium arboreum, Gossypium anaking*
Geog. Reg(s)	: Africa, Asia, Asia-China, Asia-India (subcontinent), Asia-Southeast
End-Use(s)	: Fiber, Oil
Domesticated	: Y
Ref. Code(s)	: 32, 117, 154, 170
Summary	: Tree cotton (*Gossypium arboreum*) is a perennial shrub or annual subshrub. It produces large seeds covered with white lint and is an important cotton crop in India.

Cultivated shrubs bear capsules with prominent oil glands, but there is little information concerning the oil production. Tree cotton is used in experimental crosses and has produced most of the modern cultivated cottons. Tree cotton production is limited almost exclusively to Southeast Asia, particularly the dry parts of India and Pakistan. Approximately 10% of the total cotton planted in India and Pakistan is this Asiatic cotton, and it comprises less than 4% of the total production.

There are 6 important races. The 1st, indicum, is the most primitive form grown mainly in India. Indicum lint is sparse, coarse and colored. The 2nd, burmanicum, is a perennial cotton native to India and introduced to Burma and the East Indies. Its lint varies in texture. The 3rd, cernuum, grows in the desertous, Garo hills of India. The 4th, sinense, was grown in India, but is now cultivated as a commercial cotton in China, Japan and Korea. The 5th, bengalense, has coarse, short lint that provides the bulk of India's cotton and is native to Assam. The 6th, soudanense, is not exportable. It grows in India and Africa and has very little lint.

Tussa Jute, Tossa Jute, Jew's Mallow, Nalta Jute

Sci. Name(s) : *Corchorus olitorius*
Geog. Reg(s) : Africa, America-South, Asia, Asia-China, Asia-India (subcontinent), Asia-Southeast
End-Use(s) : Fiber, Vegetable
Domesticated : Y
Ref. Code(s) : 126, 148, 154, 170
Summary : Tussa jute (*Corchorus olitorius*) is an herbaceous annual whose stems yield a commercial fiber. It is similar to white jute (*Corchorus capsularis*), but is usually taller with slender, cylindrical pods and fine, strong, lustrous fiber that is less durable than ramie, flax, or cotton because it has less cellulose and more lignin.

Commercial plantings are in Bangladesh, India, and China. Other less prominent areas of production are in Nepal, Burma, Brazil, Peru, and Thailand. The long, flexible fiber is easily spun and used in bags, carpet-backing, bale wrapping, twine, ropes, and other products. Crops are produced at relatively low costs. In the Middle East, Egypt, the Democratic Republic of the Sudan, and tropical Africa, the stems are treated as spinach.

Tussa jute is most successful in higher areas where flood threats are minimal. For quality yields, manures or fertilizers should be applied. Plants are suitable for harvesting 5 months from planting. There is little or no specific yield data for tussa jute fiber.

Upland Cotton, Cotton

Sci. Name(s) : *Gossypium hirsutum, Gossypium mexicanum*
Geog. Reg(s) : America-Central, America-North (U.S. and Canada), Eurasia
End-Use(s) : Fiber, Forage, Pasture & Feed Grain, Oil
Domesticated : Y
Ref. Code(s) : 117, 154, 170
Summary : The American upland cotton (*Gossypium hirsutum*) is a small perennial tree or shrub that provides the bulk of United States cotton. Its fiber-covered seeds are harvested as a major fiber crop and comprise 90% of the world production of 65 million bales of fiber in textile industries for cordage and other products. A short fiber from the seeds taken prior to crushing, is a source of industrial cellulose. Cotton seeds are pressed to yield cotton seed oil used in cooking and for other culinary purposes. The protein-rich, expressed oilcake is fed to livestock.

Fifteen states stretching from North Carolina to California comprise the major commercial cotton states. The main types of upland cotton are 'Acala,' 'Delta,' 'Plains,' and 'Eastern. Cotton trees require 50 cm of water during their growing season for adequate production.

In 1983, cotton production in the United States was approximately 1.7 million MT. The United States a principal exporter along with Russia and Turkey. The principal importers are Hong Kong, China, and Japan.

Vetiver, Khus-Khus
(See Essential Oil)

Vogel Fig
(See Isoprenoid Resin & Rubber)

Wheat, Bread Wheat, Common Wheat
(See Cereal)

White Jute

Sci. Name(s)	: *Corchorus capsularis*
Geog. Reg(s)	: America-North (U.S. and Canada), America-South, Asia, Asia-China, Asia-India (subcontinent), Tropical
End-Use(s)	: Fiber
Domesticated	: Y
Ref. Code(s)	: 126, 170
Summary	: White jute (*Corchorus capsularis*), is an herbaceous annual fiber crop that provides commercial jute. Most jute plantations are in India and Pakistan. Other producing countries are China, Taiwan, and Brazil. The United States is the largest importer of manufactured jute fiber.

The soft bast fibers are weaker than hemp or flax but are 2nd in importance to cotton as a textile fiber. It contains 60-65% cellulose, an amount lower than flax, ramie, and cotton. A woody substance, lignin, makes jute less durable than other fibers. About 75% of the world's jute is used for coarse woven fabrics, sacking, or burlap. It is also used in twines and carpet yarns. White jute is more commonly grown for fiber than the closely related, tussa jute (*Corchorus olitorius*).

Jute grows wild in southern China. The crop is grown experimentally in many tropical countries, but with the exception of Brazil, has seldom been successful. Plants thrive in areas with temperatures of 24-35 degrees Celsius. Crops are harvested 100-130 days after planting. In 1984, the United States imported 581,741 kg of jute products valued at $2.7 millon.

White Lupine, Egyptian Lupine
(See Forage, Pasture & Feed Grain)

White Melilot, White Sweetclover
(See Forage, Pasture & Feed Grain)

White Mulberry
(See Miscellaneous)

Wild Cotton

Sci. Name(s)	: *Gossypium anomalum*
Geog. Reg(s)	: Africa-North, Africa-South
End-Use(s)	: Fiber
Domesticated	: N
Ref. Code(s)	: 117, 170
Summary	: Wild cotton (*Gossypium anomalum*) is a small shrub to 2 m in height which grows in desert areas. Its seeds are about 6 mm long with brown fiber. There is little

information concerning economic value for this particular cotton. Its natural habitat is along the southern borders of the Sahara and into southwest Africa.

Wine Palm, Coco de Chile, Coquito Palm, Honey Palm, Wine Palm of Chile, Chile Coco Palm
(See Beverage)

Forage, Pasture & Feed Grain

Abyssinian Oat
(See Cereal)

African Locust Bean
(See Oil)

Alfalfa, Lucerne, Sativa
Sci. Name(s) : *Medicago sativa, Medicago sativa* subsp. *sativa*
Geog. Reg(s) : Ameica-North (U.S. and Canada), Mediterranean, Subtropical, Temperate, Tropical
End-Use(s) : Dye & Tannin, Erosion Control, Fiber, Forage, Pasture & Feed Grain, Medicinal, Miscellaneous, Oil
Domesticated : Y
Ref. Code(s) : 56, 154, 170, 220
Summary : Alfalfa is grown for fodder and is a major forage crop worldwide. Plant leaves are used as a commercial source of chlorophyll and the flowers attract bees. Alfalfa seeds contain a drying oil used in paints. Alfalfa is grown as a cover crop to reduce erosion. It is grown in combination with corn for sileage. Alfalfa fiber is used in manufacturing paper. In folk medicine, it is used as a cooling poultice for boils, for weight gain, peptic ulcers, and urinary and bowel problems. Seeds are reported to contain trypsin inhibitors and are used to make a yellow dye.

Plants are native to the Mediterranean and grow in most temperate, subtropical, and tropical areas. A large proportion of the alfalfa produced is grown in the United States. Alfalfa grows on a variety of soils but prefers rich, well-drained, loamy soil. It does not tolerate waterlogging or acid soils.

Alkali Sacaton
Sci. Name(s) : *Sporobolus airoides*
Geog. Reg(s) : America-North (U.S. and Canada)
End-Use(s) : Forage, Pasture & Feed Grain
Domesticated : N
Ref. Code(s) : 20, 96
Summary : The alkali sacaton is an erect, tufted perennial from 40-90 cm in height which provides good forage on alkaline soils. In the United States, it grows in Kansas, Nebraska, Montana, California, Texas, and New Mexico.

Almond
(See Nut)

Alsike Clover
Sci. Name(s) : *Trifolium hybridum*
Geog. Reg(s) : Africa-North, America-North (U.S. and Canada), Asia, Europe
End-Use(s) : Forage, Pasture & Feed Grain

Domesticated : Y
Ref. Code(s) : 56, 220
Summary : Alsike clover is a short-crowned perennial legume popular as an annual or biennial forage and hay crop. Plants adapt to high-altitudes and thrive in cool, moist climates. They are also successful on low, wet, fertile lands. Alsike clover is suggested for growing in creek bottoms, wet, natural meadows, and swales. If grown in mixtures with red clover, timothy, or other grasses, hay quality and yield quantity are improved.

Plants are grown in Europe, the Caucasus, Asia Minor, North Africa, and the United States. In the United States, yields are highest in Idaho and Oregon with 300-375 kg/ha of seeds. Elsewhere in the United States, seed yields are 160-200 kg/ha. Plants are harvested for hay while in bloom.

Alyceclover, Oneleaf-Clover

Sci. Name(s) : *Alysicarpus vaginalis, Alysicarpus nummularifolius*
Geog. Reg(s) : Africa-East, Asia-Southeast, Tropical
End-Use(s) : Erosion Control, Forage, Pasture & Feed Grain
Domesticated : Y
Ref. Code(s) : 220
Summary : Alyceclover is an herbaceous tropical legume cultivated for pasture, hay, and forage. It is used for soil improvement and conservation and provides effective erosion control on newly established terraces. Especially in Malaysia, it is a good cover crop on rubber plantations. Alyceclover grows throughout Southeast Asia and East Africa.

Angleton Bluestem

Sci. Name(s) : *Dichanthium aristatum*
Geog. Reg(s) : Africa, America-North (U.S. and Canada), Asia-India (subcontinent), Asia-Southeast, Australia
End-Use(s) : Forage, Pasture & Feed Grain
Domesticated : Y
Ref. Code(s) : 24, 25, 122, 142
Summary : Angleton bluestem is a densely tufted, creeping perennial grass up to 3 m in length. It provides excellent fodder and, in wet environments, experiences thick, lush growth. Its size depends on the conditions of the soil and amount of moisture. The grass is common in southeastern Africa, southern India, and islands throughout the Indian Ocean. In the Philippines, it is used for pasture improvement. The grass has been introduced to many other areas including North America and Australia.

Annual Bluegrass, Low Speargrass, Dwarf Meadow Gold, Six-Weeks Grass, Plains Bluegrass

Sci. Name(s) : *Poa annua*
Geog. Reg(s) : America-North (U.S. and Canada), Europe, Hawaii
End-Use(s) : Forage, Pasture & Feed Grain, Ornamental & Lawn Grass, Weed
Domesticated : Y
Ref. Code(s) : 17, 20

Summary : Annual bluegrass is a European annual grass introduced and naturalized in North America at higher-altitudes. It is palatable and nutritious forage and is often used for pasture. It grows sparingly at higher-altitudes in Hawaii. The grass is sometimes considered a troublesome weed in the continental United States. In Michigan, three or more crops are grown from seed in the span of one season. The grass is used for lawns in shady areas with regular watering.

Annual Ryegrass, Italian Ryegrass, Australian Ryegrass

Sci. Name(s) : *Lolium multiflorum*
Geog. Reg(s) : America-North (U.S. and Canada), Europe
End-Use(s) : Forage, Pasture & Feed Grain, Ornamental & Lawn Grass
Domesticated : Y
Ref. Code(s) : 23, 96
Summary : Annual ryegrass is an important winter forage grass in Europe. It is used to a limited extent in the United States for meadows, pastures, and lawns.

Antelope Grass

Sci. Name(s) : *Echinochloa pyramidalis*
Geog. Reg(s) : Africa-South, Africa-West, Tropical
End-Use(s) : Forage, Pasture & Feed Grain
Domesticated : Y
Ref. Code(s) : 24, 109, 142, 171, 220
Summary : Antelope grass is a reedlike perennial up to 5 m in height that is an excellent fodder and pasture grass when young. It is cultivated for hay in South Africa and Zimbabwe. The grass is burned at the end of the dry season and fresh growth grazed. Antelope grass withstands drought fairly well but is most successful in wet or irrigated areas. It is considered good pasture for land prone to flooding in the tropics and in the swampy areas of tropical Africa.

Apricot
(See Fruit)

Arabica Coffee, Coffee
(See Beverage)

Arrowleaf Clover

Sci. Name(s) : *Trifolium vesiculosum*
Geog. Reg(s) : America-North (U.S. and Canada), Eurasia, Europe, Mediterranean
End-Use(s) : Forage, Pasture & Feed Grain
Domesticated : Y
Ref. Code(s) : 56
Summary : Arrowleaf clover is an annual legume cultivated in the southern United States and southern Europe for forage. The clover is palatable and provides good forage during the late winter and early spring months.

Plants are native to the western and eastern Mediterranean regions, the Balkans, Greece, the western Caucasus, and southern Russia. They have been introduced to the United States. Plants thrive on silty loam soils and continue to produce on various clays. Arrowleaf clover is fairly drought-resistant. Forage yields are recorded at 7-12 MT/ha and seed yields at 100-500 kg/ha.

Arrowroot
(See Gum & Starch)

Australian Bluestem
Sci. Name(s)   : *Bothriochloa intermedia*
Geog. Reg(s)   : Asia-Southeast, Australia, Subtropical, Tropical
End-Use(s)   : Erosion Control, Forage, Pasture & Feed Grain
Domesticated   : Y
Ref. Code(s)   : 122
Summary   : Australian bluestem is a hardy, palatable fodder grass that tolerates heavy grazing. It is drought-resistant and has potential as an erosion control agent. Australian bluestem grows throughout Southeast Asia and Australia and most tropical and subtropical areas.

Australian Sheepbush, Austral Sheepbush
Sci. Name(s)   : *Pentzia virgata*
Geog. Reg(s)   : Africa-North, Africa-South, Tropical
End-Use(s)   : Forage, Pasture & Feed Grain
Domesticated   : Y
Ref. Code(s)   : 220
Summary   : The Australian sheepbush is a South African shrub used as fodder. It provides fodder in tropical areas and parts of North Africa. There is no available data on the yields or economics of this plant.

Azuki Bean, Adzuki Bean
(See Vegetable)

Babassu, Orbignya speciosa
(See Oil)

Babul
(See Gum & Starch)

Bahia Grass, Bahiagrass
Sci. Name(s)   : *Paspalum notatum*
Geog. Reg(s)   : Africa-Central, America-Central, America-North (U.S. and Canada), America-South, Subtropical, Tropical
End-Use(s)   : Erosion Control, Forage, Pasture & Feed Grain, Ornamental & Lawn Grass
Domesticated   : Y
Ref. Code(s)   : 17, 24, 171
Summary   : Bahia grass is a perennial grass that forms a dense cover and is valued as a grazing grass and soil erosion control agent. The grass is highly productive and fairly easy to establish. It is valued in the tropics and subtropics as a pasture grass but produces low herbage yields of 3-8 MT/ha of dry matter. Bahia grass withstands cold

weather. It was introduced into Uganda as pasture and forage but was found most useful as erosion control on bunds. In the United States, bahia grass is grown for lawns. It is native to South and Central America.

Ball Clover

Sci. Name(s) : *Trifolium nigrescens*
Geog. Reg(s) : America-North (U.S. and Canada), Mediterranean, Worldwide
End-Use(s) : Erosion Control, Forage, Pasture & Feed Grain
Domesticated : Y
Ref. Code(s) : 56, 220
Summary : Ball clover is an annual winter legume grown in the southern United States for pasture and soil improvement. Ball clover is often mixed with abruzzi rye for longer and better production. This clover has a high seed yield and is able to reseed itself. It is grown for permanent grass sods in parts of the southeastern United States. Ball clover is not heat- or drought-hardy.

Plants are native to the western Mediterranean and have naturalized throughout most of the world. They thrive on loam or clay soils. Dry forage yields are recorded at 2.5-7.5 MT/ha.

Bard Vetch, Monantha Vetch

Sci. Name(s) : *Vicia monantha, Vicia monanthos*
Geog. Reg(s) : Asia-China, Europe, Mediterranean
End-Use(s) : Cereal, Erosion Control, Forage, Pasture & Feed Grain
Domesticated : Y
Ref. Code(s) : 24, 138, 141, 220
Summary : The bard vetch is a Mediterranean herb cultivated as a cover or forage crop. It has edible seeds that are added to soups. The vetches are grown for hay, silage, and grazing in mixtures with spring and winter cereal crops. Plowing plants into the soil helps moisture retention and improves the nitrogen-fixing qualities of the root. It can grow with common vetch or wooly pod vetch but is less winter-hardy than the hairy vetch.

Barley
(See Cereal)

Barnyardgrass, Barnyard Millet

Sci. Name(s) : *Echinochloa crusgalli, Panicum crusgalli*
Geog. Reg(s) : America-North (U.S. and Canada), Asia, Asia-Southeast, Tropical
End-Use(s) : Cereal, Erosion Control, Forage, Pasture & Feed Grain, Vegetable, Weed
Domesticated : Y
Ref. Code(s) : 35, 54, 122, 171
Summary : Barnyardgrass is an annual tropical grass about 1 m in height and provides a useful fodder. In Asia and California, barnyardgrass is considered one of the worst possible weeds of rice fields. In other countries, the grain is used for food in times of scarcity. In Java, the young shoots are eaten as vegetables. In Egypt, the grass is used for reclaiming saline areas.

The grass grows from sea level to 2,000 m in altitude and thrives in wet, sandy or clay soils along coastal and inland areas.

Basin Wildrye, Giant Wildrye

Sci. Name(s)	: *Elymus cinereus*
Geog. Reg(s)	: America-North (U.S. and Canada)
End-Use(s)	: Forage, Pasture & Feed Grain
Domesticated	: Y
Ref. Code(s)	: 153, 164, 206
Summary	: Basin wildrye is an important perennial forage grass, 0.6-2 m in height, of the dry plains, sand hills and ditches of the western United States. It provides adequate forage but is less robust than the giant wild rye (*Elymus condensatus*). Its growth begins in the early spring, becomes unpalatable by summer, matures in August, and then regenerates itself. It is most successful in dry areas below 3,046 m.

Beggarlice, Tick Clovers, Greenleaf Desmodium

Sci. Name(s)	: *Desmodium intortum*
Geog. Reg(s)	: Africa-East, Africa-South, Hawaii, Tropical
End-Use(s)	: Forage, Pasture & Feed Grain
Domesticated	: Y
Ref. Code(s)	: 5, 56, 170
Summary	: Beggarlice is a perennial tropical subshrub with economic potential as tropical fodder and pasture for beef cattle. In the semitropical conditions of Hawaii, beggarlice production reaches 27 MT/ha. Beef weight gain of 450-675 kg/ha/year has been demonstrated. In Zimbabwe, pastures withstand grazing for 6 years if fertilized with phosphorous. In Uganda, pastures are grazed for 3 years. Total yields increase with longer cutting intervals of 4-12 weeks.

The grass grows wild in a variety of areas up to 2,400 m in altitude. It prefers moist to wet soils but will grow in rocky areas. Beggarlice does not tolerate frost. It requires temperatures of 7.3-27.1 degrees Celsius.

Benoil Tree, Horseradish-Tree
(See Oil)

Bermudagrass, Star Grass, Bahama Grass, Devilgrass
(See Erosion Control)

Berseem Clover, Egyptian Clover

Sci. Name(s)	: *Trifolium alexandrinum*
Geog. Reg(s)	: America-North (U.S. and Canada), Mediterranean, Subtropical, Tropical
End-Use(s)	: Forage, Pasture & Feed Grain
Domesticated	: Y
Ref. Code(s)	: 56
Summary	: Berseem clover is an annual legume with stems from 3-6 dm in height widely cultivated as a palatable forage plant, particularly in the Mediterranean area. It is used as green chop or ensilage. Plants have about half the crude fiber content and

more carbohydrates and fats than alfalfa. Berseem is the least winter-hardy species of the Trifolium genus.

Plants have adapted to the Gulf coast areas and the southwestern United States. In Florida, berseem clover is grown as a winter annual. Plants grow in other tropical and subtropical regions. If berseem is planted in October or November, harvests are made 4 times. Fresh weight forage yields are 25-37.5 MT/ha.

Big Bluegrass
(See Ornamental & Lawn Grass)

Big Bluestem, Turkeyfoot Grass
Sci. Name(s) : *Andropogon gerardii*
Geog. Reg(s) : America-North (U.S. and Canada)
End-Use(s) : Forage, Pasture & Feed Grain
Domesticated : Y
Ref. Code(s) : 164
Summary : Big bluestem is a prairie grass of North America. It provides good pasture and hay.

Big Trefoil
Sci. Name(s) : *Lotus uliginosus, Lotus pedunculatus*
Geog. Reg(s) : America-North (U.S. and Canada), America-South, Australia, Europe, Pacific Islands
End-Use(s) : Forage, Pasture & Feed Grain
Domesticated : Y
Ref. Code(s) : 56
Summary : Big trefoil is a perennial legume with a creeping, shallow rhizome. It is valued as a pasture, hay, and nonbloating forage crop in the United States. This legume has a high-protein content with no prussic acid. It also contains flavonol polymers in its roots and leaves. This trefoil has a tolerance to disease, fungus, and waterlogging. It is suited for marshes and wet grasslands but is not drought-resistant.

The plant has been introduced to Australia, New Zealand, South America, and the United States and does well on the acidic coastal soils of the Pacific Northwest, especially areas frequently flooded. In New Zealand, it thrives in pastures that are infertile and in peat areas where the grazing is light. It is native to Central and Atlantic Europe. Yields vary yearly and insects are necessary for fertilization.

Billion Dollar Grass, Japanese Barnyard Millet, Japanese Millet
(See Cereal)

Birdsfoot Trefoil
(See Erosion Control)

Bitter Vetch, Ervil
Sci. Name(s) : *Vicia ervilia, Ervilia sativa, Ervum ervillia*
Geog. Reg(s) : America-North (U.S. and Canada), Asia, Europe, Mediterranean
End-Use(s) : Cereal, Forage, Pasture & Feed Grain

Domesticated : Y
Ref. Code(s) : 27, 138, 141, 179, 220
Summary : The bitter vetch is a Mediterranean herb widely cultivated for its edible seeds, which are fed to livestock or eaten in soups. The plant is grown for fodder and soil improvement. The seeds are considered poisonous to pigs and are dangerous to cattle at the time when the pods are nearing maturity and are still moist.

Bitter vetch is said to have economic importance and is considered a valuable pasture plant. Vetches are good for folding sheep and as silage and hay. To make good hay, plants should be cut while in flower. It is important to mix the seed of upright-growing plants with vetches, such as winter vetch, with winter rye or oats and spring vetch with spring rye or oats. Vetches grow on most soils if there is enough lime. The addition of phosphatic and potash fertilizer increases fodder yield. Bitter vetch is found in parts of Europe, Asia, and the United States.

Black Gram, Urd, Wooly Pyrol
(See Vegetable)

Black Grama
Sci. Name(s) : *Bouteloua eriopoda*
Geog. Reg(s) : America-North (U.S. and Canada)
End-Use(s) : Forage, Pasture & Feed Grain
Domesticated : Y
Ref. Code(s) : 96, 164
Summary : Black grama is one of the best perennial forage grasses in the southwestern United States, especially Arizona. It recovers quickly from heavy grazing, spreads rapidly and provides pasture throughout the year. Black grama thrives on open, dry plains and hills.

Black Locust, False Acacia
(See Timber)

Black Medic, None-Such
Sci. Name(s) : *Medicago lupulina*
Geog. Reg(s) : Africa-South, America-Central, America-North (U.S. and Canada), America-South, Asia, Europe, Temperate
End-Use(s) : Forage, Pasture & Feed Grain, Weed
Domesticated : Y
Ref. Code(s) : 56, 154, 214
Summary : In Europe, black medic is used as green manure. Cattle and poultry are fed the empty pods. It is often mixed with other grasses and clovers and used as forage. Plants are considered weeds eaten by wild and some domestic animals. This plant is often confused with shamrocks.

Black medic grows primarily in the temperate areas of Asia, North Africa, the United States, Central and South America, and Africa. Plants grow on a variety of well-limed soils. They do not tolerate salinity but can withstand frost if well-established.

Blackpod Vetch, Narrow Leaf Vetch

Sci. Name(s)	: *Vicia sativa* subsp. *nigra*
Geog. Reg(s)	: America-North (U.S. and Canada), America-South, Asia-China, Europe, Mediterranean
End-Use(s)	: Erosion Control, Forage, Pasture & Feed Grain
Domesticated	: Y
Ref. Code(s)	: 17, 154, 166, 214
Summary	: The blackpod vetch is an herb grown for pasture, food, green manure, and as a cover crop in orchards. Plowing green vetch into the soil helps the soil hold its moisture and improves its nitrogen-fixing qualities. Crops can be plowed under in about 3 months. Plants do not thrive under hot, dry conditions. They are cultivated in the United States, South America, the Mediterranean, and China.

Blue Clitoria, Butterfly Pea, Asian Pigeon-Wings, Butterfly Bean, Kordofan Pea

Sci. Name(s)	: *Clitoria ternatea*
Geog. Reg(s)	: Africa, America, Australia, Tropical
End-Use(s)	: Dye & Tannin, Forage, Pasture & Feed Grain, Ornamental & Lawn Grass, Vegetable
Domesticated	: Y
Ref. Code(s)	: 56, 170, 220
Summary	: The blue clitoria, also known as the butterfly pea, is a twining perennial. In Australia, plants are grown for forage. They have 11% digestible proteins, about 50% total digestible nutrients, and 39% starch. In the Democratic Republic of the Sudan, blue clitoria produces well under irrigation and is used for forage and for its purgative seeds and roots.

Blue clitoria grows rapidly and its dense foliage provides good ground cover. The pods are eaten as vegetables in some tropical countries. In Ambon, an island in the Moluccas, the seeds and flowers are used to dye rice blue. The seeds and bark yield tannin. Blue clitoria is often grown as an ornamental.

Plants are most successful in rich soils from sea level to 1,600 m in altitude. They need full sunlight and an average annual rainfall from 3.8-42.9 dm and annual temperatures from 19.4-27.9 degrees Celsius.

Blue Grama

Sci. Name(s)	: *Bouteloua gracilis*
Geog. Reg(s)	: America-North (U.S. and Canada)
End-Use(s)	: Forage, Pasture & Feed Grain
Domesticated	: Y
Ref. Code(s)	: 96, 164
Summary	: Blue grama is a perennial grass that provides excellent forage during summer and winter. It has erect culms reaching from 20-50 cm. This grass is found on the open plains and lightly wooded mountains of the western United States. Blue grama often grows in the same fields with buffalograss and the two are sometimes confused. Blue grama lacks the creeping stems of buffalograss.

Blue Lupine, European Blue Lupine, New Zealand Blue Lupine, Narrow-Leaved Lupine

Sci. Name(s) : *Lupinus angustifolius*
Geog. Reg(s) : Africa-South, America, America-North (U.S. and Canada), Australia, Europe, Mediterranean, Pacific Islands
End-Use(s) : Forage, Pasture & Feed Grain, Medicinal, Miscellaneous
Domesticated : Y
Ref. Code(s) : 56, 154
Summary : Blue lupine is used as forage and silage and for late winter, early spring grazing. According to some sources, it is poisonous to sheep. Sweet cultivar seeds are used as a protein additive in animal feed. Bitter cultivars are grown mainly for soil improvement. Blue lupine flowers are useful in honey production. In India, the plant is grown as green manure between crops. The germinating seeds are rich in asparagine and are used in the culture media for commercial production of tuberculin.

A native of the Mediterranean Basin and France, blue lupine is now cultivated in Australia, Tasmania, New Zealand, South Africa, northern Europe, and the United States. It grows as a winter annual on well-drained soil in the southern United States.

Blue lupine can produce as much as 5-7.5 MT/ha of dry herbage, 7.5-12.5 MT/ha of green manure and seed yields of 500-600 kg/ha.

Blue Panicgrass, Blue Panicum, Giant Panicum

Sci. Name(s) : *Panicum antidotale*
Geog. Reg(s) : Asia-India (subcontinent)
End-Use(s) : Cereal, Forage, Pasture & Feed Grain
Domesticated : Y
Ref. Code(s) : 17, 24, 154
Summary : Blue panicgrass is a hardy perennial grass grown for forage and grain production in dry areas. It grows at low altitudes in sun and shade and reaches heights of 2.7 m. It tolerates most well-drained soils but is slow to establish.

First cuttings are made 50-60 days from sowing. Yields range from 10-50 MT/ha of fresh fodder. Blue panicgrass has fairly good nutritive value for livestock. It is grown in parts of India.

Blue Wildrye
(See Ornamental & Lawn Grass)

Bluebunch Wheatgrass

Sci. Name(s) : *Agropyron spicatum*
Geog. Reg(s) : America-North (U.S. and Canada)
End-Use(s) : Forage, Pasture & Feed Grain
Domesticated : Y
Ref. Code(s) : 164
Summary : Bluebunch wheatgrass is an important pasture grass of the northwestern United States and thrives particularly on the dry plains and open mountain slopes. The grass is highly palatable and eaten by all kinds of livestock in any season.

This tufted perennial has culms 60-100 cm in length with wide leaf blades about 1-4 mm broad.

Boer Lovegrass
(See Erosion Control)

Brazilian Lucerne

Sci. Name(s)	: *Stylosanthes guianensis*
Geog. Reg(s)	: America-North (U.S. and Canada), America-South, Australia, Tropical
End-Use(s)	: Erosion Control, Forage, Pasture & Feed Grain
Domesticated	: Y
Ref. Code(s)	: 56, 170
Summary	: Brazilian lucerne is an herbaceous perennial with stems up to 1 m in length used mainly as a pasture legume on infertile soils or in the humid tropics. In Australia, plants are grown as a form of soil erosion control while in the United States, it is grown in mixed pastures in southern Florida.

Plants are native to South America and can be grown in most open, rocky environments and in fields at elevations approaching 2,000 m. Pastures are utilized when fully established, which is usually about 3-5 months from planting. In southern Florida, yield data taken from the two cultivars 'Cook' and 'Endeavor,' show that the average is about 15,000 MT/ha of green matter with 14% crude protein. Dry weight yields are from 3.5-10 MT/ha.

Breadfruit
(See Vegetable)

Broadbean, Horsebean, Field Bean, Tick Bean, Windsor Bean
(See Vegetable)

Browntop Millet

Sci. Name(s)	: *Brachiaria ramosa*
Geog. Reg(s)	: Asia-Southeast
End-Use(s)	: Cereal, Forage, Pasture & Feed Grain
Domesticated	: Y
Ref. Code(s)	: 122, 126
Summary	: Browntop millet is a highly palatable annual or perennial fodder grass found wild in forests of Southeast Asia. This millet is similar to finger millet (*Eleusine coracana*) in terms of climatic requirements. When grown as a grain crop, browntop millet should be planted in rocky, shallow soils, fertilized and kept free of weeds for best results. These plants need moist climates at altitudes of 2,000-2,500 m.

Millets mature in 3-5 months. Crops grown under natural rainfall yield from 2,000-4,000 kg/ha with higher yields possible under controlled irrigation.

Buckwheat
(See Cereal)

Buffalo Gourd, Wild Gourd
(See Oil)

Buffalograss

Sci. Name(s)	: *Buchloe dactyloides*
Geog. Reg(s)	: America-North (U.S. and Canada)
End-Use(s)	: Erosion Control, Forage, Pasture & Feed Grain
Domesticated	: Y
Ref. Code(s)	: 96, 164
Summary	: Buffalograss is a perennial forage grass important in the Great Plains states of the United States. It furnishes excellent forage year-round. The foliage cures on the ground, providing nutritious winter feed. Plants are good cover for exposed dry banks.

The grass is frequently grown in mixture with blue grama grass and the two are often confused. Plants usually bloom in the spring.

Buffelgrass

Sci. Name(s)	: *Cenchrus ciliaris*
Geog. Reg(s)	: Africa, Asia-India (subcontinent), Mediterranean
End-Use(s)	: Forage, Pasture & Feed Grain, Ornamental & Lawn Grass
Domesticated	: Y
Ref. Code(s)	: 122
Summary	: Buffelgrass is a very productive and nutritious pasture and hay grass. It is a hardy perennial that tolerates hot, dry conditions. The grass is most often considered wild but is occasionally cultivated for turf. Buffelgrass is native to Africa, India, and the Mediterranean.

Bulbous Bluegrass

Sci. Name(s)	: *Poa bulbosa*
Geog. Reg(s)	: Africa-North, America-North (U.S. and Canada), Eurasia
End-Use(s)	: Forage, Pasture & Feed Grain
Domesticated	: Y
Ref. Code(s)	: 17, 96
Summary	: There is not much information on bulbous bluegrass except that it is native to Eurasia and North Africa and has been introduced and naturalized throughout the mainland United States. The grass grows in fields and meadows and can be propagated by its bulblets.

Burnet, Garden Burnet, Small Burnet
(See Weed)

Butterfly Pea
(See Erosion Control)

Buttonclover

Sci. Name(s)	: *Medicago orbicularis*
Geog. Reg(s)	: America-North (U.S. and Canada), Asia, Mediterranean
End-Use(s)	: Forage, Pasture & Feed Grain
Domesticated	: Y
Ref. Code(s)	: 56, 220

Summary : In the Mediterranean region and in sections of Asia, buttonclover is grown as low-quality forage. Sheep prefer this clover to burclover. It does have some value as green manure.

In California, yields of the buttonclover are 1,000 kg/ha of seeds. It does best on fairly well-drained loamy soil.

Cabbage, Savoy Cabbage
(See Vegetable)

California Brome Grass, Mountain Brome
Sci. Name(s) : *Bromus carinatus*
Geog. Reg(s) : America-North (U.S. and Canada)
End-Use(s) : Forage, Pasture & Feed Grain
Domesticated : Y
Ref. Code(s) : 96, 164
Summary : California brome grass is grown for forage. It has culms standing 50-100 cm tall with flat blades. These plants are common on open ground and in lightly wooded areas of the western United States where they provide a great deal of range forage.

The seeds of this grass are now available commercially and are sold for revegetation purposes or range grass stock.

California Burclover, Toothed Burclover
Sci. Name(s) : *Medicago polymorpha*
Geog. Reg(s) : Africa-North, America-North (U.S. and Canada), America-South, Asia-Central, Australia, Europe, Temperate
End-Use(s) : Erosion Control, Forage, Pasture & Feed Grain
Domesticated : Y
Ref. Code(s) : 56
Summary : California burclover is a forage and pasture grass valuable as concentrated fodder during dry seasons. It is useful for soil renovation, green manure, and winter cover for erosion control. For hay production it is mixed with oats and wheat. Unless the stand is dense, California burclover lies close to the ground and is hard to mow.

Plants grow throughout southern Europe, western and Central Asia, North Africa, and other warm, dry temperate areas. It is grown in Argentina, Australia, and the Pacific Coast of the United States and is adapted to regions with mild winters and moist, well-drained, slightly alkaline soils. Its natural habitat is in shady places along streams but it does not tolerate salt-charged soils or frost. Clover yields are from 5-7.5 MT/ha of hay.

Camelthorn, Camill Thorn
Sci. Name(s) : *Alhagi pseudalhagi*
Geog. Reg(s) : Eurasia
End-Use(s) : Forage, Pasture & Feed Grain, Medicinal
Domesticated : Y
Ref. Code(s) : 57, 213
Summary : Camelthorn (*Alhagi pseudalhagi*) is a shrub grown in the steppe region of southeastern Russia and the western portions of the Kazakh Republic. Its leaves are

edible and have unspecified medicinal qualities. The shrub is often used as fodder for livestock.

Its classification is difficult because it is often associated with 2 similar legume species, *Alhagi camelorum* in India, and *Alhagi maurorum* in Arabia and Syria.

Canada Wildrye

Sci. Name(s)	: *Elymus canadensis*
Geog. Reg(s)	: America-North (U.S. and Canada)
End-Use(s)	: Erosion Control, Forage, Pasture & Feed Grain
Domesticated	: Y
Ref. Code(s)	: 33, 164
Summary	: Canada wildrye is a perennial tufted grass 0.75-1.8 m in height planted experimentally for forage production in the midwestern United States. This grass is fairly tall and coarse but provides good forage. It is used for erosion control on inland sand dunes. Canada wildrye is more commonly found in its natural state on both dry and moist soils throughout the Great Plains and Pacific Northwest of the United States.

Canarygrass, True Canarygrass
(See Miscellaneous)

Candletree

Sci. Name(s)	: *Parmentiera cerifera*
Geog. Reg(s)	: America-Central, Tropical, West Indies
End-Use(s)	: Forage, Pasture & Feed Grain, Fruit
Domesticated	: Y
Ref. Code(s)	: 17, 154, 171, 199
Summary	: The candletree reaches heights of approximately 6 m and produces yellow, candlelike fruits that are fed to cattle. Candletrees are grown mainly in the West Indies and other tropical areas as an oddity. However, there is little information concerning the importance of the fruit. Trees are native to Panama.

Cape Aloe, Bitter Aloe, Red Aloe
(See Medicinal)

Caribgrass

Sci. Name(s)	: *Eriochloa polystachya*
Geog. Reg(s)	: America-South, Tropical
End-Use(s)	: Forage, Pasture & Feed Grain
Domesticated	: Y
Ref. Code(s)	: 17, 24, 96, 220
Summary	: Caribgrass is an annual or short-lived perennial grass grown in tropical America for fodder and quality hay. It does not tolerate cold climates and adverse conditions and produces good quality hay with a medium- to high-nutritive content. In South America it is often found in wet soil.

Carob, St. John's-Bread
(See Gum & Starch)

Carpetgrass

Sci. Name(s) : *Axonopus affinis*
Geog. Reg(s) : America-South, Asia-Southeast, Tropical
End-Use(s) : Forage, Pasture & Feed Grain, Ornamental & Lawn Grass, Weed
Domesticated : Y
Ref. Code(s) : 122
Summary : Carpetgrass is a mat-forming perennial grass used for turfing lawns and airstrips and as a pasture grass. It is common in Southeast Asia, South America, and in other tropical countries and warmer regions of the world.

This grass flourishes in waterlogged soils at higher elevations. It can become a troublesome weed.

Cassava, Manioc, Tapioca-Plant, Yuca, Mandioca, Guacomole
(See Energy)

Cassia
(See Timber)

Castorbean
(See Oil)

Catjang, Jerusalem Pea, Marble Pea, Catjan
(See Vegetable)

Caucasian Bluestem

Sci. Name(s) : *Bothriochloa caucasica*
Geog. Reg(s) : Asia-India (subcontinent), Asia-Southeast, Eurasia, Subtropical, Tropical
End-Use(s) : Erosion Control, Forage, Pasture & Feed Grain
Domesticated : Y
Ref. Code(s) : 122
Summary : The caucasian bluestem is a fodder grass able to withstand heavy stocking densities. It is highly drought-resistant. This grass grows in India, Russia, and most tropical, and subtropical regions of the world. It was introduced to Thailand as an experimental soil binding and fodder grass.

Chewings Fescue

Sci. Name(s) : *Festuca rubra* subsp. *commutata*
Geog. Reg(s) : America-North (U.S. and Canada), Europe, Pacific Islands
End-Use(s) : Forage, Pasture & Feed Grain, Ornamental & Lawn Grass
Domesticated : Y
Ref. Code(s) : 17, 96
Summary : Chewings fescue is cultivated as a forage and meadow grass in New Zealand but is rarely grown for fodder in the United States. It is used for shaded lawns in Europe and is found in meadows, hills, and marshes in the cooler parts of the northern hemisphere.

Chicory
(See Beverage)

Chinese Tallow-Tree, Tallow-Tree
(See Oil)

Chlabato, Ghiabato
(See Vegetable)

Chufa, Ground Almond, Tigernut, Yellow Nutsedge

Sci. Name(s)	:	*Cyperus esculentus*
Geog. Reg(s)	:	Africa-West, Asia, Asia-India (subcontinent), Temperate, Tropical
End-Use(s)	:	Beverage, Cereal, Forage, Pasture & Feed Grain, Gum & Starch, Oil, Tuber
Domesticated	:	Y
Ref. Code(s)	:	111, 171, 220
Summary	:	Chufa, or ground almond, is a perennial grasslike herb that reaches 30-90 cm in height. It grows in most tropical and warm temperate areas, provided there is sufficient water. In the southern parts of the United States, chufas are grown for pig food. The herb is cultivated in tropical Africa and India for its tubers, which are eaten as vegetables. Tubers contain approximately 38% starch, 27% fat, 3.4% protein, and 2.5% ash. In India, the tubers are a source of flour and can be used as an adulterant of cocoa and coffee. In Spain and Italy, the tubers are harvested and eaten as vegetables and are used to make a beverage, horchuta. One kg of chufas yields about 5.5 liters of juice. In Sicily, a cultivar has been developed with a high saccharose content. It is a commercial source of alcohol.

The tubers also yield 20-28% of a yellow, nondrying oil that is used for culinary purposes. In Spain and Italy, the oil is used like olive or sweet almond oil in cooking, and in the manufacture of scented soaps. The leaves show possible use for paper making; digestion with soda lye produces about 35-40% of a deep yellow paper pulp.

Chufas are most successful on light, well-drained sandy loams in areas with moderately high temperatures. They are grown from sea level to elevations approaching 2,000 m. No fertilizer or cultivation is required. Tubers mature 3-4 months from planting.

Cocoa, Cacao
(See Beverage)

Coconut
(See Oil)

Collards, Boreocole, Kale
(See Vegetable)

Colonial Bentgrass, Rhode Island Bentgrass, Browntop, Common Bentgrass

Sci. Name(s)	:	*Agrostis tenuis, Agrostis vulgaris*
Geog. Reg(s)	:	America-North (U.S. and Canada), Asia-Central, Australia, Europe, Pacific Islands

End-Use(s)   : Forage, Pasture & Feed Grain, Ornamental & Lawn Grass
Domesticated : Y
Ref. Code(s)   : 15, 42, 164
Summary   : Colonial bentgrass is a slender, erect perennial that grows 20-40 cm in height. In the United States, it is cultivated as a lawn and pasture grass. It grows throughout Britain, Europe, and northern Asia. It thrives on acidic soils. Colonial bentgrass has been introduced to Australia and New Zealand.

There are many developed strains prized as lawn or pasture grasses in the cooler and humid regions of the United States.

Columbus Grass, Almum Sorghum, Almum

Sci. Name(s)   : *Sorghum X almum*
Geog. Reg(s)   : America-South, Australia
End-Use(s)   : Forage, Pasture & Feed Grain
Domesticated : Y
Ref. Code(s)   : 171
Summary   : Columbus grass is used for grazing and silage. It is native to Argentina where it is still grown extensively. There are several experimental plantings in Australia.

This grass is favored as pasture because of its easy propagation and eradication. Pastures should not be grazed too early or during drought because of possible prussic acid poisoning.

Comagueyana, Hurricane Grass, Pitted Bluestem

Sci. Name(s)   : *Bothriochloa pertusa*
Geog. Reg(s)   : Africa, Asia-India (subcontinent), Asia-Southeast
End-Use(s)   : Forage, Pasture & Feed Grain
Domesticated : Y
Ref. Code(s)   : 122
Summary   : Comagueyana is a pasture grass that grows throughout Southeast Asia, India, and Africa.

Common Bean, French Bean, Kidney Bean, Runner Bean, Snap Bean, String Bean, Garden Bean, Green Bean, Haricot Bean
(See Vegetable)

Common Bent-Grass, Redtop

Sci. Name(s)   : *Agrostis gigantea, Agrostis nigra*
Geog. Reg(s)   : America-North (U.S. and Canada), Asia, Asia-China, Australia, Eurasia, Europe, Pacific Islands
End-Use(s)   : Forage, Pasture & Feed Grain, Weed
Domesticated : Y
Ref. Code(s)   : 15, 42, 164
Summary   : Common bent-grass (*Agrostis gigantea*) is a perennial meadow grass reaching 0.6-1.2 m in height with vigorous underground stems growing on arable lands in Britain. It grows in Central and southern Russia, China, and Japan and has been introduced to New Zealand and Australia. This grass prefers light soils, is common in damp, woody areas and thrives in grassy places. It is regarded usually as a weed.

In the United States, common bent-grass is called 'redtop.' This creates some confusion because the Agrostis variety known as 'redtop' is *Agrostis alba*. *Agrostis alba* is the grass most commonly used in the United States as a lawn and meadow grass, not *Agrostis gigantea* (common bent-grass).

Common Fig, Adriatic Fig
(See Fruit)

Common Indigo, Indigo
(See Dye & Tannin)

Common Licorice
(See Spice & Flavoring)

Common Oat, Oat, Oats
(See Cereal)

Common Persimmon, Persimmon, American Persimmon
(See Fruit)

Common Reed Grass, Carrizo
(See Miscellaneous)

Common Vetch

Sci. Name(s)	: *Vicia sativa* subsp. *sativa*
Geog. Reg(s)	: Africa, America, America-South, Asia, Australia, Europe, Mediterranean, Pacific Islands, Temperate, Worldwide
End-Use(s)	: Erosion Control, Forage, Pasture & Feed Grain
Domesticated	: Y
Ref. Code(s)	: 24, 138, 213, 220
Summary	: The common vetch is widely cultivated as a cover crop and for fodder. It is less winter-hardy than hairy vetch. Common vetch is best adapted to a well-drained, loam soil and will tolerate sandy soils. The common, purple, and hairy vetches are used for orchard crops and are edible and palatable for livestock. It is an herb native to the Mediterranean and western Asia.

Cooba
(See Timber)

Cowpea, Southern Pea, Black-Eyed Pea, Crowder Pea

Sci. Name(s)	: *Vigna unguiculata* subsp. *unguiculata*
Geog. Reg(s)	: Africa, America-North (U.S. and Canada), Asia-Southeast, Temperate, Tropical
End-Use(s)	: Beverage, Forage, Pasture & Feed Grain, Vegetable
Domesticated	: Y
Ref. Code(s)	: 126, 132, 220

Summary   : The cowpea (*Vigna unguiculata* subsp. *unguiculata*) is a forage or green manure crop grown in warmer regions. The seeds are edible and used as stock food or in soups. As a pulse, they are ground into meal or roasted as a substitute for coffee. Plant leaves are high in vitamin A. Dry beans yield 1-3 MT/ha of seed.

Compared with *Phaseolus vulgaris*, cowpeas are more drought-resistant and can tolerate lower rainfall and humidity. They are adapted to a wide range of soils, from sandy to heavy loams, and from fertile to poor. Cowpeas are one of the staple food crops in the southern United States and are grown in Southeast Asia and Africa in both temperate and tropical areas.

Crambe, Colewort
(See Oil)

Creeping Bent-Grass, Fiorm, Red Top
Sci. Name(s)   : *Agrostis stolonifera* var. *palustris, Agrostis palustris, Agrostis alba*
Geog. Reg(s)   : Africa-South, America-North (U.S. and Canada), Asia, Australia, Europe, Pacific Islands
End-Use(s)   : Forage, Pasture & Feed Grain, Ornamental & Lawn Grass
Domesticated : Y
Ref. Code(s)   : 15, 42, 164
Summary   : Creeping bent-grass is grown in the United States for golf greens and lawns. It requires frequent cutting and watering to produce good turf. This grass has erect culms about 0.5-1 m high and is common in lowland areas of Britain and grassy and marshy areas of Europe, Asia, and North America. It is subject to several serious diseases. It has been introduced to Australia, New Zealand, and South Africa.

Several cultivated clones are the Metropolitan and Washington bents, and the Cocoos and Coos bents, which are cultivated for lawns and golf greens.

Creeping Foxtail
Sci. Name(s)   : *Alopecurus arundinaceus*
Geog. Reg(s)   : America-North (U.S. and Canada), Eurasia, Temperate
End-Use(s)   : Forage, Pasture & Feed Grain
Domesticated : Y
Ref. Code(s)   : 25
Summary   : Creeping foxtail is a tall perennial grass occasionally cultivated for pasture and hay. It is usually found in alpine meadows at elevations of 2,000-3,000 m in the United States, eastern Canada, and temperate Eurasia.

Crimson Clover
Sci. Name(s)   : *Trifolium incarnatum*
Geog. Reg(s)   : Africa-North, America-North (U.S. and Canada), Europe
End-Use(s)   : Erosion Control, Forage, Pasture & Feed Grain
Domesticated : Y
Ref. Code(s)   : 56, 220
Summary   : Crimson clover is an important winter annual legume grown in the southern United States, which provides excellent pasturage and hay, protection for the soil, and green manure. Plants are good seed producers.

154

This clover is grown throughout North America, Europe, and North Africa. It thrives in areas with cool, humid climates on sandy or clay soils. Best quality hay is cut while plants are in bloom. Hay yields were recorded at 4.5-5 MT/ha, with higher yields from good soils. A typical seed yield is 340-410 kg/ha of seed.

Crowfoot Grass
(See Weed)

Crownvetch, Trailing Crownvetch
Sci. Name(s) : *Coronilla varia*
Geog. Reg(s) : America-North (U.S. and Canada), Eurasia, Europe, Hawaii, Mediterranean, Subtropical, Temperate
End-Use(s) : Erosion Control, Forage, Pasture & Feed Grain, Ornamental & Lawn Grass, Weed
Domesticated : Y
Ref. Code(s) : 15, 56
Summary : Crownvetch is a hardy, long-lived perennial herb which is an important forage plant in most temperate and subtropical areas. The leaves are easily digested and are comparable in food value to alfalfa. Best results are obtained through regular grazing.

The plants have deep root systems which make them useful for erosion control. In the United States and Hawaii, 3 cultivars are used extensively for soil conservation: 'Chemung', 'Emerald', and 'Penngift'. In Europe, crown vetch is cultivated as a greenhouse plant.

Crownvetch is native to Central and southern Europe and grows throughout the Mediterranean, Central Russia, the Caucasus, western Syria, and Iran. It was introduced into North America and other temperate areas. It can become a troublesome weed. Plants thrive on most well-drained soils. They are fairly drought-resistant and frost-hardy.

Cuddapah Almond, Chirauli Nut
(See Oil)

Curly Mesquite
Sci. Name(s) : *Hilaria belangeri*
Geog. Reg(s) : America-Central, America-North (U.S. and Canada)
End-Use(s) : Forage, Pasture & Feed Grain
Domesticated : N
Ref. Code(s) : 96
Summary : Curly mesquite is the dominant "short grass" of the United States Texas plains. It is an important range grass because it withstands close grazing. This grass grows low, its culms are usually 1.2-2.4 m tall and it produces firm sods with curly tufts. Curly mesquite is found on the mesas and plains from Texas to Arizona and into northern Mexico.

Dallisgrass
Sci. Name(s) : *Paspalum dilatatum*

Geog. Reg(s)   : America-North (U.S. and Canada), Hawaii, Subtropical, Temperate, Tropical
End-Use(s)   : Erosion Control, Forage, Pasture & Feed Grain, Weed
Domesticated   : Y
Ref. Code(s)   : 24, 154, 171
Summary   : Dallisgrass is a hardy perennial grass that provides pasture and forage in subtropical, tropical, and warm temperate areas. It shows potential as pasture in southern Australia because of its persistent growth, drought-hardiness, and ability to withstand heavy grazing. In Hawaii, dallisgrass supplies good pasture but is also considered a weed.

Yields of fresh herbage range from 39-65 MT/ha. Dallisgrass is most successful
on moist, fertile soils; it is common in areas receiving over 900 mm of rainfall/year.
The grass is also frost-resistant. It establishes quickly in mixtures of other rapid-
growing grasses. Dallisgrass is used for erosion control.

Damas
(See Timber)

Danicha
(See Fiber)

Diaz Bluestem
Sci. Name(s)   : *Dichanthium annulatum*
Geog. Reg(s)   : Asia-India (subcontinent), Asia-Southeast, Subtropical, Temperate, Tropical
End-Use(s)   : Forage, Pasture & Feed Grain
Domesticated   : Y
Ref. Code(s)   : 24, 25, 122
Summary   : Diaz bluestem is a tufted perennial grass with stems approximately 1 m in
height, which provides excellent fodder. It has been introduced to many tropical,
subtropical, and warm temperate regions of the world. In India and Burma, this grass
is considered reliable pasture if it is not overgrazed. Herbage yields range from
1.8-18 MT/ha.

The grass grows in fairly dry to moist conditions in areas receiving an annual
rainfall from 300-1,500 mm.

Eastern Elderberry, American Elderberry, American Elder
(See Fruit)

Edible Canna, Gruya, Queensland Arrowroot
(See Gum & Starch)

Einkorn, One-Grained Wheat
Sci. Name(s)   : *Triticum monococcum*
Geog. Reg(s)   : Europe
End-Use(s)   : Beverage, Cereal, Forage, Pasture & Feed Grain, Miscellaneous
Domesticated   : Y
Ref. Code(s)   : 17, 161, 171

Summary : Einkorn is a hardy cereal that is seldom cultivated for its grain because of its poor quality, late ripening habits, and small yields. The grain is used husked, in place of barley, as a fodder corn for livestock. The grains have been used in the manufacture of certain beers and vinegars. There are both the winter and spring forms of einkorn, but the winter form is the most commonly cultivated.

Plants are resistant to frost, rust, and poor soils. Einkorn can be found in some economic gardens and breeding collections. Plants are cultivated in parts of Yugoslavia and Turkey.

Elephant Bush, Spekboom
(See Ornamental & Lawn Grass)

Elephant Grass, Napier Grass

Sci. Name(s) : *Pennisetum purpureum*
Geog. Reg(s) : Africa, Tropical
End-Use(s) : Erosion Control, Fiber, Forage, Pasture & Feed Grain
Domesticated : Y
Ref. Code(s) : 24, 171
Summary : Elephant grass is a tall clumped perennial pasture grass native to tropical Africa and is valued for its high leaf yields, vigorous growth, and palatability. The grass does best in areas with a mean annual rainfall of 1,000-1,500 mm. Substantial yields are obtained with nitrogen or organic manure. Fields should be replanted every 5-6 years. Four to six cuts/year have produced 45.3-136 MT/ha of green matter. In its early stages, elephant grass is used for silage. It makes good mulch and ground cover for soil conservation as well as a windbreak. Dry stems are used for fencing and thatching throughout the tropics.

This grass has been crossed with bulrush millet (*Pennisetum americanum*) to give a sterile triploid that produces better fodder than either parent. Elephant grass thrives in tropical areas from sea level to 2,000 m in altitude in most fertile soils.

Emmer
(See Cereal)

Fairway Crested Wheatgrass
(See Erosion Control)

Fat Hen, Lamb's-Quarters, White Goosefoot, Lambsquarter
(See Cereal)

Fenugreek
(See Spice & Flavoring)

Flax
(See Fiber)

Fox Grape
(See Fruit)

Foxtail Millet, Italian Millet, German Millet, Hungarian Millet, Siberian Millet
(See Cereal)

Galleta Grass

Sci. Name(s)	: *Hilaria jamesii*
Geog. Reg(s)	: America-North (U.S. and Canada), Asia-India (subcontinent)
End-Use(s)	: Forage, Pasture & Feed Grain
Domesticated	: Y
Ref. Code(s)	: 96
Summary	: Galleta grass is an erect, grazable range grass with culms 20-40 cm in height that grows best in deserts, canyons, and on dry plains. In the United States, its habitat extends from Wyoming to Utah and into Texas. Galleta grass has been introduced into India.

Gamba Grass

Sci. Name(s)	: *Andropogon gayanus*
Geog. Reg(s)	: Africa, America-South, Asia-India (subcontinent), Australia, Tropical
End-Use(s)	: Erosion Control, Forage, Pasture & Feed Grain
Domesticated	: Y
Ref. Code(s)	: 171
Summary	: Gamba grass is an adaptable, persistent grass used in pastures and for erosion control. The grass must be grazed before flowering. It grows in a variety of soils and withstands a long, dry season. It provides good results as pasture in northern Nigeria and Ghana.

The 2 m tall perennial is native to tropical Africa and introduced into Brazil, India, and Queensland.

Garbanzo, Chickpea, Gram
(See Vegetable)

Garden Orach, Butler Leaves, Orach, Mountain Spinach, Orache
(See Vegetable)

Giant Wildrye

Sci. Name(s)	: *Elymus condensatus*
Geog. Reg(s)	: America-North (U.S. and Canada)
End-Use(s)	: Cereal, Forage, Pasture & Feed Grain
Domesticated	: Y
Ref. Code(s)	: 153, 164
Summary	: The giant wildrye is a tall pasture grass about 1.5-3.5 m high that, when young, makes good hay. It grows in dense clumps and has short, thick culms. The grass can tolerate slightly salinic soils and is fairly drought-resistant. At one time, wildrye seeds were ground into flour by Native Americans.

In the United States, giant wildrye grows in the Central and southern states and along the California coast. It is most successful in dry areas below 1,523 m in altitude.

Golden Timothy Grass

Sci. Name(s) : *Setaria sphacelata*
Geog. Reg(s) : Africa-East, Africa-South, Africa-West, Tropical
End-Use(s) : Forage, Pasture & Feed Grain
Domesticated : Y
Ref. Code(s) : 96, 171
Summary : Golden timothy is a palatable, nutritious perennial grass with culms from 0.5-1.5 m in height. It is used for hay, silage, and soil improvement. Golden timothy is native to tropical Africa and its natural habitat extends to South Africa.

The grass grows well on fertile, planted lays. Because of its growth potential, golden timothy is becoming increasingly important as a pasture grass in East and South Africa and Zimbabwe.

Goosegrass, Fowl Foot Grass, Yardgrass
(See Weed)

Grass Pea, Chickling Vetch, Chickling Pea
(See Cereal)

Green Needlegrass, Feather Bunchgrass

Sci. Name(s) : *Stipa viridula*
Geog. Reg(s) : America-North (U.S. and Canada)
End-Use(s) : Forage, Pasture & Feed Grain
Domesticated : Y
Ref. Code(s) : 20, 220
Summary : Green needlegrass (*Stipa viridula*) has culms from 40-70 cm in height and is grown as forage on the Rocky Mountains and prairies of North America. This grass is valued because it does not produce the pointed rachilla that make other members of its genus unpalatable to livestock.

Guar, Clusterbean
(See Gum & Starch)

Guineagrass

Sci. Name(s) : *Panicum maximum*
Geog. Reg(s) : Africa, America-North (U.S. and Canada), America-South, Asia-India (subcontinent), Hawaii, Tropical
End-Use(s) : Forage, Pasture & Feed Grain
Domesticated : Y
Ref. Code(s) : 24, 154, 171
Summary : Guineagrass is a dense-growing, perennial grass reaching 0.5-4.5 m in height, which is one of the most important cultivated range and fodder grasses of lowland tropical America. It also grows well in the drier pastures of Hawaii. Young guineagrass provides excellent forage and remains palatable to livestock throughout most stages of its growth.

The grass grows throughout tropical Africa and has been introduced into other tropical areas including parts of India and North America. It grows from sea level to altitudes approaching 1,800 m in nearly any area with warm, tropical climates. Guineagrass needs fairly fertile soil and adequate moisture for good growth. The best soils are well-drained, light-textured, preferably sandy loams.

Guineagrass is difficult to propagate because its ears ripen unevenly and the seeds are released early. The best method of propagation is by root division. For cut fodder, the first harvest occurs at 10 weeks and later ones in 6-8 weeks. Yields of 35-100 MT/ha/annum of green matter are produced. Much higher yields are obtained with proper fertilization.

Gum Arabic, Senegal Gum, Kher
(See Gum & Starch)

Hairy Indigo
(See Dye & Tannin)

Hairy Vetch, Winter Vetch, Russian Vetch

Sci. Name(s)	: *Vicia villosa*
Geog. Reg(s)	: Africa-North, Asia, Europe, Mediterranean
End-Use(s)	: Erosion Control, Forage, Pasture & Feed Grain, Vegetable
Domesticated	: Y
Ref. Code(s)	: 17, 24, 27, 138, 213, 220
Summary	: The hairy vetch is an annual or biennial herb grown as a cover crop and for hay and silage in mixtures with spring and winter cereals. It is also eaten as a vegetable. It is winter-hardy, adapted to light-sandy or heavy soils, and withstands sub-zero temperatures. It grows in the United States, Europe, the Mediterranean, North Africa, western Asia, and most tropical and subtropical areas.

Hard Fescue

Sci. Name(s)	: *Festuca longifolia, Ovina* var. *duriuscula, Festuca trachyphylla*
Geog. Reg(s)	: America-North (U.S. and Canada), Europe
End-Use(s)	: Forage, Pasture & Feed Grain
Domesticated	: Y
Ref. Code(s)	: 96
Summary	: Hard fescue is cultivated as a fodder grass. It is cultivated in Maine, parts of Iowa, and Virginia and is native to Europe where it provides pastures, particularly in the steppe and mountainous regions.

Harding Grass, Toowoomba Canary Grass

Sci. Name(s)	: *Phalaris stenoptera, Phalaris aquatica, Phalaris tuberosa*
eog. Reg(s)	: Africa-South, America-North (U.S. and Canada), Australia, Europe
End-Use(s)	: Forage, Pasture & Feed Grain
Domesticated	: Y
Ref. Code(s)	: 35, 96, 152, 218
Summary	: Harding grass is a perennial grass which grows to 1.5 m in height. It is a very promising forage grass in experimental cultivation at the California Experimental Station. It is cultivated in Oregon, Washington, and North Carolina for forage as well as in Europe, South Africa, and Australia.

Hirta Grass
(See Erosion Control)

Hog-Plum, Yellow Mombin, Jobo
(See Fruit)

Hop Clover

Sci. Name(s)	: *Trifolium* spp.
Geog. Reg(s)	: America-North (U.S. and Canada), Europe, Subtropical, Temperate
End-Use(s)	: Forage, Pasture & Feed Grain
Domesticated	: Y
Ref. Code(s)	: 163, 220
Summary	: Trifolium species are a genus of about 300 herbal clovers. Most clovers thrive in temperate and subtropical regions and many are grown as winter forage when most grasses are dormant.

There are 3 species of clover included in the Trifolium genus and these are the hop clovers. They are known individually as *Trifolium procumbens*, *Trifolium dubium*, and *Trifolium agrarium*. Hop clovers are native to Europe and have naturalized throughout most areas of the southern United States. *Trifolium agrarium* is the least economical member. Most clovers are highly nutritious and are capable of giving substantial yields.

Hops
(See Beverage)

Horsegram
(See Cereal)

Hungarian Vetch

Sci. Name(s)	: *Vicia pannonica*
Geog. Reg(s)	: America-North (U.S. and Canada), Europe, Mediterranean
End-Use(s)	: Erosion Control, Forage, Pasture & Feed Grain
Domesticated	: Y
Ref. Code(s)	: 138, 141, 220
Summary	: The Hungarian vetch is a commonly cultivated forage and cover crop in Europe. It is somewhat more winter-hardy than the common vetch but not as hardy as hairy vetch. It grows on moderately fertile soils in mild temperatures in the United States. Plants withstand wet soils. This herb is native to the Mediterranean region.

Hungry-Rice, Fonio, Fundi, Acha
(See Cereal)

Idaho Fescue

Sci. Name(s)	: *Festuca idahoensis*
Geog. Reg(s)	: America-North (U.S. and Canada)
End-Use(s)	: Forage, Pasture & Feed Grain
Domesticated	: N

Ref. Code(s)   : 96, 104
Summary   : Idaho fescue is a grass used to provide supplemental forage in pasture lands. It is grazed throughout the spring and summer months and is highly nutritious. Although the grass loses palatability as it ages, it is valued for its persistence and hardiness in pastures and for its ability to form dense stands.

This grass grows in the open woods and on the rocky slopes of the United States from Colorado to central California northward.

Indian Bael, Bael Fruit, Bengal Quince, Bilva, Siniphal, Bael Tree
(See Fruit)

Indian Grass
Sci. Name(s)   : *Sorghastrum avenaceum, Sorghastrum nutans*
Geog. Reg(s)   : America-North (U.S. and Canada)
End-Use(s)   : Forage, Pasture & Feed Grain
Domesticated   : N
Ref. Code(s)   : 25, 96
Summary   : Indian grass is a tall perennial grass important as prairie and range hay in the United States. It can be found in wooded areas of Florida and other southern states.

Indian Jujube, Beri, Inu-Natsume, Ber Tree
(See Fruit)

Indian Melilot, Sourclover, Indian Sweetclover, Annual Yellow Sweetclover
Sci. Name(s)   : *Melilotus indica*
Geog. Reg(s)   : Asia-Central, Asia-India (subcontinent), Mediterranean, Temperate
End-Use(s)   : Erosion Control, Forage, Pasture & Feed Grain
Domesticated   : Y
Ref. Code(s)   : 56
Summary   : Indian melilot (*Melilotus indica*) is an annual herb with erect stems reaching up to 50 cm in height that is cultivated for forage, hay, and silage. It is also used for green manure and soil improvement. This crop is generally considered less productive and less palatable than white melilot (*Melilotus alba*). In the Punjab, Indian melilot is grown as an irrigated, cool-season crop. Crops are harvested for forage during full bloom and average 2-4 cuttings/crop. This plant is often found as a winter weed on cultivated land.

Plants grow in most warm temperate areas and continue to thrive up to heights of 1,650 m. They grow well in most soils but are most productive on well-drained, neutral, or alkaline ones. Yields range from 4.5-6 MT/ha of green forage and 250 kg/ha of seed. Indian melilot is native to the Mediterranean region, Central Asia and India.

Indian Rice, Silkgrass, Indian Millet, Indian Ricegrass
(See Cereal)

Indianfig, Spineless Cactus, Prickly Pear, Indianfig Pricklypear
(See Fruit)

Intermediate Wheatgrass, Grenar Intermediate Wheatgrass

Sci. Name(s) : *Agropyron intermedium*
Geog. Reg(s) : America-North (U.S. and Canada)
End-Use(s) : Erosion Control, Forage, Pasture & Feed Grain
Domesticated : Y
Ref. Code(s) : 201
Summary : Intermediate wheatgrass is a high producing variety of wheatgrass often used for pasture and erosion control because of its vigorous growth and mild sod-forming qualities. It thrives at elevations of 300-1,066 m on well-drained soils in areas with rainfall of 35-75 cm.

This grass produces well in the northwestern United States wherever dryland alfalfa grows.

Jackbean, Horsebean, Swordbean
(See Vegetable)

Jackfruit, Jack
(See Fruit)

Japanese Lespedeza, Japanese Clover, Common Lespedeza, Striate Lespedeza

Sci. Name(s) : *Lespedeza striata*
Geog. Reg(s) : America-North (U.S. and Canada), Asia, Asia-China
End-Use(s) : Erosion Control, Forage, Pasture & Feed Grain
Domesticated : Y
Ref. Code(s) : 56
Summary : Japanese lespedeza is an annual herb cultivated for pasture, hay, erosion control, soil improvement, and green manure. It provides good permanent pasture on soils too acidic for most other clovers. This herb contains less moisture than alfalfa or red clover and is cured as hay more quickly. The best quality hay is from cuttings made just prior or during the first bloom.

Plants originate in Manchuria, China, Korea, the Ryukyu Islands, and Taiwan. They were introduced to North America and are cultivated in the southern United States. About 2.5-7.5 MT/ha of hay and 100-250 kg/ha of seeds are possible.

Jaragua Grass

Sci. Name(s) : *Hyparrhenia rufa*
Geog. Reg(s) : Africa-West, America-North (U.S. and Canada), Tropical
End-Use(s) : Forage, Pasture & Feed Grain, Weed
Domesticated : Y
Ref. Code(s) : 25, 96, 171
Summary : Jaragua grass is a fast-growing, highly-palatable, tufted perennial pasture grass with culms up to 1 m in height. It recovers quickly from grazing and must be cut or mowed frequently to control flowering. In parts of Africa, it is considered excellent fodder and silage. The grass grows throughout tropical Africa in high rainfall areas at low to medium altitudes. It is also cultivated to a limited extent in Florida and along the Gulf Coast but is most often found as a weed.

Jerusalem-Artichoke, Girasole
(See Tuber)

Job's Tears, Adlay, Adlay Millet
(See Miscellaneous)

Johnson Grass, Johnsongrass

Sci. Name(s) : *Sorghum halepense*
Geog. Reg(s) : America-North (U.S. and Canada), Eurasia, Mediterranean
End-Use(s) : Forage, Pasture & Feed Grain, Weed
Domesticated : Y
Ref. Code(s) : 163, 171
Summary : Johnson grass is an erect grass, with culms from 0.6-1.5 m in height, grown for fodder. Rootstocks are also used for livestock feed. This grass forms extensive underground root systems and can become a weed. If grown on rich, black soils, 3 or more cuttings can be made.

The grass grows in the United States where it has adapted to the cotton-producing areas and to the climates of New Mexico, Arizona and California. Plants are native to the Mediterranean and Near East.

Kapok, Silk-Cotton-Tree, Ceiba
(See Fiber)

Kentucky Bluegrass, June Grass
(See Ornamental & Lawn Grass)

Kidney-Vetch, Ladies' Fingers

Sci. Name(s) : *Anthyllis vulneraria*
Geog. Reg(s) : Africa-North, Eurasia, Europe
End-Use(s) : Forage, Pasture & Feed Grain
Domesticated : Y
Ref. Code(s) : 42, 220
Summary : The kidney-vetch is an erect or decumbent perennial herb or shrub cultivated as fodder for sheep and goats. It is covered with fine, soft hairs. Plants grow on dry, shallow soils but are more abundant on calcareous soils. The kidney-vetch grows throughout Europe, the Caucasus and North Africa.

Kikuyu Grass

Sci. Name(s) : *Pennisetum clandestinum*
Geog. Reg(s) : Africa-East, America-South, Subtropical, Tropical
End-Use(s) : Erosion Control, Forage, Pasture & Feed Grain, Ornamental & Lawn Grass
Domesticated : Y
Ref. Code(s) : 24, 171
Summary : Kikuyu grass is a densely matted perennial grass that produces valuable pasturage capable of withstanding heavy grazing. It has a high leaf yield, is rich in protein and low in fiber. Kikuyu grass makes good lawns and is planted for soil conservation alongside roads, on bunds, and hills.

The grass is native to the East African highlands and thrives at higher-altitudes in most fertile to acidic soils in areas with an annual rainfall of at least 1,000 mm. Kikuyu grass was introduced throughout the tropics and moist regions of the sub-tropics and has naturalized in the Andes mountains of Colombia.

Kodo Millet, Kodo, Kodra
(See Cereal)

Korean Lespedeza, Korean Clover
(See Erosion Control)

Kudzu Vine, Kudzu
(See Tuber)

Kura Clover, Kura Gourd, Pellett Clover, Honeyclover, Caucasian Clover

Sci. Name(s)	: *Trifolium ambiguum*
Geog. Reg(s)	: America-North (U.S. and Canada), Asia, Australia, Europe
End-Use(s)	: Forage, Pasture & Feed Grain, Miscellaneous
Domesticated	: Y
Ref. Code(s)	: 56
Summary	: Kura clover is a long-lived perennial legume with creeping stems from 1-4 dm in length that provides excellent grazing. Its flowers attract bees and are good for honey production. This clover is resistant to pests, cold winter conditions, and drought. It is most often grazed in the northern United States and southern Canada. Kura clover is being tested in the United States as a possible forage crop. Reports from Australia show that this clover is less productive than white clover.

Plants are native to Asia Minor and southeastern Europe. They thrive in northern humid regions at altitudes approaching 3,000 m.

Lablab, Lablab Bean

Sci. Name(s)	: *Lablab purpureus, Dolichos lablab*
Geog. Reg(s)	: Africa, America-Central, America-South, Asia-Southeast, Tropical
End-Use(s)	: Cereal, Forage, Pasture & Feed Grain, Vegetable
Domesticated	: Y
Ref. Code(s)	: 56, 126
Summary	: Lablab beans are short-term perennials grown as annual, palatable forage for livestock. In southern Asia, Latin America, and Africa, these beans are produced in 75-300 days. Plants have high yields of vegetative mass and are used for green manure and soil improvement. Young pods are eaten as snapbeans or cooked and eaten as vegetables. Dried seeds are edible. Fresh seeds contain poisonous prussic acid and must be thoroughly cooked before eating.

Plants thrive in semiarid to subhumid regions with an annual rainfall of 60-90 cm on infertile but well-drained soils. Plants show fairly good resistance to certain pests and diseases. There are both bush and climbing varieties of lablab.

Dry seed production is from 400-500 kg/ha with higher yields up to 1,150 kg/ha. About 5-10 MT of fodder are produced.

Lanceleaf Crotalaria
(See Erosion Control)

Leadtree, Lead Tree

Sci. Name(s)   : *Leucaena latisliqua*
Geog. Reg(s)   : America-Central, America-North (U.S. and Canada)
End-Use(s)   : Beverage, Erosion Control, Forage, Pasture & Feed Grain, Gum & Starch, Vegetable
Domesticated   : Y
Ref. Code(s)   : 56
Summary   : The leadtree provides an excellent source of protein for cattle fodder. It is used in erosion control, water conservation, soil improvement, and as a green manure crop. Its seeds have 25% gum. The dried seeds are strung for jewelry. Young pods and shoots can be eaten as vegetables, while the ripe seeds make a coffee substitute. The tree wood is very hard and heavy. Leadtrees are often used as shade for black pepper, coffee, cocoa, quinine, and vanilla.

    The tree is native to Mexico and northern Central America. It needs long, warm growing seasons.

Leucaena-Hawaiian Type
(See Timber)

Leucaena-Salvadorian Type
(See Timber)

Limpo Grass

Sci. Name(s)   : *Hemarthria altissima*
Geog. Reg(s)   : Africa, America-North (U.S. and Canada), Asia-India (subcontinent), Asia-Southeast, Subtropical, Tropical
End-Use(s)   : Forage, Pasture & Feed Grain
Domesticated   : Y
Ref. Code(s)   : 118, 122
Summary   : Limpo grass is a hardy perennial pasture grass grown in tropical and subtropical areas. It can withstand heavy grazing and is found wild in Indochina, Thailand, India, and Africa. In the United States, primarily Florida, limpo grass is cultivated as commercial pasturage.

Little Bluestem

Sci. Name(s)   : *Schizachyrium scoparium, Andropogon scoparius*
Geog. Reg(s)   : America-North (U.S. and Canada)
End-Use(s)   : Forage, Pasture & Feed Grain
Domesticated   : Y
Ref. Code(s)   : 18, 20, 146
Summary   : Little bluestem is a tufted perennial grass with slender culms from 60-90 cm in height. It provides good pasture on dry plains, prairies, and hills at elevations from 1,500-2,400 m. Little bluestem is occasionally cut for hay. Plants are sensitive to frost.

This grass is variable in growth but usually starts late in spring and flowers from late August to early October. It is most successful in sandy soils. Plants are found in North America.

Maize, Corn, Indian Corn
(See Cereal)

Mango
(See Fruit)

Manila Grass, Zoysia, Japanese Carpet Grass
(See Ornamental & Lawn Grass)

Meadow Fescue

Sci. Name(s)	: *Festuca pratensis, Festuca elaitor*
Geog. Reg(s)	: America-North (U.S. and Canada), Asia, Eurasia, Europe
End-Use(s)	: Forage, Pasture & Feed Grain
Domesticated	: Y
Ref. Code(s)	: 14, 17, 183
Summary	: Meadow fescue is cultivated as a tall meadow or pasture grass grown in the cooler areas of North America as well as meadows and forest borders throughout Europe and Asia. It is native to Eurasia.

This grass is considered a valuable fodder plant, is highly nutritious, and valuable as fresh forage or hay. It revives quickly from mowing and can be cut 2-3 times a year. Its hay yield is 3,000-6,000 kg/ha. It is a tall grass and rejuvenates early in the spring. This grass thrives in wet climates but withstands brief drought, waterlogging, and light-frost. Meadow fescue thrives in rich, moist, gravelly soil.

Meadow Foxtail

Sci. Name(s)	: *Alopecurus pratensis*
Geog. Reg(s)	: America-North (U.S. and Canada), Europe
End-Use(s)	: Forage, Pasture & Feed Grain
Domesticated	: Y
Ref. Code(s)	: 164
Summary	: Meadow foxtail is a tufted perennial grass similar to timothy and is cultivated for forage in the northern United States. It grows wild in meadows and on waste ground. It was introduced to the United States from Europe and is naturalized in both countries.

Mesquite, Algorrobo
(See Timber)

Molasses Grass
(See Insecticide)

Moth Bean, Mat Bean
(See Vegetable)

Mountain Brome Grass

Sci. Name(s) : *Bromus marginatus*
Geog. Reg(s) : America-Central, America-North (U.S. and Canada)
End-Use(s) : Forage, Pasture & Feed Grain
Domesticated : Y
Ref. Code(s) : 18, 146
Summary : Mountain brome grass is a stout, erect perennial pasture grass grown in mixtures of alfalfa and sweet clover. It is commonly found in canyons, meadows, and waste grounds of the northwestern United States and Mexico at elevations of 2,100-2,900 m.

Mung Bean, Green Gram, Golden Gram
See Vegetable)

Mutton Grass, Mutton Bluegrass

Sci. Name(s) : *Poa fendleriana*
Geog. Reg(s) : America-North (U.S. and Canada)
End-Use(s) : Forage, Pasture & Feed Grain
Domesticated : Y
Ref. Code(s) : 96, 186
Summary : Mutton grass is a perennial bunch grass reaching 30.5-61 cm in height and producing a very nutritious and palatable pasture, particularly popular with sheep raisers.
It is a North American range grass found on mesas, in open dry woods, on rocky hillsides, and fairly cool, moist habitats at altitudes approaching 3,657 m.

Narrowleaf Trefoil

Sci. Name(s) : *Lotus tenuis*
Geog. Reg(s) : America-North (U.S. and Canada), Asia-Central, Europe
End-Use(s) : Forage, Pasture & Feed Grain
Domesticated : Y
Ref. Code(s) : 56
Summary : The narrowleaf trefoil is a perennial herb with a shallow root system grown for pasture and hay in areas with dry soil. The hay is leafy, nutritious, and comparable to other good legume hays. It is native to Europe and Central Asia and has been introduced into the United States. Narrowleaf trefoil thrives in rich, wet meadows, sands, and heavy-textured clay soils but can adapt to dry soils with poor drainage.

Natal Indigo
(See Dye & Tannin)

Needle and Thread Grass

Sci. Name(s) : *Stipa comata*
Geog. Reg(s) : America-North (U.S. and Canada)
End-Use(s) : Forage, Pasture & Feed Grain
Domesticated : Y
Ref. Code(s) : 20, 220

Summary : Needle and thread grass grows from 30-120 cm in height and is a fairly important pasture grass of North America particularly in the Rocky Mountain areas and on prairies. There is little or no information concerning its physical appearance or economic potential.

Neem
(See Timber)

Niger-Seed
(See Oil)

Nigerian Lucerne

Sci. Name(s) : *Stylosanthes erecta, Stylosanthes guineensis*
Geog. Reg(s) : Africa-West, Tropical
End-Use(s) : Forage, Pasture & Feed Grain, Medicinal, Miscellaneous
Domesticated : N
Ref. Code(s) : 56, 170, 220
Summary : Nigerian lucerne is a woody herb from 0.1-1.5 m in height cultivated as a tropical pasture legume. Plants are usually grazed or cut for fodder according to immediate demand. Nigerian lucerne plants are a common ingredient in herbal medicines. Parts of the plant are smoked like tobacco with the smoke acting as a healing agent.

Plants are most successful on sandy ground in coastal areas. They are adapted to savannah conditions and survive on dry, gravelly hills and in sandy pastures. There is no yield data available.

Nile Grass

Sci. Name(s) : *Acroceras macrum*
Geog. Reg(s) : Africa-South
End-Use(s) : Forage, Pasture & Feed Grain
Domesticated : Y
Ref. Code(s) : 220
Summary : Nile grass is a creeping perennial cultivated for hay and as substantial permanent pasture. In Africa, the grass grows in Botswana and is a prominent pasture grass of the northeastern parts of South Africa's Transvaal.

Oil-Bean Tree, Owala Oil
(See Oil)

Olive
(See Oil)

Orchardgrass, Cocksfoot

Sci. Name(s) : *Dactylis glomerata*
Geog. Reg(s) : Africa-North, America-North (U.S. and Canada), Asia, Europe, Temperate
End-Use(s) : Forage, Pasture & Feed Grain
Domesticated : Y
Ref. Code(s) : 42, 54, 80, 220

Summary   : Orchardgrass is a medium-sized perennial grass reaching from 0.45-1.5 m in height and cultivated in most temperate countries for winter pasture and hay. It is most commonly grown in mixtures with clover or alfalfa or as permanent pasture. The grass is fairly drought-hardy and adapts well in rich or poor soil and shaded areas. In the United States, it grows wild along roadsides and in waste areas. Orchardgrass also grows throughout the British Isles, Europe, temperate Asia, and North Africa.

Ouricury Palm, Licuri, Nicuri
(See Fat & Wax)

Pangolagrass
Sci. Name(s)   : *Digitaria decumbens*
Geog. Reg(s)   : America-Central, West Indies
End-Use(s)   : Forage, Pasture & Feed Grain
Domesticated   : Y
Ref. Code(s)   : 24, 80, 171
Summary   : Pangolagrass is a perennial grazing grass in the humid parts of Central America and the Caribbean. It forms a dense, leafy, trailing mat that grows up to 60 cm in height, is highly nutritious and palatable. It is valuable for pasture because it does not produce fertile seed and is therefore not successful as a weed. It is extremely competitive and crowds out all the less valuable grasses and weeds.

This grass is most successful in wet, coastal lowland areas receiving 500-900 mm/year of rain with a pronounced dry season. Yields are from 10-20 MT/ha of dry matter from moderately fertilized pastures.

Para Grass, Paragrass
Sci. Name(s)   : *Brachiaria mutica*
Geog. Reg(s)   : Africa-West, America-South, Tropical
End-Use(s)   : Forage, Pasture & Feed Grain
Domesticated   : Y
Ref. Code(s)   : 16, 171
Summary   : Para grass is a trailing perennial approximately 2 m tall used mostly for silage. It provides good fodder in various tropical areas but is poor pasture because it is not drought-resistant and requires careful management. Para grass grows at low altitudes on poorly-drained soil in South America and West Africa.

Pataua, Seje Ungurahuay
(See Oil)

Pea, Garden Pea, Field Pea
(See Vegetable)

Peanut, Groundnut, Goober, Mani
(See Oil)

Pearl Millet, Bulrush Millet, Spiked Millet, Cat-tail Millet
(See Cereal)

Perennial Ryegrass, English Ryegrass

Sci. Name(s)   : *Lolium perenne*
Geog. Reg(s)   : America-North (U.S. and Canada), Europe, Temperate
End-Use(s)   : Forage, Pasture & Feed Grain, Ornamental & Lawn Grass
Domesticated   : Y
Ref. Code(s)   : 17, 23, 74, 212, 220
Summary   : Perennial ryegrass is a valuable grazing and hay grass grown extensively in most temperate countries. In the United States, this grass grows in Texas as a cool-season forage or winter lawn. Perennial ryegrass is native to Europe.

Perennial Soybean, Pempo

Sci. Name(s)   : *Glycine wightii*
Geog. Reg(s)   : Africa, America-South, Asia, Asia-India (subcontinent), Tropical
End-Use(s)   : Forage, Pasture & Feed Grain, Spice & Flavoring
Domesticated   : Y
Ref. Code(s)   : 56
Summary   : The perennial soybean is a climbing or trailing plant grown for forage, green manure, and cover crops in the hot, tropical climates of Africa, India, and Sri Lanka where they are adapted to elevations from sea level to 2,000 m. In Malawi, plant leaves are used as potherbs. Production yields in these countries approach 7 MT/ha. When grown with other grasses, yields reach 8-10 MT/ha.

Plants are native to tropical Africa and Asia and introduced to tropical America with limited production there.

Perilla
(See Oil)

Persian Clover, Shaftal, Birdseye Clover, Reversed Clover

Sci. Name(s)   : *Trifolium resupinatum*
Geog. Reg(s)   : Africa-North, America-North (U.S. and Canada), Asia, Europe, Mediterranean, Temperate
End-Use(s)   : Forage, Pasture & Feed Grain, Ornamental & Lawn Grass
Domesticated   : Y
Ref. Code(s)   : 56
Summary   : Persian clover is a coarse, annual or biennial legume grown for its nutritious winter forage, hay, green manure, seeds, and as an ornamental. Plants are native to parts of Asia, Iran, Greece, and Egypt. This clover was introduced to England and the United States and its cultivation has spread to many temperate countries. It thrives on wet, heavy soils in low areas. In the United States, the clover has adapted to the climates of the southeastern states and along the Pacific coast. Plants are cut for hay just prior to full bloom. Recorded green manure yields are at 37.5 MT/ha with hay yields at 2.5-5 MT/ha. Seed yields are 150-300 kg/ha. Crops may be grazed from winter through late spring.

Pigeon Pea, Red Gram, Congo Pea, No-Eye Pea
(See Vegetable)

Potato, European Potato, Irish Potato, White Potato
(See Tuber)

Poulard Wheat, Cone Wheat, Rivet Wheat
(See Cereal)

Prickly Comfrey, Comfrey, Quaker Comfrey, Russian Comfrey, Blue Comfrey

Sci. Name(s)	: *Symphytum peregrinum, Symphytum uplandicum, Symphytum orientale*
Geog. Reg(s)	: Asia, Europe, Mediterranean
End-Use(s)	: Forage, Pasture & Feed Grain
Domesticated	: Y
Ref. Code(s)	: 220
Summary	: Prickly comfrey is a hybrid of *Symphytum asperum* and *Symphytum officinale*. Parts of the plant are fed to livestock. There is very little information about its economics or characteristics. Plants are found in Europe, the Mediterranean and Asia.

Proso Millet, Hog Millet, Common Millet, Brown Corn Millet, Broomcorn Millet
(See Cereal)

Purple Vetch
(See Erosion Control)

Purslane, Pusley, Wild Portulaca, Akulikuli-Kula
(See Ornamental & Lawn Grass)

Quack Grass, Torpedograss
(See Weed)

Quackgrass, Twitchgrass, Couchgrass, Couch, Twitch

Sci. Name(s)	: *Agropyron repens*
Geog. Reg(s)	: America-North (U.S. and Canada), Europe
End-Use(s)	: Forage, Pasture & Feed Grain, Weed
Domesticated	: Y
Ref. Code(s)	: 16, 33
Summary	: Quackgrass is a perennial used for grazing and hay production but its long, creeping rootstocks make it a troublesome weed. It is common along roadsides and meadows. It has naturalized in the United States and is native to Europe.

This genus is closely related to wheat (*Triticum aestivum*). Agropyron differs from Triticum structurally as it is made up primarily of perennials.

Raintree, Saman
(See Timber)

Ramie, Rhea, China Grass
(See Fiber)

Ramon, Breadnut
(See Fruit)

Rape, Colza
(See Oil)

Red Clover

Sci. Name(s)	: *Trifolium pratense*
Geog. Reg(s)	: America-North (U.S. and Canada), Temperate
End-Use(s)	: Dye & Tannin, Erosion Control, Forage, Pasture & Feed Grain, Medicinal
Domesticated	: Y
Ref. Code(s)	: 56, 220
Summary	: Red clover is a perennial legume usually grown in mixtures with rye or timothy.

It is grown as a winter annual in the northeastern United States and Canada and at higher altitudes in the southeastern and western parts of the United States. The flowers yield a yellow dye. They can be dried and used in a decoction to stimulate the appetite or as a sedative.

Plants are usually grazed in their first year and cut 5-15 days after the first bloom. The second crop is used as pasture, is harvested for seed, or used for soil improvement and green manure. Red clover shows some resistance to the potato leaf hopper. Plants thrive on well-drained loam soils and will adapt to wet soils. Seed yields have been recorded at 70-100 kg/ha, without irrigation. With irrigation, seed yields are from 600-800 kg/ha.

Red Fescue

Sci. Name(s)	: *Festuca rubra* subsp. *rubra*
Geog. Reg(s)	: America-North (U.S. and Canada), Asia, Europe, Temperate
End-Use(s)	: Forage, Pasture & Feed Grain, Ornamental & Lawn Grass
Domesticated	: Y
Ref. Code(s)	: 96, 183, 220
Summary	: Red fescue is a durable pasture grass grown on poor, hilly soils, meadows and damp, sandy places in most of Europe and temperate Asia. In the northern United States, it is used in grass mixtures for pastures.

This grass has value on poor soils and marshy meadows as fodder, hay and pasture particularly for sheep. Its yield is 2,500-3,000 kg/ha. In ornamental horticulture, it is considered one of the best grasses for lawns and grows well on gravelly soils or in dry, shady places.

Red Mulberry
(See Ornamental & Lawn Grass)

Red Oat, Mediterranean Oat, Algerian Oat
(See Cereal)

Reed Canarygrass

Sci. Name(s)   : *Phalaris arundinacea*
Geog. Reg(s)   : America-North (U.S. and Canada)
End-Use(s)   : Erosion Control, Forage, Pasture & Feed Grain
Domesticated   : Y
Ref. Code(s)   : 16, 20, 102, 163
Summary   : Reed canarygrass is a tall long-lived perennial with fairly good potential as a meadow and pasture grass in both cool and warm areas. In the United States, there has been increased plantings of the grass along the coasts of Washington and Oregon. Its seed can be used for fish food.

   This grass is native to North America. It is recommended for planting in parks, along stream banks, and artificial ponds. It succeeds in most soils but prefers moist, clay loams. Grass should be cut prior to 1st bloom to insure subsequent cuttings. About 200 kg/ha of seed is obtained.

Rescue Grass

Sci. Name(s)   : *Bromus unioloides*
Geog. Reg(s)   : Africa-East, America-North (U.S. and Canada), America-South, Tropical
End-Use(s)   : Forage, Pasture & Feed Grain
Domesticated   : Y
Ref. Code(s)   : 96, 171
Summary   : Rescue grass is an annual or biennial with culms from 70-100 cm in height grown in mixtures of rhodesgrass (*Chloris gayana*) and white clover (*Trifolium repens*) to produce high-quality, durable pasture. This mixture is called the Kachwekano mixture. In the southern United States, this grass provides hardy winter forage.

   It is native to South America and has been introduced to higher-altitudes in the tropics. In the Kigezi District in southwestern Uganda at altitudes above 1,800 m, it produces a promising ley. Rescue grass seeds are easy to collect and germinate.

Rhodesgrass, Rhodes Grass

Sci. Name(s)   : *Chloris gayana*
Geog. Reg(s)   : Africa-Central, Africa-East, Africa-South
End-Use(s)   : Forage, Pasture & Feed Grain
Domesticated   : Y
Ref. Code(s)   : 171
Summary   : Rhodesgrass is a highly adaptable, perennial grass that provides valuable grazing, fodder and hay and grows from 0.5-1.5 m in height to provide a temporary ley. It has several forms cultivated as annuals. It is native to the higher-elevations of East and South Africa and is important in East and Central Africa as pasture grass. Grass seed can be purchased in stores.

Rice
(See Cereal)

Rice Bean, Ohwi, Ohashi
(See Vegetable)

Rose Clover

Sci. Name(s)   : *Trifolium hirtum*
Geog. Reg(s)   : America-North (U.S. and Canada), Eurasia, Europe, Mediterranean
End-Use(s)   : Forage, Pasture & Feed Grain
Domesticated   : Y
Ref. Code(s)   : 56, 220
Summary   : Rose clover is an annual legume with stems from 1-4 dm in height cultivated as range fodder in the United States, especially California and in southern Europe and Russia. It is palatable either fresh or dried. Carefully grazed clover reseeds itself and consequently there is no yield data available. It is not actively harvested in the United States.

   Plants are native to the Mediterranean region. They thrive on most well-drained soils in areas below 1,000 m in elevation.

Rough Bluegrass, Roughstalk Bluegrass, Roughish Bluegrass

Sci. Name(s)   : *Poa trivialis*
Geog. Reg(s)   : Africa-North, America-North (U.S. and Canada), Eurasia, Europe, Hawaii
End-Use(s)   : Forage, Pasture & Feed Grain
Domesticated   : Y
Ref. Code(s)   : 17, 20, 184
Summary   : Rough bluegrass is a perennial grass with culms from 40-60 cm in height cultivated as meadow and pasture. It is often grown in mixtures with other grasses, usually redtop and fowl meadow grass.

   This grass is common in European pastures, especially in Great Britain. It grows in North Africa, Siberia, North America, and Hawaii in deep, moist loams. The grass has slow growth and is easily crowded out by more vigorous grasses. Rough bluegrass is native to Eurasia and North Africa.

Rough Pea

Sci. Name(s)   : *Lathyrus hirsutus*
Geog. Reg(s)   : America-North (U.S. and Canada), Asia, Europe
End-Use(s)   : Forage, Pasture & Feed Grain, Weed
Domesticated   : Y
Ref. Code(s)   : 56
Summary   : Rough peas are important winter annual legumes cultivated for pasture, hay and soil improvement. Peas provide valuable forage for the winter and early spring months. Rough peas are adapted to the Cotton Belt area of the United States.

   Plants thrive in well-drained soils but can grow on soils too wet for clover and small grains or too heavy for most other annual legumes. Rough peas are native to southern Europe and southwestern Asia.

   Rough peas flower in the 1st part of spring and produce rich, palatable pasturage that can be grazed until pod formation begins. When pods ripen, plants are cut. Seed yields range from 300-400 kg/ha in the United States. Up to 1,000-2,000 kg/ha have been produced experimentally. About 10,500 MT/ha/annum of seed are obtained.

Russian Wildrye

Sci. Name(s)   : *Elymus junceus, Psathyrostachys juncea*

Geog. Reg(s) : America-North (U.S. and Canada), Asia, Asia-China, Eurasia, Europe
End-Use(s) : Erosion Control, Forage, Pasture & Feed Grain
Domesticated : Y
Ref. Code(s) : 22, 52, 211, 220
Summary : Russian wildrye is a hardy perennial grass native to the steppe and desert regions of the USSR and China where it grows at low and high altitudes. In the United States, it is important in rejuvenating depleted rangeland in the northern Great Plains and intermountain regions.

It is an economically important, drought-resistant, and salt tolerant forage grass used primarily for range and pasture seedlings. There are 30 year old stands that remain productive in spite of heavy grazing. This species is palatable and has broad leaves with a high level of crude protein available throughout all stages of its development. Together with native rangeland flora, it provides quality forage for late summer and fall grazing in the northern Great Plains and provinces of Canada.

Rutabaga, Swede
(See Vegetable)

Rye
(See Cereal)

Sacaton
Sci. Name(s) : *Sporobolus wrightii*
Geog. Reg(s) : America-Central, America-North (U.S. and Canada)
End-Use(s) : Forage, Pasture & Feed Grain
Domesticated : N
Ref. Code(s) : 20, 96
Summary : Sacaton is an erect perennial with clumps from 1.2-1.8 m height used for grazing. Its leaves are tough but palatable. This grass is often grown along water courses. It is grazed to about 30-60 cm above the ground. Sacaton is found on mesas and in valleys in Central and North America.

Sainfoin, Holyclover, Esparret
Sci. Name(s) : *Onobrychis viciaefolia, Onobrychis viciifolia, Onobrychis sativa*
Geog. Reg(s) : America-North (U.S. and Canada), Eurasia, Europe, Temperate
End-Use(s) : Forage, Pasture & Feed Grain, Miscellaneous
Domesticated : Y
Ref. Code(s) : 17, 56
Summary : Sainfoin is an erect, perennial herb from 10-80 cm in height grown for forage. It can be used as dryland pasture or hay in regions receiving limited rainfall or irrigation. As forage and hay, sainfoin is palatable to livestock and does not cause bloating if fed green. It is considered a good feed supplement for pigs. Hay yields are high but delays in harvest reduce forage value.

Plants are native to temperate Eurasia. There has been interest in Europe and Canada for using sainfoin in honeybee farming. Plants grow on banks, wasteland, and grassy areas in most soils. For best results, they should be planted in spring or autumn. Yields are 1 MT/ha of hay in the 1st year; 8.5 MT/ha of hay in the 2nd year and seed yields of 2,900-5,800 kg/ha.

Saltbush, Fourwing Saltbush, Chamisa

Sci. Name(s) : *Atriplex canescens*
Geog. Reg(s) : America-Central, America-North (U.S. and Canada)
End-Use(s) : Forage, Pasture & Feed Grain, Vegetable
Domesticated : N
Ref. Code(s) : 220
Summary : The saltbush is a shrub with edible seeds whose habitat ranges from the western United States to Mexico. The shrub is occasionally fed to livestock.

Sand Lovegrass
(See Ornamental & Lawn Grass)

Sandberg Bluegrass

Sci. Name(s) : *Poa secunda*
Geog. Reg(s) : America-South
End-Use(s) : Forage, Pasture & Feed Grain
Domesticated : N
Ref. Code(s) : 17
Summary : Sandberg bluegrass (*Poa secunda*) is native to Chile. This grass has often been incorrectly identified as *Poa sandbergii*. There is little information concerning its characteristics or economic potential. This grass is not presently cultivated.

Sea Island Cotton, American Egyptian Cotton, Extra Long Staple Cotton
(See Fiber)

Senegal Rosewood, Barwood, West African Kino, Red Barwood
(See Dye & Tannin)

Sericea Lespedeza

Sci. Name(s) : *Lespedeza cuneata, Lespedeza sericea*
Geog. Reg(s) : America-North (U.S. and Canada), Asia-Central, Subtropical, Temperate, Tropical
End-Use(s) : Erosion Control, Fiber, Forage, Pasture & Feed Grain
Domesticated : Y
Ref. Code(s) : 56
Summary : Sericea lespedeza is a perennial herb from 60-100 cm in height that provides fair pasture but is considered less nutritious than Korean or Japanese clover. Plants show more potential as soil improving and erosion control agents. Plant stems are made into paper pulp, used alone for cheap paper, or blended with longer fibered pulps to be used in specialty papers. Seeds are used in wild bird seed mixes.

Plants are tolerant of heat and poor soils and thrive in most tropical, subtropical, and warm temperate areas of the world. Sericea lespedeza is native to the Himalayas and is grown mainly in parts of the Orient and the United States.

Hay is cut twice a season with yields of 3,360-7,620 kg/ha. Forage yields exceed 9,000 kg/ha from cuttings taken at 9 week intervals. Seed yields are 335-1,000 kg/ha.

Serradella

Sci. Name(s)	: *Ornithopus sativus*
Geog. Reg(s)	: Africa-East, Africa-North, America-North (U.S. and Canada), Europe, Temperate
End-Use(s)	: Erosion Control, Forage, Pasture & Feed Grain
Domesticated	: Y
Ref. Code(s)	: 17, 205
Summary	: Serradella is an annual herb from 20-70 cm tall found mainly in Spain and Morocco and cultivated in other temperate areas for green manure and cover crops. It has little value as an economic forage crop.

In Kenya, serradella is grown for seed. In Europe, seed yields are 350-1,200 kg/ha. In the southeastern United States, the herb is grown mainly as a winter annual. It has potential in the northern United States if planted with oats and barley as a cover or forage crop. It thrives in light, sandy soils. This herb is not yet established commercially in the United States but forage yields range from 2-2.5 MT/ha.

Sesame, Simsim, Beniseed, Gingelly, Til
(See Oil)

Sesbania

Sci. Name(s)	: *Sesbania exaltata*
Geog. Reg(s)	: America-North (U.S. and Canada), Temperate
End-Use(s)	: Erosion Control, Fiber, Forage, Pasture & Feed Grain
Domesticated	: Y
Ref. Code(s)	: 56, 214
Summary	: Sesbania is a perennial, woody herb grown mainly as a cover or green manure crop. It provides smooth, strong fiber used for nets and fish lines. Plants are an important winter food source for wildlife. They are also used for soil improvement on irrigated land.

Plants thrive on moist soils in the warm temperate areas of the southern United States. They are grown as spring annuals. Plants are susceptible to frost. Sesbania flowers from April to October and fruits throughout the winter.

Sheep Fescue
(See Ornamental & Lawn Grass)

Shittim-Wood

Sci. Name(s)	: *Acacia seyal*
Geog. Reg(s)	: Africa, Mediterranean
End-Use(s)	: Dye & Tannin, Forage, Pasture & Feed Grain, Gum & Starch
Domesticated	: N
Ref. Code(s)	: 170
Summary	: The shittim-wood tree is used as a browse plant and fodder for livestock. It has very little practical industrial use. The timber is weak and often twisted. The tree

yields an edible gum considered inferior to that of the gum Arabic tree. A tannin is obtained from the bark but is inadequate unless mixed with other tannins.

The small, thorny tree grows in the drier parts of western Sudan and over into Egypt.

Showy Crotalaria
(See Erosion Control)

Siberian Wheatgrass
Sci. Name(s)	: *Agropyron sibiricum*
Geog. Reg(s)	: America-North (U.S. and Canada)
End-Use(s)	: Forage, Pasture & Feed Grain
Domesticated	: Y
Ref. Code(s)	: 201
Summary	: Siberian wheatgrass is a pasture grass more productive than its relative, crested wheatgrass. Siberian wheatgrass is an ashless form of the crested variety. It has slow seedling vigor but thrives on dry, light soils in the United States.

Sickle Medic, Yellow-Flowered Alfalfa
Sci. Name(s)	: *Medicago falcata*
Geog. Reg(s)	: Asia-Central, Asia-India (subcontinent), Europe, Temperate
End-Use(s)	: Erosion Control, Forage, Pasture & Feed Grain, Miscellaneous, Weed
Domesticated	: Y
Ref. Code(s)	: 56, 220
Summary	: Sickle medic (*Medicago falcata*) is grown periodically as a forage crop. It is used mainly in breeding to improve crop frost-resistance. Sickle medic is native to Europe and Central Asia and grows in a variety of ways, often as a weed. It is used as pasture and hay. Plants provide cover for banks, slopes, and borders and are often planted for erosion control.

Yields are higher if hybrids are used. Hybridization occurs with Alfalfa lucerne (*Medicago sativa*). This hybrid grows in India and throughout the temperate and cold regions of the world.

Side-Oats Grama
Sci. Name(s)	: *Bouteloua curtipendula*
Geog. Reg(s)	: America-North (U.S. and Canada)
End-Use(s)	: Forage, Pasture & Feed Grain
Domesticated	: Y
Ref. Code(s)	: 33, 164
Summary	: Side-oats grama is a valuable perennial forage grass in the dry woods and prairies of the western United States where it furnishes feed year round. This grass grows from 0.3-1 m in height and thrives at altitudes up to 2,700 m.

Signal Grass, Palisade Grass
Sci. Name(s)	: *Brachiaria brizantha*
Geog. Reg(s)	: Africa-South, Africa-West, Asia-India (subcontinent), Tropical
End-Use(s)	: Forage, Pasture & Feed Grain, Ornamental & Lawn Grass

Domesticated : Y
Ref. Code(s) : 171
Summary : Signal grass is a valuable pasture and hay grass in tropical and South Africa and Sri Lanka. It is cultivated for ornamental hedges in Equatorial Africa. This grass grows from 0.6-1.5 m in height. It is palatable and thrives under controlled grazing. Signal grass does not seed well and is usually propagated by root divisions. It grows well at altitudes up to 2,000 m in areas receiving over 750 mm of rain.

Single-Flowered Vetch, Monantha Vetch, One-Flowered Vetch
(See Erosion Control)

Siratro, Purple Bean, Conchito
Sci. Name(s) : *Phaseolus atropurpureus, Macroptilium atropurpureum*
Geog. Reg(s) : Africa-South, America-North (U.S. and Canada), Tropical
End-Use(s) : Forage, Pasture & Feed Grain
Domesticated : Y
Ref. Code(s) : 24, 191
Summary : Siratro is a perennial, trailing, tropical pasture legume with a high-crude protein content. It is grown in mixtures with other tropical grasses to produce a balanced protein diet for livestock. Siratro is valued for its drought-resistance, frost-hardiness, and high productivity. Plants grow in most soils but preferably light-textured, well-drained ones. Siratro is popular in parts of South Africa and has been introduced to Central and southern Florida as summer pasture.

Siris, East Indian Walnut, Koko
(See Timber)

Slender Wheatgrass
Sci. Name(s) : *Agropyron trachycaulum*
Geog. Reg(s) : America-North (U.S. and Canada)
End-Use(s) : Forage, Pasture & Feed Grain
Domesticated : Y
Ref. Code(s) : 164
Summary : Slender wheatgrass is an important perennial forage grass in the mountainous areas of the western United States. It thrives on moist grasslands and in open woods. The 2 variations of slender wheatgrass are *Agropyron panciflorum* (awnless) and *Agropyron subsecundum* (awned). Slender wheatgrass has stems 50-100 cm tall.

Slenderleaf Crotalaria
Sci. Name(s) : *Crotalaria brevidens* var. *brevidens*
Geog. Reg(s) : Africa, Subtropical, Tropical
End-Use(s) : Erosion Control, Forage, Pasture & Feed Grain, Gum & Starch
Domesticated : Y
Ref. Code(s) : 56, 170
Summary : Slenderleaf crotalaria is an annual herb with pale yellow flowers that is cultivated for green manure and fodder in parts of tropical Africa. This crotalaria is considered nontoxic, unlike other crotalarias that have alkaloids toxic to stock. Plants are

hardy and adaptable. They thrive in most tropical or subtropical soils. Slenderleaf crotalaria has potential for erosion control. Plant seeds contain a mucilage gum that is obtained through a dry milling process.

Crotalaria grows rapidly and requires little attention. It is most common in grassy or lightly wooded areas from 600-2,700 m in elevation.

Sloughgrass

Sci. Name(s)	: *Beckmannia syzigachne*
Geog. Reg(s)	: America-North (U.S. and Canada), Asia
End-Use(s)	: Forage, Pasture & Feed Grain
Domesticated	: Y
Ref. Code(s)	: 164
Summary	: Sloughgrass is an annual tufted grass reaching 100 cm in height grown in Asia and North America as a highly palatable fodder and hay. In the western United States, sloughgrass grows in wet meadows and along ditches and lakes.

Small Buffalo Grass, Blue Panicgrass, Kleingrass

Sci. Name(s)	: *Panicum coloratum*
Geog. Reg(s)	: Africa-South
End-Use(s)	: Forage, Pasture & Feed Grain
Domesticated	: N
Ref. Code(s)	: 24, 184, 220
Summary	: Small buffalo grass is a tufted perennial with stems from 40-150 cm tall that provides excellent pasture and hay. This grass grows at altitudes from 500-2,000 m in areas seasonally waterlogged but does not tolerate heavy, waterlogged clays. There are 2 main varieties of small buffalo grass: variety makarikariense and variety coloratum. Small buffalo grass is native to South Africa.

Smilograss

Sci. Name(s)	: *Oryzopsis miliacea*
Geog. Reg(s)	: America-North (U.S. and Canada), Mediterranean
End-Use(s)	: Erosion Control, Forage, Pasture & Feed Grain, Ornamental & Lawn Grass
Domesticated	: Y
Ref. Code(s)	: 17, 76
Summary	: Smilograss is a perennial herb used as forage, a sand-binder, and for reseeding burned areas. It reaches heights up to 1.2 m and is grown as an ornamental. Plants thrive on open ground in any good, well-drained soil in full sunlight. Smilograss originated in the Mediterranean region and was introduced into the United States in California, New Jersey, and Pennsylvania.

Smooth Brome Grass

Sci. Name(s)	: *Bromus inermis*
Geog. Reg(s)	: America-North (U.S. and Canada), Eurasia
End-Use(s)	: Forage, Pasture & Feed Grain
Domesticated	: Y
Ref. Code(s)	: 146, 164

Summary   : Smooth brome is a perennial grass with culms to 1.2 m in height native to Eurasia where it is cultivated for hay. In the United States, it has become one of the most successful forage crops and is widely planted for pasture and hay production in all states with the exception of some southern states. Smooth brome grass often grows wild along roadsides, ditches, and in moist, woody areas.

Smooth Crotalaria, Striped Crotalaria

Sci. Name(s)   : *Crotalaria pallida, Crotalaria mucronata, Crotolaria striata*
Geog. Reg(s)   : Africa, Africa-North, America-North (U.S. and Canada), America-South, Asia-India (subcontinent), Australia, Hawaii, Tropical
End-Use(s)   : Beverage, Erosion Control, Forage, Pasture & Feed Grain
Domesticated   : Y
Ref. Code(s)   : 56, 170
Summary   : Smooth crotalaria is a short-lived perennial growing up to 3 m. It is cultivated throughout the tropics for pasture and as a cover crop. In Hawaii, the southeastern United States, and North Africa, the grass is usually grown for pasture. The plants are also valuable in erosion control. Crotalaria seeds are used as a coffee substitute.

Plants are common along river banks and in grasslands and woody areas from sea level to 1,500 m. They thrive in tropical, frost-free climates, and are found growing in Africa, India, Australia, South America, Hawaii, and the southeastern United States. Replanting occurs from 1.5-2 years, with average green forage yields from 10.5-10.7 MT/ha.

Snail Medic

Sci. Name(s)   : *Medicago scutellata*
Geog. Reg(s)   : Asia, Mediterranean
End-Use(s)   : Forage, Pasture & Feed Grain
Domesticated   : N
Ref. Code(s)   : 56
Summary   : Snail medic is an annual or perennial herb that reaches 15-60 cm in height and is grown for forage in the Mediterranean region and southwestern Asia. Plants occur naturally on heavy loam soils in meadows and forest margins in areas with warm temperatures and mild winters. Plants are cultivated to a limited extent. There is limited data on yields and economic potential.

Sorghum, Great Millet, Guinea Corn, Kaffir Corn, Milo, Sorgo, Kaoling, Durra, Mtama, Jola, Jawa, Cholam Grains, Sweets, Broomcorn, Shattercane, Grain Sorghums, Sweet Sorghums
(See Cereal)

Soybean
(See Oil)

Spanish Sainfoin

Sci. Name(s)   : *Hedysarum coronarium*
Geog. Reg(s)   : Africa-North, America-North (U.S. and Canada), Asia-India (subcontinent), Australia, Mediterranean

182

End-Use(s) : Forage, Pasture & Feed Grain, Ornamental & Lawn Grass
Domesticated : Y
Ref. Code(s) : 56
Summary : Spanish sainfoin is a small, deep rooting perennial or biennial herb with stems from 30-120 cm tall grown for forage and green manure. This herb is considered equal to red clover in nutritional value. It is fed to livestock fresh or dried as hay. It has a deep root system that improves the soils while providing rich pasturage. It is often grown as an ornamental in India.

This herb grows in North America, Australia, and North Africa. In Tunisia, it grows well on marls and clay limestone soils in areas receiving an annual rainfall of 45-60 cm. Plants withstand frost and adapt easily to most deep, well-drained calcareous soils. Countries bordering the Mediterranean and parts of Australia are the principal areas of production. Spanish sainfoin is grown experimentally in the United States. When irrigated, 100-125 MT/ha of green forage are produced. Rainfed crops yield 15-40 MT/ha.

Spelt
(See Cereal)

Spotted Burclover
(See Erosion Control)

St. Augustine Grass, Buffalo Grass of Australia, Pimento Grass of Jamaica
(See Ornamental & Lawn Grass)

Standard Crested Wheatgrass, Crested Wheatgrass
(See Erosion Control)

Stargrass, Giant Stargrass, Starr Grass
Sci. Name(s) : *Cynodon plectostachyus*
Geog. Reg(s) : Africa-East
End-Use(s) : Forage, Pasture & Feed Grain
Domesticated : Y
Ref. Code(s) : 24, 171
Summary : Stargrass is a perennial grass with no underground rhizomes, and with stems from 30-90 cm in height. It provides valuable, palatable pasture in the rift valleys of East Africa and areas from 800-2,000 m in elevation in Uganda, Kenya, Tanzania, and Ethiopia. It grows on dry, light-textured soils and is a good seed producer.

Sterile Oat, Animated Oat
(See Cereal)

Strawberry Clover
Sci. Name(s) : *Trifolium fragiferum*
Geog. Reg(s) : America-North (U.S. and Canada), Australia, Pacific Islands
End-Use(s) : Forage, Pasture & Feed Grain, Miscellaneous, Ornamental & Lawn Grass
Domesticated : Y

Ref. Code(s)   : 56

Summary   : Strawberry clover is a perennial with creeping stems from 1-3 cm in length grown as lawn and pasture in New Zealand and Australia. In the United States, it is an important pasture legume on saline and alkaline soils. This clover is used for green manure but does not grow tall enough for hay. It is an appropriate crop for large, poorly drained, irrigated areas. The plant is also used for honey production.

Plants are fairly resistant to short droughts and close grazing. Their natural habitats are wet meadows, river valleys, and bogs. Strawberry clover is widely cultivated in the western United States. Seed yields are recorded at 40-300 kg/ha with an average yield of 100 kg/ha.

Streambank Wheatgrass, Sodar Streambank Wheatgrass

Sci. Name(s)   : *Agropyron riparium*
Geog. Reg(s)   : America-North (U.S. and Canada)
End-Use(s)   : Forage, Pasture & Feed Grain
Domesticated   : Y
Ref. Code(s)   : 201
Summary   : Streambank wheatgrass is grown in conjunction with tall wheatgrass to provide durable range pasture and good ground cover in North America. It is drought-resistant, alkali-tolerant, and low-growing with rapid sod-forming ability. This grass is easy to establish, aggressive, spreads rapidly, but is low in palatability. It suppresses weeds and forms a smooth, long-lived, protective sod.

Sub Clover, Subterranean Clover

Sci. Name(s)   : *Trifolium subterraneum*
Geog. Reg(s)   : Africa, America-North (U.S. and Canada), Asia, Australia, Europe, Pacific Islands
End-Use(s)   : Erosion Control, Forage, Pasture & Feed Grain
Domesticated   : Y
Ref. Code(s)   : 56
Summary   : Sub clover is an annual winter pasture legume that provides excellent pasturage when mixed with other grasses, hay, and silage. Its stems form a dense mat that provides good erosion control. In the southern United States, this clover makes permanent pastures that last for about 25 years.

Plants are native to southern Europe, Asia, Africa, and southern England. They were introduced to, and are cultivated in, Australia, New Zealand, and the western and southern United States. Sub clover succeeds in areas with moist, warm winters and dry summers on light soils. It is harvested for forage 4-8 times each spring. When harvested with perennial rye grass, silage yields are 19.4-21.3 MT/ha and hay yields are 5.6-5.9 MT/ha.

Sudangrass, Sudan Grass

Sci. Name(s)   : *Sorghum sudanense*
Geog. Reg(s)   : Africa-South, America-Central, America-South, Australia, Hawaii, West Indies
End-Use(s)   : Forage, Pasture & Feed Grain
Domesticated   : Y
Ref. Code(s)   : 36, 163

184

Summary   : Sudangrass is a tall, annual grass from 1.8-3 m in height grown for fodder, hay, and summer pasture. It is among the first of the Sorghum genus to ripen, is the most drought-resistant, but does not tolerate extreme humidity.

     Plants are grown under irrigation in Australia, South Africa, Argentina, Brazil, Cuba, Puerto Rico, and Hawaii. In northern areas, 1 grass cutting/season is made, while in southern areas 2-3 cuttings are made. Sudangrass grows well in mixtures. Most of the commercially produced seed comes from drier regions. Hay yields in humid areas are recorded at 3.4-15.7 MT/ha while yields in semiarid regions are 2.2-7.8 MT/ha. Under irrigation, hay yields show a significant jump from 6.7-13 MT/ha.

Sugarcane
(See Sweetener)

Sunflower
(See Oil)

Sunn Hemp, Sann Hemp
(See Fiber)

Surinam Grass

Sci. Name(s)	: *Brachiaria decumbens*
Geog. Reg(s)	: Africa-West, Tropical, West Indies
End-Use(s)	: Forage, Pasture & Feed Grain
Domesticated	: Y
Ref. Code(s)	: 171

Summary   : Surinam grass (*Brachiaria decumbens*) is a natural pasture grass in tropical Africa. In Jamaica, it has potential as an alternative pasture grass to pangola grass (*Digitaria decumbens*). Surinam grass is also a valuable bottom grass in large, heavily grazed areas.

     The grass is very palatable and recovers quickly under rotational grazing. It is a trailing perennial reaching about 25-65 cm in height. It does not seed well and must be planted by root divisions or stem cuttings.

Sweet Orange, Orange
(See Fruit)

Sweet Potato
(See Tuber)

Sweet-Pitted Grass

Sci. Name(s)	: *Bothriochloa insculpta*
Geog. Reg(s)	: Africa-East
End-Use(s)	: Forage, Pasture & Feed Grain
Domesticated	: N
Ref. Code(s)	: 171

Summary   : Sweet-pitted grass is a highly palatable, tufted perennial 1 m tall that has potential as a ley grass in the higher-altitudes of East Africa. It thrives at elevations of

1,000-2,000 m in areas of low to medium rainfall. Heavily grazed grass forms a close sward. Sweet-pitted grass is propagated by seed.

Sweetclover

Sci. Name(s)	: *Melilotus suaveolens*
Geog. Reg(s)	: Asia, Asia-India (subcontinent), Temperate
End-Use(s)	: Forage, Pasture & Feed Grain
Domesticated	: Y
Ref. Code(s)	: 56
Summary	: Sweetclover is an herb grown for forage. It is native to Asia and India and is introduced to many temperate areas. In the 1st year, hay yields are 1.2-1.4 MT/ha while in the 2nd year, yields are 1.8-2.9 MT/ha.

Switchgrass

Sci. Name(s)	: *Panicum virgatum*
Geog. Reg(s)	: America-Central, America-North (U.S. and Canada), West Indies
End-Use(s)	: Erosion Control, Forage, Pasture & Feed Grain, Ornamental & Lawn Grass
Domesticated	: Y
Ref. Code(s)	: 17, 76, 220
Summary	: Switchgrass is a perennial grass grown for fodder, as an ornamental and erosion control agent. In North America, it grows naturally on prairies from Mexico to Florida and Arizona. It is also found in the West Indies and Central America.

There are 2 varieties: rubrum with less vigorous growth and strictum considered inferior to true switchgrass. Switchgrass grows best on light, sandy soil in full sun.

Swordbean

Sci. Name(s)	: *Canavalia gladiata*
Geog. Reg(s)	: Africa, Asia, Asia-India (subcontinent), Tropical
End-Use(s)	: Erosion Control, Forage, Pasture & Feed Grain, Vegetable
Domesticated	: Y
Ref. Code(s)	: 126, 170
Summary	: The swordbean is a climbing plant widely cultivated across Asia, particularly in India as green manure, cover, and forage. The young pods and beans are eaten as vegetables in tropical Asia and must be specially processed in salt water before eating. Unripe fresh beans contain 88.6% water, 2.7% protein, 0.2% fat, 6.4% carbohydrate, 1.5% fiber and 0.6% ash.

This plant is closely related to the jackbean (*Canavalia ensiformis*). Swordbeans grow well in the humid tropical areas of Africa and Asia. In subhumid regions, the necessary annual rainfall is 900-1,200 mm.

Tall Fescue
(See Ornamental & Lawn Grass)

Tall Meadow Oat Grass

Sci. Name(s)	: *Arrhenatherum elatius*
Geog. Reg(s)	: Africa-North, America-North (U.S. and Canada), Asia, Australia, Europe
End-Use(s)	: Forage, Pasture & Feed Grain

Domesticated : Y
Ref. Code(s) : 42, 220
Summary : Tall meadow oat grass is an erect, perennial grass cultivated as pasture and hay on the rough land of Britain, other parts of Europe, the mountainous areas of North Africa and western Asia, Australia, and North America. In the United States, it is found in meadows at altitudes from 1,500-2,750 m.

Tall Wheatgrass

Sci. Name(s) : *Agropyron elongatum*
Geog. Reg(s) : America-North (U.S. and Canada)
End-Use(s) : Forage, Pasture & Feed Grain
Domesticated : Y
Ref. Code(s) : 18
Summary : Tall wheatgrass is a perennial grass reaching heights of 1-5 m. It was introduced to the United States for forage but now grows wild in the southwestern states, especially New Mexico.

Tarwi, Tarhui, Chocho, Pearl Lupine, Tarin Altramuz, Muti, Ullus
(See Oil)

Teff
(See Cereal)

Teosinte

Sci. Name(s) : *Zea mays* subsp. *mexicana, Euchlaena mexicana*
Geog. Reg(s) : America-Central, America-North (U.S. and Canada)
End-Use(s) : Cereal, Forage, Pasture & Feed Grain, Oil, Ornamental & Lawn Grass, Sweetener
Domesticated : Y
Ref. Code(s) : 17, 84, 132, 171, 187
Summary : Teosinte is an annual grass that crosses easily with maize (*Zea mays* subsp. *maize*) and used in Mexico, Guatemala, and the southern United States as fodder and for its edible seeds. These seeds are processed for starch, sugar, and a cooking oil. This oil is gaining importance.

These plants resemble maize in habit. Taller forms are grown to produce silage and harvested before the grain ripens. It is also cultivated as an ornamental.

Tepary Bean
(See Vegetable)

Texas Bluegrass
(See Ornamental & Lawn Grass)

Thickspike Wheatgrass

Sci. Name(s) : *Agropyron dasystachyum*
Geog. Reg(s) : America-North (U.S. and Canada)
End-Use(s) : Forage, Pasture & Feed Grain

Domesticated : N
Ref. Code(s) : 18
Summary : Thickspike wheatgrass is an upright, rhizomatous perennial. It prefers sandy or gravelly soils and is found mainly in the north-central plains and mountains of New Mexico at elevations of 1,800-2,450 m.

Tikus, Tikug
(See Fiber)

Timothy, Common Timothy, Mountain Timothy, Herd's Grass
Sci. Name(s) : *Phleum pratense, Phleum nodosum*
Geog. Reg(s) : America-Central, America-North (U.S. and Canada), America-South, Asia, Europe, Temperate
End-Use(s) : Forage, Pasture & Feed Grain
Domesticated : Y
Ref. Code(s) : 17, 186
Summary : Timothy is a perennial meadow grass from 15-61 cm in height cultivated for hay and pasture in mixtures with red clover or alone. It is palatable to most livestock. Temporary pastures are quite often successful.

This grass grows in most temperate climates and in moist alpine or subalpine regions. In North America, it ranges from coast to coast. The grass is also found in parts of Mexico, Chile, northern Europe, and Asia.

Tobosa Grass
Sci. Name(s) : *Hilaria mutica*
Geog. Reg(s) : America-Central, America-North (U.S. and Canada)
End-Use(s) : Forage, Pasture & Feed Grain
Domesticated : Y
Ref. Code(s) : 96
Summary : Tobosa grass is a durable, perennial range grass important in the drier areas of North America. It grows in dense clumps on dry plains and hills from Texas to Arizona and into Mexico. In Arizona, this grass is called black gramma and is considered one of the most valuable grasses in that state.

Tomato
(See Fruit)

Townsville Stylo, Townsville Lucerne
Sci. Name(s) : *Stylosanthes humilis*
Geog. Reg(s) : America-Central, America-North (U.S. and Canada), America-South, Subtropical, Tropical
End-Use(s) : Erosion Control, Forage, Pasture & Feed Grain
Domesticated : Y
Ref. Code(s) : 56
Summary : Townsville stylo is an annual legume, with stems reaching 90 cm in height, which provides excellent pasture when mixed with permanent grasses. These mixtures are also used for green chop or silage. Townsville stylo is often grown as a cover crop

in citrus groves. Plants are valuable because of their ability to absorb nutrients under low fertility conditions.

Plants are native to Central Mexico and Central and northern South America. They provide important pasturage in most tropical and subtropical areas. In the United States, they are considered good for growing in southern Florida and other Gulf Coast states. Pastures are grazed from summer to fall and after the 1st frost. Mechanical harvest yields are 1,250 kg/ha of clean seed.

Tropical Carpetgrass, Savanna Grass

Sci. Name(s)	:	*Axonopus compressus*
Geog. Reg(s)	:	America-Central, Tropical, West Indies
End-Use(s)	:	Forage, Pasture & Feed Grain, Ornamental & Lawn Grass
Domesticated	:	Y
Ref. Code(s)	:	171
Summary	:	Tropical carpetgrass is a perennial lawn grass growing up to 45 cm in height. It

withstands heavy grazing during rainy seasons, which makes it a popular tropical pasture grass. It thrives in areas with well-distributed, medium to heavy rainfall and tolerates shade.

This grass is native to the Caribbean and Central America. While it has been a principle pasture grass in tropical America, it is being replaced by the more productive pangolagrass (*Digitaria decumbens*) in some areas.

Tropical Kudzu

Sci. Name(s)	:	*Pueraria phaseoloides*
Geog. Reg(s)	:	Asia-Southeast, Tropical
End-Use(s)	:	Erosion Control, Forage, Pasture & Feed Grain
Domesticated	:	Y
Ref. Code(s)	:	24, 56, 170
Summary	:	Tropical kudzu is a climbing or trailing perennial herb that is an important

cover and green manure crop in the tropics and is being considered as possible tropical fodder and pasturage because of its dense, vigorous growth. Plants grow well in most soils and thrive in areas with adequate rainfall and moderate to high temperatures. They do not tolerate heavy grazing, close cutting, or prolonged shade.

Typical yields range from 5-10 MT/ha of dry matter with seed yields from 50-100 kg/ha. Tropical kudzu is indigenous to the lowlands of Malaysia.

Upland Cotton, Cotton
(See Fiber)

Vasey Grass, Vaseygrass

Sci. Name(s)	:	*Paspalum urvillei*
Geog. Reg(s)	:	America-North (U.S. and Canada), America-South
End-Use(s)	:	Forage, Pasture & Feed Grain
Domesticated	:	Y
Ref. Code(s)	:	17, 24
Summary	:	Vasey grass is a perennial grass that forms large tufts from 1-2 m in height and

provides good pasture and fodder. In the southern and southwestern United States, it is used primarily for hay. The best hay is from the 1st cutting while 2nd cuttings gives the best seed yields. This grass grows on wet soils and tolerates drought. It is native to South America and has naturalized in the United States.

Veldtgrass, Perennial Veldtgrass

Sci. Name(s)	: *Ehrharta calycina*
Geog. Reg(s)	: Africa-South, America-North (U.S. and Canada)
End-Use(s)	: Erosion Control, Forage, Pasture & Feed Grain
Domesticated	: Y
Ref. Code(s)	: 25, 142, 147, 220
Summary	: Veldtgrass is a variable perennial with culms from 30-70 cm in height. It provides good grazing on inland sand plains or dunes and coastal areas and is used as a sand binder in the southern United States. This grass grows along the Cape of Good Hope in South Africa.

Velvet Bentgrass, Brown Bent-Grass
(See Ornamental & Lawn Grass)

Velvetbean

Sci. Name(s)	: *Mucuna deeringiana*
Geog. Reg(s)	: America-North (U.S. and Canada), Asia, Asia-Southeast, Hawaii, Tropical
End-Use(s)	: Beverage, Cereal, Erosion Control, Forage, Pasture & Feed Grain, Vegetable
Domesticated	: Y
Ref. Code(s)	: 36, 56, 132, 220
Summary	: The velvetbean is a bushy annual or perennial vine grown as cover, green manure, or forage crops in warm areas. Plants are used primarily in animal feeds and young pods and seeds are eaten as vegetables. Mature seeds are processed for flour or parched and used as a coffee substitute. Plants are harvested 4-8 months after planting.

Plants are native to Malaysia and southern Asia and are grown throughout the tropics at low to medium elevations in most well-drained soils. In the southern United States, drought-resistant cultivars have been developed that show potential for dryland farming. In 1981, velvetbean yields in the United States were 1,000-2,000 kg/ha. In Hawaii, yields ranged from 900-1,500 kg/ha.

Vine Mesquite, Vine Mesquitegrass

Sci. Name(s)	: *Panicum obtusum*
Geog. Reg(s)	: America-North (U.S. and Canada)
End-Use(s)	: Cereal, Forage, Pasture & Feed Grain
Domesticated	: Y
Ref. Code(s)	: 17, 20, 18, 96, 119
Summary	: Vine mesquite is a perennial forage and hay grass reaching 15-40 cm in height grown in Texas and the southwestern United States. It produces starchy seeds processed for flour. Vine mesquite grows in sandy or gravelly soils in moist areas from 1,830-2,440 m in elevation.

Vogel Tephrosia
(See Erosion Control)

Water Canarygrass, Bulbous Canarygrass
Sci. Name(s) : *Phalaris aquatica, Phalaris tuberosa*
Geog. Reg(s) : America-North (U.S. and Canada), Australia, Mediterranean
End-Use(s) : Forage, Pasture & Feed Grain
Domesticated : Y
Ref. Code(s) : 80
Summary : Water canarygrass is cultivated in Australia and the United States as a pasture grass. It is native to the Mediterranean. There is limited information available concerning individual characteristics or economic potential for this grass.

Water Spinach, Swamp Morning-Glory, Kangkong
(See Vegetable)

Watermelon
(See Fruit)

Weeping Lovegrass
Sci. Name(s) : *Eragrostis curvula*
Geog. Reg(s) : Africa-South, America-North (U.S. and Canada)
End-Use(s) : Erosion Control, Forage, Pasture & Feed Grain
Domesticated : Y
Ref. Code(s) : 82, 171, 220, 224
Summary : Weeping lovegrass is a hardy, durable range and pasture grass of South Africa. In South Africa, it provides important pasture. It is drought-resistant, withstands intensive grazing, adverse conditions and is useful for stabilizing sand dunes. In the southern Plains of the United States it is useful for increasing beef production on sandy-upland soils. It can be used in mixed plantings and does well on a range of soils. It is adapted to deep sands, clays and rocky soils but forage quality is best on richer soils.

Western Wheatgrass
Sci. Name(s) : *Agropyron smithii*
Geog. Reg(s) : America-North (U.S. and Canada)
End-Use(s) : Forage, Pasture & Feed Grain
Domesticated : Y
Ref. Code(s) : 164
Summary : Western wheatgrass is a perennial forage grass reaching 30-60 cm in height, and grown in the moist alkaline soils of the western United States. Further east, it has adapted to dry upland meadow conditions.

Wheat, Bread Wheat, Common Wheat
(See Cereal)

White Clover, Ladino Clover

Sci. Name(s)	: *Trifolium repens*
Geog. Reg(s)	: America-North (U.S. and Canada), Mediterranean, Worldwide
End-Use(s)	: Erosion Control, Forage, Pasture & Feed Grain
Domesticated	: Y
Ref. Code(s)	: 56
Summary	: White clover is a long-lived, nutritious forage legume grown throughout the world and used alone or in mixtures as pasturage, hay, and silage. It is also grown as a cover crop and for soil-improvement and protection. In the United States, about half of the 45 million hectares of humid or irrigated pastureland is estimated to have differing amounts of white clover.

This clover is native to the eastern Mediterranean region of Asia Minor. It grows on most nonsaline, cool, moist, limey soils. In the United States, yields from 1,500-2,500 MT/ha of seeds are produced. In humid areas, seed yields are from 30-200 kg/ha. In the western parts of the United States, from 150-600 kg/ha of seed are produced. Forage yields are variable.

White Lupine, Egyptian Lupine

Sci. Name(s)	: *Lupinus albus, Lupinus termis*
Geog. Reg(s)	: Africa-South, America-North (U.S. and Canada), America-South, Australia, Europe, Mediterranean, West Indies
End-Use(s)	: Cereal, Erosion Control, Fiber, Forage, Pasture & Feed Grain, Medicinal
Domesticated	: Y
Ref. Code(s)	: 56
Summary	: White lupine is a winter annual used for grazing, green manure, and general soil improvement. The seeds contain toxins and trypsin inhibitors and need prolonged boiling before eating. A flour is made from the seeds. They have medicinal qualities and a wide range of potential use in liquors, soaps, and fiber-work.

This plant is thought to be from the Balkan Peninsula. It is widely grown in the Mediterranean, the Canary Islands, Madeira, and the Upper Nile. It has been introduced to Europe, South Africa, Australia, South America, and the United States. Plants grow on mildly acidic, sandy loams. White lupine is considered the most winter hardy of the lupines. About 5-7.5 MT/ha of dry herbage are obtained. Seed yields are approximately 800-1,000 kg/ha.

White Melilot, White Sweetclover

Sci. Name(s)	: *Melilotus alba*
Geog. Reg(s)	: Asia, Asia-India (subcontinent), Europe, Temperate
End-Use(s)	: Fiber, Forage, Pasture & Feed Grain, Miscellaneous, Oil
Domesticated	: Y
Ref. Code(s)	: 56, 154, 205
Summary	: White melilot (*Melilotus alba*) is an aromatic annual or biennial plant grown for forage, hay, silage, cover, and green manure. Plants are also used for pasture and soil improvement. The flowers are among the most important for honey production. Fiber from the plant rind is used for paper pulp production. Oil from the seeds is used in paints and varnishes. The expressed seed meal is added as a protein supplement to cattle feeds if toxic substances have been removed.

Plants are native to Europe, Asia, and India. They have been introduced to other temperate areas where they tolerate heavy clays and light sands but are most successful in well-drained lime soils. Plants have adapted to dry conditions and have been planted in the Great Plains areas of the United States. They produce from sea level to 2,000 m in altitude. In the 1st year, hay yields are 2.2-3.5 MT/ha. In the 2nd year, yields increase from 2.2-8.1 MT/ha.

White Mustard

Sci. Name(s)	: *Sinapis alba, Brassica alba, Brassica hirta*
Geog. Reg(s)	: Tropical
End-Use(s)	: Forage, Pasture & Feed Grain, Oil, Spice & Flavoring, Vegetable
Domesticated	: Y
Ref. Code(s)	: 31, 72, 135, 170
Summary	: White mustard (*Sinapis alba*) is an annual plant that reaches 30 cm in height grown primarily a catch-crop for grazing or green manure. Young seedlings are used in salads. White mustard tends to have more vegetative growth than its relative, black mustard (*Brassica nigra*), and the leaves are considered more palatable. The seeds contain the glucoside sinalbin that yields 30% mustard oil. Prepared seeds produce common table mustard. These seeds and oil have a milder flavor than those of black mustard. Plants are grown throughout the year in greenhouses. Seedlings are collected when 1 cm high. White mustard is native to the tropics.

White Tephrosia
(See Erosion Control)

Whitetip Clover

Sci. Name(s)	: *Trifolium variegatum*
Geog. Reg(s)	: America-North (U.S. and Canada)
End-Use(s)	: Forage, Pasture & Feed Grain
Domesticated	: Y
Ref. Code(s)	: 56
Summary	: Whitetip clover is an annual legume highly valued as forage in the coastal and mountainous regions of North America. Crops are grazed in summer. The clover produces thick stands and experiences good growth under variable conditions. It thrives in moist areas below 2,600 m in elevation.

Wild Oat
(See Cereal)

Wine Grape, Grape, European Grape, California Grape
(See Beverage)

Winged Bean, Goa Bean, Asparagus Pea, Four-Angled Bean, Manilla Bean, Princess Pea
(See Vegetable)

Woolypod Vetch, Winter Vetch, Hairy Vetch

Sci. Name(s) : *Vicia dasycarpa, Vicia villosa*
Geog. Reg(s) : Africa, America-North (U.S. and Canada), Asia-China, Europe
End-Use(s) : Forage, Pasture & Feed Grain
Domesticated : Y
Ref. Code(s) : 17, 138, 141, 220
Summary : The woolypod vetch is grown in the southern United States as a cover crop, for soil improvement and fodder. It is found in North America, Africa, and China and is native to Europe. It does not tolerate hot, dry climates.

Yague
(See Cereal)

Yellow Flame, Yellow Poinciana, Soga
(See Ornamental & Lawn Grass)

Yellow Highstem, Yellow Bluestem

Sci. Name(s) : *Bothriochloa ischaemum*
Geog. Reg(s) : Africa-North, Asia, Europe, Pacific Islands
End-Use(s) : Forage, Pasture & Feed Grain
Domesticated : Y
Ref. Code(s) : 122
Summary : Yellow highstem is a pasture grass found in Papua New Guinea, Asia, North Africa, and southern Europe.

Yellow Lupine, European Yellow Lupine

Sci. Name(s) : *Lupinus luteus*
Geog. Reg(s) : Africa-South, America-North (U.S. and Canada), Australia, Europe, Mediterranean
End-Use(s) : Cereal, Forage, Pasture & Feed Grain, Miscellaneous, Ornamental & Lawn Grass
Domesticated : Y
Ref. Code(s) : 56
Summary : Yellow lupine is used for grazing, forage and silage, and in fruit orchards, for soil improvement, and green manure. The seeds are ground to flour and provide a protein supplement. In Australia, this is used as a protein concentrate in pet foods. Fresh seeds contain the alkaloid lupinine and are thought to be poisonous. Yellow lupine is used in honey production and is also grown as an ornamental.

    Plants are native to the Mediterranean basin and are now widely cultivated in northern Europe, South Africa, Australia, and the southern United States. They require a mild-temperatured growing season and sandy soils but will tolerate temporary water-logging. Yields range from 15-50 MT/ha of forage and 3-8.7 MT/ha of roots.

Yellow Melilot, Yellow Sweetclover

Sci. Name(s) : *Melilotus officinalis*
Geog. Reg(s) : Africa-North, America-North (U.S. and Canada), America-South, Asia, Europe

194

End-Use(s) : Erosion Control, Forage, Pasture & Feed Grain, Medicinal, Miscellaneous, Spice & Flavoring, Tuber
Domesticated : Y
Ref. Code(s) : 56, 205

Summary : Yellow melilot (*Melilotus officinalis*) is a biennial herb to 130 cm in height cultivated for forage, hay, and pasture. Its roots contain nitrogen-fixation nodules, making plants excellent for soil-improvement as well as erosion control. Its flowers attract bees and are used as honey pasture on bee farms. Parts of the plant are used to flavor cheese and tobacco snuff and an antithrombotic preparation can be obtained. Melilot roots are edible.

Plants are native to Europe, Asia, and North Africa and have been introduced to North and South America where they thrive from sea level to 4,000 m in neutral, well-drained soils. They are fairly drought-resistant.

Yellow melilot matures 10-14 days before white melilot (*Melilotus alba*). They must be heavily grazed to retain palatability. About 4.2-4.5 MT/ha of hay is obtained in the 1st year, increasing in the 2nd year to 5.5-8.5 MT/ha of hay.

Fruit

Aceituna, Olivo, Paradise-Tree
(See Oil)

African Oil Palm, Oil Palm
(See Oil)

African Star-Apple

Sci. Name(s) : *Chrysophyllum delevoyi*
Geog. Reg(s) : Africa, Tropical
End-Use(s) : Fruit, Oil, Timber
Domesticated : Y
Ref. Code(s) : 220
Summary : The African star-apple is an evergreen tree cultivated in tropical Africa for its sweet, acidic fruit (odara pears). An oil is extracted from the seeds and used in parts of Africa for soap making. The wood is used for turning, carving, and furniture.

Akee

Sci. Name(s) : *Blighia sapida*
Geog. Reg(s) : Africa-West, West Indies
End-Use(s) : Fruit
Domesticated : Y
Ref. Code(s) : 170
Summary : Akee is a tree planted for its edible, fully-ripened fruit aril. Unripe or damaged fruit should not be eaten because of a toxic peptide, hypoglycin A. Carefully selected fruit arils are eaten fresh or cooked.

Trees bear both male and female flowers and reach 7-25 m in height. Seedlings fruit in 5 years. Akee is grown in Jamaica and grows wild in the forests of West Africa.

Ambarella, Otaheite Apple, Golden-Apple, Jew-Plum

Sci. Name(s) : *Spondias dulcis, Spondias cytherea*
Geog. Reg(s) : America-North (U.S. and Canada), Asia-Southeast, Tropical
End-Use(s) : Fruit, Gum & Starch, Miscellaneous
Domesticated : Y
Ref. Code(s) : 17, 36, 65, 170
Summary : Ambarella is a fast-growing tree from 9-18 m tall that produces plum-sized, edible fruits. The fruit is eaten fresh, cooked, and as preserves or pickles. In Java, young tree leaves are eaten steamed or cooked with tough meat as a tenderizer. A gum can be obtained from the tree bark.

Trees have been grown with some success in southern Florida but are too tender to be grown in California. Ambarella fruits are found in some tropical markets.

American Gooseberry, Currant Gooseberry, Hairy Gooseberry

Sci. Name(s) : *Ribes hirtellum*
Geog. Reg(s) : America-North (U.S. and Canada)

End-Use(s)   : Fruit, Miscellaneous
Domesticated : N
Ref. Code(s)   : 15, 16, 17
Summary   : The American gooseberry is a shrubby, North American plant from 0.6-1.2 m
in height. It produces edible berries used in pies, jams, and jellies. Shrubs are suc-
cessful in producing hybrids.

American Red Raspberry

Sci. Name(s)   : *Rubus idaeus* var. *strigosus*
Geog. Reg(s)   : America-North (U.S. and Canada)
End-Use(s)   : Fruit
Domesticated : Y
Ref. Code(s)   : 17
Summary   : The American red raspberry is a robust, hardy shrub that produces edible
berries popular in pies, jams, and jellies. Shrubs are most common in the United
States and grow in the northern parts of North America. American red raspberries
remain productive for as long as 20 years. Plants thrive in deep, well-drained soil.

Angled Luffa, Angled Loofah, Singkwa Towelgourd, Seequa, Dishcloth Gourd

Sci. Name(s)   : *Luffa acutangula*
Geog. Reg(s)   : Asia-India (subcontinent), Tropical
End-Use(s)   : Fruit, Medicinal, Vegetable
Domesticated : Y
Ref. Code(s)   : 154, 170
Summary   : Angled luffa is a plant that produces edible fruits used as vegetables. This
plant is grown in India for domestic purposes. The fruit is used in chop suey. Shoots,
flowers, and young leaves are cooked and eaten as greens. The leaves and fruits are
used in unspecified medicinal preparations.

Luffas grow best in rich soils in the low humid tropics. Fruit must be harvested
while still young and tender, usually at 2 months. Fully mature fruit is bitter and
inedible. Each plant produces 15-20 fruits.

Apple

Sci. Name(s)   : *Malus sylvestris, Malus pumila*
Geog. Reg(s)   : America-North (U.S. and Canada), America-South, Europe, Mediterranean
End-Use(s)   : Beverage, Energy, Fruit, Gum & Starch, Timber
Domesticated : Y
Ref. Code(s)   : 61, 220
Summary   : The apple tree is native to the Mediterranean and has been introduced to Eu-
rope and North and South America. Europe is the leading apple producer. Most of
the apples produced are eaten raw. The rest are used for making cider, soft drinks,
and vinegars. Pectin is a valuable by-product of the cider. Its hard, strong wood
makes quality tool handles and excellent firewood.

There are three main varieties of apple trees: 1. The cider variety, with bittersweet apples, 2. The cooking variety, with moderately acidic apples, and 3. The dessert
variety with apples having a high sugar content.

An apple tree can bear fruits for about 100 years although the fruits may become commercially unprofitable. About 3.8 million MT of apples were produced in the United States in 1983. Fifty-five percent of the crop was sold fresh, 24% was processed for juice and cider, 15% was canned, 3% was dried, 2% was frozen, and 1% was used for other apple products.

Apricot

Sci. Name(s)	: *Prunus armeniaca*
Geog. Reg(s)	: America-North (U.S. and Canada), Hawaii
End-Use(s)	: Forage, Pasture & Feed Grain, Fruit, Oil
Domesticated	: Y
Ref. Code(s)	: 17, 36, 116, 154, 192, 196
Summary	: The apricot is a small, deciduous tree growing to 6 m in height that produces edible fruit. Apricots enter commercial markets as fresh, canned, or dried fruit. Fruit kernels are the source of a useful culinary oil. The expressed oilseed cake is used as cattle feed. Tree leaves provide fodder for sheep and goats. Apricot trees are hardy and their fruits ripen earlier than peaches. Trees bear after 5 years and continue for approximately 30 years. Yields are from 30-50 kg/tree of fruit. Apricots thrive on friable loams with well-drained subsoil.

In the United States, trees are grown commercially along the Pacific Coast and in some Rocky Mountain states. Apricots can be grown successfully in Hawaii at altitudes above 1,050 m. About 95% of United States apricot production is in California. In 1983, apricot production in the United States reached 86,256 MT; 39,899 MT were canned, 26,394 MT were dried, 10,703 MT were frozen, and 9,260 MT were sold fresh. Production was valued at just under $29.9 million.

Archucha, Wild Cucumber

Sci. Name(s)	: *Cyclanthera pedata*
Geog. Reg(s)	: America-Central, America-South
End-Use(s)	: Fruit, Vegetable
Domesticated	: Y
Ref. Code(s)	: 170, 220
Summary	: Archucha is a strong-smelling annual vine sometimes cultivated in tropical America for its edible fruits. These fruits are especially popular in Peru as vegetables. Plants grow in the highlands of the Andes mountains in South America. Archucha is native to Mexico.

Argus Pheasant-Tree

Sci. Name(s)	: *Dracontomelon mangiferum*
Geog. Reg(s)	: Asia-India (subcontinent), Asia-Southeast
End-Use(s)	: Fruit, Medicinal, Spice & Flavoring, Timber
Domesticated	: Y
Ref. Code(s)	: 36, 159, 170, 220
Summary	: The argus pheasant-tree is a large tree grown in India and the Malay Pennisula for its edible fruits. The fruit is sold in local markets and in the Malay Pennisula where it is eaten as a sour relish with fish. The flowers are sometimes used as condiments. The bark has been used medicinally. Timber is utilized in India and the

Malay Pennisula for making matchsticks and for house construction. The wood does not have much market potential as it lacks strength and durability.

Australian Desert Lime

Sci. Name(s) : *Eremocitrus glauca*
Geog. Reg(s) : Australia
End-Use(s) : Beverage, Fruit
Domesticated : N
Ref. Code(s) : 17, 220
Summary : The Australian desert lime is a shrubby tree whose tart fruit can be made into jams and beverages. Trees grow primarily in Australia.

Australian Nightshade

Sci. Name(s) : *Solanum aviculare*
Geog. Reg(s) : Australia
End-Use(s) : Fruit, Medicinal
Domesticated : N
Ref. Code(s) : 49, 220
Summary : Australian nightshade (*Solanum aviculare*) is a shrub reaching 4 m in height often confused with another member of its genus, the kangaroo apple (*Solanum laciniatum*). *Solanum aviculare* has pale stems and smaller fruit than the kangaroo apple. This plant produces sweet, acidic berries eaten fresh. Leaves and unripe berries are poisonous. Fruit flesh contains sucrose. This plant is being researched for the manufacture of steroid hormone drugs used in birth control and rheumatoid arthritis.

Shrubs are native to Australia where they grow along the edges of lowland forests. Plants bear year-round but fruit most heavily during the summer and autumn months.

Avocado, Alligator Pear

Sci. Name(s) : *Persea americana, Persea gratissima*
Geog. Reg(s) : Africa-South, America-Central, America-North (U.S. and Canada), America-South, Australia, Europe, Hawaii, Mediterranean, West Indies
End-Use(s) : Fruit, Oil
Domesticated : Y
Ref. Code(s) : 132, 170, 199, 220
Summary : The avocado tree reaches 16 m in height and produces a highly nutritious, high-energy content fruit. Avocados are large fruits that vary in weight, shape, skin texture, and color. Most types of avocados have soft yellow pulp with a large seed. The pulp is a popular salad fruit in most markets and is generally consumed fresh. Avocado pulp can be processed as frozen pulp used in guacamole and various sauces. It is also used to flavor ice creams. About 3-30% of an edible oil is obtained from the pulp. This oil is similar to olive oil and is very digestible.

Trees originated in Central America and spread throughout the tropics, parts of Africa, Australia, and North America. Avocados remain important food crops in Central America. Commercial production is limited mainly to parts of the United

States--particularly California and Hawaii--Cuba, Argentina, Brazil, South Africa, Israel, Mexico, and Australia.

Individual trees produce 136 kg/year of fruit. The usual yield for commercial orchards ranges from 9-18.5 MT/ha/year. Trees grown from seed take 5-7 years until harvest. Grafted trees give earlier yields usually after 3-4 years. In 1983, the United States produced 222,215 MT of avocados, valued at $108 million. In 1982, the United States imported 936 MT of avocados.

Avocado trees grow in a variety of soils with good drainage and low saline conditions. They are most successful in sandy loams. In general, trees need cool-climates for successful growth.

There are 3 main ecological races: 'Mexican' is a hardy tree able to withstand cold better than other members of its genus; 'Guatemalan' is less resistant to cold climates than the 'Mexican' and is native to the highlands of Central America; 'West Indian' is the most tender of all the races and is native to the lowlands of Central America.

Bacury

Sci. Name(s)	: *Platonia esculenta, Platonia insignis*
Geog. Reg(s)	: America-South
End-Use(s)	: Fruit, Oil, Timber
Domesticated	: N
Ref. Code(s)	: 219
Summary	: Bacury is a South American tree with edible fruits used in pastries and preserves. They are popular market items in Brazil. Bacury wood is heavy, durable, and used for flooring, finish work, and building construction. The fruit seeds provide a nondrying oil (bacury kernel oil) used in candles and soap. This tree grows in the Amazon regions of Brazil and is found in Guiana.

Balsam-Apple, Wonder-Apple

Sci. Name(s)	: *Momordica balsamina*
Geog. Reg(s)	: Asia-India (subcontinent), Asia-Southeast, Pacific Islands, Tropical
End-Use(s)	: Fruit, Medicinal
Domesticated	: Y
Ref. Code(s)	: 132, 220
Summary	: The balsam-apple tree (*Momordica balsamina*) is grown for its fruits which are eaten raw, boiled, or fried. Sap from the leaves has been used to treat stomach problems and intestinal worms. The tree is a minor member of the Momordica genus. It grows in tropical areas from India to the Philippines and New Guinea.

Banana, Plantain

Sci. Name(s)	: *Musa X paradisiaca, Musa paradisiaca* var. *sapientum, Musa sapientum*
Geog. Reg(s)	: Asia-Southeast, Tropical
End-Use(s)	: Fruit
Domesticated	: Y
Ref. Code(s)	: 17, 154, 171, 220
Summary	: The banana (*Musa X paradisiaca*) is a tropical tree grown for its edible fruits. This genus includes both the dessert and cooking bananas. They are generally

distinguished according to cultivar name. The ripe fruit is eaten fresh, boiled, baked, fried, or roasted and is made into a type of vinegar. This fruit is a good source of carbohydrates and potassium in the tropics. *Musa* X *paradisiaca* is the result of a cross between *Musa acuminata* and *Musa balbisiana*. It is an interspecific hybrid with naturally occurring clones that do not produce seed.

Banana trees are native to parts of Southeast Asia and about 200 different varieties grow in other humid, tropical areas. Bananas require fertile soils and need adequate rainfall for successful growth and fruiting. This fruit is generally larger than those of *Musa acuminata*.

Worldwide, approximately 40.7 million MT of bananas are produced each year, of which 20.8 million MT are exported, mainly to temperate countries. In 1983, the United States imported 2.1 million MT of bananas, valued at approximately $600 million.

Banana Passion Fruit, Curuba, Banana Fruit

Sci. Name(s)	: *Passiflora mollissima, Tacsonia mollissima*
Geog. Reg(s)	: America-South, Hawaii, Pacific Islands
End-Use(s)	: Beverage, Fruit, Weed
Domesticated	: Y
Ref. Code(s)	: 119, 132, 170
Summary	: The banana passion fruit (*Passiflora mollissima*) is a woody vine with edible fruit eaten raw or in sherbets and beverages. The pulp is subacid in flavor and is a good source of vitamin C, vitamin A, niacin, and carbohydrates. Vines are suited to colder conditions and do well at higher-elevations. They grow wild in the Andes mountains and have been introduced to New Zealand. The vines grow in Hawaii at elevations of 1,200-1,675 m and are noxious weeds that threaten higher elevation native flora.

Baniti

Sci. Name(s)	: *Garcinia dulcis*
Geog. Reg(s)	: Asia-Southeast
End-Use(s)	: Dye & Tannin, Fruit
Domesticated	: Y
Ref. Code(s)	: 4, 170, 220
Summary	: Baniti is a medium-sized tree with a short trunk and large leaves. The tree produces pulpy, edible fruit. The bark yields a white latex that gives a brown color when mixed with indigo. The latex from the fruit is yellow. Trees grow in Southeast Asia from the Philippines to Java.

Barbados Cherry, Acerola, West Indian-Cherry, Barbados-Cherry, West Indian Cherry

Sci. Name(s)	: *Malpighia punicifolia*
Geog. Reg(s)	: America-North (U.S. and Canada), America-South, Tropical, West Indies
End-Use(s)	: Beverage, Fruit, Ornamental & Lawn Grass
Domesticated	: Y
Ref. Code(s)	: 170, 220

Summary   : The Barbados cherry tree is grown for its fruits which are used in juice production. The cherries are also made into preserves and jellies. They are rich in ascorbic acid and are often used to enrich other fruit juices with a low vitamin C content. The unripe fruit has the highest amount of ascorbic acid. The tree grows in the West Indies and in South America up to Texas. It has been introduced into the tropics and subtropics and it makes a good hedge.

Beach Plum, Shore Plum
(See Erosion Control)

Bignay, Chinese Laurel, Salamander Tree
Sci. Name(s) : *Antidesma bunius*
Geog. Reg(s) : Asia-India (subcontinent), Asia-Southeast, Australia
End-Use(s)   : Beverage, Fruit, Medicinal, Timber
Domesticated : Y
Ref. Code(s) : 36, 170, 220
Summary   : Bignay is a small tree reaching 9 m in height. It grows wild in India and Australia. In Malaysia, the tree is cultivated for its fruit made into preserves, jellies, and a sauce for fish. The fruit is also used in brandy and syrups. The tree provides hard, reddish timber made into paper pulp. Tree bark is poisonous but has various uses in folk medicines.

Bilimbi
Sci. Name(s) : *Averrhoa bilimbi*
Geog. Reg(s) : Asia-Southeast, Tropical
End-Use(s)   : Fruit
Domesticated : Y
Ref. Code(s) : 170
Summary   : Bilimbi is a small Malaysian tree that reaches 15 m in height and is grown for its acidic, edible fruits, which are used for making pickles, curries, and preserves. They resemble cucumbers in appearance. Bilimbi trees are found throughout the tropics.

Bira Tai
Sci. Name(s) : *Garcinia multiflora*
Geog. Reg(s) : Asia-Southeast
End-Use(s)   : Fruit, Oil
Domesticated : Y
Ref. Code(s) : 220
Summary   : The fruit of the bira tai tree is used as a lemon substitute, and an oil expressed from the seeds is used for lighting. Bira tais grow throughout Southeast Asia. There is little additional information concerning the economic potential of the fruit or the expressed oil.

Black Currant, European Black Currant
Sci. Name(s) : *Ribes nigrum*
Geog. Reg(s) : America-North (U.S. and Canada), Asia, Asia-Central, Europe

End-Use(s)   : Beverage, Fruit, Medicinal
Domesticated : Y
Ref. Code(s)   : 15, 16, 17, 69, 220
Summary   : The black currant is a shrub whose edible fruit is cultivated as a garden and fresh produce market crop. The fruit is popular in jams and can be processed into drinks with a high vitamin C content. The fruit is also fermented to make a liqueur called cassis. Black currants are rarely eaten raw. The dried leaves have been used in home cough remedies.

   Shrubs thrive on heavy, nitrogen-rich soils. They are found in Europe, northern and Central Asia, the Himalayas, and North America. Plants have a very strong and unpleasant odor.

Black Mulberry

Sci. Name(s)   : *Morus nigra*
Geog. Reg(s)   : Asia, Asia-China, Eurasia, Europe, Mediterranean, Tropical
End-Use(s)   : Beverage, Dye & Tannin, Fruit, Medicinal
Domesticated : Y
Ref. Code(s)   : 17, 170, 205, 220
Summary   : The black mulberry (*Morus nigra*) tree reaches 10 m in height and is cultivated for its fruit, which contains 10% sugar, 2% malic acid, and vitamin C. Mulberries are mostly used in homemade products such as, wines, jams, conserves, and a dyestuff. The fruit is made into a syrup for sore throats.

   Trees thrive at higher-altitudes in the tropics. They require protection from cold winds and frosts and need warm, loam soils. They originated in China and Japan and are cultivated in Europe and the Near East.

Black Raspberry

Sci. Name(s)   : *Rubus occidentalis*
Geog. Reg(s)   : America-North (U.S. and Canada)
End-Use(s)   : Fruit
Domesticated : Y
Ref. Code(s)   : 15, 17, 65
Summary   : The black raspberry is a North American shrub with edible berries eaten raw or in pies, jams, jellies, and ice creams. Plants prefer deep soils but tolerate lighter soils. A good planting of black raspberries fruits for 5-10 years.

Black Sapote, Zapte Negro

Sci. Name(s)   : *Diospyros digyna*
Geog. Reg(s)   : America-Central, Hawaii, West Indies
End-Use(s)   : Fruit, Timber
Domesticated : Y
Ref. Code(s)   : 1, 167
Summary   : The black sapote tree grows in Mexico and is cultivated for its ebony wood and sweet fruit. The wood is hard, close-grained, and valuable for cabinet making and furniture. Black sapotes were cultivated on a small scale in the West Indies and Hawaii. If grown in deep, rich, moist soil, black sapotes can reach heights of 15-18 m. Trees mature in 5-6 years and fruit regularly and heavily.

Blackberry

Sci. Name(s) : *Rubus* spp.
Geog. Reg(s) : America-North (U.S. and Canada)
End-Use(s) : Fruit
Domesticated : Y
Ref. Code(s) : 17
Summary : The blackberry is from a genus of berry-producing shrubs native to North America. Most shrubs are wild but several cultivated varieties have been produced. Blackberries can be described in 5 basic groups: plants with erect growth; eastern trailing shrubs with or without hairy canes; southeastern shrubs with trailing hairy canes; shrubs common along the Pacific Coast; and semievergreen trailing shrubs. Shrubs provide good windbreaks for other crops and large yields of edible berries.

In the United States, blackberry culture is limited by the extreme climates of the northern and plains states and by the droughts and dry heat of the southwestern states. They grow in most well-drained, humus-rich soil. Plants respond to fertilising. Canes are biennial. Fruits grown for commercial use can be harvested mechanically.

Boston Marrow, Pumpkin, Winter Squash, Squash, Marrow
(See Vegetable)

Brazilian Guava, Guinea Guava

Sci. Name(s) : *Psidium guineense*
Geog. Reg(s) : America-South, Tropical, West Indies
End-Use(s) : Fruit
Domesticated : Y
Ref. Code(s) : 17, 167, 170, 207
Summary : The Brazilian guava is a low-growing bush or tree indigenous to the West Indies and parts of tropical America. Trees produce small, edible fruits occasionally collected and used to make jelly. The fruit quality is too poor to consider it as a promising horticultural crop.

Breadfruit
(See Vegetable)

Bullock's-Heart, Corazon, Custard-Apple

Sci. Name(s) : *Annona reticulata*
Geog. Reg(s) : America-Central, Tropical, West Indies, Worldwide
End-Use(s) : Fruit, Miscellaneous
Domesticated : Y
Ref. Code(s) : 148, 170
Summary : Bullock's-heart is a small tree native to Central America and the West Indies cultivated in tropical areas for its bland fruit. This fruit is of good quality but not as highly-favored as the cherimoya. Trees withstand diverse conditions and thrive through long, dry periods. Best results are from trees planted in well-drained soils. Bullock's-heart is valued as a vigorous, resistant rootstock for other members of the Annona genus.

Cacao Blanco, Nicaraguan Cacao, Patashiti, Bacao
(See Spice & Flavoring)

Cainito, Star-Apple

Sci. Name(s)	: *Chrysophyllum cainito*
Geog. Reg(s)	: America-Central, Tropical, West Indies
End-Use(s)	: Fruit, Ornamental & Lawn Grass
Domesticated	: Y
Ref. Code(s)	: 132, 170
Summary	: Cainito is an ornamental tree whose fruits have edible pulp that is a good source of carbohydrates. Fresh pulp is eaten after the skin is removed.

This tree is native to the West Indies and Central America. It grows to 12 m in height and makes an attractive ornamental. Cainito thrives in hot, tropical lowland climates and is fairly frost-tolerant.

Camel's-Foot, Gemsbok Bean
(See Beverage)

Candletree
(See Forage, Pasture & Feed Grain)

Cantaloupe, Melon, Muskmelon, Honeydew, Casaba

Sci. Name(s)	: *Cucumis melo*
Geog. Reg(s)	: America-North (U.S. and Canada), Asia, Asia-China, Asia-India (subcontinent), Europe, Temperate, Tropical
End-Use(s)	: Fruit, Oil, Vegetable
Domesticated	: Y
Ref. Code(s)	: 148, 170
Summary	: The cantaloupe is an annual vine with edible melons eaten as dessert fruits, preserves, or vegetables. Melon seeds are edible and yield an edible oil. The most popular cultivars are: the cantaloupe melon of Europe; the musk melon of the United States; the casaba or winter melon (includes the 'Honeydew melon.'); and other cucumber-shaped melons grown mostly in India, China and Japan as vegetables.

Melons are grown across a wide climatic range but the bulk of production is in warm temperate countries. They require moderate rain, full sun, and rich, loamy soils with low acidity. They are susceptible to frost and mildews. Melons are harvested 3-4 months after planting.

Cape-Gooseberry, Peruvian-Cherry, Peruvian Groundcherry

Sci. Name(s)	: *Physalis peruviana*
Geog. Reg(s)	: America, Tropical
End-Use(s)	: Fruit
Domesticated	: Y
Ref. Code(s)	: 36, 92, 170
Summary	: The cape-gooseberry is a shrubby herb from 61-91 cm in height that produces edible berries stewed and made into jam. These berries are an excellent source of

vitamins A and C, a good source of vitamin B complex, high in protein, phosphorous and iron, but fairly low in calcium.

Herbs are native to tropical America and are cultivated in other tropical countries. Plants are most successful on well-drained, sandy soils, in sheltered, sunny areas. First season yields are highest but plants continue to grow and fruit regularly for 2-3 years. Fruit ripens best in areas with dry winters.

Cashew
(See Nut)

Ceriman, Monstera
(See Ornamental & Lawn Grass)

Ceylon-Gooseberry, Ketembilla, Quetembila
Sci. Name(s) : *Dovyalis hebecarpa*
Geog. Reg(s) : America-North (U.S. and Canada), Asia, Asia-India (subcontinent), Tropical
End-Use(s) : Fruit
Domesticated : Y
Ref. Code(s) : 167, 220
Summary : The Ceylon-gooseberry is a bushy shrub from 4.5-6.1 m in height that grows throughout tropical Asia, particularly India and Sri Lanka, and is cultivated for its fruit eaten fresh or in preserves. In the United States, there is small scale commercial cultivation in southern Florida and California. Shrubs are not drought-hardy. They require plenty of moisture for satisfactory growth and fruit production. Shrubs are injured by high temperatures.

Chayote
(See Vegetable)

Chempedale, Chempedak
Sci. Name(s) : *Artocarpus integer*
Geog. Reg(s) : Asia-Southeast
End-Use(s) : Fruit
Domesticated : Y
Ref. Code(s) : 170
Summary : Chempedale is a wild Malaysian tree with edible fruit pulp. This pulp and immature fruit are eaten in soups. Chempedale seeds are edible if cooked. This tree is under minor cultivation in parts of the Malay Archipelago.

Cherimoya
Sci. Name(s) : *Annona cherimola*
Geog. Reg(s) : America-North (U.S. and Canada), America-South
End-Use(s) : Beverage, Fruit
Domesticated : Y
Ref. Code(s) : 148, 170
Summary : Cherimoya is a small tree common in the Andean valleys of Ecuador and Peru grown for its edible fruits, which contain 18% sugar. They are used in ice creams,

sherbets, and exotic drinks. Cherimoyas are seldom grown in commercial orchards because of their perishability.

Trees do not survive the low, hot tropics and are only grown above 900 m in elevation. In the United States, a few varieties of cherimoya trees have been grown with some success in southern California. In areas with temperatures of 24-29 degrees Celsius, the best fruit crops are obtained by hand pollination and light pruning.

Cherry Plum, Myrobalan Plum
(See Ornamental & Lawn Grass)

Chicle, Sapodilla, Naseberry, Nispero, Chicle Tree
(See Gum & Starch)

Chilean Strawberry

Sci. Name(s)	: *Fragaria chiloensis*
Geog. Reg(s)	: America-North (U.S. and Canada), America-South
End-Use(s)	: Fruit
Domesticated	: Y
Ref. Code(s)	: 154, 220
Summary	: The Chilean strawberry is an herb cultivated in the Andean regions of South America for its edible fruit. Chilean strawberrys ripen from June to September. They are native to the Pacific regions of North and South America.

Chinese Jujube

Sci. Name(s)	: *Ziziphus jujuba*
Geog. Reg(s)	: America-North (U.S. and Canada), Asia-China, Mediterranean, Temperate
End-Use(s)	: Fruit, Medicinal
Domesticated	: Y
Ref. Code(s)	: 36, 170, 236
Summary	: Chinese jujube is a temperate tree cultivated in the relatively dry parts of the world for its edible fruits. This fruit is an excellent source of vitamin C and is eaten fresh, as a dried dessert fruit, or as preserves. Dried jujubes are exported from China. Bark and seeds are used as a remedy for diarrhea and the roots for treating fevers. The powdered roots and a leaf poultice are used to treat wounds. The bark is used as a digestive tonic. The fruits are similar to dates in appearance and taste and are grown in parts of North America and the Mediterranean.

Cimarrona, Mountain Soursop

Sci. Name(s)	: *Annona montana*
Geog. Reg(s)	: Tropical, West Indies
End-Use(s)	: Beverage, Fruit
Domesticated	: Y
Ref. Code(s)	: 148, 170
Summary	: The cimarrona tree is grown in the West Indies for its edible fruit. This fruit is soft and perishable, which makes extensive transportation difficult. For easier transportation, most fruits are processed and sold in nectar form. Only limited information has been found concerning this tree's potential as a tropical fruit crop.

Citron

Sci. Name(s) : *Citrus medica*
Geog. Reg(s) : Europe, Mediterranean, Tropical
End-Use(s) : Fruit, Spice & Flavoring
Domesticated : Y
Ref. Code(s) : 170
Summary : The citron is a shrub or small tree reaching 3 m in height which produces edible fruit economically unimportant as a fresh fruit. The fruit peel is fermented in brine and candied or used as flavoring for cakes and confectionery. It contains the glucoside, hesperidin. Citrons are the most tender of the Citrus genus. Trees are grown in most tropical countries but commercial plantings are limited to Italy, Greece, and Corsica.

Clausena dentata

Sci. Name(s) : *Clausena dentata, Clausena willdennovii, Amyris dentata*
Geog. Reg(s) : Asia, Asia-India (subcontinent), Tropical
End-Use(s) : Fruit, Timber
Domesticated : N
Ref. Code(s) : 29, 67, 170, 220
Summary : Clausena dentata is a shrub or small tree with cherry-sized fruit similar in taste to the black currant berry. These fruits are eaten in tropical Asia and India. The timber is white, hard, close-grained, and resembles boxwood. Tree leaves have an odor similar to anise.

Cochineal Cactus
(See Dye & Tannin)

Cocona

Sci. Name(s) : *Solanum hyporhodium*
Geog. Reg(s) : America-South, Tropical, West Indies
End-Use(s) : Fruit
Domesticated : Y
Ref. Code(s) : 92, 170
Summary : Cocona is a shrub that reaches 1.5 m in height and is grown for its edible fruits. It is native to the upper Amazon region in South America, is grown experimentally in Puerto Rico, and shows promise in other geographical areas.

Shrubs are successful on light, well-drained soils in exposed, sunny areas. In the tropics, they are grown from sea level to 1,219 m in elevation. Plants reproduce readily by seed. Fruit is harvested 7 months after planting.

Coconut
(See Oil)

Common Fig, Adriatic Fig
Sci. Name(s) : *Ficus carica*
Geog. Reg(s) : Europe, Mediterranean, Temperate

End-Use(s) : Beverage, Forage, Pasture & Feed Grain, Fruit, Medicinal, Ornamental & Lawn Grass
Domesticated : Y
Ref. Code(s) : 132, 148, 154, 170, 220
Summary : The common fig tree bears sweet, edible fruit--popular fresh, dried or in confectionery. Figs are high in calcium, sugar, iron, and copper. Fruit pulp is a good source of carbohydrates, potassium, and vitamin A. It is also brewed as an alcoholic beverage or used as a laxative. The leaves are used as potherbs or fed to livestock. The tree is also cultivated for shade. Many varieties are known but only a few are adopted to the tropics. Because of insect damage by entering the "eye" near maturity, varieties with closed "eye" are often selected. There are 2 major varieties of commerce. One type produces a brown fruit crop twice a year while the 2nd type produces white and golden fruit crops once a year. Trees have been known to live as long as 200 years.

Although most fig trees survive a wide range of temperatures, trees grown in warm climates are most successful. They do not survive the low, humid tropics. For best quality fruit, temperatures should not exceed 38 degrees Celsius. Trees tolerate deep, dry soil but prefer moisture. Nitrogen increases the yield while phosphorus and potash do not. Fruit is pollinated by hand or by wasps. Trees are native to the Mediterranean and are grown in temperate Europe.

Common Guava, Lemon Guava

Sci. Name(s) : *Psidium guajava*
Geog. Reg(s) : America-North (U.S. and Canada), America-South, Asia-India (subcontinent), Hawaii, Pacific Islands, Subtropical, Tropical
End-Use(s) : Beverage, Dye & Tannin, Fruit, Medicinal, Weed
Domesticated : Y
Ref. Code(s) : 132, 170
Summary : The guava is a tree growing to 10 m in height that bears edible fruits that are popular raw, stewed in pies, and as jams, jellies, and preserves. Fruit pulp is processed into a fresh juice, juice concentrate, and a nectar that contains nearly twice the amount of vitamin C as fresh orange juice. One of its prinicipal commercial uses is jelly. In some countries, guava leaves have been used to control diarrhea, and in the leather industry for dyeing and tanning.

Trees are indigenous to the American tropics where they range from sea level to 1,525 m, as both wild and cultivated plants. Because trees are very adaptable, they have been declared troublesome weeds in some areas. Trees are most successful in warm to hot, frost-free, tropical and subtropical climates in areas receiving low to medium rainfall.

Guava trees bear fruit about 2 years after transplanting and fruit may be picked for as long as 30 years. Fruits mature about 5 months after flowering. A good yield is from 25-30 MT/ha/year of fruit. Guavas are grown commercially in India, Hawaii, Florida, Brazil, and Guyana. Guavas are considered very adaptive.

Common Persimmon, Persimmon, American Persimmon

Sci. Name(s) : *Diospyros virginiana*
Geog. Reg(s) : America-North (U.S. and Canada)

End-Use(s) : Forage, Pasture & Feed Grain, Fruit, Miscellaneous, Ornamental & Lawn
Grass
Domesticated : Y
Ref. Code(s) : 59, 65, 192, 199, 220
Summary : The common persimmon is a slow-growing tree of the eastern and southeastern United States which reaches 15-30 m in height grown for its edible fruits. The fruit is eaten fresh when fully ripe as a dessert or for other culinary purposes. Immature fruit is too bitter for consumption. Poor quality fruit is fed to livestock.

Persimmon trees have a hard, dark brown wood inferior as lumber. They are grown as ornamentals, shade trees, and root stock for kaki (*Diospyros kaki*). They do well under varying conditions and in sandy, clay, and stony soils. The quality of the fruit diminishes in cool-climates. In the United States, better quality persimmons grow in the Mississippi River Valley area.

Common Red Ribes, Aka-Suguri, Garden Currant

Sci. Name(s) : *Ribes sativum*
Geog. Reg(s) : America-North (U.S. and Canada), Asia, Europe
End-Use(s) : Fruit
Domesticated : Y
Ref. Code(s) : 15, 188, 213, 220
Summary : Common red ribes is a shrub from 1.2-1.8 m in height cultivated for its edible berries. The berries are sold as domestic fresh market fruit. Fruits are used to make jams and preserves and are eaten fresh. Shrubs are propagated by hardwood cuttings. Healthy shrubs may yield at least 10-12 fruit crops. Common red ribes is native to Europe and has been introduced into the United States and Japan.

Costa Rican Guava, Wild Guava

Sci. Name(s) : *Psidium friedrichsthalianum*
Geog. Reg(s) : America-Central, America-North (U.S. and Canada), Asia-Southeast
End-Use(s) : Fruit, Ornamental & Lawn Grass
Domesticated : Y
Ref. Code(s) : 36, 167, 170, 207
Summary : The Costa Rican guava tree reaches 7.6 m in height and has edible, acidic fruits used in jellies and pies. Trees are grown as ornamentals. Costa Rican guavas are slow in growth habit and do not fruit heavily. Successful fruit crops are obtained after a cold winter.

Trees are native to Central America and introduced to Singapore, Malaysia, and the United States, especially southern Florida and California. In the United States, Costa Rican guavas are economically unpromising.

Cowberry, Red Whortleberry, Mountain Cranberry, Lingon Berry, Rock Cranberry, Lingen

Sci. Name(s) : *Vaccinium vitis-idaea*
Geog. Reg(s) : America-North (U.S. and Canada), Europe
End-Use(s) : Fruit, Ornamental & Lawn Grass
Domesticated : Y
Ref. Code(s) : 17, 42, 220, 236

Summary   : The cowberry is an evergreen shrub that grows at high-elevations in Europe and North America. It is difficult to establish in the garden. It produces edible berries that are occasionally used in pies, jams, and jellies. Plants are also valued as ornamentals because of their small white flowers and red fruit. Cowberrys persist on acidic soils and usually dominate their surroundings.

Cranberry, Large Cranberry, American Cranberry

Sci. Name(s) : *Vaccinium macrocarpon, Oxycoccus macrocarpon*
Geog. Reg(s) : America-North (U.S. and Canada), Temperate
End-Use(s)   : Beverage, Fruit
Domesticated : Y
Ref. Code(s) : 17, 154, 220, 236
Summary   : The cranberry is a shrub whose edible berries are of considerable economic value. Cranberries are eaten as a sauce with turkey or chicken and can be eaten in tarts and pies. They are also used for juice, relishes, and eaten fresh. Cranberries require a large initial capital investment as well as careful surveillance of weather and insects; plants tend to yield a low profit margin.

Plants are cultivated in temperate countries. Cranberries have brilliant autumn foliage. They require acidic soil and plenty of moisture. Most varieties are self-pollinating. They are grown commercially in carefully maintained bogs where plants will produce berries for a century or longer.

In 1983, the United States produced 137,500 MT of cranberries, valued at $154 million. Eighty-seven percent of the cranberries were used for processing.

Cuajilote, Cuachilote

Sci. Name(s) : *Parmentiera aculeata*
Geog. Reg(s) : America-Central
End-Use(s)   : Fruit, Medicinal
Domesticated : N
Ref. Code(s) : 220, 232
Summary   : Cuajilote is a tree that produces edible fruits that are eaten raw, cooked, preserved, and pickled. A decoction of the tree roots is considered to be diuretic and has been used in the treatment of dropsy. Trees are found mainly in Central America, especially Mexico.

Cupuacu

Sci. Name(s) : *Theobroma grandiflorum*
Geog. Reg(s) : America-South, Tropical
End-Use(s)   : Beverage, Fruit
Domesticated : N
Ref. Code(s) : 170
Summary   : Cupuacu is a tree with edible fruit eaten fresh or used in drinks. Trees thrive in the rain forests of tropical America from sea level to 990 m in elevation. Cupuacu is native to Brazil.

Currant Tomato

Sci. Name(s) : *Lycopersicon pimpinellifolium*

Geog. Reg(s) : America-Central, America-South, Hawaii
End-Use(s) : Fruit, Miscellaneous
Domesticated : Y
Ref. Code(s) : 132, 154, 170, 214
Summary : The currant tomato (*Lycopersicon pimpinellifolium*) is an herb used primarily in South America for preserves and canning. It grows wild in Hawaii. Plants thrive in hot climates with cool nights on rich, well-drained soil. It is often crossed with the tomato (*Lycopersicon esculentum*) to produce a disease-resistant cultivar.

Curuba, Carua, Casabanana
(See Ornamental & Lawn Grass)

Damson Plum, Tart Damson Plum, Bullace Plum
Sci. Name(s) : *Prunus domestica* subsp. *insititia*
Geog. Reg(s) : America-North (U.S. and Canada), Asia, Europe, Temperate
End-Use(s) : Fruit, Miscellaneous
Domesticated : Y
Ref. Code(s) : 17, 89, 94, 178
Summary : The damson plum (*Prunus domestica* subsp. *institia*) is a small tree reaching 7.5 m in height grown for its tart, edible fruits used mainly in jams and jellies. Trees are often used as rootstocks for more commercially important plums. Trees grow wild in parts of Europe and Asia and are common in many temperate areas of the United States.

Date Palm
Sci. Name(s) : *Phoenix dactylifera*
Geog. Reg(s) : Africa-North, America-South, Asia, Subtropical
End-Use(s) : Beverage, Fiber, Fruit, Sweetener
Domesticated : Y
Ref. Code(s) : 132, 148, 171
Summary : The date palm reaches to 30 m in height and has a variety of uses. Its soft, brown fruits are about 5 cm in length and are usually eaten fresh or dried. Date pulp is a good source of calcium, phosphorous, and carbohydrates. Palm trunks can be tapped for a sap that is made into palm wine. Palm leaves have been used as thatch and fiber. Date yields have been recorded at approximately 20-100 kg/year. Large yields are not usually obtained before 5-8 years.

Most palms grow in dry, subtropical climates on moist soils, although they will tolerate dry, tropical soils. Northeastern Brazil appears to be a promising date palm producing region. Presently, Iraq supplies about one third of the world's dates. Palms are subtropical in origin, and may have originated in Mesopotamia. They are widely cultivated in the arid regions of the Middle East, particularly in southern Arabia, North Africa, and the areas bordering the Sahara.

Desert Date
Sci. Name(s) : *Balanites aegyptiaca*
Geog. Reg(s) : Africa-West, Asia-Southeast
End-Use(s) : Energy, Fruit, Medicinal, Miscellaneous, Oil

Domesticated : Y
Ref. Code(s) : 67, 220
Summary : The desert date is a small evergreen tree that produces edible, bittersweet fruit. In African medicine, the fruit is used to treat liver and spleen problems. The gray bark is used as a fish poison, and when combined with the desert date fruit and roots, is lethal to snails, some types of fish, and tadpoles. Date seeds yield a fixed oil. Fruit kernels have been filled with gunpowder and used as fireworks. Fruit pulp has been used to clean silk in Rajputana. Its moderately hard, yellowish-white wood is used for walking sticks and fuel.

Desert dates grow on the dry, wooded grasslands and alluvial flats of tropical Africa and Burma.

Dewberry

Sci. Name(s) : *Rubus* spp.
Geog. Reg(s) : America-North (U.S. and Canada), Temperate
End-Use(s) : Fruit
Domesticated : Y
Ref. Code(s) : 17, 65
Summary : The dewberry is a shrub with ground-trailing canes that produce edible fruit. Fruit ripens quickly and is picked when glossy black in color. Because they do not remain fresh more than 3 days, their market potential is limited.

Cultivated shrubs need support for best production. Most planting is done in the spring. Dewberries are propagated by cane tips and root cuttings. Shrubs are found throughout North America and other temperate areas.

Durian

Sci. Name(s) : *Durio zibethinus*
Geog. Reg(s) : Asia, Asia-India (subcontinent), Asia-Southeast, Tropical
End-Use(s) : Fruit, Vegetable
Domesticated : Y
Ref. Code(s) : 4, 87, 167, 170
Summary : Durian is a tree that reaches 37 m in height in Malaysia and Southeast Asia. It bears large, heavy fruits with an unpleasant odor but sweet taste. Ripe fruit is made into sauces and unripe fruits are eaten as vegetables. Fruit seeds are eaten dried, roasted, or fried in coconut oil. Since seeds quickly lose their viability, durian is difficult to establish.

The tree fruits after 7 years and the fruit takes 3 months to ripen. There are 2 crops/year for 50 years. Production in India is 100 fruits/tree. These trees are most successful in tropical climates on deep, rich, moist soil.

Dwarf Banana, Pisong Jacki

Sci. Name(s) : *Musa acuminata, Musa cavendishii*
Geog. Reg(s) : Africa-East, Asia-India (subcontinent), Asia-Southeast, Australia, Pacific Islands, Tropical
End-Use(s) : Fiber, Fruit
Domesticated : Y
Ref. Code(s) : 17, 132, 171, 220

Summary : The dwarf banana tree is one of the parents of many edible varieties of banana but lacks the quality of the improved hybrids. This tree produces edible fruit eaten fresh, boiled, steamed, roasted, baked, or fried. In Malaysia, fishermen use fiber from the leaves for cordage.

Dwarf bananas grow in most tropical areas and are common from Burma to the Malay Pennisula and Archipelago, New Guinea, Australia, and Samoa. Other forms are found in East Africa, southern India, and the Philippines. Trees are perennial herbs from 2-9 m in height and thrive in hot, lowland tropical climates to 1,000 m in altitude. Bananas prefer fertile, well-drained soil. Fruit is harvested 12-18 months from planting. Dwarf bananas are often crossed with *Musa balbisiana* to produce a better adapted and flavored fruit. All edible fruited clones are derived from *Musa acuminata*.

Eastern Elderberry, American Elderberry, American Elder

Sci. Name(s) : *Sambucus canadensis*
Geog. Reg(s) : America-North (U.S. and Canada)
End-Use(s) : Forage, Pasture & Feed Grain, Fruit, Insecticide, Medicinal, Spice & Flavoring
Domesticated : Y
Ref. Code(s) : 17, 119, 192, 223
Summary : The eastern elderberry is a shrub from 3-3.6 m in height that produces large, sweet, edible berries that are used for making wines, sauces, jellies, and pies. A decoction of the flowers has been used as a treatment for stomach complaints, as a diuretic and stimulant. Flowers are fried and eaten like fritters. Flower buds have been used as substitute flavoring for capers. Dried leaves have certain insecticidal properties. Elderberries are also important foodstuffs to wild birds.

Plants grow along streams and in low places on rich, alluvial soils. They flower in July and fruit in September. In North America, their natural habitat extends from Nova Scotia, south to Florida and Texas.

Egg-Fruit-Tree, Canistel

Sci. Name(s) : *Pouteria campechiana, Lucuma nervosa, Lucuma salicifolia*
Geog. Reg(s) : America-North (U.S. and Canada), America-South, West Indies
End-Use(s) : Fruit
Domesticated : Y
Ref. Code(s) : 15, 151, 170
Summary : The egg-fruit-tree is an evergreen that reaches 23 m in height, cultivated in South America for its edible fruits. This fruit is used in salads and the pulp in pies, puddings, ice creams, or jams. Fruiting season depends on when trees are planted. Trees fruit heavily and fairly regularly. They are native to northeastern South America and naturalized in the West Indies and the United States, particularly southern Florida.

Egyptian Carissa

Sci. Name(s) : *Carissa edulis*
Geog. Reg(s) : Africa-West, Tropical
End-Use(s) : Fruit, Ornamental & Lawn Grass
Domesticated : Y

214

Ref. Code(s) : 170, 207
Summary : Egyptian carissa is a tropical African shrub with edible fruits. This plant has potential as an ornamental hedge.

Emblic, Myrobalan, Emblica

Sci. Name(s) : *Phyllanthus emblica*
Geog. Reg(s) : Asia, Asia-Southeast
End-Use(s) : Dye & Tannin, Fruit, Medicinal, Ornamental & Lawn Grass
Domesticated : Y
Ref. Code(s) : 16, 154, 170
Summary : Emblic is a tall, attractive tree to 9 m in height that is popular because of its handsome foliage, which resembles that of fir or hemlock. Trees produce fruit that is eaten boiled or as preserves. Leaves and bark are rich in tannins. In the Mascarene Islands, Asia, and the East Indies, emblic fruit is dried for use in unspecified medicinal preparations.

English Gooseberry, European Gooseberry

Sci. Name(s) : *Ribes uva-crispa, Ribes grossularia, Grossularia uva-crispa*
Geog. Reg(s) : America-North (U.S. and Canada), Europe
End-Use(s) : Beverage, Fruit
Domesticated : Y
Ref. Code(s) : 16, 17, 220
Summary : The English gooseberry is a low-growing shrub with tiny, yellow fruit. English gooseberries are usually sold as domestic market crops. Fruit can be made into wine. Plants tend to escape from cultivation and become naturalized as weeds. Caution should be used when growing the English gooseberry in the United States because of 2 harmful diseases, gooseberry rust and gooseberry mildew.

European Strawberry, Strawberry

Sci. Name(s) : *Fragaria vesca*
Geog. Reg(s) : America-North (U.S. and Canada), Asia, Europe
End-Use(s) : Beverage, Dye & Tannin, Fruit, Medicinal
Domesticated : Y
Ref. Code(s) : 205, 220
Summary : The European strawberry is an herb with a pleasantly flavored, edible fruit. It prefers moist soils and grows primarily in the woody regions of eastern North America. It bears fruit twice a year. The fruits also have medicinal uses as laxatives and diuretics. Some of their constituents are tannins, flavonoids, organic acids, vitamin C, and sugars. A decoction of the roots was formerly used in the treatment of gonorrhea. Its leaves can be brewed as tea.

In 1983, the United States produced 264,954 MT of strawberries for the fresh market, valued at $309.2 million. Additionally, 139,877 MT of strawberries were produced for the processing market, valued at $97.1 million.

Feijoa

Sci. Name(s) : *Feijoa sellowiana*
Geog. Reg(s) : America-North (U.S. and Canada), America-South, Subtropical, Tropical

End-Use(s) : Fruit
Domesticated : N
Ref. Code(s) : 154, 162, 167, 170, 214, 220
Summary : Feijoa is a wild plant with fruit stewed and made into jams and jellies. Fully ripened fruit can be eaten fresh. This fruit spoils quickly in hot, humid weather. In cool storage, they last about a month.

Plants grow wild in Brazil, Uruguay, Paraguay, and Argentina in subtropical areas. In the United States, several unsuccessful attempts were made to grow feijoa in Florida. In California, plants will grow but yields are poor. Under proper conditions, feijoa fruits 4-5 years after planting. Plants thrive in sandy, humus-rich loams but tolerate the red clay adobe soils of California.

Fox Grape
Sci. Name(s) : *Vitis labrusca*
Geog. Reg(s) : America-North (U.S. and Canada)
End-Use(s) : Beverage, Dye & Tannin, Forage, Pasture & Feed Grain, Fruit, Oil, Ornamental & Lawn Grass
Domesticated : Y
Ref. Code(s) : 17, 84, 154, 213, 220
Summary : The fox grape is a climbing tree whose edible fruits are eaten raw or made into grapejuice. The fruit is popular in sweets, ice cream, jellies, and wines. The pressed fruit is fed to livestock. Grape seeds yield an economically important oil and tannin used in cream of tartar.

Plants are native to eastern North America. Vines are grown as ornamentals. Because of their musky odor, they are considered inferior to European grapes. However, fox grapes are not affected by Phylloxera, a harmful disease that usually affects grape vines.

Gamboge Tree
Sci. Name(s) : *Garcinia tinctoria, Garcinia xanthochymus*
Geog. Reg(s) : Asia-India (subcontinent), Asia-Southeast, Tropical
End-Use(s) : Fruit
Domesticated : Y
Ref. Code(s) : 220
Summary : The gamboge tree is grown for its edible fruits eaten mainly as breakfast fruits. It is native to southern India and Malaysia and is grown in most tropical areas. There is limited information concerning its commercial cultivation.

Garden Strawberry
Sci. Name(s) : *Fragaria* X *ananassa, Fragaria chiloensis* X *Fragaria virginiana*
Geog. Reg(s) : America-North (U.S. and Canada), Hawaii
End-Use(s) : Fruit
Domesticated : Y
Ref. Code(s) : 14, 154
Summary : The garden strawberry is a hybrid herb that bears an edible, large-fruited variety of strawberry eaten fresh or in confectionery. It is a good source of vitamin C. In the United States, plants grow as a small scale commercial crop in Hawaii,

particularly at elevations from sea level to 305 m. On Maui and Hawaii, plants are grown with some success at altitudes from 1,067-1,828 m. Production is hindered by the Chinese rose beetle. There is limited information on the economics of this particular variety of strawberry.

Giant Granadilla, Barbardine

Sci. Name(s) : *Passiflora quadrangularis*
Geog. Reg(s) : America-South, Hawaii, Tropical
End-Use(s) : Beverage, Fruit, Spice & Flavoring, Tuber, Vegetable
Domesticated : Y
Ref. Code(s) : 4, 155, 170, 192
Summary : Giant granadilla is a robust, perennial vine that grows in hot, moist climates. It produces edible fruit whose pulp is a good source of phosphorous and vitamin A. The pulp is used to flavor ice creams and sherbets, to make beverages, and to make jams. Green, unripe fruits can be boiled and eaten as vegetables. The tuberous root is thought to be poisonous but is purportedly eaten in Jamaica as a substitute for yams.

Vines are native to tropical South America. Outside of South America, poor fruit-set often occurs with hand-pollination the general recourse. In the Hawaiian islands, however, there are one or more insect pollinators present that promote satisfactory fruit-setting. Vines are usually propagated from seed or by cuttings. Fruit crops are grown mainly for domestic consumption.

Gilo

Sci. Name(s) : *Solanum gilo*
Geog. Reg(s) : Africa-Central, America-South
End-Use(s) : Fruit, Spice & Flavoring, Vegetable
Domesticated : Y
Ref. Code(s) : 170, 237
Summary : Gilo is a plant grown in Africa and South America as a fruit crop. Its bitter, immature fruits are cooked and eaten as vegetables or used as seasoning. In Nigeria, the young shoots are chopped and used as bitter flavoring in soups. Plants were introduced into South America from Central Africa. They are harvested green approximately 90-100 days after planting.

Governor's-Plum, Ramontchi, Rukam
(See Ornamental & Lawn Grass)

Grapefruit

Sci. Name(s) : *Citrus paradisi*
Geog. Reg(s) : Africa-South, America-North (U.S. and Canada), America-South, Mediterranean, Subtropical, Tropical, West Indies
End-Use(s) : Beverage, Essential Oil, Fruit
Domesticated : Y
Ref. Code(s) : 132, 170, 180
Summary : The grapefruit tree produces a commercially important breakfast fruit eaten fresh, as canned segments, or juice. The pulp is an excellent source of vitamin C and a good source of potassium. The fruit peel yields an essential oil used as a food

flavoring, in soft drink powders, gelatin desserts, and candies. Small amounts are used in lotions and colognes.

Trees are grown in hot, lowland tropical climates or hot, subtropical climates with a well-distributed rainfall. In the United States, crops grow mainly in Florida, California, Texas, and Arizona. Important exporters are Israel, South Africa, the West Indies, and Brazil.

Two varieties are the most commonly planted: the 'Marsh' variety with 8 seeds/fruit and the pink-fleshed 'Thompson' variety, with 3-5 seeds/fruit. There are 2 popular red-fleshed cultivars that arose as mutants from the 'Thompson' variety. These are the 'Ruby' and 'Webb' grapefruits. Grapefruit trees grow to 15 m in height.

United States grapefruit production in 1983 was 2.2 million MT valued at $190 million. Forty-seven percent of the grapefruits were processed and the remaining 53% were sold as fresh fruit.

Green Sapote

Sci. Name(s) : *Pouteria viridis, Calocarpum viride*
Geog. Reg(s) : America-South
End-Use(s) : Fruit
Domesticated : Y
Ref. Code(s) : 213, 219
Summary : The green sapote is a tree cultivated in South America for its delicate, edible fruit. The sweet, juicy pulp is eaten fresh or as preserves. Green sapote fruit is popular in South American domestic markets.

Highbush Blueberry, Swamp Blueberry

Sci. Name(s) : *Vaccinium corymbosum*
Geog. Reg(s) : America-North (U.S. and Canada)
End-Use(s) : Fruit, Ornamental & Lawn Grass
Domesticated : Y
Ref. Code(s) : 17, 220, 236
Summary : The highbush blueberry is a shrub of eastern North America. It is cultivated for its edible berries, which are made into tarts and pies. The fruit is canned and sold as domestic produce.

This variety requires acidic soil (pH 4.0-5.2) with good drainage, moisture, and aeration. It will not tolerate temperatures below -28 degrees Celsius and needs a growing season of 160 days. Plants bear in their third year. They are valued as ornamentals.

Hog-Plum, Yellow Mombin, Jobo

Sci. Name(s) : *Spondias mombin, Spondias lutea*
Geog. Reg(s) : America-Central, America-South, Asia, Asia-Southeast, Tropical, West Indies
End-Use(s) : Beverage, Forage, Pasture & Feed Grain, Fruit
Domesticated : Y
Ref. Code(s) : 36, 65, 151, 170
Summary : The hog-plum tree reaches 12-18 m in height and is occasionally cultivated in parts of the tropics as a fruit tree. Its fruit has a flavor similar to plums and are eaten fresh, stewed, canned, or as jam, jelly and wine. It is also used for a feed for pigs.

Trees grow wild in Mexico, Central America, northern South America, and the West Indies. They are cultivated in the Philippines and in the United States, are successful in southern Florida. Trees thrive on rich, moist, heavy loams.

Huckleberry, Black Huckleberry

Sci. Name(s) : *Gaylussacia baccata*
Geog. Reg(s) : America-North (U.S. and Canada)
End-Use(s) : Fruit
Domesticated : Y
Ref. Code(s) : 16, 220
Summary : The huckleberry is an erect shrub growing to 1 m in height with sweet, firm fruit. This fruit is a common black huckleberry sold in domestic markets. It is popular fresh or in pies.

Shrubs grow wild on rocky or sandy soils in shady areas. They are hardy and tolerate dry climates in full sunlight. Huckleberries are native to, and grow throughout, North America.

Icecream Bean

Sci. Name(s) : *Inga edulis*
Geog. Reg(s) : Africa-East, America-Central, America-South, Hawaii, West Indies
End-Use(s) : Fruit, Spice & Flavoring, Vegetable
Domesticated : Y
Ref. Code(s) : 56
Summary : Icecream bean trees reach 17 m in height, have broad, spreading crowns and are grown as shade for coffee, tea, and cocoa plants in Central and South America. Trees produce an edible fruit pulp used to flavor desserts. The pods are occasionally eaten as vegetables in Central and South America.

Trees thrive in tropical climates and need considerable moisture for satisfactory growth. Trees are native to Central and South America and are introduced throughout the tropics. Icecream beans are grown as domestic fruit crops in South America, Hawaii, the West Indies, and East Africa.

Ilama

Sci. Name(s) : *Annona diversifolia*
Geog. Reg(s) : America-Central, America-North (U.S. and Canada), Tropical
End-Use(s) : Fruit
Domesticated : Y
Ref. Code(s) : 148, 170, 220
Summary : The ilama tree is occasionally cultivated in Central America and Florida for its edible fruit. It thrives under a variety of climatic conditions and withstands long, dry seasons and tropical locations below 600 m in elevation. Trees are ideally suited as tropical fruit crops. They do not need hand pollination. Ilama fruits are comparable to cherimoya.

Imbe
(See Beverage)

Imbu, Hog Plum

Sci. Name(s) : *Spondias tuberosa*
Geog. Reg(s) : America-South, Tropical
End-Use(s) : Fruit
Domesticated : Y
Ref. Code(s) : 36, 65, 220
Summary : The imbu (*Spondias tuberosa*) is a plant grown for its edible fruits with a flavor similar to the sweet orange. This fruit is eaten fresh or made into a milk jelly called imbuzada. Plants are native to Brazil, are introduced throughout the tropics, and grow in some arid regions.

Indian Bael, Bael Fruit, Bengal Quince, Bilva, Siniphal, Bael Tree

Sci. Name(s) : *Aegle marmelos, Aegle correa*
Geog. Reg(s) : Asia-India (subcontinent), Asia-Southeast
End-Use(s) : Beverage, Dye & Tannin, Essential Oil, Forage, Pasture & Feed Grain, Fruit, Gum & Starch, Medicinal, Timber
Domesticated : Y
Ref. Code(s) : 170, 220
Summary : Indian bael is a tree native to India and cultivated for its light, yellowish hardwood and edible fruit throughout Southeast Asia and the East Indian Archipelago. This durable wood is used in a variety of ways, ranging from house construction to tool handles. Its hard-shelled fruit has soft, aromatic pulp that is made into beverages, or eaten fresh, dried and as sherbet. It is also used in preparations to treat dysentery and dyspepsia and can be made into a soap substitute. An oil (Marmelle oil), extracted from the rind, is used as a hair tonic and as a yellow dye in calico printing. The flowers are distilled and used in perfumed products. Indian bael leaves provide fodder and the tree bark exudes an adhesive gum.

Indian Jujube, Beri, Inu-Natsume, Ber Tree

Sci. Name(s) : *Ziziphus mauritiana*
Geog. Reg(s) : Africa, Asia, Asia-India (subcontinent), Europe, Pacific Islands, Tropical
End-Use(s) : Beverage, Dye & Tannin, Forage, Pasture & Feed Grain, Fruit, Miscellaneous
Domesticated : Y
Ref. Code(s) : 17, 26, 132, 154, 170, 213, 220
Summary : Indian jujube is an evergreen shrub or tree grown for its edible fruits high in vitamin C. These are eaten fresh, candied, or dried as a dessert fruit. They are made into a drink and, in the Democratic Republic of the Sudan, the fermented pulp is used to make cakes similar to gingerbread. The leaves and bark are used for tanning.

Plants are cultivated in hot, dry areas of tropical Africa, Asia, and India where they have a wide range of uses such as fodder, silkworm production, and fencing. Shrubs provide protection for other trees. Indian jujube withstands drought and light-frost.

Indian Wood-Apple

Sci. Name(s) : *Feronia limonia*
Geog. Reg(s) : Asia-India (subcontinent), Asia-Southeast
End-Use(s) : Dye & Tannin, Fruit, Gum & Starch, Medicinal

Domesticated : Y
Ref. Code(s) : 170, 220
Summary : The Indian wood-apple is a tree whose aromatic fruit is eaten fresh or in sherbets and jellies. Fruit production is mainly in India. Medicinally, wood-apple fruit and leaves have been used to treat indigestion. In Thailand, fruit juice is used as a yellow ink for writing on palm leaves. The tree exudes a water-soluble gum used in glues, water paints and varnishes.

Indianfig, Spineless Cactus, Prickly Pear, Indianfig Pricklypear

Sci. Name(s) : *Opuntia ficus-indica, Opuntia occidentalis*
Geog. Reg(s) : America-Central, Subtropical, Tropical
End-Use(s) : Forage, Pasture & Feed Grain, Fruit, Sweetener
Domesticated : Y
Ref. Code(s) : 17, 132, 167, 192, 199, 210, 220
Summary : Indianfig is a large bush or tree grown in tropical and subtropical areas for its edible fruit high in calcium and vitamin C that is eaten fresh, dried, or cooked, and can be made into a syrup, jam, and preserve. Plants are used as forage.

In its natural habitat, indianfig grows in warm, dry, rocky places on fairly rich soils. In Mexico, plants are cultivated for domestic consumption but a small amount of fruit is shipped to the United States. Plants begin to fruit 3 years after planting and continue to produce for many years. Fruit yields are about 20,178 kg/ha. Indianfigs decay over time and form rich organic matter.

Jaboticaba, Brazilian Grape Tree

Sci. Name(s) : *Myrciaria cauliflora*
Geog. Reg(s) : Subtropical, Tropical
End-Use(s) : Fruit
Domesticated : Y
Ref. Code(s) : 132, 220
Summary : The jaboticaba tree reaches 12 m in height and produces edible fruit. The pulp provides carbohydrates and vitamin C. Trees thrive in warm, tropical, and subtropical regions in areas with light rainfall. They are fairly frost-hardy.

Jackfruit, Jack

Sci. Name(s) : *Artocarpus heterophyllus*
Geog. Reg(s) : Asia, Asia-India (subcontinent), Tropical
End-Use(s) : Dye & Tannin, Forage, Pasture & Feed Grain, Fruit, Gum & Starch, Timber, Vegetable
Domesticated : Y
Ref. Code(s) : 170
Summary : The jackfruit is a tropical tree whose large ripe fruit is eaten in India. The fruit pulp is eaten fresh or preserved. Immature fruits are eaten as vegetables, used in soups and made into pickles. Seeds are eaten boiled or roasted and taste similar to chestnuts. The fruit rind is fed to livestock. The tree provides valuable timber and its heartwood yields a yellow dye. Latex from the tree is used to mend earthenware utensils. Although it has many uses, there is no commercial jackfruit production except in parts of tropical Asia.

Jackfruit trees tolerate higher-altitudes than the breadfruit and grows in most deep, well-drained soils. Best results are from deep alluvial soils. Trees grow up to 20 m in height and provide agricultural shade for coffee (*Coffea arabica*) and areca (*Areca catechu*). Trees are used as living supports for pepper (*Piper nigrum*) plants.

A variety of the jackfruit, 'Singapore Jack', fruits 3 years after planting and continues fruiting for about 4 months. Most jackfruits weigh 9-27 kg. Yields of up to 250 fruits/tree/annum are produced.

Jambolan, Java Plum

Sci. Name(s) : *Syzygium cumini, Eugenia cumini*
Geog. Reg(s) : America-North (U.S. and Canada), America-South, Asia-India (subcontinent), Hawaii
End-Use(s) : Energy, Fruit, Medicinal
Domesticated : Y
Ref. Code(s) : 17, 36, 170
Summary : Jambolan is an evergreen shrub grown for its edible fruits in California, Florida, and Hawaii. In India, a vinegar is made from the unripe fruit juice. Seeds have been used for the treatment of diabetes, diarrhea, and dysentery. In Europe, seeds have served as the basis of patent diabetes medicine.

The seed contains 41% starch and 6.3% proteids. The wood is often used as fuel.

Japanese Plum

Sci. Name(s) : *Prunus salicina, Prunus triflora*
Geog. Reg(s) : Africa-South, America-North (U.S. and Canada), Asia, Tropical
End-Use(s) : Fruit
Domesticated : Y
Ref. Code(s) : 16, 17, 89, 94, 148, 212
Summary : The Japanese plum tree (*Prunus salinica*) reaches 6-9 m in height and is cultivated in Japan for its edible fruit. These plums are eaten fresh, sun-dried as prunes, or used in canning and jams. Trees are valued for their hardy growth and tolerance of varying environmental conditions. They give early yields and are resistant to most diseases. Trees flower prematurely and are susceptible to frost.

Trees do not require a severe chilling for flower set and thrive in areas with hot summers and mild springs. In the tropics, trees are successful at high-altitudes, usually from 1,640-2,280 m in elevation. Japanese plums are grown in the warmer parts of the United States, Asia, and South Africa.

Java Apple

Sci. Name(s) : *Syzygium samarangense*
Geog. Reg(s) : Asia-Southeast, Tropical
End-Use(s) : Fruit
Domesticated : Y
Ref. Code(s) : 132
Summary : The Java apple tree reaches 15 m in height and produces edible fruits. The pulp is fair source of carbohydrates, potassium, and vitamin C. Trees are native to Southeast Asia. They adapt to the hot, wet, lowland tropics and do not tolerate frost.

Kaki, Kaki Persimmon, Japanese Persimmon

Sci. Name(s) : *Diospyros kaki*
Geog. Reg(s) : America-North (U.S. and Canada), Asia-China
End-Use(s) : Fruit, Natural Resin
Domesticated : Y
Ref. Code(s) : 40, 65, 192
Summary : Kaki is a deciduous Chinese tree that reaches 12-15 m in height and is cultivated for its fruit, which has a fairly high sugar content of 17% and is eaten as a fresh
dessert fruit or in pies. A dried fruit market is developing and there is good potential
for commercial production if the problems of drying the fruit are solved. In China, a
natural resin from the unripe fruit is applied as a waterproofing varnish for paper hats
and umbrellas.

Major fruit orchards in the United States are in Florida, several Gulf states, and
California. Trees are planted in deep, fairly heavy, well-drained, humus-rich soils for
best results.

Kangaroo Apple

Sci. Name(s) : *Solanum laciniatum*
Geog. Reg(s) : Australia
End-Use(s) : Fruit, Sweetener, Vegetable
Domesticated : N
Ref. Code(s) : 49, 121, 220
Summary : The kangaroo apple (*Solanum laciniatum*) is a shrub often confused with an-
other member of its genus, *Solanum aviculare*. *Solanum laciniatum* differs only
slightly in appearance; it has purplish leaf stems and slightly larger fruit with paler
seeds. Kangaroo apples are edible and in Australia are eaten fresh or as a vegetable.
This fruit contains the natural sugars sucrose, glucose, and fructose.

Plants grow throughout Australia and are common along the edges of rain
forests and in other protected, moist areas. Kangaroo apples flower summer to autumn.

Karanda
(See Ornamental & Lawn Grass)

Kariis, Karii

Sci. Name(s) : *Garcinia lateriflora*
Geog. Reg(s) : Asia-Southeast
End-Use(s) : Fruit, Miscellaneous, Timber
Domesticated : N
Ref. Code(s) : 220
Summary : Kariis is a tree that produces fruits with a variety of shapes and flavors, making
the plant a potential source for experimental breeding. The wood is used to make
pestles. Trees are found in Southeast Asia. There is limited information concerning
their habits or economic value.

Kiwi, Yangtao, Chinese-Gooseberry
Sci. Name(s) : *Actinidia chinensis*

Geog. Reg(s)   : America-North (U.S. and Canada), Asia-China
End-Use(s)   : Fruit, Ornamental & Lawn Grass
Domesticated : Y
Ref. Code(s)   : 16, 220
Summary   : Kiwi (*Actinidia chinensis*) is a woody vine climbing to heights of 7.6 m grown for its edible fruits. These are eaten fresh or used in jams and preserves and are popular in China. In the United States, a fresh kiwi market has evolved and considerable acreage of kiwi is planted in California. Kiwis are planted as ornamentals.

Vines are hardy and thrive in moist, rich soils in full or partial sunlight. Both sexes are required for fruiting. Plants are most successful when grown in southern climates but can be grown further north if protected during the autumn and winter months.

Kumquat

Sci. Name(s)   : *Fortunella* spp.
Geog. Reg(s)   : America-North (U.S. and Canada), Asia, Asia-China, Asia-Southeast
End-Use(s)   : Fruit
Domesticated : Y
Ref. Code(s)   : 148, 154, 170, 220
Summary   : Kumquats are small trees with edible fruit. There are several useful Fortunella species: the meiwa kumquat (*Fortunella crassifolia*), occasionally cultivated for its sweet fruit eaten fresh and in salads; the round kumquat (*Fortunella japonica*), with an acidic fruit made into jellies and preserves; and the oval kumquat (*Fortunella margarita*), grown commercially in China, Japan, and Florida for its acidic fruit used in jellies and preserves. Kumquats are native to eastern Asia and Malaysia. Trees bloom in May and the fruit is picked in December.

Lac-tree, Ceylon Oak, Kussum Tree, Malay Lac-tree
(See Fat & Wax)

Langsat

Sci. Name(s)   : *Lansium domesticum*
Geog. Reg(s)   : Asia-Southeast
End-Use(s)   : Fruit
Domesticated : Y
Ref. Code(s)   : 132, 154
Summary   : Langsat is a tree reaching 15 m in height that produces edible fruits. The sweet pulp is a fair source of carbohydrates and potassium. The best fruits are of a small, sweet, seedless variety. Langsat fruit bunches have 15-30 fruits/bunch. Trees adapt to the hot, wet lowland tropics and are not frost-hardy.

Lemon

Sci. Name(s)   : *Citrus limon*
Geog. Reg(s)   : America-North (U.S. and Canada), Europe, Temperate, Tropical
End-Use(s)   : Beverage, Essential Oil, Fruit, Spice & Flavoring
Domesticated : Y
Ref. Code(s)   : 132, 170, 180

224

Summary   : The lemon tree produces acidic fruits that are used extensively both fresh and processed in temperate countries. Fruit pulp is an excellent source of vitamin C and a fair source of potassium. Commercially, lemon pulp is used in the preparation of lemonades and for culinary and confectionery purposes. The fruit also provides citric acid and pectin.

Lemon peels yield an essential oil that is an important commercial product. Its main constituents are both d and dl limonene. Oil produced in Sicily is reportedly higher in citral content than oil produced in California. Lemon peel oil is used to flavor medicines, and is an important flavor in bakery products, carbonated soft drinks, soft-drink powders, gelatin desserts, extracts, candies, and ices. The oil is also used in perfumery, eaux de Colognes, lotions, and soaps. Because lemon oil alone is not very stable, its terpenes and sesquiterpenes are usually removed, leaving a stable oil that is more soluble in aqueous alcohol. This oil contains mostly citral. The peel, a good source of vitamin P, is eaten candied.

In 1983, lemon production in the United States was 860,875 MT. Fifty-four percent of the crop was processed and the remainder sold fresh. The total value of the crop was $109.7 million.

Lemon industries in the United States center in Florida and California. Lemons may be stored for as long as 6-8 months if cured during dry weather. In 1984, the United States imported 618.4 MT of lemon oil valued at $7.7 milliom.

Lime

Sci. Name(s)   : *Citrus aurantiifolia*
Geog. Reg(s)   : America-Central, America-North (U.S. and Canada), Mediterranean, Tropical, West Indies
End-Use(s)   : Beverage, Essential Oil, Fruit, Spice & Flavoring
Domesticated   : Y
Ref. Code(s)   : 132, 170

Summary   : The lime tree (*Citrus aurantiifolia*) bears a yellow or green fruit with a sour, juicy pulp. Trees are grown throughout the tropics and are the most commonly cultivated species of the acidic Citrus family. Commercial lime production is most prominent in Mexico, the West Indies, parts of the United States, and Egypt.

Lime pulp is an excellent source of vitamin C and provides a tart juice that is used fresh as flavoring or for making limeades. Commercially, lime products include limeade, lime juice, and cordials. Fruit peels yield an essential oil, while the fruit itself provides citric acid. Trees grow to 6 m in height. They are most successful in hot, lowland tropical climates in areas with well-distributed rainfall. Trees cannot withstand frost.

Lime production in the United States in 1983 was 61,744 MT valued at $22 million. Fifty-eight percent of the limes were sold as fresh fruit and 42% were processed. Fresh limes are tender and rarely imported by temperate countries because of possible transportation damage. In 1984, the United States imported 539 MT of lime oil as an essential oil valued at $14.5 million.

Limeberry

Sci. Name(s)   : *Triphasia trifolia*
Geog. Reg(s)   : Asia-Southeast, Subtropical, Tropical

End-Use(s) : Fruit, Ornamental & Lawn Grass
Domesticated : Y
Ref. Code(s) : 170
Summary : Limeberry is a thorny, evergreen shrub with edible fruits usually made into preserves. Shrubs also form excellent hedges. Limeberry is native to Southeast Asia and is introduced throughout the tropics and subtropics.

Lingaro
Sci. Name(s) : *Elaeagnus philippensis*
Geog. Reg(s) : Asia-Southeast
End-Use(s) : Fruit
Domesticated : Y
Ref. Code(s) : 36, 143, 220
Summary : Lingaro is a tree with sweet fruits used to make jelly in the Philippines. This tree grows in moderately hot climates and is common in thickets and forests at low to medium altitudes and sometimes up to 1,500 m.

Litchi, Litchi Nut, Lychee
Sci. Name(s) : *Litchi chinensis*
Geog. Reg(s) : Asia-China, Tropical
End-Use(s) : Fruit
Domesticated : Y
Ref. Code(s) : 132, 154, 170
Summary : The litchi nut tree is native to China and is grown in the tropics at higher-altitudes. Its nuts are eaten fresh, canned in syrup, or dried. It is perishable but is highly prized for its flavor and as a good source of ascorbic acid and phosphorus. Annual nut yields range from 75-225 kg/tree.

Litchi trees prefer loamy to sandy, well-drained, mildly acidic soils. They thrive in areas with wet summers and dry winters.

Longan
Sci. Name(s) : *Euphoria longana, Dimocarpus longan*
Geog. Reg(s) : Asia-China
End-Use(s) : Fruit, Ornamental & Lawn Grass
Domesticated : Y
Ref. Code(s) : 154, 214
Summary : The longan tree is grown in China for its edible fruit similar to litchis. The fruit pulp is eaten fresh, canned, or dried. Trees make attractive ornamentals and grow rapidly on a wide range of soils, doing best in shade.

Loquat
Sci. Name(s) : *Eriobotrya japonica*
Geog. Reg(s) : Africa-Central, America-North (U.S. and Canada), Asia, Asia-China, Asia-India (subcontinent), Mediterranean, Subtropical, Tropical
End-Use(s) : Fruit, Ornamental & Lawn Grass
Domesticated : Y
Ref. Code(s) : 132, 148, 154, 220

Summary : Loquat is a drought-resistant evergreen tree grown for its edible fruit eaten fresh or cooked. Its pulp is a good source of vitamin A, potassium, and carbohydrates. The varieties with a higher proportion of edible flesh are the Tanaka, Oliver, and Early Red. The smaller-fruited types are used for cooking and jellies. Loquat trees make attractive ornamentals.

The fruit is seldom included with market produce because it is usually damaged by the Mediterranean fruit fly. Best results are from grafted trees grown at higher-altitudes. Loquat trees are native to China and economically important there and to Japan, the Mediterranean region, the United States, and India. In Central America, trees thrive at 914 m or higher in elevation. They are adapted to subtropical or highland tropical climates in areas with well-distributed rainfall. Bloom is injured by low temperatures and trees are damaged by the fire blight and pear blight diseases.

Lukrabao, Chaulmogra Tree
(See Medicinal)

Mammy-Apple, Mammee-Apple
Sci. Name(s) : *Mammea americana*
Geog. Reg(s) : America-Central, America-South, Tropical, West Indies
End-Use(s) : Beverage, Fruit, Insecticide, Medicinal, Timber
Domesticated : Y
Ref. Code(s) : 154, 167, 170
Summary : The mammy-apple is an evergreen tree that originates in tropical America and the West Indies. It bears fruit which is cooked or made into preserves. It is thought to taste similar to the apricot. The fruit, sap, and flowers of the mammy-apple tree are also used to make beverages, such as a liquor made from the flower which is called creme de creole. In Mexico, the gum and seeds are used as insecticides, and the leaves are used to treat fevers. The hard wood is used for making posts and in construction. This wood has a beautiful grain and takes a high polish.

This tree is tropical in its requirements and cannot be grown where there is any danger of frost. It likes a rich, well-drained, sandy loam. Seedlings bear in 6-7 years.

Mango
Sci. Name(s) : *Mangifera indica*
Geog. Reg(s) : Africa-South, America, America-Central, America-South, Asia-India (subcontinent), Asia-Southeast, Australia, Hawaii, Pacific Islands, Tropical
End-Use(s) : Beverage, Cereal, Forage, Pasture & Feed Grain, Fruit, Miscellaneous, Spice & Flavoring, Timber
Domesticated : Y
Ref. Code(s) : 132, 167, 170
Summary : The mango tree provides an important fruit in India and Asia. Trees are native to the Indo-Burma area where it grows wild in forests. It is now introduced into most tropical areas.

Mangoes are eaten raw or made into juice, squash, jams, jellies, or preserves. Unripe fruits are used in pickles, chutneys, and culinary preparations. The seeds are used as food or ground for flour. The fruit is a good source of vitamin A. The unripe fruit contains considerable quantities of malic and tartaric acid.

The tree leaves are an alternative food supply for cattle in times of shortage. The timber is used in boat construction. The tree and leaves are important in Hindu myths and culture.

Trees grow best in the tropics and adapt to a wide variety of soils. They prefer deep rich, alluvial loams with good drainage and produce 1 good crop every 3-4 years. Trees mature 2-5 months after fertilization and require a dry season to fruit heavily. In India, most fruit is for domestic consumption with a limited amount exported to Kuwait, Bahrain, Singapore, and Malaya. In the 10th year, a tree bears 400-600 fruits. This amount increases until the 40th year when yields begin to decline. Mangoes are grown in South and Central America, parts of the United States, Europe, and the Mediterranean as well as the Pacific.

Mangosteen

Sci. Name(s)   : *Garcinia mangostana*
Geog. Reg(s)   : Asia-Southeast, Tropical, West Indies
End-Use(s)   : Fruit
Domesticated   : Y
Ref. Code(s)   : 16, 132, 148, 170
Summary   : Mangosteen is a slow-growing tree reaching 12-15 m in height. It bears fruits with sweet, carbohydrate-rich pulp popular throughout the tropics as a fresh fruit. Mangosteen is cultivated extensively in the West Indies.

Trees are most successful in the hot, wet, lowland tropics in well-drained, clay or loam soils. They are not frost-hardy and are fairly difficult to establish because of poor seed viability. Successful plantings of trees fruit after 8-15 years, depending on growth conditions. A typical yield ranges from 200-500 fruits/year. Good yields range from 1,000-2,000 fruits/year. Mangosteens are native to Malaysia.

Marking-Nut Tree
(See Dye & Tannin)

Menteng, Kapundung

Sci. Name(s)   : *Baccaurea racemosa*
Geog. Reg(s)   : Asia-India (subcontinent), Asia-Southeast
End-Use(s)   : Dye & Tannin, Fiber, Fruit
Domesticated   : N
Ref. Code(s)   : 36
Summary   : Menteng is an Indian tree whose acidic fruits must be eaten sparingly, or vomiting will occur. The leaves are thought to contain an alkaloid. Menteng yields a fiber used in Java for paper manufacture. The leaves also yield a mauve colored dye.

Mexican Husk Tomato, Tomatillo, Jamberry, Tomatillo Ground Cherry, Husk-Tomato
(See Vegetable)

Mexican-Apple, White Sapote

Sci. Name(s)   : *Casimiroa edulis*
Geog. Reg(s)   : America-Central, Subtropical

End-Use(s)   : Fruit
Domesticated : Y
Ref. Code(s)   : 148, 170
Summary   : The Mexican-apple tree produces edible fruits with pulp that is a good source
of vitamin C, carbohydrates, and protein. This tree is native to the Mexican and
Central American highlands and is introduced throughout the subtropics. It is fast-
growing and easily propagated. In Central America, trees thrive in deep, well-drained
soils at medium elevations. The Mexican-apple can probably be grown at lower altitudes if irrigated.

Morula

Sci. Name(s)   : *Sclerocarya caffra*
Geog. Reg(s)   : Africa-South, Pacific Islands
End-Use(s)   : Beverage, Fruit, Nut, Oil
Domesticated : Y
Ref. Code(s)   : 154, 158
Summary   : Morula is a South African tree that produces fruit with edible seeds. The fruit
is highly nutritious with 4 times the vitamin C of fresh orange juice. Fruit is made into
alcoholic beverages, conserves, and jelly. Fruit seeds produce a protein-rich oil and
are eaten raw or cooked as porridge. Processed seeds can be made into a type of
cosmetic.

These trees are a protected species in South Africa. In Tonga, they are a highly
valued indigenous species. Morulas are fairly rare specimens. They are propagated
by truncheons or seed. Trees are susceptible to frost.

Mountain Papaya

Sci. Name(s)   : *Carica pubescens*
Geog. Reg(s)   : America-South
End-Use(s)   : Fruit
Domesticated : Y
Ref. Code(s)   : 148, 170
Summary   : The mountain papaya is a soft-wooded tree grown at high-altitudes in the
tropics for its small, edible fruits. This fruit is eaten stewed or in preserves. The tree
contains latex in its fruit, leaves, and trunk. The mountain papaya is native to the An-
des of South America.

Muriti
(See Fiber)

Muscadine Grape, Southern Fox Grape

Sci. Name(s)   : *Vitis rotundifolia*
Geog. Reg(s)   : America-North (U.S. and Canada)
End-Use(s)   : Beverage, Fruit
Domesticated : Y
Ref. Code(s)   : 17, 84, 154, 220

Summary : Muscadine grapes (*Vitrus rotundifolia*) are presently grown in the southeastern portion of the United States. The grapes are sold as fresh table grapes, jellies, grapejuice, and wine. The vast majority of grapes go into the production of wine.

The wines are often low in alcohol content. Quality is difficult to maintain and dependent on a number of factors, especially the degree of ripeness. The low sugar content requires the addition of sugar or other sweeteners to stabilize the wine. The wine does not hold its color well. Storage temperature and the total amount of heat used in processing greatly impacts color.

Muscadine grapes represent a small percentage of all grapes grown in the United States. Production is centered in the states of North and South Carolina, Georgia, Florida, Alabama, and Mississippi.

Naranjilla, Lulo
Sci. Name(s) : *Solanum quitoense*
Geog. Reg(s) : America-South
End-Use(s) : Beverage, Fruit
Domesticated : Y
Ref. Code(s) : 92, 154, 170, 237
Summary : Naranjilla is a shrub from 1.2-1.8 m in height cultivated for its fruit, made into preserves and pies. Its acidic fruit pulp is used to make a juice rich in proteins and mineral salts.

Plants are grown at high-altitudes on well-drained slopes. They are common in Andean valleys from southern Colombia to northern Peru. Necessary annual rainfall is 152 cm.

Natal-Palm
Sci. Name(s) : *Carissa macrocarpa, Carissa grandiflora*
Geog. Reg(s) : Africa-South
End-Use(s) : Fruit, Ornamental & Lawn Grass
Domesticated : Y
Ref. Code(s) : 170
Summary : The Natal-palm is a large South African shrub with fruits made into jelly or used as a substitute for cranberry sauce. The palm is grown as an ornamental shrub.

Olive
(See Oil)

Otaheite Gooseberry, Otaheite Gooseberry-Tree, Star Gooseberry, Jimbling, Indian Gooseberry
Sci. Name(s) : *Phyllanthus acidus, Cicca distichus, Phyllanthus distichus*
Geog. Reg(s) : Africa, Asia-India (subcontinent), Tropical
End-Use(s) : Fruit
Domesticated : Y
Ref. Code(s) : 154, 170
Summary : The otaheite gooseberry is a tropical tree reaching 9 m in height with edible fruit. The acidic fruit pulp is eaten cooked or as preserves. The otaheite gooseberry is native to Madagascar and India. It is propagated by seeds or green wood cuttings.

Ouricury Palm, Licuri, Nicuri
(See Fat & Wax)

Papaya

Sci. Name(s)	: *Carica papaya*
Geog. Reg(s)	: Africa-East, Hawaii, Subtropical, Tropical
End-Use(s)	: Dye & Tannin, Fruit, Gum & Starch, Medicinal
Domesticated	: Y
Ref. Code(s)	: 148, 170
Summary	: The papaya is a short-lived, soft-wooded tropical tree about 2 m in height

whose fleshy fruit is a good source of vitamin C and a fair source of vitamins A and B. The fruit is eaten fresh, canned, or cooked as a substitute for marrow and applesauce. Immature fruits are tapped for a latex that produces papain, a proteolytic enzyme used in meat tenderizers, in chewing gum manufacture, in cosmetics, and as a drug for digestive ailments. Tanning industries use the papain for bating hides, and it is used in the degumming of natural silks and to give shrink resistance to wool. Throughout most producing areas, papaya fruits are grown primarily for domestic consumption.

Transporting the tender fruit to temperate countries is difficult. This difficulty has given rise to alternative forms of marketing, that is, canned or processed fruit and fruit juice. Papain production is centered in Tanzania and the major importer of the enzyme is the United States. Papain collection requires special care in collecting and drying and because of this, papain has not experienced substantial commercial expansion. Hawaii has a fresh papaya industry that has demonstrated sustained growth. The U.S. fresh papaya market is sensitive to overproduction.

In tropical and subtropical regions, papaya crops produce well at altitudes up to 1,500 m. Plants are susceptible to frost and a virus, cucurbit mosaic, which is transmitted by green peach aphids. To safeguard against the virus, trees must be started in weed-free soils and later transplanted to well-drained land and grown with large amounts of nitrogen fertilizer and water. Plantations should be rotated with other crops and renewed every few years.

Papaya trees fruit after 9-14 months. Fresh fruits are harvested when traces of yellow appear on the skin and the fruits are fully ripened 4-5 days later. Papaya fruit yields vary from 30-150 fruits/tree/annum. If grown for papain production, the productive life of the tree is usually 3 years. In 1983, 21,020 MT of papayas were produced in Hawaii for the fresh market and 6,855 MT of fruit were used for processing. Total value of production was $11.6 million.

Passion Fruit, Purple Granadilla
(See Beverage)

Pawpaw

Sci. Name(s)	: *Asimina triloba*
Geog. Reg(s)	: America-North (U.S. and Canada), Temperate
End-Use(s)	: Fruit, Medicinal, Spice & Flavoring
Domesticated	: Y
Ref. Code(s)	: 220

Summary : Pawpaw is a North American shrub whose bottle-shaped fruits are favored by some and found extremely objectionable by others. This particular crop should not be confused with the economic fruit crop, the papaya (*Carica papaya*). The flesh is similar to that of a banana and is eaten raw or baked, as a filling for pies and desserts, or a flavoring in ice cream. An emetic obtained from the seeds and the leaves is applied to ulcers and boils. The fruit is sold as domestic market produce in the southern and southeastern parts of the United States.

  This deciduous shrub or tree reaches 12 m in height and produces after 6-8 years. Pawpaw is well-adapted to the temperate climate of North America. Trees bear for 50-80 years.

Peach, Nectarine

Sci. Name(s) : *Prunus persica*
Geog. Reg(s) : America-North (U.S. and Canada), Asia-Southeast, Hawaii, Temperate
End-Use(s) : Fruit, Ornamental & Lawn Grass
Domesticated : Y
Ref. Code(s) : 17, 36, 101, 116, 148, 154, 212
Summary : The peach tree is an important commercial fruit tree. It can be grown throughout the world but fruit production is best in areas with cool temperatures necessary for fruit set. In the colder parts of the United States, peach orchards provide fresh market peaches. In warmer regions like California, peaches are grown mainly for canning. Trees grow in Hawaii at lower altitudes but are severely affected by the Mediterranean fruit fly. In Malaysia, trees are grown as ornamentals and in Java, trees are mostly wild with a few cultivated for flowers and fruit.

  Trees grow in most well-drained soils and prefer fairly cold, exposed places. They bear in their 3rd year and peak at 8-12 years. In 1983, the United States produced 812,524 MT of peaches valued at approximately $259 million. Fifty-four percent of these peaches were sold fresh and the remainder sold canned, dried, or frozen.

Peach Palm, Pejibaye

Sci. Name(s) : *Bactris gasipaes, Guilielma gasipaes*
Geog. Reg(s) : America-Central, America-South
End-Use(s) : Beverage, Fruit, Nut
Domesticated : Y
Ref. Code(s) : 171
Summary : The peach palm is cultivated in Central and northern South America for its edible fruits. This fruit is eaten boiled, ground into flour, or fermented and made into a beverage called chicha. The kernels are oily but edible. Palms bear in 5-8 years and continue for 50-75 years. Each season, 5-6 fruit bunches/stem are produced. Fruit ripens in 6 months. Palms grow from sea level to 1,200 m in elevation.

Pear, Common Pear

Sci. Name(s) : *Pyrus communis*
Geog. Reg(s) : America-North (U.S. and Canada), Europe
End-Use(s) : Beverage, Fruit, Timber
Domesticated : Y
Ref. Code(s) : 17, 36, 154, 189, 220

Summary   : The pear tree is grown for its commercially important, edible fruits. A well-known commercial cultivar is the 'Bartlett.' Commercially, pear trees are best grown in areas with equitable climates. In the United States, commercial production centers in New England, around the Great Lakes and along the Pacific Coast. Pears are also produced in Europe, mainly in Italy, Germany, France, Switzerland, and the Balkan States. Many European pears are made into a popular drink called perry. Pear wood is heavy, durable, and used for cutlery and turnery.

Trees grown in favorable climates and good, humus-rich sand or clay loams are relatively long-lived and regular producers. They require 900-1,000 hours of temperatures below 7.2 degrees Celsius for adequate fruit set. About 89 cm of rain is necessary for successful growth. Fruit ripens from early August till winter.

In 1983, the United States produced 702,607 MT of pears valued at $132 million. Fifty percent of the crop was sold fresh while the remainder was used for canning or dried fruit.

Pepino, Melon-Pear

Sci. Name(s)   : *Solanum muricatum*
Geog. Reg(s)   : America-South, Tropical
End-Use(s)   : Fruit
Domesticated   : Y
Ref. Code(s)   : 132, 170, 220, 237
Summary   : Pepino is a perennial herb that reaches up to 50 cm in height cultivated in parts of tropical America for its edible fruits eaten fresh. These fruits are high in vitamin C. Plants are most successful in areas with low to moderate temperatures and rainfall. Plants do well on rich, well-drained soils. Fruit are harvested 8 weeks after planting. Pepino is probably native to Peru.

Phalsa
(See Beverage)

Pineapple

Sci. Name(s)   : *Ananas comosus*
Geog. Reg(s)   : Africa-South, America-South, Asia-Southeast, Hawaii, Subtropical, Tropical
End-Use(s)   : Beverage, Fiber, Fruit, Sweetener
Domesticated   : Y
Ref. Code(s)   : 148, 171
Summary   : Pineapples are herbs grown throughout the tropics and subtropics. They are popular as fresh or canned dessert fruits which contain 14% sugar, provide vitamins A, B, and C, and bromelin, a protein digesting enzyme. Hawaii, Thailand, Malaysia, and Brazil are the chief producers.

The majority of commercially grown pineapples is processed in the country of production. The quick-frozen flesh from fully ripe fruits and the canned juice are popular market items. Sugar-syrup is obtained from the mill juice. The pineapple is a good source of alcohol and citric acid. Plant leaves yield 2-3% of a strong, silky fiber used in a fabric called pina cloth in the Philippines and Taiwan. This textile fiber cannot be produced economically from plants grown primarily for fruit.

Pineapples are perennial herbs growing to 100 cm in height. In Hawaii, increased yields are obtained through the use of bituminized paper or black polythene. Plants are inserted through this covering and are protected from chilling, weeds, and excessive rain. Coverings also protect soil nutrients and result in better fruit. Herbs are planted 40,000-50,000/ha. They are damaged by nematodes and mealy-bug wilt. The most popular canning cultivar is "cayenne." Most pineapple breeders use this "cayenne" as a starter crop because of its consistently good qualities.

In the United States, 654.9 million MT of pineapples were produced in 1983 with a total value of over $100 million. In the same year, the United States exported approximately 12,675 MT of canned pineapples, valued at $10.5 million, imported 68,345 MT of fresh pineapple, valued at $10.1 million, and 183,665 MT of canned pineapples, valued at $114.7 million.

Plum, Common Plum, European Plum, Garden Plum, Prune Plum

Sci. Name(s) : *Prunus domestica*
Geog. Reg(s) : America-North (U.S. and Canada), Asia, Eurasia, Europe
End-Use(s) : Beverage, Fruit, Timber
Domesticated : Y
Ref. Code(s) : 17, 65, 116, 178, 220
Summary : The common plum (*Prunus domestica*) is a tree from 9-12 m in height that provides an important commercial fruit available canned or as sun-dried prunes. Plums are also popular as fresh fruit produce in markets. Prunes are from varieties of fruit with a high sugar content. Fruit is made into various alcoholic beverages and liqueurs. Tree wood is hard and occasionally used for furniture manufacture.

Trees are grown throughout most of Europe--particularly in Yugoslavia, Rumania, Italy--and the United States, in California and Oregon. In the eastern United States, plum yields are from 1-4 bushels/mature tree. Higher yields are obtained from trees in the western United States. Trees grow in most light, well-drained soils. They fruit 4-8 years after planting and are native to southwestern Asia.

Pomegranate

Sci. Name(s) : *Punica granatum*
Geog. Reg(s) : Mediterranean, Subtropical, Tropical
End-Use(s) : Beverage, Dye & Tannin, Fruit, Medicinal, Ornamental & Lawn Grass
Domesticated : Y
Ref. Code(s) : 132, 167, 170, 192, 196, 220
Summary : The pomegranate is a shrub or tree reaching 6 m in height whose red fruit is filled with an edible, juicy pulp. The fruit is sometimes used as a salad or table fruit, and the pulp can be processed into certain beverages. Fruit skin contains about 26% tannin, which has been used in the commercial tanning of leather. Dried bark, roots, seeds, and fruit rinds have been used in certain unspecified medicines as treatment for intestinal worms, diarrhea, and fevers. In the United States, pomegranates are grown mainly as ornamentals.

Trees are native to Iran. Pomegranates have spread in cultivation throughout most of the tropics and subtropics in areas having cool winters, and hot, dry summers with some humidity. They thrive on deep, heavy loams at elevations approaching 1,600 m. The best fruit is produced in semiarid regions where high temperatures can aid fruit in ripening. Fruit yields vary from 50-200 fruits/year. Trees will continue to bear for 3-4 years.

Pomerac, Malay Apple

Sci. Name(s) : *Syzygium malaccense, Eugenia malaccensis*
Geog. Reg(s) : Asia-Southeast
End-Use(s) : Fruit, Medicinal, Ornamental & Lawn Grass
Domesticated : Y
Ref. Code(s) : 36, 132, 170
Summary : Pomerac is a tree growing to 12 m in height that bears edible fruit whose sweet pulp is a fair source of potassium and vitamin C. Pomerac fruit is eaten fresh or cooked. In the Moluccas, preparations of the tree bark have been used as a mouthwash. Malays have used the powdered leaves to treat cracked tongues. Concoctions of the tree roots have been applied to itchy areas. In Cambodia, the fruits, leaves, and seeds have been prepared for use in treating fevers. Trees are planted throughout the tropics as ornamentals and windbreaks. Pomeracs are native to Malaysia.

Pulasan

Sci. Name(s) : *Nephelium mutable*
Geog. Reg(s) : Asia-Southeast
End-Use(s) : Beverage, Fruit, Medicinal, Oil, Timber
Domesticated : Y
Ref. Code(s) : 36, 132, 170
Summary : Pulasan is a tree reaching 10 m in height grown primarily in Indonesia, Malayasia, and Thailand for its edible fruit aril. Its pulp is a good source of carbohydrates, vitamin C, and calcium. Fruit seeds are boiled, roasted, and made into a beverage similar to cocoa. Seeds contain an oil used in lamps. The hard timber is useful but rarely exploited. Preparations of pulasan roots are used to treat fevers.

Pummelo, Shaddock

Sci. Name(s) : *Citrus grandis*
Geog. Reg(s) : Asia-Southeast, Subtropical, Temperate, Tropical
End-Use(s) : Fruit
Domesticated : Y
Ref. Code(s) : 132, 170
Summary : The pummelo tree is a tree from 5-15 m in height grown for its edible fruit. The fruit pulp is segmented and provides an excellent source of vitamin C and a fair source of carbohydrates, potassium and calcium. It is used primarily as a dessert fruit and the peel candied and eaten as marmelade.

Pummelos are native to Thailand and Malaysia and are grown in Thailand and other Southeast Asian countries. Trees grow in hot, tropical lowland or subtropical areas with well-distributed rainfall. Although these trees are fairly tolerant of frost, they have little or no commercial value in temperate countries.

Purple Raspberry

Sci. Name(s) : *Rubus occidentalis* X *Rubus idaeus, Rubus neglectus*
Geog. Reg(s) : America-North (U.S. and Canada)

End-Use(s) : Fruit
Domesticated : Y
Ref. Code(s) : 15, 192
Summary : The purple raspberry is a shrub that produces edible berries, most of which are used for domestic use. The shrub is the result of a cross between *Rubus strictus* and *Rubus occidentalis*. They are found in the wild but can be propagated by tip layering. Several cultivated varieties exist that are the Colombian, the Erksine Park, the Haymaker, and the Shaffer. Raspberries are grown primarily in the United States and Canada.

Quince
(See Gum & Starch)

Rabbit-Eye Blueberry
Sci. Name(s) : *Vaccinium ashei*
Geog. Reg(s) : America-North (U.S. and Canada)
End-Use(s) : Fruit
Domesticated : Y
Ref. Code(s) : 17, 188, 203, 236
Summary : The rabbit-eye blueberry is a shrubby plant of the southern United States. It produces edible berries that are used in jams, jellies, and pies. The blueberry has been successfully planted as a commercial berry crop in North Carolina, Georgia, and Florida. The produce is sold as a fresh market item. Blueberry yields from 3 year old plants are about 3 liters/plant. Berry yields increase as the plant matures, averaging 16 liters for each 12-16 year old plant. The rabbit-eye blueberry is very adaptable and will sustain growth under mild drought conditions. The productivity of rabbit-eye blueberries has been greatly improved by selection and breeding.

Rambai
Sci. Name(s) : *Baccaurea motleyana*
Geog. Reg(s) : Asia-Southeast
End-Use(s) : Beverage, Dye & Tannin, Fruit
Domesticated : Y
Ref. Code(s) : 36, 170
Summary : Rambai is a tree of medium height that is cultivated throughout western Malaysia for its fruit and bark. The sweet fruit is made into conserves or fermented into a liquor. The bark yields a dye. Rambai wood is considered inferior for use in construction.

Rambutan
Sci. Name(s) : *Nephelium mutabile*
Geog. Reg(s) : Asia-Southeast
End-Use(s) : Fruit
Domesticated : Y
Ref. Code(s) : 132, 170

Summary : Rambutan is an evergreen tree reaching up to 15 m in height grown for its edible fruit high in vitamin C and carbohydrates. Rambutan is a native to the lowlands of Malaysia where it grows successfully. Seedling trees begin fruiting in 5-6 years.

Ramon, Breadnut

Sci. Name(s) : *Brosimum alicastrum* subsp. *alicastrum*
Geog. Reg(s) : America, America-North (U.S. and Canada), Tropical
End-Use(s) : Beverage, Forage, Pasture & Feed Grain, Fruit, Miscellaneous, Timber
Domesticated : Y
Ref. Code(s) : 16, 170, 220
Summary : Ramon is a tree of tropical America that has edible fruits about 2.5 cm in diameter. The fruit contains a single large seed (ramon seed or breadnut) that is eaten roasted or used as a coffee substitute. The leaves are used as livestock fodder. The hard wood is white, fine-grained, and is used in general carpentry. Growth in the United States has been limited to botanical gardens.

Red Currant, Northern Red Currant, White Currant, Garden Currant

Sci. Name(s) : *Ribes rubrum*
Geog. Reg(s) : America-North (U.S. and Canada), Europe
End-Use(s) : Beverage, Fruit
Domesticated : Y
Ref. Code(s) : 17, 69, 220
Summary : The red currant is a shrub cultivated in Europe and North America for its edible fruits. The currants are made into jellies and wine. There are both red and white-fruited varieties. White-fruited currants are used in making Bar de Luc jelly.

Shrubs will continue fruiting for 20 years and are not bothered by pests or diseases, except birds. Red currants thrive on deep, light, humus-rich soils. Shrubs are planted during winter months, preferably October through December.

Red Mombin, Spanish Plum

Sci. Name(s) : *Spondias purpurea*
Geog. Reg(s) : America, America-Central, Tropical, West Indies
End-Use(s) : Fruit
Domesticated : Y
Ref. Code(s) : 17, 151, 170
Summary : The red mombin is a tree cultivated for its edible fruits eaten fresh, cooked, and sometimes dried. Slightly unripe fruit is made into a jelly.

Trees are native to tropical America. They grow wild and are cultivated in the West Indies, Mexico and Central America.

Red Mulberry

(See Ornamental & Lawn Grass)

Rose Apple, Jambos

Sci. Name(s) : *Syzygium jambos, Eugenia jambos*
Geog. Reg(s) : Asia-Southeast, Tropical
End-Use(s) : Dye & Tannin, Fruit, Medicinal

Domesticated : Y
Ref. Code(s) : 36, 170
Summary : The rose apple is a small tree reaching up to 10 m in height. It produces edible, rose-scented fruits that can be eaten fresh or made into preserves.

Cultivated for years in the Indo-Malaysian region, the rose apple has gained popularity and has spread throughout the tropics. The bark contains 7% tannin. Seeds have been used to treat problems with diarrhea, dysentery, and catarrh.

Roselle
(See Fiber)

Sand Pear, Chinese Pear, Japanese Pear, Asian Pear, Oriental Pear
Sci. Name(s) : *Pyrus pyrifolia, Pyrus serotina*
Geog. Reg(s) : America-North (U.S. and Canada), Asia, Asia-China
End-Use(s) : Fruit
Domesticated : Y
Ref. Code(s) : 15, 17, 154
Summary : The sand pear is a tree reaching from 9-15 m in height, grown for its juicy, edible fruit. This tree is native to China and naturalized in Japan. It tolerates more adverse climates than the European pear and, in the United States, grows further north and south. Sand pear trees prefer light soils.

Sapote, Mamme Sapote, Marmalade Plum, Marmalade Fruit
Sci. Name(s) : *Pouteria sapota*
Geog. Reg(s) : America-Central, America-North (U.S. and Canada), Asia-Southeast, Tropical, West Indies
End-Use(s) : Fruit
Domesticated : Y
Ref. Code(s) : 15, 17, 167
Summary : The sapote is a tree reaching 20 m in height that provides an important fruit in the Caribbean. It is eaten fresh and made into preserves. They do not tolerate limey soils. Sapotes grown in Florida and California are not generally successful. Seedling trees bear at 7-8 years regularly and heavily under appropriate soil and climate conditions. Plants are also found in Central America and Southeast Asia.

Sapucaja, Sapucaia Nut
(See Ornamental & Lawn Grass)

Seagrape
Sci. Name(s) : *Coccoloba uvifera*
Geog. Reg(s) : America-Central, America-North (U.S. and Canada), America-South, West Indies
End-Use(s) : Beverage, Fruit, Timber
Domesticated : Y
Ref. Code(s) : 59, 214, 220
Summary : Seagrape trees are about 7.6 m tall with grapelike clusters of sour, purple fruit that are made into a pink jelly. The heavy, close-grained wood is used for cabinet and

furniture-making. In the West Indies, the fruit is made into an alcoholic drink. Sea-grape tree bark is the source of Jamaican kino.

Trees are native to southern Florida, Bermuda, the Bahamas, the West Indies, and Central and South America. They grow along coastal areas and are seldom found inland. Trees are cultivated in Florida. Seagrape trees flower almost continually throughout the year.

Serendipity-Berry, Diel's Fruit
(See Sweetener)

Siberian Crabapple
Sci. Name(s) : *Malus baccata*
Geog. Reg(s) : Asia, Asia-China
End-Use(s) : Fruit, Miscellaneous, Ornamental & Lawn Grass
Domesticated : Y
Ref. Code(s) : 40, 61, 220, 236, 238
Summary : The Siberian crabapple tree is native to northern China and Siberia, grown for its edible fruit eaten fresh or dried. It is most often grown as an ornamental plant. Since it is one of the hardiest and tallest of its genus, it is often used in experimental breeding.

Sierra Plum, Pacific Plum, Klamath Plum
Sci. Name(s) : *Prunus subcordata*
Geog. Reg(s) : America-North (U.S. and Canada)
End-Use(s) : Fruit, Ornamental & Lawn Grass
Domesticated : Y
Ref. Code(s) : 15, 16, 17, 89
Summary : The Sierra plum is a tree reaching 7.6 m in height and producing edible fruit generally used in jams and jellies. Trees are often grown as ornamentals. They grow mainly in the highlands and mountains of Oregon and northern California and are successful in fresh, fertile, sandy soils. Wild trees fruit more heavily than domesticated and cultivated ones.

Smooth Loofah, Sponge Gourd, Dishcloth Gourd, Vegetable Sponge, Luffa, Loofah
(See Miscellaneous)

Sour Cherry, Pie Cherry
Sci. Name(s) : *Prunus cerasus*
Geog. Reg(s) : America-North (U.S. and Canada), Temperate
End-Use(s) : Beverage, Fruit, Gum & Starch, Oil, Timber
Domesticated : Y
Ref. Code(s) : 16, 17, 88, 94, 116, 220
Summary : The sour cherry (*Prunus cerasus*) is a tree 7-9 m in height that produces commercially valuable edible fruit. The cherries are used for culinary purposes and in certain liqueurs. Cherry seeds are the source of a refined semidrying oil (cherry kernel oil) used as a salad oil and in cosmetics. A gum obtained from fruit stems is used

in cotton printing. Leaves are brewed as tea. Cherry trees have hard, durable wood used for turning, inlay work, furniture, and instruments.

Trees are hardy and grown on a wide range of climates and soils. In the United States, sour cherry trees thrive in both north and south, but most trees are grown in New York, Michigan, and Wisconsin.

Sour Orange, Seville Orange
(See Essential Oil)

Soursop
Sci. Name(s) : *Annona muricata*
Geog. Reg(s) : America-Central, Tropical
End-Use(s) : Beverage, Fruit
Domesticated : Y
Ref. Code(s) : 170
Summary : The soursop (*Annona muricata*) is a small evergreen tree widely cultivated as a fruit crop in the coastal valleys of Central America and the lowland areas of the tropics. It is grown for its acidic, juicy fruit, which is made into drinks and ice cream. Because this fruit is highly perishable, it is usually transported in nectar form.

Spanish Lime
Sci. Name(s) : *Melicoccus bijugatus*
Geog. Reg(s) : America, Tropical
End-Use(s) : Fruit, Medicinal
Domesticated : N
Ref. Code(s) : 36, 132, 220
Summary : The Spanish lime is a tree reaching 20 m in height that produces a juicy, slightly acidic fruit. Its pulp is a good source of carbohydrates and a fair source of vitamin A. Trees are grown in tropical America and the fruits are collected and sold locally. Preparations of the bark have been used to treat dysentery. Trees thrive in lowland areas receiving heavy rainfall. They will not tolerate frost.

Spanish Tamarind, Voavanga
Sci. Name(s) : *Vangueria madagascariensis, Vangueria edulis*
Geog. Reg(s) : Africa-West, Asia-China, Hawaii, Pacific Islands, West Indies
End-Use(s) : Fruit
Domesticated : Y
Ref. Code(s) : 61, 154, 220
Summary : The Spanish tamarind is a small, tropical African tree or shrub cultivated for its fruit edible when over-ripe. Trees thrive in full sun on well-drained soils but can survive on dry soils. They are propagated by seed or cuttings and are found in China, the Pacific islands, the West Indies, and Hawaii.

Star-Anise Tree, Star Anise
(See Essential Oil)

Starfruit, Carambola

Sci. Name(s)	: *Averrhoa carambola*
Geog. Reg(s)	: Asia-Southeast, Tropical
End-Use(s)	: Beverage, Fruit, Miscellaneous
Domesticated	: Y
Ref. Code(s)	: 170
Summary	: The starfruit tree reaches 5-12 m in height and grows wild in Indonesia and throughout the tropics. Its edible fruit is 5-angled and star-shaped if cut in half. Ripe fruit has juicy, aromatic flesh used in salads, tarts, preserves, and drinks. The fruit may be used for cleaning brassware.

Strawberry Guava, Cattley Guava, Waiawi-'ula'ula, Wild Guava, Purple Guava

Sci. Name(s)	: *Psidium cattleianum, Psidium littorale*
Geog. Reg(s)	: America-North (U.S. and Canada), America-South, Hawaii, Tropical
End-Use(s)	: Beverage, Fruit, Ornamental & Lawn Grass
Domesticated	: Y
Ref. Code(s)	: 17, 145, 154, 167, 170, 207
Summary	: The strawberry guava is a bushy shrub or small tree that produces edible fruit eaten fresh or in jams, jellies, custards, and sherberts. Fruit pulp is processed into a type of wine. The fruit is a poor source of calcium, phosphorous, and iron but is a fair source of niacin and a good source of ascorbic acid. It is grown as an ornamental.

Hawaii is a naturalized home of the strawberry guava. It is native to Brazil and has spread throughout the tropics. Trees thrive at elevations from 60-300 m. In the United States, strawberry guavas fruit satisfactorily in Florida. Trees need a dry climate and rich, sandy loams or red clay ones for best production.

Sugar-Apple, Sweetsop

Sci. Name(s)	: *Annona squamosa*
Geog. Reg(s)	: Africa-Central, Mediterranean, Tropical, West Indies
End-Use(s)	: Beverage, Fruit
Domesticated	: Y
Ref. Code(s)	: 148, 170
Summary	: The sugar-apple tree is widely cultivated as a fruit crop throughout the tropics at low and medium altitudes. Sugar-apples contain 16-18% sugar and are mixed with wine, ice cream, or milk. In India, this fruit is called custard apple and is a dessert fruit. Sugar-apple trees are native to the West Indies. Trees tolerate hot climates such as the Nile and Jordan valleys.

Surinam-Cherry, Pitanga Cherry

Sci. Name(s)	: *Eugenia uniflora, Eugenia michellii*
Geog. Reg(s)	: America, America-South, Tropical
End-Use(s)	: Beverage, Dye & Tannin, Fruit, Insecticide, Ornamental & Lawn Grass
Domesticated	: Y
Ref. Code(s)	: 167, 170, 220
Summary	: The Surinam-cherry tree has a variety of uses. The fully ripened fruit is edible. In Brazil, jellies and sherbets are made from the cherries. Liqueurs, syrups, and

wines can also be made from the fruit. Trees are grown as hedges and windbreaks and the leaves are used for decoration. The bark is used for tanning and crushed leaves are spread in barns to repel insects.

Surinam-cherry trees experience optimum growth in light, sandy soils and require moisture. They can adapt to a wide range of soils. Trees bear fruit in 4-5 years and will continue fruiting once a year. These trees are important in Buddhist iconography.

Sweet Blueberry, Lowbush Blueberry, Low Sweet Bush, Late Sweet Bush, Late Sweet Blueberry

Sci. Name(s)	: *Vaccinium angustifolium*
Geog. Reg(s)	: America-North (U.S. and Canada)
End-Use(s)	: Fruit, Medicinal
Domesticated	: Y
Ref. Code(s)	: 17, 59, 236
Summary	: The sweet blueberry is a wild plant whose edible berries are collected and sold in fresh fruit markets or as canned fruit. The sweet berries are a valuable food source for wildlife. At one time, the berries were used in herbal medicines. It is widely distributed in woodland and sunny places along the northeastern coast of the United States.

These plants require an acidic, peaty or sandy soil, and a good supply of moisture for optimum production. They do not thrive on limey soils. Plants are grown commercially in large fields in Maine. They make good ground cover. Plants grow from 0.3-0.6 m high and tend to thrive more readily after being cut and burned.

Sweet Cherry, Bird Cherry, Mazzard, Gean

Sci. Name(s)	: *Prunus avium*
Geog. Reg(s)	: America-North (U.S. and Canada), Eurasia
End-Use(s)	: Fruit
Domesticated	: Y
Ref. Code(s)	: 16, 17, 88, 94, 220
Summary	: The sweet cherry (*Prunus avium*) is a tree reaching up to 10.6 m in height that is grown for its edible fruits eaten fresh, cooked, canned, or dried. There are two main varieties: one produces fruit with soft, tender flesh called Guigne; the other produces fruit with firm, breaking flesh, and called Bigarreaus. There are approximately 600 varieties of sweet cherry now in cultivation.

Trees are grown in regions with equitable climates. In the United States, trees thrive in the Hudson river valley of New York, along the Great Lakes, and Pacific Coast. Trees do not endure prolonged heat and need deep, well-drained, sandy or gravelly loams. Sweet cherry trees are native to Eurasia and naturalized in North America.

Sweet Granadilla, Granadilla
(See Beverage)

Sweet Orange, Orange

Sci. Name(s)	: *Citrus sinensis*

Geog. Reg(s)	: Africa-South, America-North (U.S. and Canada), Asia-China, Asia-Southeast, Europe, Subtropical, West Indies
End-Use(s)	: Beverage, Essential Oil, Forage, Pasture & Feed Grain, Fruit
Domesticated	: Y
Ref. Code(s)	: 132, 148, 170, 180
Summary	: The sweet orange tree produces one of the most popular dessert fruits, the orange. The fruit is marketed fresh, canned, or as frozen orange juice concentrate. The pulp contains 4-13% carbohydrates and is a good source of vitamin C. After juice processing, the resulting pulp is often used as cattle feed. The orange peel produces an essential oil used in flavoring and essences and is a good source of pectin. The oil is produced commercially in California, Florida, Sicily, and the West Indies. Smaller areas of production are in Spain, Zimbabwe, Israel, India, Indonesia, China, Japan, Greece, and Pakistan.

Orange trees grow from 6-8 m in height. Trees are most successful in subtropical areas from 800-1,200 m in altitude and where fruit can be exposed to cool temperatures during maturation. The tree is native to southern China.

In 1984, the United States imported 3,165.8 MT of orange oil as an essential oil, valued at $3.3 million. The fresh orange market is exceeded only by bananas in international trade. The largest fresh orange producers are the United States, Brazil, Spain, Italy, and Mexico. In 1983, orange production in the United States was 8.6 million MT, valued at $1.4 billion. Seventy-three percent of the oranges were processed, while the remaining 27% were sold as fresh fruit.

Tachibana Orange, Tachibana

Sci. Name(s)	: *Citrus tachibana*
Geog. Reg(s)	: Asia, Asia-China
End-Use(s)	: Fruit, Medicinal
Domesticated	: N
Ref. Code(s)	: 170, 225
Summary	: The tachibana orange tree (*Citrus tachibana*) provides fruits that are eaten fresh. The peel has certain unspecified medicinal qualities. Trees are usually wild, growing mainly in China and Japan. There is little or no information on economic possibilities for tachibana orange production.

Tamarind

Sci. Name(s)	: *Tamarindus indica*
Geog. Reg(s)	: Africa, Tropical
End-Use(s)	: Beverage, Cereal, Energy, Fruit, Gum & Starch, Miscellaneous, Ornamental & Lawn Grass, Spice & Flavoring, Vegetable
Domesticated	: Y
Ref. Code(s)	: 170
Summary	: The tamarind is a semievergreen tree growing to 20 m high that is grown for its edible fruit. Fruit pulp is eaten fresh or mixed with sugar as a sweet meat (tamarind balls). It is used for seasonings, in curries, and can be made into an acidic drink and sherbet. Seeds are eaten roasted, boiled, or as flour. In India, seeds are used as a carbohydrate source for sizing cloth, paper, and jute products. They are also made into a vegetable gum and used in food processing. Tree flowers and leaves are eaten

in salads. Overly ripe fruits have been used as copper and brass polishers. Tamarinds make attractive ornamental and shade trees because of their height and compact, rounded crown. Tree wood makes good charcoal.

Trees grow wild in the drier regions of tropical Africa and throughout the tropics. They are suited to the semiarid tropical regions and grow in any well-drained soil. Fruit is harvested primarily for domestic consumption. There has been a small export market for the dried pulp from India to Europe and North America, for use in chutneys and meat sauces.

Tangerine, Mandarin Orange

Sci. Name(s)   : *Citrus reticulata*
Geog. Reg(s)   : America-North (U.S. and Canada), Asia, Asia-China, Subtropical, Tropical
End-Use(s)   : Fruit
Domesticated   : Y
Ref. Code(s)   : 132, 148, 170
Summary   : Tangerine trees reach up to 8 m in height and are grown for their edible fruits, which are used as dessert fruits. Commercially, they are sold fresh or canned. The pulp is a good source of vitamin A and carbohydrates and a fair source of vitamin C. In 1983, United States production was 199,193 MT, 60% of which was used fresh and 40% processed. Total production value was about $41.6 million.

Trees have a long history of cultivation in China and Japan. They are introduced into the United States, and grow in most tropical and subtropical countries.

There are several horticultural varieties: the King group has large, seedy fruits with thick skin, and the Satsuma group, popular in Japan. These fruits are cold-hardy and adapted to tropical areas. Another group is the hybrids, mainly Kara, Kinnow, and Wilking.

Tomato

Sci. Name(s)   : *Lycopersicon esculentum*
Geog. Reg(s)   : America-North (U.S. and Canada), America-South, Europe, Hawaii, Mediterranean
End-Use(s)   : Forage, Pasture & Feed Grain, Fruit, Oil, Vegetable
Domesticated   : Y
Ref. Code(s)   : 154, 170, 220
Summary   : The tomato produces a fruit that is an important vegetable. The plant originated in Peru and Ecuador and spread to tropical America, Europe, and the Mediterranean. The United States is the largest producer. In 1983, tomato production in the United States reached 381.2 million MT valued at $1.5 billion. Seventy percent was used for processing. Italy is the largest exporter of tomato products. Tomatoes are eaten cooked or raw, in a variety of ways. Green tomatoes are pickled or preserved. Seeds contain 24% oil used as a salad oil and for making margarine and soap. The residual press cake is utilized as stock food or fertilizer.

There are several varieties: the cherry tomato, pear tomato, common tomato, potato-leaved tomato, and the upright tomato. In Hawaii, a number of disease resistant cultivar have been developed.

The tomato withstands various climatic conditions but prefers areas with long, sunny periods, light rainfall, and night temperatures of 10-20 degrees Celsius. In 1983, average yields in the United States were 44,818 kg/ha.

Tree-Tomato

Sci. Name(s)	: *Cyphomandra betacea*
Geog. Reg(s)	: America-South, Tropical
End-Use(s)	: Fruit
Domesticated	: Y
Ref. Code(s)	: 132, 170
Summary	: The tree-tomato reaching up to 3-6 m in height is grown for its edible fruits eaten fresh, stewed, or made into preserves. The fruit pulp is a good source of carbohydrates and minerals.

This tree is native to Peru and is cultivated extensively in the Andes. It requires cool, highland climates, well-distributed rainfall, and no frost. The tree tomato has been introduced to the higher-altitudes of the tropics.

Virginia Strawberry

Sci. Name(s)	: *Fragaria virginiana*
Geog. Reg(s)	: America-North (U.S. and Canada)
End-Use(s)	: Beverage, Fruit
Domesticated	: Y
Ref. Code(s)	: 9, 213, 214, 219, 238
Summary	: The Virginia strawberry is a perennial herb that produces edible fruits eaten fresh. Plants grow wild in eastern North America from Texas to Arizona. The leaves can be made into a tealike drink. These plants are self-sterile. Commercially, they are limited by their perishability.

Vogel Fig
(See Isoprenoid Resin & Rubber)

Wampi

Sci. Name(s)	: *Clausena lansium*
Geog. Reg(s)	: Asia-China, Tropical
End-Use(s)	: Fruit, Medicinal
Domesticated	: Y
Ref. Code(s)	: 36, 170
Summary	: Wampi is a small tree whose edible, slightly acidic fruit pulp is eaten fresh or as jam. Immature fruits are used in preparations for treating bronchitis. This tree is native to southern China and now grows in several tropical countries.

Water Lemon, Jamaica Honeysuckle, Belle Apple, Pomme de Liane, Yellow Granadilla

Sci. Name(s)	: *Passiflora laurifolia*
Geog. Reg(s)	: America-South, Asia-Southeast, West Indies
End-Use(s)	: Beverage, Fruit
Domesticated	: N

Ref. Code(s) : 4, 170
Summary : Water lemon is a woody, climbing vine that grows wild in the West Indies and northeastern South America. It produces edible fruit with pulp rich in calories and ascorbic acid, eaten fresh, and processed into a beverage. Plant leaves are poisonous.

Vines are cultivated and distributed throughout the tropical lowlands. In Malaysia, water lemons are considered the best of all the passion fruits.

Watermelon

Sci. Name(s) : *Citrullus lanatus*
Geog. Reg(s) : America-North (U.S. and Canada), Europe, Hawaii, Subtropical, Tropical
End-Use(s) : Forage, Pasture & Feed Grain, Fruit, Oil
Domesticated : Y
Ref. Code(s) : 170
Summary : The watermelon is a viny annual herb that produces large, sweet fruit. The ripe flesh is eaten fresh and the seeds are ground and used in baking. About 20-45% of an edible semidrying oil is extracted from the seeds and used in cooking and as an illuminant. The seeds also contain 30-40% protein and are rich in the enzyme urease which promotes the hydrolysis of urea. The expressed oilseed cake is fed to livestock.

Melons are grown throughout the drier areas of the tropics and subtropics, in southern Europe, Central and southern United States, and Hawaii as domestic market produce.

Watermelons grow from seed in fertile, sandy soil. They are fairly drought-resistant but are easily killed by waterlogging and frost. Fruit is harvested 4-5 months after sowing. It is fragile, requires careful handling, and cannot be stored for more than 2-3 weeks. A typical harvest yields approximately 1200 marketable fruits/ha.

In 1983, watermelon production in the United States reached 1.2 million MT. In addition, the United States imported 84,550 MT of fresh watermelon, valued at $12.5 million.

Waxgourd, White Gourd
(See Vegetable)

Western Elderberry, Blue Elderberry, Blueberry Elder

Sci. Name(s) : *Sambucus glauca, Sambucus cerulea*
Geog. Reg(s) : America-North (U.S. and Canada)
End-Use(s) : Fruit, Ornamental & Lawn Grass
Domesticated : Y
Ref. Code(s) : 16, 17, 192
Summary : The western elderberry is a shrub or tree reaching 4.5-9.1 m in height with edible fruit used in wines, pies, sauces, and jellies. In the United States, plants are grown as ornamentals along the Pacific Coast. They are fairly pest and disease free and grow in most moist soil.

White Mulberry
(See Miscellaneous)

Wild Guava, Montain Guava

Sci. Name(s)	: *Psidium montanum*
Geog. Reg(s)	: West Indies
End-Use(s)	: Fruit
Domesticated	: N
Ref. Code(s)	: 2, 170
Summary	: The wild guava tree produces an edible fruit collected as a fresh fruit for domestic markets in Jamaica. Trees grow wild on wooded hillsides. Fruiting occurs in January, April, June, and September.

Wine Grape, Grape, European Grape, California Grape
(See Beverage)

Wine Palm, Coco de Chile, Coquito Palm, Honey Palm, Wine Palm of Chile, Chile Coco Palm
(See Beverage)

Wintergreen
(See Essential Oil)

Wonderberry, Black Nightshade, Poisonberry, Stubbleberry
(See Miscellaneous)

Wooly Pawpaw, Flag Pawpaw

Sci. Name(s)	: *Asimina grandiflora, Asimina speciosa, Pityothamus incanus, Annona grandiflora, Orchidocarpum grandiflorum, Porcella grandiflora*
Geog. Reg(s)	: America-North (U.S. and Canada)
End-Use(s)	: Fruit
Domesticated	: N
Ref. Code(s)	: 60, 197, 213
Summary	: The wooly pawpaw (*Asimina grandiflora*) is a shrub about 1.5 m in height that bears edible fruit. This plant grows on the coastal plains of the southeastern United States.

Gum & Starch

Ajipo, Tuberosus Yam Bean, Yam Bean, Potato Bean
(See Insecticide)

Almond
(See Nut)

Ambarella, Otaheite Apple, Golden-Apple, Jew-Plum
(See Fruit)

Apple
(See Fruit)

Arrowroot

Sci. Name(s)	: *Maranta arundinacea*
Geog. Reg(s)	: Africa, America-South, Tropical, West Indies
End-Use(s)	: Forage, Pasture & Feed Grain, Gum & Starch, Medicinal
Domesticated	: Y
Ref. Code(s)	: 154, 171, 205, 220
Summary	: Arrowroot plants provide a starch used in jellies or pastes. It is often part of infant and invalid food because it is highly digestible. The starch is used as a base for face powders and in certain types of glue. It counteracts insect stings. A fibrous debris from the rhizomes is fed to cattle or returned to the land. Indians use the rhizomes to treat wounds and ulcers, sooth the stomach, and aid digestion. It is used in pill manufacture and in barium meals for X-ray treatment of the gastrointestinal system.

Arrowroot is found in northern South America and the Lesser Antilles. It has been introduced throughout the tropics. St. Vincent Island is the major producer of the starch, while Bermuda produces the highest quality starch. The principal starch importers are the United States and the United Kingdom.

Plants require 1,500 to 2,000 mm of rain a year and grow best on sandy loams. The rhizomes mature in 10-12 months. The maximum amount of starch can be obtained at 12 months but the rhizome is more fibrous and the starch harder to extract at that time. Crops can be harvested for 5-6 years and can yield 7.5-37 MT/ha of rhizomes. About 2.5-7.5 MT/ha of starch is possible.

Babul

Sci. Name(s)	: *Acacia nilotica, Acacia arabica*
Geog. Reg(s)	: Africa, Asia-India (subcontinent)
End-Use(s)	: Dye & Tannin, Energy, Erosion Control, Fiber, Forage, Pasture & Feed Grain, Gum & Starch, Timber
Domesticated	: N
Ref. Code(s)	: 50, 170, 173
Summary	: The babul tree of India exudes an industrial gum called babul gum. This water-soluble gum was probably the original gum arabic. In the Democratic Republic of the Sudan and western Pakistan, it is a valuable economic resource.

Also valuable is the hard, fine-grained babul timber. Although difficult to work, it can be brought to a high polish, making it popular in eastern Sudan for general carpentry and boat building. The wood is also used for tool handles, pestles and mortars, door frames, bowls, and other small domestic items. Babul timber makes good firewood. Trees have been used to forest flood plains in the eastern Sudan and upper Nile regions.

Tree pods, fruit, and bark provide varieties of tannins. The pods and leaves of the young plant are used as fodder. Fiber from the stems is made into mats.

Babul trees are fairly large trees, some reaching heights of 24-30.5 m. Trees grow wild throughout tropical Africa and as far east as India.

Betel Pepper
(See Miscellaneous)

Buffalo Gourd, Wild Gourd
(See Oil)

Burweed
(See Fiber)

Carob, St. John's-Bread

Sci. Name(s)	: *Ceratonia siliqua*
Geog. Reg(s)	: Mediterranean
End-Use(s)	: Beverage, Forage, Pasture & Feed Grain, Gum & Starch, Sweetener, Timber
Domesticated	: Y
Ref. Code(s)	: 47, 220
Summary	: The carob tree is cultivated in the Mediterranean for its pods, which contain about 50% sugar and protein. The pods are valuable forage for stock.

A gum (locust gum) extracted from the endosperm of the pods is made into a fine powder that has extensive use in the food industry. It stabilizes ice cream, is used in the preparation of cheeses and extruded meat products, improves the water retention properties of doughs, and is used as an emulsifier and stabilizer in dressings and sauces. In the textile industry, the gum is an excellent film-former and is often used alone or in combination with starch in textile sizing and dye thickeners and in textile printing. In the paper-making industries, the gum is used for fiber bonding and beater additives. The gum is used in drilling mud because it controls water losses and mud viscosity.

Carob seeds are used as a coffee substitute (carob coffee). In Cyprus, a molasses (pasteil) is made from the beans. The workable wood is used to make furniture and carts.

Gum yields are about 60 kg of gum per 453 kg of pods. In 1986, prices were in the $4.80-$5.50/kg range.

Cassava, Manioc, Tapioca-Plant, Yuca, Mandioca, Guacomole
(See Energy)

Chayote
(See Vegetable)

Chicle, Sapodilla, Naseberry, Nispero, Chicle Tree

Sci. Name(s)	: *Manilkara zapota, Manilkara ochras, Manilkara zapotilla*
Geog. Reg(s)	: America-Central, Tropical
End-Use(s)	: Fruit, Gum & Starch, Timber
Domesticated	: N
Ref. Code(s)	: 61, 132, 154
Summary	: Chicle is an evergreen tree native to Central America. It can grow up to 38 m but usually does not exceed 22 m. Its latex is the main source of chicle, which is used in chewing gum. The trees are tapped once every 2-3 years and yield about 68 liters of latex. Chicle also bears a soft sweet fruit that is a favorite in tropical America. The fruit is eaten fresh, used to flavor ice cream, and made into syrup and jam. Its fine hard wood is used to make various wooden articles.

Plants can grow in ordinary soils but prefer limestone and humid tropical atmospheres. They are frost hardy. Chicle needs moisture and shade from the sun.

Chinese Waterchestnut, Waternut
(See Tuber)

Chufa, Ground Almond, Tigernut, Yellow Nutsedge
(See Forage, Pasture & Feed Grain)

Cimaru, Tonka Bean
(See Spice & Flavoring)

Colchicum, Meadow Saffron, Autumn Crocus
(See Medicinal)

Convolvulacea Yam
(See Tuber)

Cow-Tree

Sci. Name(s)	: *Brosimum utile, Brosimum galactodendron*
Geog. Reg(s)	: America
End-Use(s)	: Beverage, Fiber, Gum & Starch
Domesticated	: N
Ref. Code(s)	: 170, 220
Summary	: The cow-tree is distributed throughout tropical America. A latex from the bark can be drunk raw like milk. It is also used as a base for chewing gum. The bark has been used as a cloth for making sails and blankets on a noncommercial basis.

Cuddapah Almond, Chirauli Nut
(See Oil)

Dasheen, Elephant's Ear, Taro, Eddoe
(See Tuber)

Edible Canna, Gruya, Queensland Arrowroot

Sci. Name(s) : *Canna edulis*
Geog. Reg(s) : America-Central, America-South, Asia, Australia, Pacific Islands, Tropical
End-Use(s) : Forage, Pasture & Feed Grain, Gum & Starch, Weed
Domesticated : Y
Ref. Code(s) : 148, 171
Summary : Edible canna is grown commercially in Australia for its starchy, underground stems. The stems, which contain 25% starch, are baked and eaten or used as feed for livestock. In Central America, edible canna is considered a weed.

Plants grow in clumps with stems reaching heights of 1-2.5 m. The fleshy, branched underground stem grows to about 60 cm in length.

The plant has a long history of cultivation in the Andes of South America. Today, edible canna is cultivated in Queensland where yields range from 100-161 MT/ha of tubers. It is also cultivated in Asia, parts of the Pacific, and other tropical areas.

Fenugreek
(See Spice & Flavoring)

Fishtail Palm, Toddy Palm
(See Sweetener)

Giant Taro, Giant Alocasia

Sci. Name(s) : *Alocasia macrorrhiza*
Geog. Reg(s) : Asia-India (subcontinent), Asia-Southeast, Pacific Islands
End-Use(s) : Gum & Starch
Domesticated : Y
Ref. Code(s) : 36, 171
Summary : The giant taro is a large herb whose edible, stout, permanent, above-the-ground stem is a valuable starch source. The stem contains calcium oxalate raphides, which must be removed during cooking. Giant taro is used as a famine food. Plants are native to Sri Lanka. Giant taro was cultivated throughout tropical Southeast Asia and was introduced to many Pacific islands.

Golden Wattle
(See Dye & Tannin)

Guar, Clusterbean

Sci. Name(s) : *Cyamopsis tetragonoloba, Cyamopsis psoralioides*
Geog. Reg(s) : America-North (U.S. and Canada), Asia-India (subcontinent)
End-Use(s) : Forage, Pasture & Feed Grain, Gum & Starch, Natural Resin, Vegetable
Domesticated : Y
Ref. Code(s) : 47, 126, 170
Summary : Guar is a bushy annual that is grown for gum production in India and the southwestern United States. In India and Pakistan, the young tender pods are eaten

as vegetables and the seeds are used as livestock feed. Plants are also grown as green manure crops and are used for fodder. Guar seed flour possesses 5-8 times the thickening power of starch and is used to improve the grade strength of certain types of paper and for the gum backing of postage stamps. It is also used in textile sizing and as a thickener in ice creams and salad dressings. A water-soluble natural resin is obtained from the seed extract and is used in the food, paper-making, mining, and petroleum industries.

Plants are grown on alluvial or sandy soil in semiarid regions with high temperatures and rainfall not exceeding 500-600 mm/year. When grown in drier climates, guar is fairly resistant to pests and diseases. Guar is native to tropical Africa and Asia. Plants are grown on a commercial scale in Pakistan and India, and to a lesser extent in the southwestern parts of the United States. In 1984, the United States imported 30,540 MT of gum, valued at $4.5 million.

In India, guar crops are usually grown in mixed cultivation. Crops will begin producing 3-3.5 months after sowing. Dry seed yields average 400-600 kg/ha, sometimes reaching 1,600 kg/ha.

Gum Arabic, Senegal Gum, Kher

Sci. Name(s)   : *Acacia senegal*
Geog. Reg(s)   : Africa, Asia-India (subcontinent)
End-Use(s)   : Fiber, Forage, Pasture & Feed Grain, Gum & Starch
Domesticated   : N
Ref. Code(s)   : 170
Summary   : The gum arabic tree produces one of the major industrial gums of the same name. The water-soluble gum is colorless and odorless, and is used in textiles, mucilage, paste, polishes, and confectionery. It is also used in lithographic work, as a glue in painting, and as an emulsifying and demulcent agent in pharmacy. In parts of Africa, a fiber called "me'di" is stripped from the bark. The young tree leaves are used as fodder.

Gum arabic trees grow throughout the drier regions of Africa, especially in the Democratic Republic of the Sudan and southern Sahara. Trees also grow in the dry hills of the Sind, and on the Aravalli ranges of India. Gum is collected from the tree bark 1 month after incisions have been made. Collection is from wild and semicultivated trees. In 1984, the United States' gum imports were just over 11,574 MT, valued at $19.0 million.

Gum Ghatti, India Gum

Sci. Name(s)   : *Anogeissus latifolia*
Geog. Reg(s)   : Africa, Asia-India (subcontinent)
End-Use(s)   : Gum & Starch
Domesticated   : N
Ref. Code(s)   : 47, 220, 230
Summary   : Gum ghatti is a tree that grows in India and Sri Lanka and provides a natural exudate gum. Often, the yield of gum can be increased by artificial incisions. The gum, which occurs as a calcium magnesium salt, is allowed to dry in the sun and is then sorted according to color and impurities and ground into a fine powder. The largest crop is picked in April. Yields are greatest after the monsoon season.

This gum has a viscosity intermediate between gum arabic and gum karaya. Its emulsification and adhesive properties are often equal or superior to gum arabic's.

Gum ghatti is used as an emulsifier for oil and water emulsions in foods and pharmaceutical products, such as in preparation of stable, powdered, oil-soluble vitamins. Other uses are as a waterproofing agent in liquid explosives and as a thickener in dying and sizing in the textile industry. It does not form a true gel, and films prepared with it are relatively soluble and brittle. It is also used in acidizing oil wells and in stabilizing certain colors in photoelectric work. It has potential in forming protective hydration layers around clay particles. The use of gum ghatti has declined in recent years.

Gum Tragacanth, Tragacanth

Sci. Name(s)	: *Astragalus gummifer*
Geog. Reg(s)	: Africa-West, Asia, Mediterranean
End-Use(s)	: Gum & Starch, Medicinal
Domesticated	: Y
Ref. Code(s)	: 47, 170
Summary	: Gum tragacanth is a West African plant that exudes a useful gum from its incised stems. The gum is dried and collected as ribbons, which are considered the highest quality. Lower-quality gum is collected in flakes.

The gum contains a soluble fraction, tragacanthin. Sixty to 70% of the gum is an insoluble fraction, bassorin, which swells to a gellike form. The gel is used in cosmetics, calico printing, confectionery, medicinally in insoluble powders, and in emulsifying oils and resins. Because the gum has a very high and stable viscosity, it is valued in the food industries, where it is used as a thickening or stabilizing agent in ice creams, salad dressings, and candies. It is also used in textile industries as a sizing agent and print paste thickener, and as an emulsifier in furniture, floor, and auto polishes. The gum is being replaced by propylene glycol alginate and xantham gum.

Plants are grown commercially in the dry mountains of Iran, Syria, and Turkey. Because of continued industrial expansion and political unrest in Iran, gum tragacanth prices have increased. In 1978, #1 ribbon-grade tragacanth was $57-$62/kg. In New York, United States imports in 1984 amounted to 188 MT with a value of $2 million. By 1986, prices had risen to the $79-$88/kg range.

Imbe
(See Beverage)

Indian Bael, Bael Fruit, Bengal Quince, Bilva, Siniphal, Bael Tree
(See Fruit)

Indian Wood-Apple
(See Fruit)

Jackfruit, Jack
(See Fruit)

Jelutong

Sci. Name(s) : *Dyera costulata*
Geog. Reg(s) : Asia-Southeast
End-Use(s) : Gum & Starch, Isoprenoid Resin & Rubber
Domesticated : N
Ref. Code(s) : 31, 116, 170, 202
Summary : Jelutong is a tree found in the forests of Malaysia. This tree contains 70-80% latex, which includes gums and resins. Only 20% of the total dry weight is rubber. Jelutong is now chiefly used as a substitute for chicle. About 85% of some chewing gums consists of jelutong. Each year, approximately 1 million kg is produced.

Jerusalem-Artichoke, Girasole
(See Tuber)

Kadjatoa

Sci. Name(s) : *Eugeissona utilis*
Geog. Reg(s) : Asia-Southeast
End-Use(s) : Dye & Tannin, Gum & Starch
Domesticated : Y
Ref. Code(s) : 171, 220
Summary : Kadjatoa is a plant occasionally cultivated for its starch-filled stems. A bright purple pollen is also collected and used to color rice dishes. Kadjatoa is native to Borneo.

Karaya Gum, Kateera Gum

Sci. Name(s) : *Sterculia urens*
Geog. Reg(s) : Asia-India (subcontinent), Tropical
End-Use(s) : Gum & Starch
Domesticated : Y
Ref. Code(s) : 47, 170, 220, 230
Summary : Gum karaya is a large bushy tree cultivated on the dry rocky hills in Central and northern India for its gum. It reaches heights of 9 m. Trees are tapped about 5 times in their lifetime. The exuded gum is dried on the trees, and is then collected and sorted on the basis of color and purity. Gum karaya is the least soluble gum exudate and tends to swell in cold water, producing a viscous solution.

Karaya gum is used in foods as a thickening, suspending, and stabilizing agent, and as an emulsifier and binder in cheese spreads and sausages. When used in conjunction with carageenans or alginates, karaya gum helps preserve the freshness of bakery goods. It is also used for such pharmaceutical products as dental adhesives and bulk laxatives. It is used as a pulp binder, as a thickening agent for printing dyes, and in textile sizing. It is a substitute for gum tragacanth.

In 1986, the price range of karaya gum was $4.29-$4.95/kg, depending on the grade. United States imports in 1984 were 2,401 MT of natural gum at a value of $6.5 million.

Leadtree, Lead Tree
(See Forage, Pasture & Feed Grain)

Maize, Corn, Indian Corn
(See Cereal)

Okra, Lady's Finger, Gumbo
(See Vegetable)

Papaya
(See Fruit)

Pepper Tree, Brazilian Pepper Tree, California Peppertree

Sci. Name(s)	: *Schinus molle*
Geog. Reg(s)	: America-North (U.S. and Canada), America-South, Tropical
End-Use(s)	: Beverage, Essential Oil, Gum & Starch, Medicinal, Ornamental & Lawn Grass, Spice & Flavoring
Domesticated	: Y
Ref. Code(s)	: 77, 152, 154, 170, 220
Summary	: The pepper tree is a medium-sized evergreen 5-15 m in height that is a source of an industrial gum called American mastic. The berrylike fruit can be ground and made into a beverage or can be dried and used as a pepper substitute. The fruit is aromatic and has a sweet, spicy, pepperlike flavor. About 3.35-5.2% of an essential oil may be obtained from the fruit. Tree bark can be ground and used as a purgative.

In the tropics, trees are planted at higher elevations for best productivity. In the United States, trees are cultivated in California. They flower and fruit from late spring to early autumn. South American pepper trees are often cultivated as attractive ornamentals.

Peruvian-Carrot, Arracacha
(See Tuber)

Pili Nut, Elemi
(See Nut)

Potato, European Potato, Irish Potato, White Potato
(See Tuber)

Quince

Sci. Name(s)	: *Cydonia oblonga*
Geog. Reg(s)	: Asia-Central, Europe, Mediterranean
End-Use(s)	: Fruit, Gum & Starch, Medicinal
Domesticated	: Y
Ref. Code(s)	: 42, 47
Summary	: Quince is a tree that is cultivated in warm temperate countries for its fruit, from which a gum is obtained. The gum is soluble in cold water but hydrates faster in hot water to give a highly viscous solution. The solution is used as a laxative and in cosmetics and demulcent lotions. It has also been used as an emulsifying and stabilizing agent in some medicinal preparations. The fruit is made into preserves.

Iran is a major producing country, followed by Spain, southern France, Portugal, and Russia. Quince is native to Central Asia. The 1984 New York price for quince seed in bags was $4.40-$6.05/kg, depending upon the quality.

Rice
(See Cereal)

Sacred Lotus, Lotus, East Indian Lotus
(See Tuber)

Salsify, Oyster-Plant, Vegetable Oyster
(See Tuber)

Senegal Rosewood, Barwood, West African Kino, Red Barwood
(See Dye & Tannin)

Shittim-Wood
(See Forage, Pasture & Feed Grain)

Siris, East Indian Walnut, Koko
(See Timber)

Slenderleaf Crotalaria
(See Forage, Pasture & Feed Grain)

Sorghum, Great Millet, Guinea Corn, Kaffir Corn, Milo, Sorgo, Kaoling, Durra, Mtama, Jola, Jawa, Cholam Grains, Sweets, Broomcorn, Shattercane, Grain Sorghums, Sweet Sorghums
(See Cereal)

Sour Cherry, Pie Cherry
(See Fruit)

Sugar Palm, Gomuti Palm

Sci. Name(s)	: *Arenga pinnata, Arenga saccharifera*
Geog. Reg(s)	: Asia-India (subcontinent), Asia-Southeast
End-Use(s)	: Fiber, Gum & Starch, Miscellaneous, Sweetener
Domesticated	: Y
Ref. Code(s)	: 171
Summary	: The sugar palm (*Arenga pinnata*) contains large quantities of starchy sap in its trunk. The sap is converted to sugar when palms flower at 7-10 years of age. Male inflorescence stalks are then tapped daily for 2-3 months, yielding 3-5 liters/day of sap. The sap contains about 15% sucrose. Gomuti fiber is stripped from the leaf sheaths. Sago starch is also obtained from this palm and others of the Arenga species.

Sugar palms grow wild from northeastern India through the Malay Archipelago and into the Philippines. They reach 7-12 m in height. Sugar palm sap was probably one of man's first sources of sugar. The palm is seldom cultivated except in parts of Southeast Asia and the Philippines.

Sweet Potato
(See Tuber)

Tamarind
(See Fruit)

Wheat, Bread Wheat, Common Wheat
(See Cereal)

Winged Bean, Goa Bean, Asparagus Pea, Four-Angled Bean, Manilla Bean, Princess Pea
(See Vegetable)

Yam Bean, Erosus Yam Bean, Jicama
(See Tuber)

Zedoary, Shoti
(See Essential Oil)

Insecticide

Ajipo, Tuberosus Yam Bean, Yam Bean, Potato Bean

Sci. Name(s) : *Pachyrhizus tuberosus*
Geog. Reg(s) : America-South, Asia-China, West Indies
End-Use(s) : Gum & Starch, Insecticide, Tuber
Domesticated : Y
Ref. Code(s) : 56, 170
Summary : Ajipo (*Pachyrhizus tuberosus*) is an herbaceous, perennial vine with edible tuberous roots that contain a starch used in custards and puddings. Plants contain the insecticide rotenone. Ajipo is similar to the erosus yam bean (*Pachyrhizus erosus*).

These vines are cultivated in Venezuela, Trinidad, Jamaica, Puerto Rico, and China. They thrive in loose, sandy, well-drained soil in areas with an annual rainfall of 6.4-41 dm and annual temperatures of 21.3-27.4 degrees Celsius. The average seed yield is 600 kg/ha and the average tuber yield is 7,000-10,000 kg/ha. Tubers are harvested 4-8 months after planting.

Avens
(See Spice & Flavoring)

Barbasco

Sci. Name(s) : *Lonchocarpus utilis*
Geog. Reg(s) : America-South
End-Use(s) : Insecticide
Domesticated : N
Ref. Code(s) : 220
Summary : The barbasco plant is native to Peru. Its roots are a minor source of the insecticide rotenone. There is limited information concerning physical characteristics or economic potential for this plant.

Camphor-Tree, Camphor
(See Medicinal)

Crown Daisy, Tansy
(See Medicinal)

Eastern Elderberry, American Elderberry, American Elder
(See Fruit)

European Pennyroyal
(See Essential Oil)

Indian Mulberry
(See Dye & Tannin)

Longleaf Pine
(See Essential Oil)

Luba Root, Tuba Root

Sci. Name(s) : *Derris* spp.
Geog. Reg(s) : Asia-Southeast
End-Use(s) : Insecticide
Domesticated : Y
Ref. Code(s) : 170, 174, 220
Summary : Luba root is a tropical liane that has gained increasing importance as it is a source of the insecticide rotenone, found in its roots. This insecticide is valuable because it seems to have no effect on warm-blooded animals. Plants are grown primarily in Southeast Asia. Harvest occurs in 3 years. A single root contains up to 20% rotenone. See macaquim ho timbo for import information regarding rotenone.

Macaquim Ho Timbo, Cube, Timbo, White Haiari, Barbasco

Sci. Name(s) : *Lonchocarpus nicou*
Geog. Reg(s) : America-South
End-Use(s) : Insecticide
Domesticated : N
Ref. Code(s) : 56
Summary : Macaquim ho timbo is a plant whose roots provide a rotenone extract used in insecticides. It grows in the Amazon Basin of Brazil, Guyana, Surinam, Peru, and Ecuador. Brazil is the major exporter of these roots. The United States imported 230 MT of crude roots and 271 MT of rotenone extract in 1984.

This plant is found from sea level to 1,300 m in elevation and thrives in well-drained, black, humus soil. Its yields vary according to location. In Brazil, 17.5 MT/ha of green root is obtained from a 3.5 year old plant.

Mammy-Apple, Mammee-Apple
(See Fruit)

Molasses Grass

Sci. Name(s) : *Melinis minutiflora*
Geog. Reg(s) : Africa-South, America, Australia, Tropical
End-Use(s) : Forage, Pasture & Feed Grain, Insecticide, Oil
Domesticated : Y
Ref. Code(s) : 220
Summary : Molasses grass is a forage grass in Australia and Rhodesia. An oil obtained from the herb is used as a mosquito repellent. The grass is native to tropical America.

Persian Insect-Flower

Sci. Name(s) : *Chrysanthemum coccineum*
Geog. Reg(s) : Africa-North, Eurasia, Mediterranean, Tropical
End-Use(s) : Insecticide
Domesticated : Y
Ref. Code(s) : 126, 170
Summary : The Persian insect-flower (*Chrysanthemum coccineum*) is an erect, perennial herb whose dried flower heads are a source of the insecticide toxin pyrethrin. It is

considered the ornamental pyrethrum plant. It has been used as a flea and louse powder in Iran, Turkey, and the USSR. It is known to be less toxic than the insecticide from pyrethrum (*Chrysanthemum cinerariifolium*). It is used in sprays for biting insects and those thriving on stored products and is also a component of vegetable and fruit insect dusts and sprays. Pyrethrin limitations include its high cost and rapid deterioration after application. Plants are most successful when grown in well-drained loams in areas with moderate rainfall. They are grown in the tropics above 1,000 m and are not frost-hardy.

Pyrethrum

Sci. Name(s)	: *Chrysanthemum cinerariifolium, Pyrethrum cinerariifolium*
Geog. Reg(s)	: Africa-East, America-South, Asia, Europe, Pacific Islands, Temperate, Tropical
End-Use(s)	: Insecticide
Domesticated	: Y
Ref. Code(s)	: 126, 170
Summary	: Pyrethrum (*Chrysanthemum cinerariifolium*) is an erect, perennial herb reaching 30-60 cm in height whose flower heads provide the insecticide toxin pyrethrin. Fully-opened flower heads contain pyrethrin I and II and cinerin I and II.

Plants are grown at higher elevations in East Africa, Zaire, Brazil, Ecuador, Japan, and Dalmatia. Production centers are in Kenya, Tanzania, Zaire, and New Guinea. Attempts to establish production in the United States, particularly in Colorado and California, have not produced satisfactory yields.

This insecticide is not as toxic to man and other mammals as many other insecticides. It is used in livestock and storage insect sprays and in vegetable and fruit dusts and sprays. Limitations to using pyrethrin include its high cost and quick deterioration after application. Synergists such as sesame oil or piperonyl butoxide are added, which reduce the cost of application and enhance its toxicity.

Pyrethrum is grown in northern temperate countries and tropical areas at higher altitudes. Plants are susceptible to frost. In Kenya, crops are grown from 1,800 to 2,700 m in elevation. The Kenyan average of pyrethrin extraction is 1.3%. Japanese crops yield an average of 1% pyrethrin, while the Dalmatian extraction rate is lower. Commercial flowers are planted in well-drained loam soils and kept weed-free.

Harvest occurs in 4 months. About 224 kg/ha of dried flowers are produced during the 1st year, increasing to 900-1,121 kg/ha for the 2nd and 3rd years if grown at 2,437 m. The United States is the largest importer of this insecticide.

Quinine, Quinine Tree, Peruvian Bark
(See Medicinal)

Rose Geranium, Geranium
(See Essential Oil)

Sage, Garden Sage, Common Sage
(See Spice & Flavoring)

Southern Red Cedar, Eastern Red-Cedar
(See Essential Oil)

Surinam-Cherry, Pitanga Cherry
(See Fruit)

Tansy
(See Medicinal)

Texas Cedar
(See Essential Oil)

Tobacco
(See Miscellaneous)

Tuba Root, Luba Patch, Derris, Tuba Putch

Sci. Name(s)	: *Derris elliptica*
Geog. Reg(s)	: Asia-Southeast, Tropical
End-Use(s)	: Insecticide, Miscellaneous
Domesticated	: Y
Ref. Code(s)	: 56, 170
Summary	: Tuba root is a climbing perennial shrub whose roots contain the insecticide rotenone. The powdered root is used in dusting powders with talc or kaolin and in aqueous suspensions for spraying. Extracts of the powdered root are used in organic solvents such as turpentine. Pounded roots are used as a fish poison. Plant leaves are toxic to cattle.

Plants are cultivated mainly in Southeast Asia but are introduced experimentally to most tropical countries. They thrive at low altitudes below 450 m in areas receiving an annual rainfall of 20 dm. Plants do not tolerate prolonged dry periods and grow best in sandy clay loams. Roots are harvested in 2 years when full of rotenone. Yields increase with crop age, but quality decreases after the 2nd year. See macaquim ho timbo for import information regarding rotenone.

Urucu Timbo, Fish Poison

Sci. Name(s)	: *Lonchocarpus urucu*
Geog. Reg(s)	: America-South
End-Use(s)	: Insecticide, Miscellaneous, Ornamental & Lawn Grass
Domesticated	: Y
Ref. Code(s)	: 17, 56
Summary	: The urucu timbo plant is native to the Brazilian rain forests. Its roots are a major source of rotenone used as a fish poison. Rotenone is an important export from Brazil. This plant is grown as an ornamental.

Vogel Tephrosia
(See Erosion Control)

West Indian Lemongrass, Sere Grass, Lemon Grass
(See Essential Oil)

Yam Bean, Erosus Yam Bean, Jicama
(See Tuber)

Isoprenoid
Resin & Rubber

Balata

Sci. Name(s)	: *Manilkara bidentata*
Geog. Reg(s)	: America-South
End-Use(s)	: Isoprenoid Resin & Rubber, Timber
Domesticated	: N
Ref. Code(s)	: 61, 170
Summary	: The balata tree is native to South America and grows up to heights of 40 m. It provides a nonelastic rubber obtained 3 times a year from the latex of wild trees. This latex is used in the manufacture of machine belting. Its timber is called bullet wood and has many uses.

Castilloa Rubber, Panama Rubber

Sci. Name(s)	: *Castilla elastica* subsp. *elastica*
Geog. Reg(s)	: America-Central, Asia-Southeast, West Indies
End-Use(s)	: Isoprenoid Resin & Rubber
Domesticated	: Y
Ref. Code(s)	: 170
Summary	: Castilloa rubber is a fast-growing, deciduous forest tree whose trunk can be incised and tapped for rubber. As a commercial rubber source, it is of little value because it cannot compete with the better producing para rubber tree (*Hevea brasiliensis*). The castilloa rubber tree has undergone extensive cultivation in Central America, Trinidad, and Java.

Caucho Rubber

Sci. Name(s)	: *Castilla ulei*
Geog. Reg(s)	: America-South
End-Use(s)	: Isoprenoid Resin & Rubber
Domesticated	: N
Ref. Code(s)	: 170
Summary	: The caucho rubber tree produces rubber from incisions made in its trunk. It does not have much commercial importance as a rubber source because it is unable to compete with para rubber trees (*Hevea brasiliensis*). The caucho rubber tree is native to the Amazon region of Brazil.

Ceara-Rubber

Sci. Name(s)	: *Manihot glaziovii*
Geog. Reg(s)	: Africa-East, Africa-West
End-Use(s)	: Isoprenoid Resin & Rubber, Miscellaneous, Oil
Domesticated	: Y
Ref. Code(s)	: 36, 170
Summary	: The ceara-rubber tree is introduced to East Africa for natural rubber production. This small tree grows rapidly and the rubber latex yield is low. As this latex coagulates immediately, trees cannot be tapped continuously. The rubber is of good

quality but has high resin content. Seeds contain a clear, slow-drying oil. The leaves, flowers, roots, bark, and fruit produce hydrocyanic acid. The flowers attract bees. This tree is grown mainly as a temporary shade for cocoa in West Africa.

False Rubber Tree, Funtumia africana, Lagos Silk-Rubber

Sci. Name(s) : *Funtumia africana*
Geog. Reg(s) : Africa
End-Use(s) : Isoprenoid Resin & Rubber
Domesticated : Y
Ref. Code(s) : 170
Summary : The false rubber tree produces a valueless, noncoagulating latex. It is similar to the Lagos silk-rubber tree (*Funtumia elastica*) and is often grown with it in mixed plantings in Africa.

Guayule

Sci. Name(s) : *Parthenium argentatum*
Geog. Reg(s) : America-Central, America-North (U.S. and Canada)
End-Use(s) : Isoprenoid Resin & Rubber
Domesticated : Y
Ref. Code(s) : 16, 154, 170, 220
Summary : Guayule is a shrub native to Mexico and the southern United States and used as a minor source of rubber. The latex is obtained by maceration of the whole plant and was used commercially during World War II. There is a considerable research effort in the southwestern portion of the United States to commercialize guayule as the second major source of natural rubber.

Gutta-Percha

Sci. Name(s) : *Palaquium gutta*
Geog. Reg(s) : Asia-Southeast, Hawaii
End-Use(s) : Fat & Wax, Isoprenoid Resin & Rubber, Miscellaneous, Ornamental & Lawn Grass
Domesticated : Y
Ref. Code(s) : 119, 154, 170
Summary : Gutta-percha is an evergreen tree that provides a milky sap called gutta-percha. When hardened, gutta-percha is used to insulate submarine cables. It is also used in the manufacture of splints, supports, golfballs, and dentistry products. The seeds are a source of cooking fat but contain saponin, which is a poisonous substance. Gutta-percha is obtained by tapping living trees or by collecting it after the trees have been felled. Trees grow wild throughout the forests of Malaysia. Attempts have been made to establish plantations because the tree is of some economic value.

Trees were introduced to Hawaii, but specimens are rare, if still present. In other warm areas, gutta-percha is maintained for ornamental and general-interest purposes.

Huemega

Sci. Name(s) : *Micrandra minor*
Geog. Reg(s) : America-South

End-Use(s) : Isoprenoid Resin & Rubber
Domesticated : N
Ref. Code(s) : 170
Summary : Huemega is a large tree that yields a white latex. This latex provides a high-quality rubber, but the tree cannot be tapped continuously. Huemega trees are abundant in South America, primarily in the Amazon and upper Orinoco basin areas.

India Rubber Fig, Rubber-Plant
Sci. Name(s) : *Ficus elastica*
Geog. Reg(s) : America-North (U.S. and Canada), Asia-India (subcontinent), Tropical
End-Use(s) : Isoprenoid Resin & Rubber, Ornamental & Lawn Grass, Vegetable
Domesticated : Y
Ref. Code(s) : 133, 154, 170, 214
Summary : The India rubber fig grows wild in damp, tropical forests and is native to India, where it provides an important source of a noncommercial rubber. It is tapped for rubber when the tree is about 25 years, and tapping is done by wounding the tree 4 times/year for the collection of latex. Some trees are tapped for 100 years. This rubber has a high resin content. This tree is considered an ornamental in the United States. Young leaves can be eaten in salads.

Jelutong
(See Gum & Starch)

Kirk's Gum Vine
Sci. Name(s) : *Landolphia kirkii*
Geog. Reg(s) : Africa-East
End-Use(s) : Isoprenoid Resin & Rubber
Domesticated : Y
Ref. Code(s) : 165, 170, 220
Summary : Kirk's gum vine is a woody liane native to East Africa. It yields a latex that provides several types of rubber important in Africa: Mozambique Blanc, Mozambique Rogue, Nyasa rubber and pine rubber. The rubber contains about 10% resin. This vine was once the leading rubber producer in Angola, Zaire, Kenya, Mozambique, and Malawi.

Lagos Silk-Rubber, Funtumia elastica, Lagos Silkrubber Tree, Silkrubber Tree
Sci. Name(s) : *Funtumia elastica*
Geog. Reg(s) : Africa-West, Tropical
End-Use(s) : Fiber, Isoprenoid Resin & Rubber
Domesticated : Y
Ref. Code(s) : 36, 170, 220
Summary : The Lagos silk-rubber tree yields a profitable latex that coagulates easily. About 33% of the latex weight is pure, high-quality rubber, but yields are small and cannot compete with the rubber-tree (*Hevea brasiliensis*).

These trees grow in tropical West Africa. Large plantations were cultivated in Nigeria, Ghana, and the Republic of Cameroun. The seeds yield an oil whose value has not been identified. A fiber is obtained from the seeds and used like kapok fiber.

Landolphia Rubber, Guinea Gum Vine

Sci. Name(s)	: *Landolphia heudelotii*
Geog. Reg(s)	: Africa-West
End-Use(s)	: Isoprenoid Resin & Rubber
Domesticated	: N
Ref. Code(s)	: 165, 170, 220
Summary	: Landolphia rubber is a rubber-producing, climbing plant native to West Africa. Large quantities of a white or pink latex are produced from wild plants, but these are unsuitable for large-scale cultivation. These vines were the top rubber producers in Gambia, Senegal, Upper Senegal, and Niger, with yields of 10-12 liters/year of latex.

Manicoba Rubber

Sci. Name(s)	: *Manihot dichotoma*
Geog. Reg(s)	: America-South, Tropical
End-Use(s)	: Isoprenoid Resin & Rubber
Domesticated	: N
Ref. Code(s)	: 170
Summary	: The Manicoba rubber tree grows in Brazil and produces a rubber considered unsatisfactory for natural rubber production. It is cultivated in the drier parts of the tropics.

Manila, White Dammar Pine
(See Natural Resin)

Rubber

Sci. Name(s)	: *Hevea benthamiana*
Geog. Reg(s)	: America-South
End-Use(s)	: Isoprenoid Resin & Rubber
Domesticated	: N
Ref. Code(s)	: 165, 220
Summary	: *Hevea benthamiana* is a tree reaching 24 m in height and is occasionally referred to as a rubber tree. Wounded trees exude a white latex with limited commercial value. *Hevea benthamiana* is native to the Amazon Basin of Brazil, where it is common in low, alluvial, flooded land.

Rubber-Tree, Para Rubber

Sci. Name(s)	: *Hevea brasiliensis*
Geog. Reg(s)	: Africa-North, America-South, Asia-Southeast
End-Use(s)	: Isoprenoid Resin & Rubber, Medicinal, Oil
Domesticated	: Y
Ref. Code(s)	: 148, 154, 170
Summary	: The rubber-tree (*Hevea brasiliensis*) is the major source of natural rubber. Hevea rubber is used primarily in tires and tire products. This rubber has low heat-

retaining capacity which is an essential quality for high-friction airplane and truck tires. There is considerable competition between natural rubbers and synthetically produced rubber. Synthetic rubber is derived from petroleum. Hevea rubber and another natural rubber, guayule, dominate specialty product markets. Most rubber is imported from Indonesia, Brunei, Liberia, and to a lesser extent, Brazil.

Rubber-tree seed kernels contain 40-50% of a semidrying oil extracted commercially for soap-making. These seeds also contain a cyanogenetic glucoside, linamarin, and an enzyme, linase, which hydrolyses the glucoside to hydrocyanic acid.

These trees live up to 200 years. Tapping begins when each tree has reached a certain girth, usually in 5-7 years. Trees must be allowed adequate rest periods to maintain steady growth. They thrive in tropical environments in areas with relatively high rainfall. They need deep, well-drained, fertile, friable soil.

Russian Dandelion

Sci. Name(s) : *Taraxacum kok-saghyz*
Geog. Reg(s) : Eurasia
End-Use(s) : Isoprenoid Resin & Rubber
Domesticated : Y
Ref. Code(s) : 17, 170
Summary : The Russian dandelion is a perennial herb whose roots contain a valuable latex once utilized as rubber. There is little or no demand for this rubber at present. In parts of the USSR, Russian dandelions are still cultivated.

Vogel Fig

Sci. Name(s) : *Ficus vogelii*
Geog. Reg(s) : Africa, Africa-West, Tropical
End-Use(s) : Dye & Tannin, Fiber, Fruit, Isoprenoid Resin & Rubber, Medicinal
Domesticated : N
Ref. Code(s) : 36, 170, 220, 226
Summary : The vogel fig is a rubber-producing tree grown in tropical Africa. The rubber is of good quality but has a high resin content, making it virtually unmarketable. Tree bark is often pounded into cloth. Through a distillation process, the wood can be made into soap. The latex contains a tannin, which, when fermented, has been used to dress wounds. In Africa, juice from the fruit has been applied to sores. The leaf, stem, and fruit give positive tests for sterols and have been used for internal problems. In tropical Africa, the fruit is considered edible.

West African Gum Vine

Sci. Name(s) : *Landolphia owariensis*
Geog. Reg(s) : Africa, Tropical
End-Use(s) : Isoprenoid Resin & Rubber
Domesticated : N
Ref. Code(s) : 165, 170, 220
Summary : The West African gum vine is a woody liane that grows in tropical Africa. It yields a latex from which the rubbers Rouge de Congo and Rouge de Kassai are produced. This vine is the most common member of the Landolphia species and grows throughout Africa.

Medicinal

Aconite, Monkshood, Friar's-Cap

Sci. Name(s)   : *Aconitum napellus*
Geog. Reg(s)   : America-North (U.S. and Canada), Asia, Europe, Temperate
End-Use(s)   : Medicinal, Ornamental & Lawn Grass
Domesticated   : Y
Ref. Code(s)   : 15, 217
Summary   : The roots and leaves of the aconite tree (*Aconitum napellus*) are a source of the drug aconite, used for medicinal purposes. The tree produces deep blue flowers during summer and autumn, making it a popular ornamental.

     Aconites are the most important and well-known of the Aconitum species. Trees grow in Europe, temperate Asia, and to a limited extent, in the Rocky Mountains and northwestern parts of North America.

Alexandrian Senna

Sci. Name(s)   : *Cassia senna, Cassia acutifolia*
Geog. Reg(s)   : Africa-North
End-Use(s)   : Medicinal
Domesticated   : Y
Ref. Code(s)   : 170
Summary   : Alexandrian senna is a shrub whose leaves and pods produce the purgative senna. Leaf and pod yields of up to 507 kg/ha are obtained in about 70 days. Senna exports from the Democratic Republic of the Sudan are 450 MT/year. The United States and the United Kingdom are the largest importers.

     Senna crops will grow on poor, sandy soils with little irrigation, making them important economical crops in arid regions. The shrub is native to the eastern Sahara.

Alfalfa, Lucerne, Sativa
(See Forage, Pasture & Feed Grain)

American Beech
(See Timber)

American Licorice, Wild Licorice
(See Spice & Flavoring)

Ammi

Sci. Name(s)   : *Trachyspermum ammi, Trachyspermum copticum, Ammi copticum*
Geog. Reg(s)   : Asia-India (subcontinent)
End-Use(s)   : Medicinal, Oil, Spice & Flavoring
Domesticated   : Y
Ref. Code(s)   : 220
Summary   : Ammi is a plant that is cultivated for its seeds. The seeds yield an oil that contains thymol. The oil and seeds can be used as a condiment. The oil also has antiseptic properties and can be used as a digestive aid. Ammi is native to India.

Angled Luffa, Angled Loofah, Singkwa Towelgourd, Seequa, Dishcloth Gourd
(See Fruit)

Anise
(See Essential Oil)

Arabica Coffee, Coffee
(See Beverage)

Aramina Fiber, Congo Jute, Cadillo, Caesarweed, Aramina
(See Fiber)

Arbor-Vitae, Northern White-Cedar
(See Essential Oil)

Argus Pheasant-Tree
(See Fruit)

Arrowroot
(See Gum & Starch)

Australian Nightshade
(See Fruit)

Avens
(See Spice & Flavoring)

Azuki Bean, Adzuki Bean
(See Vegetable)

Balsam-Apple, Wonder-Apple
(See Fruit)

Barbados Aloe, Mediterranean Aloe, True Aloe

Sci. Name(s) : *Aloe barbadensis*
Geog. Reg(s) : Africa-North, Africa-West, Hawaii, Mediterranean
End-Use(s) : Medicinal, Ornamental & Lawn Grass
Domesticated : Y
Ref. Code(s) : 177
Summary : The Barbados aloe is a succulent plant whose leaves yield an important sappy latex that provides the drug dioscoroides. This drug is used to treat arthritis and athlete's foot. The fresh latex is used to treat irritations and burns and is used in cosmetics. Barbados aloes are also popular as ornamentals. This succulent grows wild in North Africa from Morocco eastward, and on the Cape Verde and Canary Islands. Aloe has been grown commercially in Hawaii on the island of Maui.

Barbatimao
(See Dye & Tannin)

Barwood, Camwood
(See Dye & Tannin)

Basil, Sweet Basil
(See Essential Oil)

Belladonna, Deadly Nightshade

Sci. Name(s) : *Atropa bella-donna*
Geog. Reg(s) : America-North (U.S. and Canada), Asia-India (subcontinent), Eurasia
End-Use(s) : Medicinal
Domesticated : Y
Ref. Code(s) : 170
Summary : Belladonna is a poisonous plant whose leaves and flowers contain the alkaloid
atropine. In the United States and India, belladonna is cultivated as a commercial
drug plant. In parts of southern Europe and Asia Minor, the flowers and leaves are
used to relieve body pains.

Benoil Tree, Horseradish-Tree
(See Oil)

Bergamot
(See Essential Oil)

Betel Palm, Betel Nut, Areca Palm
(See Ornamental & Lawn Grass)

Bible Frankincense, Oil of Olibanum, Essence of Olibanum
(See Natural Resin)

Bignay, Chinese Laurel, Salamander Tree
(See Fruit)

Bitter Almond
(See Essential Oil)

Bitter Lettuce

Sci. Name(s) : *Lactuca virosa*
Geog. Reg(s) : Europe
End-Use(s) : Medicinal
Domesticated : Y
Ref. Code(s) : 36, 170, 205
Summary : Bitter lettuce is a strongly scented biennial that develops a leafy rosette in its
1st year and in its 2nd year forms a sturdy stem up to 1.5 m in height. A latex taken
from the dried stem is used to prepare the sedative drug lactucarium. Today, the

drug is rarely used in medicine, and plants are no longer cultivated to any extent. Plants are found on dry, nitrogenous soils, usually on the wasteland areas and hillsides of Europe.

Black Cumin
(See Spice & Flavoring)

Black Currant, European Black Currant
(See Fruit)

Black Mulberry
(See Fruit)

Blue Lupine, European Blue Lupine, New Zealand Blue Lupine, Narrow-Leaved Lupine
(See Forage, Pasture & Feed Grain)

Borage
(See Ornamental & Lawn Grass)

Broad-Leaved Lavender
(See Essential Oil)

Buchu, Bacco
(See Essential Oil)

Buckwheat
(See Cereal)

Buffalo Gourd, Wild Gourd
(See Oil)

Burdock, Great Burdock, Edible Burdock

Sci. Name(s)	: *Arctium lappa*
Geog. Reg(s)	: America-North (U.S. and Canada), Eurasia, Europe
End-Use(s)	: Essential Oil, Medicinal, Vegetable
Domesticated	: N
Ref. Code(s)	: 42, 220
Summary	: Burdock is a tall, biennial herb whose young roots contain inulin, a bitter compound used as an appetite stimulant and diuretic. The roots yield an essential oil and in Japan are eaten as vegetables. This herb has stems 90-130 cm. It grows wild in Europe, Asia Minor, and North America.

Burweed
(See Fiber)

Butternut
(See Nut)

Cajeput

Sci. Name(s)	:	*Melaleuca cajuputi*
Geog. Reg(s)	:	Asia-Southeast, Australia
End-Use(s)	:	Essential Oil, Medicinal, Timber
Domesticated	:	N
Ref. Code(s)	:	220
Summary	:	The cajeput tree is the source of an oil used in Malaysia and Australia as an antiseptic, digestive stimulant, and treatment for intestinal worms. Its hard wood has been used for posts and ship-building.

Calendula, Pot-Marigold
(See Ornamental & Lawn Grass)

Camelthorn, Camill Thorn
(See Forage, Pasture & Feed Grain)

Camphor-Tree, Camphor

Sci. Name(s)	:	*Cinnamomum camphora*
Geog. Reg(s)	:	America-North (U.S. and Canada), Asia, Asia-China
End-Use(s)	:	Essential Oil, Insecticide, Medicinal, Miscellaneous
Domesticated	:	Y
Ref. Code(s)	:	170, 205
Summary	:	The camphor-tree is an evergreen tree reaching heights of 12 m. Its distilled wood yields camphor and white oil of camphor, both of which have medicinal qualities. Both oils contain safrole, acetaldehyde, terpineol, eugenol, cineole, d-pinene, and phellandrene. The oils are somewhat antiseptic in nature and are stimulants, carminatives, expectorants, and mild analgesics. They are also used as a rubefacient and parasiticide. The oils are most often used in external medicines for relieving rheumatic pain and inflammation, fibrositis, and neuralgia. They are usually used in combination with other substances.

Camphor oil is used in the manufacture of disinfectants, celluloid, and other chemical preparations. Safrole, a poisonous residual of the oil, is used in soap and perfume manufacture. Camphor oil has largely been replaced with synthetic substances derived from coal tar and pinene, a turpentine derivative.

The tree is native to China, Japan, and Taiwan. It has been introduced into many tropical countries, where it does best at altitudes up to 750 m. Taiwan is the main producer of camphor oil. In 1984, the United States imported 58 MT of camphor as an essential oil, valued at $184,000.

Candytuft, Rocket Candytuft
(See Ornamental & Lawn Grass)

Cape Aloe, Bitter Aloe, Red Aloe

Sci. Name(s)	:	*Aloe ferox*

Geog. Reg(s) : Africa-South
End-Use(s) : Forage, Pasture & Feed Grain, Medicinal, Miscellaneous
Domesticated : Y
Ref. Code(s) : 16, 220
Summary : The Cape aloe is a succulent whose leaves provide a latex that is dried and used as an effective laxative and in certain veterinary medicines. The latex can also be made into a jam. Cape aloes are used as fodder for livestock in times of drought. The dried leaves may be burnt as an insect repellent. Cape aloes are grown and exported from South Africa.

Caper
(See Spice & Flavoring)

Cardamom, Cardamon
(See Spice & Flavoring)

Catnip
(See Miscellaneous)

Champac
(See Ornamental & Lawn Grass)

Chaulmoogra
Sci. Name(s) : *Hydnocarpus kurzii*
Geog. Reg(s) : Asia-Southeast, Hawaii
End-Use(s) : Medicinal, Oil
Domesticated : Y
Ref. Code(s) : 154
Summary : Chaulmoogra is an evergreen tree 12-15 m in height whose seeds yield a valuable thick, yellowish oil called chaulmoogra oil, used in the treatment of skin diseases.
Trees are native to Burma and Thailand but are now cultivated in parts of Hawaii. In 1920, chaulmoogra crude oil was researched at the University of Hawaii and was discovered to have derivatives that could provide the painless curatives needed in the treatment of leprosy. Expressed oilseed cakes have been used for manure. Tree leaves and shoots are thought to be toxic to humans.

Cherry Laurel, English Laurel
(See Erosion Control)

Chervil
(See Spice & Flavoring)

Chicory
(See Beverage)

Chili Pepper, Tabasco, Bird Pepper, Tabasco Pepper
(See Spice & Flavoring)

Chinese Ephedra

Sci. Name(s) : *Ephedra sinica*
Geog. Reg(s) : Asia-China
End-Use(s) : Medicinal
Domesticated : N
Ref. Code(s) : 220
Summary : Chinese ephedra is a plant that when processed, supplies the medicinally important ephedrine antiasthmatic drug. Plants grow along the coasts of southern China.

Chinese Jujube
(See Fruit)

Chinese Rhubarb, Medicinal Rhubarb, Canton Rhubarb

Sci. Name(s) : *Rheum officinale*
Geog. Reg(s) : Asia-China, Asia-Southeast
End-Use(s) : Medicinal, Ornamental & Lawn Grass
Domesticated : Y
Ref. Code(s) : 16, 17, 21, 94, 170, 220
Summary : Chinese rhubarb is a large, perennial herb whose dried rhizomes provide a drug known as rhubarb. Rhubarb has been used in some medicines as a purgative, laxative, and tonic. In Java, rhubarb has been used in certain cosmetics. The active principles of the rhizome include glucogallin, tetrarin, and calcium oxalate. Chinese rhubarb has been grown as an ornamental.

Plants are native to China and Tibet. Roots are dug and cut into small pieces, which are then threaded on a string for sun drying.

Chinquapin, Allegheny Chinkapin
(See Nut)

Chives
(See Vegetable)

Clove
(See Spice & Flavoring)

Coca

Sci. Name(s) : *Erythroxylum coca*
Geog. Reg(s) : America-South
End-Use(s) : Medicinal
Domesticated : Y
Ref. Code(s) : 154, 170, 205
Summary : Coca is a plant whose leaves provide the narcotic alkaloid cocaine. Leaves are harvested 1-3 years after planting and are dried and prepared for the legal and illegal drug markets. Coca is used as a local anesthetic in South American countries, where it is grown mainly at higher elevations. Commercial cultivation of this herb is subject to worldwide constraint.

Colchicum, Meadow Saffron, Autumn Crocus

Sci. Name(s)	: *Colchicum autumnale*
Geog. Reg(s)	: Europe
End-Use(s)	: Dye & Tannin, Fat & Wax, Gum & Starch, Medicinal, Ornamental & Lawn Grass
Domesticated	: Y
Ref. Code(s)	: 205
Summary	: Colchicum or autumn crocus is a perennial plant whose dried corms and seeds yield the toxic alkaloid colchicine. This alkaloid has potential medicinal use for the treatment of gout, rheumatism, eczema and bronchitis. Also present are other toxic alkaloids and starches, gums, sugar, fats, and tannins that make the plant poisonous. With medical supervision, it has been used as a diuretic, sedative, and appetite stimulant.

This plant is often grown as an ornamental. It is native to Europe, where it is most successful on deep clays and rich loamy soils. Colchicum is found both wild and cultivated.

Colocynth

Sci. Name(s)	: *Citrullus colocynthis*
Geog. Reg(s)	: Africa, Asia-India (subcontinent), Tropical
End-Use(s)	: Medicinal
Domesticated	: N
Ref. Code(s)	: 170, 220
Summary	: Colocynth is a viny herb that produces a drug from the dried pulp of its mature fruit. The drug contains a bitter alkaloid and resin that cause a violent purging action. Colocynth grows wild from India to tropical Africa.

Coltsfoot

Sci. Name(s)	: *Tussilago farfara*
Geog. Reg(s)	: Africa-North, America-North (U.S. and Canada), Eurasia, Temperate
End-Use(s)	: Beverage, Dye & Tannin, Medicinal, Vegetable
Domesticated	: Y
Ref. Code(s)	: 17, 129, 220, 236
Summary	: Coltsfoot is an herb whose leaves and flowers are used in cough remedies and in the treatment of bronchitis. Leaves were smoked in herbal tobaccos to alleviate asthmatic conditions, and the stalk and leaves were brewed as a tea to relieve coughs. The leaves are eaten as vegetables. All parts of the plant contain mucilage, tannin, and traces of a bitter amorphous glucoside. The flowers contain phytosterol and faradial.

Plants can easily become weeds because of their vigorously spreading rootstalks. The herb is native to temperate Eurasia and is grown in North Africa and North America.

Common Fig, Adriatic Fig
(See Fruit)

Common Guava, Lemon Guava
(See Fruit)

Common Reed Grass, Carrizo
(See Miscellaneous)

Common Rue, Garden Rue

Sci. Name(s)	: *Ruta graveolens*
Geog. Reg(s)	: Mediterranean
End-Use(s)	: Essential Oil, Medicinal, Ornamental & Lawn Grass, Spice & Flavoring
Domesticated	: Y
Ref. Code(s)	: 64, 154, 170
Summary	: Rue is a perennial herb about 0.7 m in height, grown most often as an ornamental. Leaves yield an essential oil used medicinally and as flavoring in cooking.

Plants are usually propagated by seed and do best if planted in the early spring or late summer. Rue favors a gravelly soil.

Common Thyme
(See Spice & Flavoring)

Composite Yam, Wild Yam

Sci. Name(s)	: *Dioscorea composita*
Geog. Reg(s)	: America-North (U.S. and Canada), America-South, West Indies
End-Use(s)	: Medicinal, Tuber
Domesticated	: Y
Ref. Code(s)	: 48, 149, 171
Summary	: The composite yam (*Dioscorea composita*) is one of the main sources of sapogenins used in oral contraceptives. The tubers contain 4-6% sapogenins. Tubers reach their maximum size in 4-5 years. In one Puerto Rican experiment, wild plants produced about 6,526 kg of dry tubers, yielding about 308 kg/ha of sapogenin.

Composite yams have been introduced into the eastern and southern parts of the United States, Puerto Rico, Surinam, and Costa Rica. Plants are most successful in sandy, well-drained loams.

Copaiba
(See Natural Resin)

Coriander
(See Spice & Flavoring)

Corn Mint, Japanese Mint, Field Mint
(See Essential Oil)

Costmary, Mint Geranium, Alecost, Bible-Leaf Mace
(See Spice & Flavoring)

Crown Daisy, Tansy

Sci. Name(s)	: *Chrysanthemum spatiosum, Tanacetum vulgare, Chrysanthemum tanacetum, Chrysanthemum vulgare*
Geog. Reg(s)	: America-North (U.S. and Canada), Asia, Europe, Temperate
End-Use(s)	: Essential Oil, Insecticide, Medicinal
Domesticated	: Y
Ref. Code(s)	: 205, 220
Summary	: The crown daisy is an aromatic northern temperate herb. Its dried leaves and flowers yield an oil (oil of tansy) that has been used medicinally to treat intestinal worms, control menstruation, and as a stimulant. The oil is also used in insect repellents because it contains 70% borneol and thujone and an unspecified amount of camphor. It is a source of vitamin C, tannins, resin, citric acid, butyric acid, oxalic acid, and lipids. The oil can be dangerous if taken in excess.

The herb has been introduced into the northeastern parts of the United States, where it is often grown in herb gardens because of its beautiful and long-lived yellow flowers. The flowers are sometimes used in insect-repelling sachets or potpourris. The crown daisy is a perennial herb growing 60-120 cm. It is native to Europe and Asia. The herb thrives on nitrogen-rich, loamy soils to 1,500 m in elevation. It will not tolerate waterlogging but can withstand shade.

Cuajilote, Cuachilote
(See Fruit)

Culebra, Snakegourd, Patole

Sci. Name(s)	: *Trichosanthes cucumerina*
Geog. Reg(s)	: Asia-China, Asia-India (subcontinent), Australia
End-Use(s)	: Medicinal
Domesticated	: N
Ref. Code(s)	: 36
Summary	: The culebra is a climbing vine that produces bitter fruits. The fruit has been used for unspecified medicinal purposes in parts of India and China. Vines range in habitat from India to Australia.

Curry-Leaf-Tree
(See Spice & Flavoring)

Dandelion, Common Dandelion
(See Beverage)

Desert Date
(See Fruit)

Dill
(See Spice & Flavoring)

Dittany, Candle Plant, Gas-Plant

Sci. Name(s)	: *Dictamnus albus, Dictamnus fraxinella*

Geog. Reg(s)   : Asia-China, Europe
End-Use(s)   : Beverage, Medicinal
Domesticated : Y
Ref. Code(s)   : 220
Summary   : Dittany is a perennial herb grown in southern Europe and northern China for its roots and leaves. The root contains the alkaloid dictamine, which is used medicinally as a uterine stimulant. The plant has a lemonlike aroma, and its leaves are sometimes brewed as a tea in various parts of Europe and China.

Dyer's-Greenwood, Dyer's-Greenweed
(See Dye & Tannin)

East Indian Rhubarb, Chinese Rhubarb, Turkey Rhubarb
Sci. Name(s)   : *Rheum palmatum*
Geog. Reg(s)   : Asia-China
End-Use(s)   : Medicinal
Domesticated : Y
Ref. Code(s)   : 21, 36, 170, 220
Summary   : East Indian rhubarb (*Rheum palmatum*) is a large, perennial herb whose rhizomes contain an alkaloid used medicinally as a tonic for treatment of upset stomachs. The rhizomes of East Indian rhubarb are thought to be the botanical source of the best rhubarb, but plants are rarely cultivated.

Plants thrive on rich, forest soils, usually in shady positions up to elevations of 3,046 m.

East Indian Sandalwood
(See Essential Oil)

Eastern Elderberry, American Elderberry, American Elder
(See Fruit)

Emblic, Myrobalan, Emblica
(See Fruit)

Eucalyptus
(See Essential Oil)

Eucalyptus, Broad-Leaved Peppermint Tree
(See Essential Oil)

Eucalyptus
(See Timber)

European Arnica, Mountain Tobacco
Sci. Name(s)   : *Arnica montana*
Geog. Reg(s)   : America-North (U.S. and Canada), Asia, Europe
End-Use(s)   : Medicinal

Domesticated : Y
Ref. Code(s) : 205
Summary : European arnica is an aromatic perennial herb whose dried flowers and stems are used in external medicines as a skin stimulant and diuretic. Homeopathic doses have some potential as treatment for epilepsy and seasickness.

It grows in Europe, Asia, and western North America. It is cultivated by root division in the spring.

European Beech
(See Ornamental & Lawn Grass)

European Pennyroyal
(See Essential Oil)

European Strawberry, Strawberry
(See Fruit)

Fat Hen, Lamb's-Quarters, White Goosefoot, Lambsquarter
(See Cereal)

Fennel, Florence Fennel, Finocchio
(See Spice & Flavoring)

Fenugreek
(See Spice & Flavoring)

Field Horsetail
Sci. Name(s) : *Equisetum arvense*
Geog. Reg(s) : Temperate
End-Use(s) : Beverage, Medicinal, Miscellaneous, Ornamental & Lawn Grass, Vegetable, Weed
Domesticated : Y
Ref. Code(s) : 38, 83, 205, 214, 220
Summary : Field horsetail is a reedlike plant that grows primarily in temperate areas on sandy, moist soil. The upper parts of the plant are used for medicinal purposes. It is listed in some books as being poisonous to livestock.

Its edible tubers, roots, and stems are prepared in a variety of ways, and a powdered form is used as tea. Traditionally, field horsetail was used to make brushes, in hair rinses, and in cosmetic preparations. It is valuable in this respect because of the silicon deposited in the cell walls.

It is often grown as an ornamental but can become a persistent weed in areas with high water tables.

Fitweed, False Coriander
(See Spice & Flavoring)

Flax
(See Fiber)

Foxglove, Digitalis

Sci. Name(s)	: *Digitalis purpurea*
Geog. Reg(s)	: Europe
End-Use(s)	: Medicinal
Domesticated	: Y
Ref. Code(s)	: 21, 220
Summary	: Foxglove (*Digitalis purpurea*) is a biennial herb ranging 0.6-2 m in height that is cultivated in Central and southern Europe as a drug plant. The plant leaves contain digitalin, ditalein, and digitorum, which are components of the powerful heart stimulant, digitalis. Leaves are collected from 2 year old wild plants, and the drug is extracted by pulverization and direct treatment or by solvent extraction. Foxglove has also been an herbal medicine used in the treatment of epilepsy.

Plants grow in dry, sandy, or gravelly soils, usually in open woods, heaths, and hedgebanks. Leaves that contain digitalis also contain a nonnitrogenised neutral principle called digitalin, which is about 100 times more powerful than digitalis.

French Lavender
(See Essential Oil)

Galanga, Kent Joer
(See Ornamental & Lawn Grass)

Gambier, White Cutch
(See Dye & Tannin)

Garbanzo, Chickpea, Gram
(See Vegetable)

Garden Angelica, Angelica
(See Essential Oil)

Garden Myrrh, Sweet Scented Myrrh

Sci. Name(s)	: *Myrrhis odorata*
Geog. Reg(s)	: Europe
End-Use(s)	: Medicinal, Miscellaneous, Spice & Flavoring, Sweetener, Vegetable
Domesticated	: Y
Ref. Code(s)	: 205, 220
Summary	: Garden myrrh is an aromatic, hardy, perennial herb cultivated commercially in western Europe as a sugar substitute for diabetics. Its leaves and seeds are made into a weak diuretic and antiseptic tonic, or eaten as salad vegetables. Its flowers attract bees. Parts of the herb are used to add an aniselike flavor to liqueurs.

Plants grow in grassy soils and thrive in shaded areas. They will tolerate some exposure to the sun.

Garden Rhubarb
(See Vegetable)

Garlic
(See Spice & Flavoring)

Gbanja Kola
(See Nut)

Ginger
(See Spice & Flavoring)

Ginseng, American Ginseng

Sci. Name(s)	: *Panax quinquefolius*
Geog. Reg(s)	: America-North (U.S. and Canada)
End-Use(s)	: Beverage, Medicinal, Tuber
Domesticated	: Y
Ref. Code(s)	: 17, 154, 194
Summary	: American ginseng is a perennial herb with tuberous roots valued in a variety of

medicines. Roots are available as whole roots, cut capsules, fiber, crushed powder, tablets, and tea or extract. Ginseng root influences the central nervous system, the endocrine and adrenocortical systems, internal organs, the metabolism, blood pressure and sugar, and gonadotropic activity.

The overall quality of cultivated ginseng root is thought to be inferior to the roots of wild ginseng. Market value is determined by the appearance and flavor of the root. The herb is native to North America and ranges from Quebec to Alabama. The American ginseng plant is considered an endangered species.

American ginseng thrives in shady areas on rich, moist, well-drained soil. It requires an annual rainfall of 0.7-1.3 m for adequate growth. Seedling plants should be transplanted in their 1st or 2nd year. Roots are usually a marketable size after 5-7 years.

Globe Artichoke, Artichoke
(See Vegetable)

Governor's-Plum, Ramontchi, Rukam
(See Ornamental & Lawn Grass)

Guaiac, Guaiacum
(See Natural Resin)

Guarana
(See Beverage)

Gum Tragacanth, Tragacanth
(See Gum & Starch)

Harmala Shrub, African-Rue, Harmel Piganum

Sci. Name(s)	: *Peganum harmala*
Geog. Reg(s)	: Asia-Central, Europe, Mediterranean
End-Use(s)	: Dye & Tannin, Medicinal, Oil
Domesticated	: N
Ref. Code(s)	: 119, 204, 220, 233
Summary	: The harmala shrub produces fruit with seeds that contain certain narcotic properties. The seeds are often used as condiments. They were once used in treatment for eye diseases, as aphrodisiacs, and as appetite stimulants. In parts of Russia, decoctions of the roots and leaves are used to treat nervous disorders and rheumatism. "Turkey Red" is a red dye extracted from the fruits and used in Turkey to dye tarbooshes. "Zit-el-Harmel" oil is extracted from the seeds.

Shrubs grow at elevations of 2,530 m in Tibet and 1,524 m in the northwestern Himalayas. The harmala shrub is native to the Mediterranean and Central Asia.

Henbane, Black Henbane

Sci. Name(s)	: *Hyoscyamus niger*
Geog. Reg(s)	: Africa-North, Asia, Europe
End-Use(s)	: Medicinal
Domesticated	: Y
Ref. Code(s)	: 170, 205, 220
Summary	: Henbane is a strong-smelling, annual or biennial herb that is cultivated for its leaves, which contain medicinally important alkaloids. The most important of these alkaloids is hyoscyamine, which is used to relieve spasms and to dilate the pupil of the eye. Plants are cultivated in Germany, Russia, and the United Kingdom. Leaves are collected when the plants begin flowering. Henbane grows in Europe, Asia, and North Africa in areas with well-drained, sandy or chalky soils.

Henna
(See Ornamental & Lawn Grass)

Hoary Basil, Camphor Basil

Sci. Name(s)	: *Ocimum kilimandscharicum*
Geog. Reg(s)	: Africa-East
End-Use(s)	: Medicinal
Domesticated	: Y
Ref. Code(s)	: 194, 220
Summary	: Hoary basil, or camphor basil, is an East African herb that is a source of camphor. The leaves have a strong odor and contain 60% camphor. The camphor has not been extracted commercially since World War II.

Holy Basil

Sci. Name(s)	: *Ocimum sanctum*
Geog. Reg(s)	: Asia, Asia-India (subcontinent), Asia-Southeast, Australia
End-Use(s)	: Medicinal, Spice & Flavoring
Domesticated	: Y
Ref. Code(s)	: 36, 170, 194

Summary : Holy basil is an annual herb whose leaves can be used as a condiment and can be eaten sparingly in salads. The leaves have a strong camphor scent because of the presence of eugenol. Medicinally, the leaves may be used to treat gastric disorders, the roots to treat fevers, and leaf juice as laxatives. Holy basil is grown throughout India. It has a long history of importance in the Hindu religion. The herb is native to Malaysia, Australia, India, and western Asia.

Hops
(See Beverage)

Horehound, White Horehound
Sci. Name(s) : *Marrubium vulgare*
Geog. Reg(s) : America-North (U.S. and Canada), Europe
End-Use(s) : Dye & Tannin, Essential Oil, Medicinal, Natural Resin
Domesticated : Y
Ref. Code(s) : 205, 220
Summary : The horehound plant originated in Europe and has been introduced to North America. Its leaves and plant tops are used to treat colds, sore throats, respiratory ailments, and other maladies. It stimulates the appetite and promotes bile flow in the stomach. An oil is obtained from the leaves and used in liqueurs and to treat gall bladder and stomach disorders and menstrual pain. It is also a weak sedative and can be applied to minor cuts. In large doses it acts as a laxative. It contains tannins, vitamin C, sterols, resin, mucilage, and a volatile oil. The leaves are used as condiments and in confections. Plants are cultivated commercially in the United Kingdom.

Huang T'eng
Sci. Name(s) : *Fibraurea tinctoria*
Geog. Reg(s) : Asia-Southeast
End-Use(s) : Dye & Tannin, Medicinal
Domesticated : N
Ref. Code(s) : 220
Summary : Huang t'eng is an herb whose roots have a bitter principle used in Malaysia and southern Vietnam for medicinal purposes. Plant stems yield a yellow dye. In Malaya, parts of the plant are used to treat fevers.

Hyssop
(See Essential Oil)

Imbe
(See Beverage)

Inca-Wheat, Quihuicha, Quinoa, Love-Lies-Bleeding
(See Cereal)

Indian Bael, Bael Fruit, Bengal Quince, Bilva, Siniphal, Bael Tree
(See Fruit)

Indian Borage
(See Spice & Flavoring)

Indian Long Pepper
(See Spice & Flavoring)

Indian Mulberry
(See Dye & Tannin)

Indian Wood-Apple
(See Fruit)

Indian-Bark, Indian Cassia
(See Spice & Flavoring)

Indian-Tobacco
(See Ornamental & Lawn Grass)

Inula, Elecampane
(See Essential Oil)

Ipecac, Ipecacuanina
Sci. Name(s)	: *Cephaelis ipecacuanha*
Geog. Reg(s)	: America-South, Asia-India (subcontinent), Asia-Southeast
End-Use(s)	: Medicinal
Domesticated	: Y
Ref. Code(s)	: 45, 170
Summary	: Ipecac is a small shrub whose roots contain an ether-soluble alkaloid, emetine, a drug (ipecac) used in the treatment of amoebic dysentery. According to the Pharmacopeia of the United States, the plant yields not less than 2% of this alkaloid.

The shrub is native to Brazil and was introduced into India and western Malaysia where it is produced commercially on a small scale basis.

Jackbean, Horsebean, Swordbean
(See Vegetable)

Jambolan, Java Plum
(See Fruit)

Java Devil Pepper
Sci. Name(s)	: *Rauvolfia serpentina*
Geog. Reg(s)	: Asia-India (subcontinent), Asia-Southeast
End-Use(s)	: Medicinal
Domesticated	: Y
Ref. Code(s)	: 36, 116, 149, 154, 220
Summary	: The Java devil pepper is an evergreen subshrub 15-45 cm in height whose roots contain 1.7-3% of the alkaloid reserpine. Reserpine is used to relieve hypertension

and to depress the activity of the hypothalamus. Extracts of plant leaves have been used as treatment for opacity of the cornea. In Java, plant roots are used in certain veterinary medicines.

Plants are grown mainly in India, Sri Lanka, Burma, Thailand, and Indonesia. Java devil peppers thrive in hot, humid areas on humus-rich soils up to 1,000 m in elevation but are most successful at low to medium altitudes. Plants are started from root cuttings, because seeds do not have a good generation rate. Root harvest occurs after 2-3 years. Yields from seeded plantations are 1,175 kg/ha of air-dried roots, 345 kg/ha of root cuttings, and 175 kg/ha of stem cuttings.

Javanese Long Pepper
(See Spice & Flavoring)

Job's Tears, Adlay, Adlay Millet
(See Miscellaneous)

Jojoba, Goat-Nut
(See Fat & Wax)

Juniper, Common Juniper
(See Essential Oil)

Kamraj
(See Vegetable)

Kava, 'Awa
(See Beverage)

Khat

Sci. Name(s)   : *Catha edulis*
Geog. Reg(s)   : Africa-East, Africa-North, Mediterranean
End-Use(s)   : Medicinal
Domesticated   : Y
Ref. Code(s)   : 170
Summary   : Khat is a small tree cultivated in Ethiopia and Arabia for its leaves and buds. These are chewed fresh or dried as a stimulant. They contain an alkaloid that, if overconsumed, can cause a type of intoxication that may lead to possible coma or death.

Trees grow wild in the higher altitudes of East Africa. They reach 6 m in height.

Kinka Oil Ironweed

Sci. Name(s)   : *Vernonia anthelmintica*
Geog. Reg(s)   : Asia, Asia-India (subcontinent), Tropical
End-Use(s)   : Medicinal, Ornamental & Lawn Grass
Domesticated   : Y
Ref. Code(s)   : 17, 220

Summary   : Kinka oil ironweed is a tropical Asian plant grown in India for its leaves, which are made into a salve for leprosy and skin diseases and in decoctions as an abortive. Plants are grown in gardens or as ornamental borders. They flower in the late summer and autumn and grow easily in rich soil.

Lehmann Lovegrass
(See Erosion Control)

Lesser Galanga
(See Spice & Flavoring)

Ling, Horn-Nut, Horned Chestnut

Sci. Name(s)   : *Trapa bicornis*
Geog. Reg(s)   : Asia-China, Asia-Southeast
End-Use(s)   : Medicinal
Domesticated   : Y
Ref. Code(s)   : 36, 92
Summary   : Ling, or horn-nut, is an aquatic plant cultivated in southern China for its hard seeds. The fresh kernels contain 16% starch and 3% protein. In Cambodia, seed husks are brewed as a medicinal tea for fevers.

Lovage, Common Lovage
(See Spice & Flavoring)

Lukrabao, Chaulmogra Tree

Sci. Name(s)   : *Hydnocarpus anthelmintica*
Geog. Reg(s)   : Asia-China, Asia-India (subcontinent), Asia-Southeast, Tropical
End-Use(s)   : Energy, Fiber, Fruit, Medicinal, Oil, Timber
Domesticated   : Y
Ref. Code(s)   : 205, 220
Summary   : Lukrabao is a tree reaching 20 m in height whose seeds yield a nondrying oil used in Thailand and Cochin-China to treat leprosy and skin problems. The oil and seeds have antibacterial, alterative, and irritant properties. The seeds are used mostly as external medicine, usually as soothing ointments, and in India are crushed and used as an alterative tonic. They provide lighting. Lukrabao trees have light-colored timber, occasionally made into planks. The bark can be made into rough cordage. The fruit pulp is edible but tasteless. Trees are native to Burma and Thailand and have been introduced into other tropical countries.

Mahua
(See Oil)

Malabar Nut

Sci. Name(s)   : *Adhatoda vasica, Justicia adhatoda*
Geog. Reg(s)   : Asia-India (subcontinent)
End-Use(s)   : Dye & Tannin, Medicinal, Miscellaneous
Domesticated   : N

Ref. Code(s) : 220
Summary   : The Malabar nut tree is a small shrub found in tropical India. A decoction of the leaves is an expectorant used in treating bronchitis. The leaves may be boiled with sawdust, yielding a yellow dye. Malabar nut wood is carved into beads and also makes good gunpowder charcoal.

Mammy-Apple, Mammee-Apple
(See Fruit)

Mandrake, European Mandrake, Mandragora
Sci. Name(s) : *Mandragora officinarum*
Geog. Reg(s) : Europe, Mediterranean
End-Use(s)   : Medicinal, Ornamental & Lawn Grass
Domesticated : Y
Ref. Code(s) : 61, 170, 220
Summary   : The mandrake plant is native to the Mediterranean region. In Europe, mandrake roots have been used for medicinal purposes as narcotics, purgatives, and aids for pregnancy. They are also important for treating asthma, hay fever, and coughs. There are certain poisonous properties present. The fruit has a sweet, insipid flavor.

Plants are hardy and easy to grow in sunny, well-drained locations on deep, fertile soil. Mandrake roots often resemble the human form, making the plant an interesting ornamental.

Marijuana, Hemp, Marihuana
(See Fiber)

Maritime Wormwood, Sea Wormwood
Sci. Name(s) : *Artemisia maritima*
Geog. Reg(s) : Asia-Central, Asia-India (subcontinent), Europe
End-Use(s)   : Medicinal, Spice & Flavoring
Domesticated : Y
Ref. Code(s) : 170, 220
Summary   : Maritime wormwood (*Artemisia maritimi*) is an herb whose leaves are a main source of santonin, a substance used to treat stomach complaints. The leaves are also used as flavoring.

The herb is native to Pakistan and grows along the European and Mongolian coastlines.

Marking-Nut Tree
(See Dye & Tannin)

Melegueta-Pepper, Grains of Paradise, Guinea Grains
(See Spice & Flavoring)

Musk Okra, Musk Mallow, Ambrette
(See Essential Oil)

Nigerian Lucerne
(See Forage, Pasture & Feed Grain)

Nutmeg, Mace
(See Spice & Flavoring)

Okra, Lady's Finger, Gumbo
(See Vegetable)

Ololiqui, Christmas Pops

Sci. Name(s)	: *Turbina corymbosa*
Geog. Reg(s)	: America-Central
End-Use(s)	: Medicinal
Domesticated	: N
Ref. Code(s)	: 220
Summary	: Ololiqui is an herb whose seeds have been used in Mexico as hallucinogens and in religious ceremonies. There is very little information about this plant.

Opium Poppy

Sci. Name(s)	: *Papaver somniferum*
Geog. Reg(s)	: Asia-China, Asia-India (subcontinent), Eurasia, Europe
End-Use(s)	: Medicinal, Oil, Ornamental & Lawn Grass, Spice & Flavoring
Domesticated	: Y
Ref. Code(s)	: 171, 194
Summary	: The opium poppy is an annual herb whose flower capsules produce a latex called opium. There are about 20 different alkaloids in the latex, but the most important and the most well known are morphine, codeine, and heroin. Opium latex is used medicinally as a painkiller and sedative, but it is also misused as a narcotic with habit-forming effects. In the United States, opium poppy culture is controlled. The principle producers of opium are India, China, Asia Minor, and areas within the Balkan Peninsula.

Poppy seeds yield a nonnarcotic, fixed oil used in paints, varnishes, and soaps. Whole seeds are edible and are often used as condiments with baked goods and pastries. Opium poppies have attractive flowers and are valued as ornamentals in flower gardens.

Poppies are most successful in areas receiving 0.3-1.7 m/year of rainfall on rich, moist soils. They will not tolerate wet, tropical lowland conditions, or frost. Individual plants reach heights of 1.2 m. The opium poppy is a cultigen derived from *Papaver setigerum*, a plant from Asia Minor.

Ouabain, Cream Fruit

Sci. Name(s)	: *Strophanthus gratus, Roupellia grata*
Geog. Reg(s)	: Africa, Tropical
End-Use(s)	: Medicinal, Ornamental & Lawn Grass
Domesticated	: Y
Ref. Code(s)	: 17, 36, 157, 170

Summary : *Strophanthus gratus*, also known as ouabain, is a small shrub or tree whose seeds are the official source of a powerful glycoside, ouabain. Ouabain is used in a manner similar to strophanthin, another powerful glycoside, in the treatment of cardiac failure. Ouabain has been proven to be the most rapidly acting of the digitalis glycosides and is valuable because it can be obtained commercially in a relatively pure form. There are 3 other members of the Strophanthus genus that are minor sources of a similar glycoside, but these are considered economically unimportant. *Strophanthus gratus* is also a source of the drug cortizone. Plants can be grown as ornamentals. They are native to tropical Africa.

Pakistan Ephedra

Sci. Name(s)	: *Ephedra gerardiana*
Geog. Reg(s)	: Asia, Asia-China, Asia-India (subcontinent), Temperate
End-Use(s)	: Medicinal
Domesticated	: N
Ref. Code(s)	: 205, 220
Summary	: Pakistan ephedra is a plant with considerable medicinal value. When processed, the plant stems provide the commercial source of ephedrine alkaloids, used in a general class of antiasthmatic drugs. The drugs are used to treat bronchial asthma and hay fever. Pakistan ephedra should not be used by people with coronary thrombosis, thyrotoxicosis, or hypertension, as it is a stimulant.

Plants grow wild in dry temperate regions, primarily in the Himalayas. Pakistan ephedras also grow in parts of China and India.

Papaya
(See Fruit)

Para Cress

Sci. Name(s)	: *Spilanthes acmella*
Geog. Reg(s)	: America-North (U.S. and Canada), Worldwide
End-Use(s)	: Medicinal
Domesticated	: Y
Ref. Code(s)	: 36
Summary	: Para cress is an herb occasionally cultivated for its leaves and stems. The plant has a strong, bitter taste due to the presence of spilanthol. Spilanthol produces an anesthetic effect, and has been used to soothe toothaches. Para cress plants have also been used as a diuretic or purgative. Para cress is native to North America and has spread throughout the world.

Paroquet Bur, Pulut-Pulut
(See Fiber)

Parsnip, Wild Parsnip
(See Tuber)

Pawpaw
(See Fruit)

Pepper, Black Pepper
(See Spice & Flavoring)

Pepper Tree, Brazilian Pepper Tree, California Peppertree
(See Gum & Starch)

Peppermint
(See Essential Oil)

Periwinkle, Madagascar Periwinkle, Bright-Eyes

Sci. Name(s)	: *Catharanthus roseus*
Geog. Reg(s)	: Africa, Asia
End-Use(s)	: Medicinal, Ornamental & Lawn Grass
Domesticated	: Y
Ref. Code(s)	: 120, 170, 225
Summary	: The periwinkle is an annual or perennial shrub 30-50 cm in height. The leaves have medicinal use as an antihypoglycaemic, anticarcinogenic, and counterirritant. Derivatives of this plant have been used in treating diabetes. In Japan, it is grown as an ornamental. It is native to Madagascar.

Peru Balsam, Balsam-of-Peru
(See Natural Resin)

Pinus pumilio, Swiss Mountain Pine, Dwarf Pine
(See Essential Oil)

Pokeweed, Poke, Skoke, Pigeon Berry
(See Weed)

Pomegranate
(See Fruit)

Pomerac, Malay Apple
(See Fruit)

Princess-Feather
(See Vegetable)

Pulasan
(See Fruit)

Purging Croton

Sci. Name(s)	: *Croton tiglium*
Geog. Reg(s)	: Asia-China, Asia-India (subcontinent), Asia-Southeast
End-Use(s)	: Medicinal, Miscellaneous, Natural Resin, Oil
Domesticated	: Y
Ref. Code(s)	: 170, 205, 220

Summary   : The purging croton is a small tree or shrub about 6 m tall, cultivated for its seeds, which yield approximately 60% of a fatty oil (croton oil). This oil is one of the world's most powerful purgatives.

Plants are under commercial cultivation in China and southeast Asia. Croton oil contains croton-resin, a lactone that produces a purging effect. The oil is used commercially in soap manufacture, illumination, and medicine. It has been used as a cathartic, counterirritant, vesicant, and a rubefacient. It is rarely taken internally and is only used externally in a diluted form as a counterirritant in the treatment of gout and neuralgia. In Asia, croton seeds are used to stun fish.

Purging crotons are native to the Malabar coast region, southwest India, and Tavoy in Burma.

Quince
(See Gum & Starch)

Quinine, Quinine Tree, Peruvian Bark
Sci. Name(s)   : *Cinchona officinalis*
Geog. Reg(s)   : Africa-East, America-South, Asia-India (subcontinent), Asia-Southeast
End-Use(s)   : Dye & Tannin, Insecticide, Medicinal
Domesticated   : N
Ref. Code(s)   : 170, 205
Summary   : Quinine (*Cinchona officinalis*) is an evergreen tree from which an alkaloid providing an antimalarial drug can be obtained. The alkaloid is taken from the dried stem bark and has antipyretic, bitter, tonic, and stomachic uses. It has been used as an astringent throat gargle, for common colds, toothaches, and as a red dye for fabric. Quinine is still used in traditional medicine for relief of muscle cramps. The drug can cause vomiting, and extensive use of it can cause cinchonism, a disorder marked by temporary deafness, headaches, and dizziness. The drug is also used in sunburn lotions and insecticidal preparations.

The tree grows wild in the forests of South America, Java, Burma, Sri Lanka, India, and East Africa. Trees are most productive at 1,500-2,500 m in altitude. There are commercial plantations in Java. Synthetic drugs have now largely replaced the commercial market for quinine.

Quinine trees grow 6-25 m in height. Trees are native to South America.

Red Clover
(See Forage, Pasture & Feed Grain)

Red Sanderswood, Red Sandalwood, Calialur Wood
(See Dye & Tannin)

Ringworm Bush
Sci. Name(s)   : *Cassia alata*
Geog. Reg(s)   : Tropical
End-Use(s)   : Medicinal, Ornamental & Lawn Grass
Domesticated   : Y
Ref. Code(s)   : 170, 225

Summary : The ringworm bush is a tropical shrub whose leaves and pods are a minor source of the purgative senna. Parts of the ringworm bush have been used in the treatment of ringworm. The bush is often cultivated as an ornamental. It is an erect shrub growing up to 3 m in height.

Roman Chamomile, Chamomile
(See Essential Oil)

Roquette, Garden Rocket, Rocket Salad, Roka
(See Oil)

Rose Apple, Jambos
(See Fruit)

Rose Geranium, Geranium
(See Essential Oil)

Rubber-Tree, Para Rubber
(See Isoprenoid Resin & Rubber)

Russian Wormseed, Levant Wormseed, Tarragon, Santonica

Sci. Name(s) : *Artemisia cina*
Geog. Reg(s) : America-North (U.S. and Canada), Eurasia
End-Use(s) : Medicinal
Domesticated : Y
Ref. Code(s) : 170, 220
Summary : Russian wormseed is an herb whose dried flower heads (*flores ciriac*) contain sanonin. This has been used to combat intestinal worms. The herb is native to Russia and the western United States.

Sacred Lotus, Lotus, East Indian Lotus
(See Tuber)

Saffron, Saffron Crocus
(See Dye & Tannin)

Saigon Cinnamon, Cassia
(See Spice & Flavoring)

Sarsaparilla, Sarsaparillja

Sci. Name(s) : *Smilax aristolochiifolia*
Geog. Reg(s) : America-Central, Tropical
End-Use(s) : Medicinal
Domesticated : Y
Ref. Code(s) : 220
Summary : Sarsaparilla is a plant whose rhizomes are dried and used for the medicinal tonic sarsaparilla. The plant is cultivated in southern Mexico and in many parts of the

tropics. Sarsaparilla rhizomes have been used to treat digestive disorders, rheumatism, skin diseases, venereal diseases, and kidney complaints.

Scarlet Poppy, Oriental Poppy

Sci. Name(s)	: *Papaver bracteatum*
Geog. Reg(s)	: Africa-North, Asia, Eurasia, Mediterranean
End-Use(s)	: Medicinal, Oil, Ornamental & Lawn Grass
Domesticated	: Y
Ref. Code(s)	: 17, 119, 149
Summary	: The scarlet poppy is a perennial herb reaching 125 cm in height grown as an ornamental and for its dried roots, which are a potential source of thebaine, an element made to synthesize codeine, naloxone and other narcotic antagonists. These roots contain 0.50-1.3% thebaine and the stems contain about 0.09-0.22%.

Plants are grown from root cuttings and flower in their 2nd year. This plant thrives in dry, rocky areas at altitudes of 1,500-3,000 m. It is native to the Russian Caucasus and northern and western Iran. In Iran, poppy seeds are eaten or used as a source of a cooking and drying oil.

Combined yields of thebaine are 15-70 kg/ha. Thebaine content is highest 3-4 weeks after the flowers open.

Sesame, Simsim, Beniseed, Gingelly, Til
(See Oil)

Siam Benzoin, Siam Balsam

Sci. Name(s)	: *Styrax tonkinensis*
Geog. Reg(s)	: Asia-China, Asia-Southeast
End-Use(s)	: Medicinal, Natural Resin, Spice & Flavoring
Domesticated	: N
Ref. Code(s)	: 220
Summary	: Siam benzoin is a plant whose stems yield a resin that has been used medicinally as an antiseptic and in the perfume industry for making incense, perfumes, and scented soaps. The resin has also been used to flavor chocolate. Plants are native to Malaya. Production is centered in the Cochin region of China.

Sisal
(See Fiber)

Smooth Loofah, Sponge Gourd, Dishcloth Gourd, Vegetable Sponge, Luffa, Loofah
(See Miscellaneous)

Socotrine Aloe

Sci. Name(s)	: *Aloe perryi*
Geog. Reg(s)	: Africa-East, Africa-North, Asia-Southeast
End-Use(s)	: Medicinal
Domesticated	: N
Ref. Code(s)	: 36

Summary : The Socotrine aloe is a succulent whose medicinal latex is considered economically unimportant. It is the least cultivated of the aloes. The Socotrine aloe is native to the Socatra Islands in the Indian Ocean. It is under minor cultivation in Arabia and East Africa.

Sodom Apple
(See Vegetable)

Solanum khasianum
Sci. Name(s) : *Solanum khasianum*
Geog. Reg(s) : Asia-India (subcontinent)
End-Use(s) : Medicinal
Domesticated : Y
Ref. Code(s) : 110
Summary : Solanum khasianum is a subshrub 0.7-1.5 m in height whose fruit contains large quantities of the glyco-alkaloid solasodine. Solasodine is used in the manufacture of several drugs, the hormones of which are used as active ingredients in contraceptive pills.

Plants are usually propagated by seed. About 8,000 kg/ha of fresh fruit is considered to be a possible yield with smaller yields of dried fruit, giving an average of 2.5% solasodine content/ha. Solanum khasianum is native to India.

Sour Orange, Seville Orange
(See Essential Oil)

Sour-Relish Brinjal, Ram-Begun
Sci. Name(s) : *Solanum ferox*
Geog. Reg(s) : Asia-India (subcontinent), Tropical
End-Use(s) : Medicinal, Vegetable
Domesticated : N
Ref. Code(s) : 36, 98, 170
Summary : The sour-relish brinjal reaches heights of 0.6-1.2 m, is similar to the eggplant in general appearance, and is sometimes grown for its fruits, used as a sour relish in curries. Seeds are considered to have undocumented medicinal properties and are used to alleviate toothaches. Root decoctions ease body pains and indigestion. Plants are found in eastern and southern India and the tropics.

Southernwood, Lad's Love, Southern Wormwood
(See Ornamental & Lawn Grass)

Spanish Carrot, Khella, Toothpick Ammi
Sci. Name(s) : *Ammi visnaga*
Geog. Reg(s) : Africa-North, Mediterranean
End-Use(s) : Medicinal
Domesticated : Y
Ref. Code(s) : 220

Summary : The Spanish carrot is a Mediterranean herb of the carrot family, that provides a semisynthetic substance, khellin, used in a relatively new treatment for asthma. The plant has a long history as an antiasthmatic agent and as a treatment for urinary disorders among the Arabs. In Egypt, the fruit stalks are sold as toothpicks.

Spanish Lime
(See Fruit)

Spanish Oregano, Origanum
(See Essential Oil)

Squirting Cucumber

Sci. Name(s)	: *Ecballium elaterium*
Geog. Reg(s)	: Mediterranean
End-Use(s)	: Medicinal
Domesticated	: Y
Ref. Code(s)	: 28, 159, 199, 220

Summary : The squirting cucumber is a trailing, perennial vine whose unripe fruits and fleshy roots are used as purgatives, diuretics, and emetics. The fruit contains a poisonous juice used in the manufacture of the drug elaterium. The plant grows in the Mediterranean region.

Star-Anise Tree, Star Anise
(See Essential Oil)

Stinging Nettle, European Nettle
(See Weed)

Strophanthus, Arrow Poison

Sci. Name(s)	: *Strophanthus sarmentosus*
Geog. Reg(s)	: Africa-West
End-Use(s)	: Medicinal
Domesticated	: N
Ref. Code(s)	: 170

Summary : Strophanthus, a liane of the Strophanthus genus, is another minor source of the powerful heart stimulant glycoside ouabain. Plants are also a source of the drug cortisone. Strophanthus is native to tropical Africa.

Strophanthus kombe

Sci. Name(s)	: *Strophanthus kombe*
Geog. Reg(s)	: Africa-West
End-Use(s)	: Medicinal
Domesticated	: N
Ref. Code(s)	: 157, 170

Summary : Strophanthus kombe is a member of the Strophanthus genus and provides a minor source of the cardiac glycoside ouabain. The plant is also a source of the drug cortisone. There is little or no information concerning the economic potential for this plant.

Strychnine Tree

Sci. Name(s)   : *Strychnos nux-vomica*
Geog. Reg(s)   : Asia-India (subcontinent), Asia-Southeast
End-Use(s)   : Medicinal
Domesticated   : N
Ref. Code(s)   : 220
Summary   : The strychnine tree is a major source of the commercial substance strychnine. Strychnine is obtained from seeds collected from wild trees. There is very limited commercial production. Trees are native to the East Indies, Sri Lanka, and India.

Styrax, Benzoin Gum, Sumatra Benzoin
(See Natural Resin)

Summer Savory
(See Spice & Flavoring)

Sweet Blueberry, Lowbush Blueberry, Low Sweet Bush, Late Sweet Bush, Late Sweet Blueberry
(See Fruit)

Sweet Flag, Calamus, Flagroot
(See Essential Oil)

Sweet Marjoram, Knotted Marjoram
(See Spice & Flavoring)

Sweet Woodruff
(See Spice & Flavoring)

Tablus Spice, Baboon-Spice, Grains-of-Paradise

Sci. Name(s)   : *Aframomum sceptrum*
Geog. Reg(s)   : Africa-West
End-Use(s)   : Medicinal
Domesticated   : N
Ref. Code(s)   : 50
Summary   : The tablus spice plant is a member of the Aframomum genus whose fruit is said to relieve thirst, fatigue, and fever. Plants grow in Africa, specifically in Sierra Leone and Ghana. There is limited information on this particular plant.

Tachibana Orange, Tachibana
(See Fruit)

Tansy
Sci. Name(s)   : *Tanacetum vulgare*

Geog. Reg(s) : America-North (U.S. and Canada), Asia, Europe, Temperate
End-Use(s) : Essential Oil, Insecticide, Medicinal
Domesticated : Y
Ref. Code(s) : 205, 220
Summary : Tansy is a scented, perennial herb reaching 60-120 cm in height that is grown for its flowers and leaves. When dried, these yield an essential oil used for treating intestinal worms and menstruation problems and as a stimulant. If the oil is applied externally, it serves as an insect repellent. Tansy oil can be dangerous if taken in excess. Fresh or dried flowers and leaves are used in many sachets and similarly scented articles.

Herbs are native to Europe and Asia and have been introduced and naturalized in most temperate areas, primarily North America. Plants thrive on rich, loamy soils and will grow at altitudes approaching 1,500 m.

Tarwi, Tarhui, Chocho, Pearl Lupine, Tarin Altramuz, Muti, Ullus
(See Oil)

Terongan, Turkeyberry, Platebush
(See Vegetable)

Thyme
(See Essential Oil)

Tiberato, Indian Nightshade
(See Vegetable)

Tobacco
(See Miscellaneous)

Tolu Balsam, Balsam-of-Tolu
(See Natural Resin)

Tsi
(See Vegetable)

Turmeric
(See Spice & Flavoring)

Udjung Atup
(See Essential Oil)

Untsuti
Sci. Name(s) : *Strophanthus hispidus*
Geog. Reg(s) : Africa-West
End-Use(s) : Medicinal
Domesticated : N
Ref. Code(s) : 170, 220

Summary : Untsuti is a member of the Strophanthus genus, whose seeds provide a minor source of the heart stimulant glycoside ouabain. The liane is considered an economically unimportant source of the drug. The glycoside is only two-fifths as toxic as the glycoside from *Strophanthus gratus*. Untsuti is native to West Africa.

Upland Cress, Wintercress, Yellow Rocket
(See Vegetable)

Vanilla
(See Essential Oil)

Vogel Fig
(See Isoprenoid Resin & Rubber)

Wampi
(See Fruit)

White Lupine, Egyptian Lupine
(See Forage, Pasture & Feed Grain)

Wild Chamomile, Sweet False Chamomile, German Chamomile
(See Essential Oil)

Wild Thyme, Creeping Thyme
(See Ornamental & Lawn Grass)

Wild Yam, Floribunda Yam
Sci. Name(s) : *Dioscorea floribunda*
Geog. Reg(s) : America-North (U.S. and Canada), America-South, West Indies
End-Use(s) : Medicinal, Tuber
Domesticated : Y
Ref. Code(s) : 48, 149, 171
Summary : The wild yam (*Dioscorea floribunda*) is the most important of the sapogenin-containing yams. The plant is cultivated in parts of America for the sapogenin used in the manufacture of the drug cortisone. Tubers have a sapogenin content of 10% by dry weight. Like the composite yam (*Dioscorea composita*), tubers are collected from wild, 3-4 year old plants. They are much easier to harvest than composite yams because they develop horizontally and rarely reach depths beyond 30 cm.

In Puerto Rico, 4 year old plants produced 4,195 kg of dry tubers, which in turn yielded 331 kg of sapogenin. Wild yams are most successful in damp, woody, or thicketed areas in well-drained, sandy loam soils, at elevations of 1,500 m. The yams were introduced into the eastern and southern parts of the United States, Puerto Rico, and Costa Rica.

Winter Purslane
(See Vegetable)

Winter's-Bark

Sci. Name(s)   : *Drimys winteri*
Geog. Reg(s)   : America-South
End-Use(s)   : Medicinal, Spice & Flavoring, Timber
Domesticated   : N
Ref. Code(s)   : 220
Summary   : Winter's-bark is a South American tree whose bark is aromatic and used in home remedies. It is made into tonics and is antiscorbutic. In South America, the bark tonics are used to treat stomach complaints and dysentery. The powdered bark is used as a condiment. The wood has been used for house interiors and packing cases.

Wintergreen
(See Essential Oil)

Wonderberry, Black Nightshade, Poisonberry, Stubbleberry
(See Miscellaneous)

Wormseed, Mouse Food, Mexican Tea, Wormseed Goosefoot

Sci. Name(s)   : *Chenopodium ambrosioides*
Geog. Reg(s)   : America, Tropical
End-Use(s)   : Essential Oil, Medicinal
Domesticated   : Y
Ref. Code(s)   : 170
Summary   : Wormseed, or Mexican tea, is an aromatic herb about 3 m tall, from which a volatile oil is extracted by steam distillation. This oil is used in the treatment of hookworm and other intestinal parasites.

        The herb is native to tropical America and has naturalized in other tropical areas.

Wormwood, Absinthe, Absinthewood, Absinthium
(See Essential Oil)

Yellow Flame, Yellow Poinciana, Soga
(See Ornamental & Lawn Grass)

Yellow Foxglove

Sci. Name(s)   : *Digitalis grandiflora*
Geog. Reg(s)   : Asia, Eurasia, Europe
End-Use(s)   : Medicinal
Domesticated   : Y
Ref. Code(s)   : 149, 220
Summary   : Yellow foxglove (*Digitalis grandiflora*) is a biennial or perennial herb 0.6-1 m in height cultivated in Central and southern Europe as a drug plant. Plant leaves contain digitalin, digitalein and digitonin, which compose the heart stimulant drug digitalis. The plant has been used in herbal medicines for the treatment of epilepsy.

Plants grow wild on the rocky hills of southern Europe and western Asia. In the USSR, yellow foxglove is cultivated as a commercial drug plant. Yellow foxglove leaves are about half as potent as those of digitalis (*Digitalis purpurea*).

Yellow Gentian
(See Beverage)

Yellow Melilot, Yellow Sweetclover
(See Forage, Pasture & Feed Grain)

Yerba Mate, Paraguay Tea, Mate
(See Beverage)

Zedoary, Shoti
(See Essential Oil)

Miscellaneous

Afara, Limba
(See Timber)

African Locust Bean
(See Oil)

Alfalfa, Lucerne, Sativa
(See Forage, Pasture & Feed Grain)

Ambarella, Otaheite Apple, Golden-Apple, Jew-Plum
(See Fruit)

American Beech
(See Timber)

American Gooseberry, Currant Gooseberry, Hairy Gooseberry
(See Fruit)

American Hazelnut, Beaked Hazel

Sci. Name(s)	: *Corylus cornuta*
Geog. Reg(s)	: America-North (U.S. and Canada)
End-Use(s)	: Miscellaneous, Nut
Domesticated	: N
Ref. Code(s)	: 108, 235
Summary	: The American hazelnut is a hardy, nut-bearing shrub or small tree up to 2.4 m in height, found throughout North America. The nuts have little commercial value but the trees show potential for developing durable hybrids in harsh, cold climates.

American Oil Palm, Corozo
(See Oil)

Arabica Coffee, Coffee
(See Beverage)

Bergamot
(See Essential Oil)

Betel Pepper

Sci. Name(s)	: *Piper betel*
Geog. Reg(s)	: Africa-East, Asia-India (subcontinent), Asia-Southeast, Hawaii, Pacific Islands
End-Use(s)	: Gum & Starch, Miscellaneous, Nut
Domesticated	: Y
Ref. Code(s)	: 154, 170, 205
Summary	: The betel pepper (*Piper betel*) is a climbing shrub grown for its leaves and nuts, which have a long history as a masticatory in India, Malaysia, Indonesia, Oceania, and

many other parts of the world. These plants are native to Malaysia and India. Betel peppers also grow in Hawaii and are known as a relative of the 'awa (*Piper methysticum*). The practice of chewing the betel nut is stylistic and there are a variety of methods.

Plants are cultivated from sea level to 914 m. They require fertile soil and need protection from the wind. The leaves and nuts may be picked 18 months after planting, traditionally in the mornings. Betel pepper gardens can last up to 50 years under proper cultivation. Betel peppers do not enter world trade.

Bible Frankincense, Oil of Olibanum, Essence of Olibanum
(See Natural Resin)

Blue Lupine, European Blue Lupine, New Zealand Blue Lupine, Narrow-Leaved Lupine
(See Forage, Pasture & Feed Grain)

Borage
(See Ornamental & Lawn Grass)

Broad-Leaved Lavender
(See Essential Oil)

Broadbean, Horsebean, Field Bean, Tick Bean, Windsor Bean
(See Vegetable)

Buckwheat
(See Cereal)

Buffalo Gourd, Wild Gourd
(See Oil)

Bullock's-Heart, Corazon, Custard-Apple
(See Fruit)

Butternut
(See Nut)

Calabash Gourd, Calabash, Bottle Gourd, White Flowered Gourd
(See Vegetable)

California Black Walnut, Hind's Walnut
(See Nut)

Camphor-Tree, Camphor
(See Medicinal)

Canarygrass, True Canarygrass

Sci. Name(s)	: *Phalaris canariensis*
Geog. Reg(s)	: Africa-North, America-North (U.S. and Canada), America-South, Australia, Mediterranean
End-Use(s)	: Forage, Pasture & Feed Grain, Miscellaneous
Domesticated	: Y
Ref. Code(s)	: 20, 154, 163
Summary	: Canarygrass is grown for seed used in birdseed mixtures and as a sizing agent in textile industries. It is cultivated in the Middle East, Argentina, Australia, Canada, and the United States. Canadian farmers planted 35,000 ha of canarygrass in 1981. Productive capability in the United States far exceeds yearly production because of low grower returns brought on by stagnant demand and strong downward price pressure exerted by other supplying countries.

Annual canarygrass is often confused with reed canarygrass, a perennial forage crop and wild grass. Canarygrass heads are spikelike and resemble club wheat. In North America, this crop is propagated by seed. Little cultivation is required after planting. Dry seed yields are 1,200-2,000 kg/ha and reach 2,000-3,000 kg/ha on irrigated land.

Cape Aloe, Bitter Aloe, Red Aloe
(See Medicinal)

Cardamom, Cardamon
(See Spice & Flavoring)

Catnip

Sci. Name(s)	: *Nepeta cataria*
Geog. Reg(s)	: America-North (U.S. and Canada), Europe
End-Use(s)	: Medicinal, Miscellaneous
Domesticated	: Y
Ref. Code(s)	: 194, 205, 220
Summary	: Catnip is a perennial native European herb whose leaves are dried and manufactured into cat toys. Cats are attracted to and somewhat intoxicated by the spicy aroma. The leaves were once used as flavoring and were smoked to relieve chronic bronchitis. This practice was stopped when the smoke proved to have hallucinogenic effects. The herb has limited medicinal value and its leaves are used mainly in home remedies for treating minor cuts and scrapes. The white, purple-spotted flowers attract bees. Plants thrive in most well-drained soils.

Ceara-Rubber
(See Isoprenoid Resin & Rubber)

Cherry Plum, Myrobalan Plum
(See Ornamental & Lawn Grass)

Chinese Chestnut
(See Nut)

Chinese Hazelnut, Chinese Filbert, Chinese Hazel
(See Nut)

Clary, Clary Wort, Clary Sage
(See Essential Oil)

Common Licorice
(See Spice & Flavoring)

Common Persimmon, Persimmon, American Persimmon
(See Fruit)

Common Reed Grass, Carrizo

Sci. Name(s)	: *Phragmites australis, Phragmites communis*
Geog. Reg(s)	: Africa, America-North (U.S. and Canada), America-South, Australia, Eurasia
End-Use(s)	: Fiber, Forage, Pasture & Feed Grain, Medicinal, Miscellaneous, Vegetable
Domesticated	: Y
Ref. Code(s)	: 17, 36
Summary	: Common reed grass is a wet ground or marsh plant whose stems and leaves are used in lattice work, matting, and sandal and pen making. Young foliage is palatable to livestock. In China, young buds are eaten in salads, and the sprouts are dried and used for undocumented medicinal purposes. The grass grows in Eurasia, Australia, Africa, South America, and parts of the United States.

Common Thyme
(See Spice & Flavoring)

Coolabah
(See Timber)

Coriander
(See Spice & Flavoring)

Cork Oak

Sci. Name(s)	: *Quercus suber*
Geog. Reg(s)	: America-North (U.S. and Canada), Europe, Mediterranean
End-Use(s)	: Miscellaneous, Timber
Domesticated	: Y
Ref. Code(s)	: 17, 36, 116, 119, 154, 202, 220
Summary	: The cork oak is an evergreen tree, 12-18 m in height, whose bark is the source of commercial cork. Cork is valuable because of its extreme light weight, durability, nonflammability, and insulating qualities. It is used mainly as a flooring material and for insulating buildings. Other commercial uses are as gasket seals, linoleum, floats, and bottle-stoppers.

Trees are cultivated in the southwestern Mediterranean area, Portugal, and parts of North America. Cork oaks thrive on rocky soils on the lower slopes of mountains.

The best cork is obtained from trees already stripped. Stripped trees take up to 10 years to regenerate but have a productive cycle of more than 150 years.

Currant Tomato
(See Fruit)

Damson Plum, Tart Damson Plum, Bullace Plum
(See Fruit)

Dasheen, Elephant's Ear, Taro, Eddoe
(See Tuber)

Dendrocalamus, Bamboo
(See Timber)

Desert Date
(See Fruit)

East Indian Sandalwood
(See Essential Oil)

Einkorn, One-Grained Wheat
(See Forage, Pasture & Feed Grain)

Elephant Bush, Spekboom
(See Ornamental & Lawn Grass)

English Walnut, Persian Walnut
(See Nut)

Eucalyptus
(See Timber)

Eucalyptus urophylla
(See Timber)

European Hazel, European Filbert
(See Nut)

Field Horsetail
(See Medicinal)

Fishtail Palm, Toddy Palm
(See Sweetener)

Garden Myrrh, Sweet Scented Myrrh
(See Medicinal)

Giant Filbert, Lambert's Filbert, Filbert
(See Nut)

Giant Reed

Sci. Name(s)	: *Arundo donax*
Geog. Reg(s)	: America-North (U.S. and Canada), Europe, Mediterranean, Tropical
End-Use(s)	: Fiber, Miscellaneous, Ornamental & Lawn Grass, Weed
Domesticated	: Y
Ref. Code(s)	: 15
Summary	: The giant reed is a perennial weed that reaches heights of 1.80-6.0 m. It grows in the Mediterranean region and in the tropics, where it is frequently planted as an ornamental. In Europe, the reed is made into musical wind instruments, while in North America it is made into screens and mats.

Grass Pea, Chickling Vetch, Chickling Pea
(See Cereal)

Gutta-Percha
(See Isoprenoid Resin & Rubber)

Heartnut, Cordate Walnut, Siebold Walnut
(See Nut)

Idigbo, Framire
(See Timber)

Imbe
(See Beverage)

Indian Jujube, Beri, Inu-Natsume, Ber Tree
(See Fruit)

Indian-Almond, Country Almond, Tropical-Almond, Myrobalan, Almendro
(See Oil)

Jasmine, Poet's Jessamine
(See Essential Oil)

Jerusalem-Artichoke, Girasole
(See Tuber)

Job's Tears, Adlay, Adlay Millet

Sci. Name(s)	: *Coix lacryma-jobi*
Geog. Reg(s)	: Asia, Asia-China, Asia-Southeast, Tropical
End-Use(s)	: Beverage, Cereal, Forage, Pasture & Feed Grain, Medicinal, Miscellaneous
Domesticated	: Y
Ref. Code(s)	: 67, 214, 220, 225

Summary : Job's tears is a broad-leaved tropical grass cultivated for its hard gray seeds, which are parched and ground to flour for tea or left whole as decorative beads. In China and the Philippines, the seeds are used as a diuretic. The foliage is fed to livestock.

The grass has culms to 1 m in height. It grows in wet fields and on wasteland. It is native to tropical Asia.

Kariis, Karii
(See Fruit)

Kura Clover, Kura Gourd, Pellett Clover, Honeyclover, Caucasian Clover
(See Forage, Pasture & Feed Grain)

Malabar Nut
(See Medicinal)

Mango
(See Fruit)

Marking-Nut Tree
(See Dye & Tannin)

Mayflower, Apamate, Roble
(See Timber)

Nigerian Lucerne
(See Forage, Pasture & Feed Grain)

Nutgrass, Purple Nutsedge, Nutsedge
(See Weed)

Olive
(See Oil)

Palmyra Palm, African Fan Palm
(See Sweetener)

Pecan
(See Nut)

Purging Croton
(See Medicinal)

Radish
(See Vegetable)

Raintree, Saman
(See Timber)

Ramon, Breadnut
(See Fruit)

Rattan
(See Timber)

Red Quebracho
(See Dye & Tannin)

Rice
(See Cereal)

Rose Geranium, Geranium
(See Essential Oil)

Sainfoin, Holyclover, Esparret
(See Forage, Pasture & Feed Grain)

Sea Island Cotton, American Egyptian Cotton, Extra Long Staple Cotton
(See Fiber)

Siberian Crabapple
(See Fruit)

Sickle Medic, Yellow-Flowered Alfalfa
(See Forage, Pasture & Feed Grain)

Silky Oak
(See Timber)

Smooth Loofah, Sponge Gourd, Dishcloth Gourd, Vegetable Sponge, Luffa, Loofah

Sci. Name(s)	: *Luffa aegyptiaca, Luffa cylindrica*
Geog. Reg(s)	: America-North (U.S. and Canada), Asia, Asia-India (subcontinent), Tropical
End-Use(s)	: Fiber, Fruit, Medicinal, Miscellaneous, Oil, Vegetable
Domesticated	: Y
Ref. Code(s)	: 154, 170
Summary	: The smooth loofah produces edible fruit eaten fresh, cooked as a vegetable, and used in soups. The fibrous network of the fruit serves as a sponge used as a filter. In addition, table mats, insoles, sponges, sandals, and gloves are made from it. It is valued for its shock-absorbing properties; because of this quality, the United States Army has used loofah fiber in linings for steel helmets and armored vehicles.

An edible oil is extracted from the seeds. A stem sap is used in medicines and toilet preparations. The plant probably originated in India and has spread throughout the tropics. In the East, loofahs are used in local medicines. Japan is a principal producer.

Plants mature 4-5 months after planting. Each vine holds 20-25 fruits, and as many as 24,000 fruits/ha are obtained under ideal conditions.

Sodom Apple
(See Vegetable)

Sorghum, Great Millet, Guinea Corn, Kaffir Corn, Milo, Sorgo, Kaoling, Durra, Mtama, Jola, Jawa, Cholam Grains, Sweets, Broomcorn, Shattercane, Grain Sorghums, Sweet Sorghums
(See Cereal)

Sour Orange, Seville Orange
(See Essential Oil)

Southern Red Cedar, Eastern Red-Cedar
(See Essential Oil)

Starfruit, Carambola
(See Fruit)

Strawberry Clover
(See Forage, Pasture & Feed Grain)

Sugar Maple, Rock Maple, Hard Maple
(See Timber)

Sugar Palm, Gomuti Palm
(See Gum & Starch)

Sugarcane
(See Sweetener)

Sunchoke

Sci. Name(s)	: *Helianthus annuus* X *Helianthus tuberosus*
Geog. Reg(s)	: America-North (U.S. and Canada), Eurasia, Europe
End-Use(s)	: Miscellaneous
Domesticated	: Y
Ref. Code(s)	: 37
Summary	: The sunchoke (*Helianthus annuus* X *H. tuberosus*) is a perennial hybrid developed from the sunflower (*Helianthus annuus*) and the Jerusalem artichoke (*Helianthus tuberosus*). In Russia, plants are used for the improvement of the cultivated sunflower. It is one of the "group immunity" cultivars researched at the Pustovoit All-Union Oilseed Institute at Krasnodar, USSR. It has proven resistant to downy mildew in Europe but other studies show that this cultivar may not be resistant in North America.

A parallel example holds true with rust. The sunchoke has shown resistance to rust in Soviet experiments but rust has developed on cultivars grown in the United States. Sunchoke hybrids experience vigorous growth but are almost all female and sterile.

Sweet Balm, Lemon Balm
(See Essential Oil)

Sweet Birch, Black Birch, Cherry Birch, Mountain Mahogany
(See Essential Oil)

Tamarind
(See Fruit)

Tamarugo, Tamarugal
(See Timber)

Ti Palm, Ti

Sci. Name(s)	: *Cordyline terminalis, Cordyline fruticosa*
Geog. Reg(s)	: Asia-India (subcontinent), Australia, Hawaii, Pacific Islands
End-Use(s)	: Beverage, Miscellaneous
Domesticated	: Y
Ref. Code(s)	: 220
Summary	: The ti palm is a shrub or small tree that grows mostly in India, Australia, and the Pacific islands. It can be grown in other tropical or warm temperate areas. In Hawaii, the underground parts of the palm are fermented to make a beverage, and the leaves are used to make skirts. In other parts of Polynesia, palm leaves have been used to wrap fish for baking.

Tobacco

Sci. Name(s)	: *Nicotiana tabacum*
Geog. Reg(s)	: Africa-East, America-North (U.S. and Canada), Asia-China, Asia-India (subcontinent), Asia-Southeast
End-Use(s)	: Insecticide, Medicinal, Miscellaneous
Domesticated	: Y
Ref. Code(s)	: 126, 220
Summary	: Tobacco plants are herbaceous annuals that are important cash crops on plantations and small farms. Plants are grown for their leaves, which are cured and dried and used for tobacco cigarettes and cigars, snuff, and as a source of nicotine for insecticides. Nicotine content is highest in the uppermost leaves. Tobacco is native to tropical America and has been introduced to many other areas. The southern United States has large areas planted to tobacco.

Leaves are harvested 4 months after planting. Air curing takes 1-2 months, fire curing 3-10 weeks, and flue curing 4-6 days. Leaves are then fermented and can be stored for long periods of time. Their overall quality improves after the first 3-4 years.

Plants need medium-rich, well-drained soil. Cigar or cigarette tobacco should be planted on rich loam soil, flue-cured or bright types on sandy loams, and air-cured or brown burley tobacco on limestone silt soils. Weed control is essential for good growth. Plants are susceptible to several fungal diseases and viruses. Control is maintained by chemicals, certain cultural practices, and by planting improved, disease-resistant cultivars.

In 1983, world production was 6 million MT. The United States produced 11% of that total, China 23%, India 10%, and Russia 6%. Brazil, Japan, Turkey, East and South Africa, and eastern Europe are other minor producers. The United States exported 239,685 MT of unmanufactured tobacco at a value of $1.5 billion in 1983.

Trifoliate Orange
(See Ornamental & Lawn Grass)

Triticale
(See Cereal)

Tuba Root, Luba Patch, Derris, Tuba Putch
(See Insecticide)

Urucu Timbo, Fish Poison
(See Insecticide)

White Melilot, White Sweetclover
(See Forage, Pasture & Feed Grain)

White Mulberry
Sci. Name(s)	: *Morus alba*
Geog. Reg(s)	: Asia-China, Asia-India (subcontinent), Tropical
End-Use(s)	: Fiber, Fruit, Miscellaneous, Timber
Domesticated	: Y
Ref. Code(s)	: 170
Summary	: The white mulberry tree (*Morus alba*) is native to China, where its bark is used in paper making. It is grown in parts of the tropics for its fruit, cooked or eaten fresh. In India, white mulberries are grown for their leaves, which are fed to silkworms. The wood is used to make lightweight sporting goods, such as tennis rackets and hockey sticks.

Wine Palm, Coco de Chile, Coquito Palm, Honey Palm, Wine Palm of Chile, Chile Coco Palm
(See Beverage)

Wonderberry, Black Nightshade, Poisonberry, Stubbleberry
Sci. Name(s)	: *Solanum nigrum*
Geog. Reg(s)	: Temperate, Tropical, Worldwide
End-Use(s)	: Fruit, Medicinal, Miscellaneous, Vegetable, Weed
Domesticated	: Y

314

Ref. Code(s) : 16, 17, 36, 170

Summary : Wonderberry, or black nightshade, has a reputation for being a poisonous plant, due to the presents of the alkaloid solasonine. Grown throughout the tropics, it is often regarded as a weed. It is cooked locally as a potherb and its ripe berries are used in pies and preserves. Its tender shoots are boiled and eaten like spinach or eaten in soups. Medicinally, it acts as a laxative and diuretic.

Yellow Lupine, European Yellow Lupine
(See Forage, Pasture & Feed Grain)

Yellow Melilot, Yellow Sweetclover
(See Forage, Pasture & Feed Grain)

Natural Resin

Aleppo Pine
(See Timber)

Balsam Fir, Canada Balsam, American Silver Fir
(See Essential Oil)

Bible Frankincense, Oil of Olibanum, Essence of Olibanum

Sci. Name(s)	: *Boswellia carteri*
Geog. Reg(s)	: Africa-East, Africa-North, Eurasia, Mediterranean
End-Use(s)	: Medicinal, Miscellaneous, Natural Resin
Domesticated	: N
Ref. Code(s)	: 77, 220, 227
Summary	: The Bible frankincense tree produces a natural resin (olibanum) from incisions made in its bark. The incisions must drain for 3 months before the gum can be collected. The tree is also a source of frankincense, which contains bassarin, 6-8% arabin and a bitter substance. Frankincense burns easily and has a pleasant smell, making it a popular incense. It has been used medicinally to treat fevers and dysentery. Modern uses of frankincense are in fumigants and fixatives for perfumes. The Bible frankincense tree grows in Somalia and the Middle East.

Brown Barrel
(See Timber)

Burmese Lacquer

Sci. Name(s)	: *Melanorrhoea usitata*
Geog. Reg(s)	: Asia-India (subcontinent), Asia-Southeast
End-Use(s)	: Natural Resin
Domesticated	: Y
Ref. Code(s)	: 220
Summary	: The Burmese lacquer tree yields a natural resin that blackens on exposure to air. It is used as a lacquer to varnish woodwork and cement Burmese glass mosaics. There is a small export market for this varnish. The tree is native to Burma, Thailand, and northeastern India in the Manipur river area.

Canary Island Pine
(See Timber)

Caribbean Pine
(See Timber)

Chinese Lacquer Tree, Varnish Tree, Lacquer Tree, Japanese Lacquer Tree, Urushi

Sci. Name(s)	: *Toxicodendron vernicifluum, Rhus verniciflua*
Geog. Reg(s)	: Asia, Temperate
End-Use(s)	: Natural Resin, Ornamental & Lawn Grass

Domesticated : Y
Ref. Code(s) : 17, 94, 199, 220
Summary : The Chinese lacquer tree reaches 6.1 m in height and is grown as an ornamental. In southwestern Japan, this tree is cultivated for its latex-filled leaves, stems,
and branches, which are dried as a principal source of a lacquer. This lacquer has a
long storage life in sealed containers and remains unchanged in the presence of acids,
alkalis, and alcohol. This tree is native to Asia and is grown in several temperate
countries.

Chir Pine
(See Timber)

Copaiba
Sci. Name(s) : *Copaifera officinalis*
Geog. Reg(s) : America-South
End-Use(s) : Medicinal, Natural Resin
Domesticated : N
Ref. Code(s) : 170, 220
Summary : Copaiba is a Venezuelan tree whose trunk yields natural resins. The hardened
resin is the commercially important copaiba balsam used in medicine and industry.
The natural resin may be used in paints, varnishes, inks, and adhesives.

Damar
(See Timber)

Dandelion, Common Dandelion
(See Beverage)

Guaiac, Guaiacum
Sci. Name(s) : *Guajacum officinale*
Geog. Reg(s) : America-North (U.S. and Canada), America-South, Subtropical, Tropical,
West Indies
End-Use(s) : Medicinal, Natural Resin, Ornamental & Lawn Grass, Timber
Domesticated : Y
Ref. Code(s) : 17, 220, 227
Summary : Guaiac is an evergreen tree or shrub that has a heavy, hard, resinous wood.
The wood contains 26% of a natural resin thought to have medicinal qualities. It is
used as a mild laxative, as a test for oxygens, and as an antioxidant for commercial
preparation of lard. This tree is also the source of lignum vitae, one of the hardest
commercial woods used in shipbuilding and for small, hard objects such as hammer
handles. Trees reach 3-9 m in height. They are often grown as ornamentals.

Trees are resistant to salt spray and have potential for development in California
and Florida. They are grown in the West Indies and along the northern coast of
South America, where they thrive in the tropical and subtropical climates. In 1983,
United States gum and natural resin imports of all types were 15,742 MT, valued at $6
million.

Guar, Clusterbean
(See Gum & Starch)

Horehound, White Horehound
(See Medicinal)

Inula, Elecampane
(See Essential Oil)

Japan Wax
(See Fat & Wax)

Kaki, Kaki Persimmon, Japanese Persimmon
(See Fruit)

Khasya Pine, Khasi Pine, Benguet Pine
(See Timber)

Labdanum
(See Essential Oil)

Loblolly Pine
(See Timber)

Longleaf Pine
(See Essential Oil)

Manila, White Dammar Pine

Sci. Name(s)	: *Agathis alba*
Geog. Reg(s)	: Asia-Southeast
End-Use(s)	: Isoprenoid Resin & Rubber, Natural Resin
Domesticated	: Y
Ref. Code(s)	: 227
Summary	: The Manila tree is grown for its oleoresin in the Philippines and islands of the Indonesian Archipelago. This oleoresin is collected at intervals to produce a soft melengket or a hard, fossillike resin called boea.

Manila resin is soluble in alcohol and used in lacquer formulas. Because of its high acid value it can be neutralized by amines in water-borne coatings and polishes.

Marijuana, Hemp, Marihuana
(See Fiber)

Mastic

Sci. Name(s)	: *Pistacia lentiscus*
Geog. Reg(s)	: Mediterranean, Subtropical
End-Use(s)	: Essential Oil, Natural Resin, Oil, Spice & Flavoring
Domesticated	: Y

Ref. Code(s) : 7, 17, 75, 77, 227

Summary : Mastic is a shrubby evergreen tree reaching 3.7-4 m in height whose bark provides mastic resin, considered one of the oldest known high-grade resins. This tree yields 3% of an essential oil with limited use in flavoring and liqueurs. Mastic resin is soluble in aromatics and alcohols and often used as a clear coating over paintings. An edible oil from the seeds contains some volatile materials removed in the distilling process.

This tree is grown on the Greek island of Chios and is native to the bordering Mediterranean areas. It prefers an arid, subtropical climate. In 1983, gum and gum resin imports of all types to the United States were 15,742 MT, valued at $6 million.

Mexican White Pine
(See Timber)

Michoacan Pine
(See Timber)

Montezuma Pine
(See Timber)

Peru Balsam, Balsam-of-Peru

Sci. Name(s) : *Myroxylon pereirae, Myroxylon balsamum* var. *pereirae*

Geog. Reg(s) : America-Central

End-Use(s) : Essential Oil, Medicinal, Natural Resin, Ornamental & Lawn Grass, Spice & Flavoring, Timber

Domesticated : Y

Ref. Code(s) : 36, 56, 154, 170

Summary : Peru balsam is a tree growing to 15 m in height whose trunk contains a gum--Peru balsam. An essential oil is also obtained from the tree and used in perfumery, cosmetics, soap, or as a flavoring. The close-grained, reddish mahogany wood is excellent for cabinet work. Trees are occasionally grown for shade or ornamental purposes. Balsam was once burned as a popular incense.

Trees have localized along the Pacific Coast jungles of Central America, El Salvador, Nicaragua, and Guatemala. These countries are the main suppliers of Peru balsam products. A healthy tree of about 6 years yields approximately 112-170 g of balsam at each extraction. Trees are tapped twice a year until about 40-60 years of age.

Pili Nut, Elemi
(See Nut)

Pino Blanco
(See Timber)

Pinus brutia
(See Timber)

Pinus merkusii
(See Timber)

Pinus oocarpa
(See Timber)

Pinus strobus var. chiapensis
(See Timber)

Ponderosa Pine
(See Timber)

Purging Croton
(See Medicinal)

Siam Benzoin, Siam Balsam
(See Medicinal)

Slash Pine
(See Timber)

Styrax, Benzoin Gum, Sumatra Benzoin

Sci. Name(s) : *Styrax benzoin*
Geog. Reg(s) : Asia-Southeast, West Indies
End-Use(s) : Medicinal, Natural Resin
Domesticated : Y
Ref. Code(s) : 205, 220
Summary : Styrax is a tree reaching 7 m in height cultivated for a resin obtained from the bark. This resin contains 60% balsamic acids, specifically esters of cinnamic and benzoic acids, benzoresinol, benzaldehyde, styrol, and vanillin. Their combined action is considered to have antiseptic properties. This resin has been used as a carminative, antiseptic, and diuretic, and internally as a genito-urinary antiseptic and expectorant in chronic bronchitis. Trees are tapped at 7 years and on up to 20 years. Styrax is found both wild and cultivated. Trees are native to Southeast Asia and the West Indies.

Tenasserim Pine
(See Timber)

Tolu Balsam, Balsam-of-Tolu

Sci. Name(s) : *Myroxylon balsamum*
Geog. Reg(s) : America-South, West Indies
End-Use(s) : Essential Oil, Medicinal, Natural Resin, Ornamental & Lawn Grass, Spice & Flavoring
Domesticated : Y
Ref. Code(s) : 56, 170

320

Summary   : Tolu balsam is a South American tree reaching 15 m in height. A gum (tolu balsam) is collected from incised trees and used as flavoring in cough syrups, soft drinks, and confectionery. Oleoresins are good fixatives. An essential oil extracted from the seeds combines well in cosmetics, soaps, oriental perfumes, and heavier floral perfumes. In the United States, the oil is used as an ingredient in tincture of benzoin. Trees are grown for shade and ornamental purposes. Venezuela, Colombia, and the West Indies provide most of the commercial tolu balsam products.

About 8-10 kg/tree of balsam is obtained during a given season. About 20 incisions are made in the bark for gum collection. Gum is collected over an 8 month period. Trees grow wild in the forests of northern South America in areas receiving an average annual rainfall of 200 mm.

West Indian Pine
(See Timber)

Nut

Abata Kola
Sci. Name(s) : *Cola acuminata*
Geog. Reg(s) : Africa-West, America-South
End-Use(s) : Beverage, Nut
Domesticated : N
Ref. Code(s) : 170
Summary : Abata kola (*Cola acuminata*) is a tree whose bitter-tasting fruit seeds are made into a stimulating beverage in parts of Africa. In Nigeria, the seeds are used ceremoniously and socially. Trees grow wild in the rain forests of West Africa and reach heights of 7-13 m.

African Locust Bean
(See Oil)

African Pumpkin, Zanzibar Oil Vine, Oysternut
(See Oil)

Almond
Sci. Name(s) : *Prunus dulcis, Prunus amygdalus, Prunus communis*
Geog. Reg(s) : Africa-South, America-North (U.S. and Canada), Asia, Australia, Europe, Hawaii, Mediterranean, Subtropical, Temperate
End-Use(s) : Energy, Forage, Pasture & Feed Grain, Gum & Starch, Nut, Oil
Domesticated : Y
Ref. Code(s) : 17, 101, 116, 154, 182
Summary : The almond is a tree reaching up to 10 m in height that produces edible nuts used mainly in candies, baked goods, and confectionery. Roasted sweet almond kernels are popular snack foods. Almonds are fairly high in protein.

Commercially important oils are obtained from both sweet and bitter almonds. Sweet almond oil has a fairly high fatty content and is used in cosmetic creams and lotions. Bitter almond oil has wider use as a flavoring agent and ingredient in cosmetic skin preparations.

Several other types of processed nut products have entered the almond market. Almond powder, which contains vitamin E and is used in almond macaroon powder, almond marzipan powder, almond paste powder and an almond drink powder. Almond butter is processed like peanut butter. Shells are burned to produce electricity and are used as roughage in cattle feed. Trees yield a gum substituted for gum tragacanth.

Almond trees are native to Asia and are productive in temperate and subtropical areas with moist, cool climates. They tolerate annual rainfalls of 2-14.7 dm and annual temperatures of 10.5-19.5 degrees Celsius. Almonds thrive in the hot, dry valleys of California. Trees grow on most soils but prefer deep, fertile, well-drained soils for highest yields.

Their productive life is about 5 years. Leading almond producers are Spain, the United States, and Italy. They are cultivated to a lesser extent in the Mediterranean, Australia, and South Africa. California is the most important almond-producing state in the United States and supplies the domestic market. Trees grown in Hawaii seldom fruit. Projected 1984 almond crop production in California was expected to reach 236,080 MT from an area of 149,850 ha.

American Beech
(See Timber)

American Black Walnut, Black Walnut, Eastern Black Walnut
(See Timber)

American Chestnut
(See Timber)

American Hazelnut, Beaked Hazel
(See Miscellaneous)

Benoil Tree, Horseradish-Tree
(See Oil)

Betel Palm, Betel Nut, Areca Palm
(See Ornamental & Lawn Grass)

Betel Pepper
(See Miscellaneous)

Bitter Almond
(See Essential Oil)

Brazil-Nut, Brazilnut

Sci. Name(s)	: *Bertholletia excelsa*
Geog. Reg(s)	: America-South
End-Use(s)	: Nut
Domesticated	: N
Ref. Code(s)	: 148, 170
Summary	: The Brazil-nut is a large tree that grows wild in the Amazon forests of South America. Commercial Brazil-nuts are collected from these wild trees. Major nut exporters are the South American countries, and the main importers are Europe and the United States.

Nuts are found in large seed pods, each containing 25 or more of the triangular-shaped nuts. Nut kernels contain 60-70% fatty oils, 87% carbohydrates, and 13-17% protein.

Trees reach heights of 40 m and bear nuts at 10-20 years of age. Fruit clusters take over a year to ripen.

Broadbean, Horsebean, Field Bean, Tick Bean, Windsor Bean
(See Vegetable)

Butternut

Sci. Name(s)	: *Juglans cinerea*
Geog. Reg(s)	: America-North (U.S. and Canada)
End-Use(s)	: Dye & Tannin, Medicinal, Miscellaneous, Nut, Sweetener, Timber
Domesticated	: Y
Ref. Code(s)	: 182, 220
Summary	: The butternut is a deciduous tree 27-55 m in height that produces edible nuts.

Butternuts are hard to crack. Their kernels have a mild, pleasant, but very distinctive taste. They are nutritious and contain 23.7% protein. Trees produce a sap sometimes used like maple syrup. Immature green fruit is occasionally pickled. The wood is close-grained, soft, light-colored, and used for furniture and house interiors. The fuzz from young twigs, leaves, buds, and fruit can be boiled to produce a light brown dye. The inner root bark has been used to treat fevers.

In North America, these trees are the most cold-resistant members of the walnut family. There is some commercial production of the nuts, but demand is generally much lower than demand for black and English walnuts. Trees are not planted commercially because of their slow growth, short life, and susceptibility to fungus attack. Most butternuts are picked for domestic use.

California Black Walnut, Hind's Walnut

Sci. Name(s)	: *Juglans hindsii*
Geog. Reg(s)	: America-North (U.S. and Canada)
End-Use(s)	: Miscellaneous, Nut
Domesticated	: Y
Ref. Code(s)	: 59, 131, 182
Summary	: The California black walnut is a tree grown for its edible nuts, produced on a

small-scale commercial basis in Missouri and Indiana. Trees are used mainly in California as a vigorous, disease-resistant rootstock for grafting with the more commercially important English walnut.

Trees grow wild along the rocky, coastal regions of Central California. Black walnuts are resistant to most diseases and pests and tolerate serious droughts. Squirrels and other rodents that scavenge the nuts pose the greatest danger. For information on the commercial production of walnuts, refer to the english walnut.

Candlenut, Candleberry, Indian Walnut
(See Oil)

Cashew

Sci. Name(s)	: *Anacardium occidentale*
Geog. Reg(s)	: Africa-East, America, America-South, Asia-India (subcontinent), Tropical
End-Use(s)	: Beverage, Dye & Tannin, Fruit, Nut, Oil
Domesticated	: Y
Ref. Code(s)	: 148, 170
Summary	: The cashew tree is an evergreen reaching 12 m in height. It provides edible

nuts used in confectionery and desserts. In India and East Africa, these nuts are important commercial items.

The nuts have 15% protein and yield 40% of a high-quality, edible oil. This oil is not generally extracted, because whole nuts have a higher value. Nut shells yield a vesicant oil used as a preservative and waterproofing agent. Distilled and polymerized cashew shell oil is used in insulation, varnishes, in the manufacture of typewriter rolls, and in oil- and acid-proof cements. Tree bark produces a dye used as indelible ink. Cashew fruits are fleshy and edible and contain vitamin C, 7-9% sugar, and 0.5% tannin. Ripe fruit is made into beverages and wine and the pulps into preserves.

Trees grow in stony, sandy, or hilly soils at 600 m in elevation. They are hardy, fairly drought-resistant, but susceptible to frost. Cashews thrive in areas with an annual rainfall of 50-380 cm. They are native to tropical America and have naturalized in other tropical countries, particularly along coastlines. Top nut-producing regions are in southern India, Brazil, Mozambique, Kenya, and Tanzania.

In 1984, world production of cashews was 397,252 MT. The major producing countries were India with 180,000 MT, Brazil with 65,000 MT, and Tanzania with 48,800 MT.

Chinese Chestnut

Sci. Name(s)	: *Castanea mollissima*
Geog. Reg(s)	: America-North (U.S. and Canada), Asia, Asia-China
End-Use(s)	: Dye & Tannin, Miscellaneous, Nut, Timber
Domesticated	: Y
Ref. Code(s)	: 39, 140
Summary	: The Chinese chestnut tree produces commercially important nuts. The straight-grained, heavy timber has been used for railroad ties, inexpensive furniture, and telephone poles. The bark yields tannin. Chestnut leaves are often fed to silkworms.

Trees grow throughout China and Korea in areas with rich, dry soil. Production centers are in the Chian province. The Chinese chestnut tree is hardy and more resistant to the devastating chestnut blight that affects the American chestnut. Chinese chestnuts have been introduced into the United States as a replacement. They begin bearing nuts at 5-6 years of age. The nuts are considered to be of high quality. In Georgia, large plantings of trees have been made in areas formerly producing peaches. The trees show promise as orchard trees.

Chinese Hazelnut, Chinese Filbert, Chinese Hazel

Sci. Name(s)	: *Corylus chinensis*
Geog. Reg(s)	: America-North (U.S. and Canada), Asia-China
End-Use(s)	: Miscellaneous, Nut
Domesticated	: Y
Ref. Code(s)	: 40, 108, 235
Summary	: The Chinese hazelnut is a tree reaching 35 m in height, grown for its commercially important edible nuts, eaten fresh, roasted, or pounded and drunk with tea. This tree is cultivated to a limited extent in China, where its nuts are used in confections.

Trees are also grown in the United States. They appear to be resistant to the Eastern filbert blight, which hampers nut production of other hazelnut trees, and show potential for hybrid development.

Chinquapin, Allegheny Chinkapin

Sci. Name(s) : *Castanea pumila*
Geog. Reg(s) : America-North (U.S. and Canada)
End-Use(s) : Medicinal, Nut, Timber
Domesticated : Y
Ref. Code(s) : 59, 140
Summary : Chinquapin (*Castanea pumila*) is an evergreen tree reaching 15-30 m. It produces sweet, edible nuts. The trees provide excellent agricultural cover. The strong, hard timber is used as fence material, but because of the tree's small stature, is not considered an economically feasible source of timber. Trees are fairly resistant to chestnut blight. Chinquapin roots have astringent qualities and have been used to treat fevers. There are two varieties, Allegheny chinkapin and the ashe chinkapin, both of which grow on the dry, sandy coastal soils of Virginia and Florida and along the Gulf coast.

The trees grow throughout the southeastern United States in dry woodlands and mountainous areas at altitudes up to 1,500 m. They bloom in June, and the brown nuts can be harvested in September. The nuts are in spiny burs and are hard to shell. They contain 5% fat, 5% protein, and 40% starch. They have about 1,800 calories/pound. Nuts are eaten raw or roasted in the shell.

Cuddapah Almond, Chirauli Nut
(See Oil)

English Walnut, Persian Walnut

Sci. Name(s) : *Juglans regia*
Geog. Reg(s) : America-North (U.S. and Canada), Asia-China, Asia-India (subcontinent), Europe
End-Use(s) : Dye & Tannin, Miscellaneous, Nut, Oil, Ornamental & Lawn Grass, Sweetener, Timber
Domesticated : Y
Ref. Code(s) : 131, 182, 220
Summary : The English walnut tree stands 37-46 m in height and produces the commercially important walnut. The dark brown nuts are rich in fat, vitamin B, iron, and phosphorous and are high in unsaturated fatty acids. Walnuts are used in baked goods, candy, ice cream, frozen foods, and packaged cake and cookie mixes. A nut meal is used to thicken savory and sweet dishes. Walnut oil is used in southern Europe, especially France, as a cooking oil and substitute for olive oil. Filtered oil is used in paint industries, and the residual oil cakes are used in dairy feeds. Trees are tapped for their sugary sap, and their roots and mature or green husks are used as a minor dye source. In England, green walnuts are pickled and eaten, or made into marmalade, jams, or syrups. The hard, heavy, and beautifully grained wood is valuable for veneer, furniture, and gunstocks.

California produces virtually all of the United States' walnuts, at an average of 0.91 MT/ha of nuts. Thirty percent of the nuts are sold in shell; 70% as "cracking stock," or shelled nuts, which are sold to oil mills; and the kernels are sent to commercial food product industries. The United States is presently the leading nut producer, but the world market is growing rapidly with competition from China and India.

Trees thrive in mild climates, coastal areas, and on most soils. However, the best nuts are from trees grown in deep, well-drained, fertile, light loams. English walnuts are ornamental in appearance and are often grown as shade, lawn, and orchard trees. Trees grown for timber are less attractive.

European Beech
(See Ornamental & Lawn Grass)

European Chestnut, Spanish Chestnut

Sci. Name(s) : *Castanea sativa*
Geog. Reg(s) : Mediterranean, Temperate
End-Use(s) : Cereal, Nut, Timber
Domesticated : Y
Ref. Code(s) : 216, 220
Summary : The European chestnut is a tree cultivated in the Mediterranean for its nuts, eaten roasted or boiled. These nuts are ground to flour, used in soup or fried in oil, and used in confectionery. The wood is fairly easy to work and often used for broad carving. It splits with little difficulty, is light in color and is used for split rail fences, general carpentry, and the manufacture of cellulose. Its heartwood is light to dark brown in color, straight-grained, durable, but damaged by wood borers. Trees tolerate some fungal attack. European chestnuts resemble the oak in appearance. They grow throughout northern temperate areas.

European Hazel, European Filbert

Sci. Name(s) : *Corylus avellana*
Geog. Reg(s) : America-North (U.S. and Canada), Europe, Temperate
End-Use(s) : Essential Oil, Miscellaneous, Nut, Oil, Timber
Domesticated : Y
Ref. Code(s) : 220
Summary : The European hazelnut tree grows in temperate areas and produces a commercially important hazelnut. Commercial nut production centers in Turkey, Spain, and the United States.

Nut kernels are eaten raw, in oil, and in certain culinary items. They are crushed for an oil used in food, soap manufacture, perfumery, paints, and lubricants. Hazel wood is soft and light but lacks durability. It is used as charcoal for gunpowder. Tree leaves are occasionally substituted for tobacco. Trees yield nuts at 4 years of age.

Gbanja Kola

Sci. Name(s) : *Cola nitida*
Geog. Reg(s) : Africa-West
End-Use(s) : Beverage, Essential Oil, Medicinal, Nut
Domesticated : N
Ref. Code(s) : 170
Summary : Gbanja kola (*Cola nitida*) is a tree about 10-13 m tall. Its nuts are chewed as a stimulant in Africa, or powdered nuts are made into refreshing beverages. The nuts

contain about 2% caffeine, 9% protein, 2% fat, 75% carbohydrates, and 2% fiber. They also have a trace of theobromine, and a glucoside, kolanin, that is used as a heart stimulant. An essential oil may also be extracted.

The tree is native to the rain forests of West Africa. Trees are cultivated primarily in Nigeria. They are most successful on fertile, well-drained soil that is rich in humus. They can be grown on soil that is considered too poor for cocoa. Trees grow slowly at first and begin fruiting in about 7 years, reaching full production at 20 years. Average yields are about 210 marketable nuts/tree or 589 kg/ha. The nuts are exported mainly to the northern parts of West Africa and the Democratic Republic of the Sudan.

Geocarpa, Kersting's Groundnut, Geocarpa Groundnut
(See Cereal)

Giant Filbert, Lambert's Filbert, Filbert

Sci. Name(s) : *Corylus maxima, Corylus tubulosa*
Geog. Reg(s) : America-North (U.S. and Canada), Eurasia
End-Use(s) : Miscellaneous, Nut
Domesticated : Y
Ref. Code(s) : 95, 131, 213, 218, 219
Summary : The giant filbert is a native Eurasian tree grown for its edible commercial nuts. Trees bear at 4 years. They thrive in full sunlight on rich, well-drained soils in areas with cool, moist climates and do not survive the tropics. This tree and the European hazelnut (*Corylus avellana*) are used in hybrid development.

In 1983, the United States produced 7,256 MT of filberts, valued at $4.6 million. Additional imports included 114 MT of unshelled filberts and 2,450 MT of shelled, blanched, or otherwise prepared or preserved filberts. The total value of these imports was approximately $5.9 million. Italy and Turkey were the main suppliers of filberts to the United States.

Heartnut, Cordate Walnut, Siebold Walnut

Sci. Name(s) : *Juglans ailantifolia, Juglans sieboldiana*
Geog. Reg(s) : America-North (U.S. and Canada), Asia, Asia-China
End-Use(s) : Dye & Tannin, Miscellaneous, Nut, Ornamental & Lawn Grass, Timber
Domesticated : Y
Ref. Code(s) : 182, 220
Summary : The heartnut is a medium-sized tree with a broad, symmetrical crown that produces edible, heart-shaped nuts. Trees are not usually grown for commercial nut production in the United States but are valued as ornamentals and occasionally for their light, soft, dark brown wood. The wood has been used in China and Japan for gunstocks and cabinetmaking. The pericarp and bark produce a brown dye, and the fruits are edible. Hulls have been used as fish poison.

Production of heartnuts is increasing in the United States. The nuts crack easily. They have a flavor similar to that of the butternut. Heartnut trees and butternut trees are easily crossbred to produce vigorous hybrids.

Heartnuts were introduced from Japan into the United States in the 19th Century and produced well until nearly all of the trees were killed by the walnut bunch disease. Present production is still limited as a result. In general, heartnuts live for about 30 years in the United States.

Hickory Nut

Sci. Name(s) : *Carya* spp.
Geog. Reg(s) : America-North (U.S. and Canada), Asia
End-Use(s) : Nut, Timber
Domesticated : N
Ref. Code(s) : 59, 216
Summary : The hickory nut tree produces nuts with commercial value as feed for game animals. The timber is strong, dark-colored, resilient and valued for furniture and cabinetry. Although it is difficult to work, it is flexible and used for making tool handles, agricultural tools, chairs, and various other articles. This wood is fairly durable and easily polished and reacts well to normal finishing treatments. It is grown in North America as well as Asia.

Himalayan Hazel, Curri

Sci. Name(s) : *Corylus ferox*
Geog. Reg(s) : Asia-Central, Asia-India (subcontinent)
End-Use(s) : Nut
Domesticated : N
Ref. Code(s) : 176, 213
Summary : The Himalayan hazel (*Corylus ferox*) is a tree which grows at higher elevations in northern India and Central Asia. It produces edible nuts with limited commercial value. This tree is similar in appearance to the Tibetan hazelnut (*Corylus tibetica*).

Japanese Chestnut
(See Ornamental & Lawn Grass)

Java-Almond, Kanari

Sci. Name(s) : *Canarium indicum, Canarium moluccanum, Canarium commune*
Geog. Reg(s) : Asia-Southeast, Pacific Islands
End-Use(s) : Nut, Oil, Ornamental & Lawn Grass
Domesticated : Y
Ref. Code(s) : 36
Summary : The Java-almond is a tree of considerable size. Its nuts are its most important product. The kernel has been used as a substitute for almonds in confectionery. The fresh oil from the nuts is used in cooking and as an illuminant. Java-almonds provide good shade. Trees are native to eastern Malaysia and New Guinea.

Jesuit Nut, Caltrop, Waterchestnut

Sci. Name(s) : *Trapa natans*
Geog. Reg(s) : Asia, Europe
End-Use(s) : Nut
Domesticated : Y
Ref. Code(s) : 90

Summary : The Jesuit nut is a plant with seed kernels containing starches and fats. The seeds have been a major food source in parts of Asia. Plants are distributed throughout Europe and Asia and grow best in warm, standing waters during summer months. Seeds are collected in autumn and stored in cold water during the winter.

Jojoba, Goat-Nut
(See Fat & Wax)

Macadamia
Sci. Name(s) : *Macadamia* spp.
Geog. Reg(s) : Africa, Asia-Southeast, Australia, Hawaii, Pacific Islands
End-Use(s) : Nut
Domesticated : Y
Ref. Code(s) : 132, 154, 170, 214, 220
Summary : The macadamia nut tree produces commercially important, edible nuts that provide a good source of protein, carbohydrates, calcium, thiamine, riboflavin, niacin, phosphorous, iron, fat, and vitamin B1.

There are 10 different species of macadamia. One type grows in Madagascar, 1 in Celebes, 5 in eastern Australia, and 3 in New Caledonia. The trees are introduced to many highland tropical areas but do not thrive in the lowland tropics. They prefer moist, fertile soils.

In Hawaii, the macadamia nut has become important as a commercial crop. Trees generally bear in 7 years. In 1983, nut yields in Hawaii were 3,811 kg/ha, to produce a total of 16,535 MT of nuts. This crop was valued at $23.9 million.

Marrow, Pumpkin, Squash, Vegetable Marrow
(See Vegetable)

Monkey-Pod
(See Timber)

Morula
(See Fruit)

Mozinda, African Breadfruit
Sci. Name(s) : *Treculia africana*
Geog. Reg(s) : Africa-West
End-Use(s) : Cereal, Nut
Domesticated : N
Ref. Code(s) : 170
Summary : Mozinda is a plant that produces edible seeds, eaten boiled or roasted, in soups, or ground to flour. Plants are grown in West Africa.

Niger-Seed
(See Oil)

Owe Kola

Sci. Name(s) : *Cola verticillata*
Geog. Reg(s) : Africa-West
End-Use(s) : Nut
Domesticated : N
Ref. Code(s) : 170
Summary : Owe kola (*Cola verticillata*) is a tree whose fruit is considered inferior to that of its relative, gbanja kola (*Cola nitida*). The tree is native to African forests along the Ivory Coast, Ghana, and southern Nigeria, to the lower parts of Zaire.

Peach Palm, Pejibaye
(See Fruit)

Peanut, Groundnut, Goober, Mani
(See Oil)

Pecan

Sci. Name(s) : *Carya illinoensis, Carya pecan, Carya viliriformis*
Geog. Reg(s) : America-North (U.S. and Canada)
End-Use(s) : Cereal, Dye & Tannin, Energy, Miscellaneous, Nut, Oil, Timber
Domesticated : Y
Ref. Code(s) : 140, 220
Summary : Pecan trees are cultivated in the southern United States for their commercially important nuts used in confectionery and cakes or sold raw and salted. An edible oil (pecan oil) is extracted from the seeds and used in the manufacture of cosmetics and certain drugs. Pecan wood is coarse and brittle but is sometimes used for fuel and tool handles.

Nut shells have some commercial value as paving for walks and driveways, fuel, mulches for ornamental plants, soil conditioners, soft abrasives in hand soap, nonskid paints and metal polishes, and for tannin and charcoal. They can also be ground into flour or used as fillers in plastic wood, adhesives, and dynamite.

The tree often reaches heights above 30 m. It is thought to be the most important native nut-bearing tree in North America and is often planted in orchards. In 1983, the United States produced 122,580 MT of pecans. This harvest was valued just under $160 million.

Pili Nut, Elemi

Sci. Name(s) : *Canarium luzonicum*
Geog. Reg(s) : Asia-Southeast
End-Use(s) : Essential Oil, Gum & Starch, Natural Resin, Nut, Oil
Domesticated : N
Ref. Code(s) : 77, 107, 227
Summary : The pili nut tree grows throughout the Philippines and is a source of an oleoresin and volatile oil used as a solvent and a pale plasticizer in lacquers. Its edible nut kernels contain 70% of a cooking oil with good keeping qualities. About 20-25% of the resin weight is the volatile oil.

Mature trees yield 4-5 kg/year of a gum used in pharmaceutical plasters, ointments, and varnishes. It is used to scent soaps and inexpensive technical preparations but is not used widely in the perfume industry.

Pinyon Pine, Silver Pine, Nut Pine, Pinyon

Sci. Name(s) : *Pinus edulis*
Geog. Reg(s) : America-North (U.S. and Canada), Hawaii
End-Use(s) : Erosion Control, Nut
Domesticated : N
Ref. Code(s) : 154
Summary : The pinyon pine or silver pine (*Pinus edulis*) produces nuts that were once a valuable food source to Native Americans. Pinyon nuts are sold in some candy stores and gourmet shops.

In the United States, trees are most often found in the western and southwestern states, particularly Wyoming and Texas. In Hawaii, about 50 of the 80 known members of the Pinus genus have been grown successfully as reforestation crops, but the nut pine is one of the few species that has never proven successful.

Pinyon Tree, Parry Pinyon Pine, Mexican Nut Pine, Pinyon

Sci. Name(s) : *Pinus quadrifolia, Pinus cembroides* var. *parryana, Pinus cembroides* var. *edulis*
Geog. Reg(s) : America-Central, America-North (U.S. and Canada)
End-Use(s) : Nut
Domesticated : Y
Ref. Code(s) : 17, 95, 212
Summary : The pinyon tree is a small, short-stemmed tree or large bush that produces roundish cones with edible seeds. The seeds are called "pinyon nuts," and there is limited commercial demand for them. Pinyon pines are native to North America. Trees range in habitat from southern Arizona to California and Mexico.

Pistachio Nut

Sci. Name(s) : *Pistacia vera*
Geog. Reg(s) : Africa-North, America-North (U.S. and Canada), Asia-India (subcontinent), Eurasia, Europe, Mediterranean
End-Use(s) : Nut, Spice & Flavoring
Domesticated : Y
Ref. Code(s) : 17, 95, 101, 116, 154, 170
Summary : The pistachio (*Pistachia vera*) is a tree reaching 6-9 m in height that produces edible fruits. Pistachio nuts are used in the food industry as flavoring and coloring for cakes, ice creams, snack foods, and certain confections. Nuts contain less than 10% sugar and have a protein and oil content of 20-40%. The major nut-producing countries are Iran, Turkey, and the Syrian Arab Republic, while minor ones are Greece, Italy, France, Spain, Israel, Lebanon, Afghanistan, and Pakistan. Pistachios are also grown in the United States, particularly in California, Texas, and Arizona.

Trees thrive in deep, moist, fertile soils and produce well in areas with long, hot summers and low humidity. They bear in 4-5 years and yield 22.7 kg/ha/year. In 1983, the United States imported 271 MT of shelled pistachios, valued at approximately $1 million, and 2,594 MT of unshelled nuts, valued $12.7 million.

The pistachio is relatively free from surface diseases but is less resistant to root fungus than other members of its genus.

Sand Dropseed
(See Erosion Control)

Seaving, Souari, Achotillo
(See Oil)

Shea-Butter, Butterseed
(See Fat & Wax)

Siberian Hazel

Sci. Name(s)	: *Corylus heterophylla*
Geog. Reg(s)	: Asia
End-Use(s)	: Nut
Domesticated	: N
Ref. Code(s)	: 39, 108
Summary	: The Siberian hazelnut (*Corylus heterophylla*) is a small tree or shrub about 6.3 m in height that grows throughout northeastern Asia and Japan. Its nuts are edible and compose a portion of the hazelnut market.

Tibetan Hazelnut, Tibetan Filbert

Sci. Name(s)	: *Corylus tibetica*
Geog. Reg(s)	: Asia-Central, Asia-China
End-Use(s)	: Nut
Domesticated	: N
Ref. Code(s)	: 108, 176
Summary	: The Tibetan hazelnut is a shrubby tree that reaches up to 7.6 m in height and bears fruit clusters similar to chestnuts'. Trees are native to Central and western China. There is limited information about these nuts and their economic potential.

Turkish Hazel, Turkish Filbert

Sci. Name(s)	: *Corylus colurna*
Geog. Reg(s)	: Asia-China, Europe
End-Use(s)	: Nut, Ornamental & Lawn Grass
Domesticated	: Y
Ref. Code(s)	: 30, 53
Summary	: The Turkish hazel tree is grown as an ornamental and produces edible nuts that do not enter commercial trade in significant volumes. Trees are native to and common in southeastern Europe and China. They thrive in areas with hot summers and cold winters and are fairly tolerant of adverse environmental conditions. Trees do well in well-drained loam soils in full sun. Turkish hazelnuts are relatively free from diseases and pests.

Waxgourd, White Gourd
(See Vegetable)

Oil

Aceituna, Olivo, Paradise-Tree

Sci. Name(s) : *Simarouba glauca*
Geog. Reg(s) : America-Central, America-North (U.S. and Canada), America-South, Hawaii, Tropical, West Indies
End-Use(s) : Fruit, Oil
Domesticated : Y
Ref. Code(s) : 17, 59, 150, 154
Summary : Aceituna is a tree reaching 15.2 m in height, grown in El Salvador for its oil-filled seeds and edible fruit pulp. The fruit is eaten raw and is a popular produce item in Central American markets. Trees are native to tropical America and are grown at low elevations along coastlines. Aceituna is grown in the West Indies from Mexico to Costa Rica and in the United States, primarily southern Florida and Hawaii. Trees fruit from April to May.

African Locust Bean

Sci. Name(s) : *Parkia filicoidea*
Geog. Reg(s) : Africa-West, Tropical
End-Use(s) : Beverage, Dye & Tannin, Fiber, Forage, Pasture & Feed Grain, Miscellaneous, Nut, Oil, Vegetable
Domesticated : Y
Ref. Code(s) : 17, 36, 50, 220, 222, 226
Summary : The African locust bean is a tall tree whose fruit seeds are a food source in parts of West Africa. The seeds contain 26% protein and a high degree of calcium. They yield 16.8% of a fixed oil. Ripe seeds are considered to be an aphrodisiac. They are used as coffee bean substitutes. Unripe seeds and leaves are eaten as vegetables. The fruit is considered good fodder for livestock. Bark contains 12-14% tannin. Seed pods are used for fish poison. Prepared tree roots yield a fibrous mass made into sponges. Trees grow in most tropical areas of the world.

African Oil Palm, Oil Palm

Sci. Name(s) : *Elaeis guineensis*
Geog. Reg(s) : Africa-West, America-Central, America-South, Asia-Southeast, Europe
End-Use(s) : Beverage, Fruit, Oil, Sweetener
Domesticated : Y
Ref. Code(s) : 31, 171
Summary : The African oil palm can reach heights of 18 m and produces 2 distinct oils. The mesocarp produces an oil that yields up to 45% by weight, and the kernel yields up to 50% by weight. Both palm oil and palm kernel oil are important in world trade. The major oil exporters are Malaysia and the countries of West Africa. Latin America has established a growing oil palm industry. The United Kingdom is the largest importer of oil palm products.

Palm oil has saturated palmitic acids, as well as oleic and linoleic acids, giving it a higher unsaturated acid content than coconut and palm kernel oils. Palm oil was used in soap and candle manufacturing and in the tin plate industry. Recently, improvements in quality resulted in a decline in its technical uses and an increase in its edible uses.

Palm kernel oil is used in edible fats, in the preparation of ice-cream and mayonnaise, and in the manufacture of soaps and detergents. By tapping the male inflorescence, a wine is made from the sap. This practice reduces the fruit yield. The central shoot or cabbage of the oil palm is edible.

Oil palms are grown from seed in nurseries and then transplanted. The first fruit ripens 3-4 years after planting. This first yield is small and considered uneconomical to harvest. Fruit yields peak with palms that are 12-15 years old, and production can continue for up to 50 years. From a single palm, as many as 2,000 fruits can be harvested in 1 year. In 1984, world production of palm oil was 7 million MT. World imports of palm oil in the same year were 4 million MT, valued at $2.7 billion. The United States imported 148,134 MT, valued at $9 million. The major exporting countries were Indonesia, with 285,800 MT valued at $150 million, and Malaysia, with about 3 million MT valued at approximately $2 billion.

Palms should be planted in areas receiving an annual rainfall of about 3 m with high humidity. Soils should be porous and well-drained.

African Pumpkin, Zanzibar Oil Vine, Oysternut

Sci. Name(s)	: Telfairia pedata
Geog. Reg(s)	: Africa-Central, Africa-East
End-Use(s)	: Nut, Oil
Domesticated	: Y
Ref. Code(s)	: 92, 132, 220
Summary	: The African pumpkin is a climbing plant cultivated for its edible seeds, which are eaten dried, roasted, in soups, or pickled. They can be made to express an oil, called castanah oil, used in soap and candle making.

Plants are cultivated in Kenya and parts of East and Central Africa from sea level to altitudes approaching 1,830 m. African pumpkin is fairly drought-resistant.

African Star-Apple
(See Fruit)

Alfalfa, Lucerne, Sativa
(See Forage, Pasture & Feed Grain)

Almond
(See Nut)

American Oil Palm, Corozo

Sci. Name(s)	: Elaeis oleifera, Elaeis melomococca
Geog. Reg(s)	: America-Central, America-South, Tropical
End-Use(s)	: Miscellaneous, Oil
Domesticated	: Y
Ref. Code(s)	: 85, 171, 220
Summary	: The American oil palm grows in tropical South and Central America and produces fruit and seeds that contain an oil used for lighting, soap manufacture, and as a

machine lubricant. The seed fat is edible and used in Colombia for hair treatments. It contains an oil with a high proportion of unsaturated fatty acids.

Research is being conducted on the improvement and hybridization of this palm, specifically in crosses with the oil-bearing African oil palm (*Elaeis guineensis*). This palm shows considerable resistance to bud rot prevalent in Colombia, as well as other diseases, and adaptivity to adverse climates. An experimental hybrid of the American and African oil palms gives higher yields than the parent American oil palm.

American oil palms grow in most tropical countries and particularly in the lowlands of South and Central America.

Ammi
(See Medicinal)

Apricot
(See Fruit)

Avocado, Alligator Pear
(See Fruit)

Babassu, Orbignya speciosa

Sci. Name(s)	: *Orbignya speciosa, Orbignya barbosiana*
Geog. Reg(s)	: America-Central, America-South, Subtropical, Tropical
End-Use(s)	: Energy, Forage, Pasture & Feed Grain, Oil, Ornamental & Lawn Grass
Domesticated	: N
Ref. Code(s)	: 17, 107, 169, 220
Summary	: Babassu is a slow-growing palm that yields palm kernel oil. It reaches heights of 18 m and is grown as an ornamental. Each palm bears 2-4 fruit bunches, which produce approximately 200-600 fruits twice a year. The shells are used for fuel or charcoal. An oil press cake and meal contains 20-23% protein and provides livestock feed. Palm oil is used in domestic soap industries. Babassu oil imported from Brazil in 1983 was valued at $436,000 for 619.2 MT.

Bacury
(See Fruit)

Baker's Meadowfoam

Sci. Name(s)	: *Limnanthes bakeri*
Geog. Reg(s)	: America-North (U.S. and Canada)
End-Use(s)	: Oil
Domesticated	: Y
Ref. Code(s)	: 70, 93, 106, 231
Summary	: Baker's meadowfoam (*Limnanthes bakeri*) is a swampy meadow plant native to North America with potential as an oil seed crop. Experimental plantings in Alaska and Oregon showed it inferior to white meadowfoam (*Limnanthes alba*) and Douglas's meadowfoam (*Limnanthes douglasii*) with respect to seed yield. In Alaska and Oregon, average yields were 405 kg/ha. More experimentation is needed for this crop.

Baobab, Monkey-Bread
(See Vegetable)

Benefing

Sci. Name(s)	: *Hyptis spicigera*
Geog. Reg(s)	: Africa-West, Tropical
End-Use(s)	: Cereal, Oil
Domesticated	: Y
Ref. Code(s)	: 220
Summary	: Benefing is an herb cultivated for its seeds, which yield about 30% of a quality drying oil. Despite the oil's potential, the small size of the seed creates difficulties for its commercial extraction. The oil is used as a substitute for linseed oil. In Africa, the seeds are eaten as a paste in stews. Benefing grows in tropical Africa.

Benoil Tree, Horseradish-Tree

Sci. Name(s)	: *Moringa oleifera, Moringa pterygosperma*
Geog. Reg(s)	: Asia-India (subcontinent), Tropical
End-Use(s)	: Forage, Pasture & Feed Grain, Medicinal, Nut, Oil, Spice & Flavoring, Sweetener, Vegetable
Domesticated	: Y
Ref. Code(s)	: 36, 132, 220
Summary	: The benoil, or horseradish-tree, is a small, short-lived but fast-growing tree cultivated throughout the tropics for its leaves and seeds. The leaves are thought to be one of the best plant sources of calcium and are rich in other vitamins, minerals, and protein. Leaves are picked after 8 weeks. Seeds yield an oil used in paints. They also have a nonsugar sweetener that is probably an ascorbic acid analogue. Tree roots are often used as a substitute for horseradish. Pods, flowers, twigs, and leaves are cooked and served as milder flavorings. Young seed pods are eaten as vegetables or used in curries, but are less valuable than the leaves. Leaves are also fed to cattle. Preparations of the roots, bark, and leaves have medicinal value.

Trees are native to India and are now found throughout the tropics in dry climates at low to moderate elevations. They are fairly drought-resistant and will thrive in most light, well-drained sandy soils. Benoil trees resemble legumes in appearance. They are considered excellent backyard vegetables.

Bira Tai
(See Fruit)

Bird Rape, Field Mustard

Sci. Name(s)	: *Brassica campestris, Brassica rapa*
Geog. Reg(s)	: Asia-China, Asia-India (subcontinent)
End-Use(s)	: Oil
Domesticated	: Y
Ref. Code(s)	: 132, 170
Summary	: Bird rape (*Brassica campestris*) is an annual herb important as an oil seed crop used in edible vegetable oils. Plants are grown in pure stands or mixed with cereals. Seed yields are usually 780-1200 kg/ha. China, India, Pakistan, Bangladesh, and

Canada are major rapeseed oil producers. Two main varieties of bird rape are variety sarson (India colza), which is mucilaginous, and variety toria (Indian rape), which is nonmucilaginous.

This plant is most successful if grown in areas with temperatures of 15-20 degrees Celsius and an annual rainfall of 200-500 cm. It thrives in fertile, well-drained, sandy or clay loams and withstands moderately saline soils.

Black Mustard
(See Spice & Flavoring)

Boston Marrow, Pumpkin, Winter Squash, Squash, Marrow
(See Vegetable)

Brown Mustard, Indian Mustard, Mustard Greens

Sci. Name(s)   : *Brassica juncea*
Geog. Reg(s)   : Africa, Asia-China, Asia-India (subcontinent), Eurasia
End-Use(s)   : Oil, Spice & Flavoring, Vegetable
Domesticated   : Y
Ref. Code(s)   : 132, 148, 170, 194
Summary   : Brown mustard is an annual herb grown as an oil seed crop in India. The seeds contain 25% of an edible oil used in cooking as an olive oil substitute. This oil is important in the food industries for its aroma from the glucoside sinigrin. The pungent leaves are used as aromatic potherbs, in mixed salads, and as cooked vegetables.

Plants thrive in rich, medium-textured soils and areas with distinct dry and wet seasons and moderate to cool temperatures. Leaves are harvested 4 weeks after transplanting and yield 45-55 MT/ha. Brown mustard is also cultivated in Africa and from eastern Europe to China.

Buffalo Gourd, Wild Gourd

Sci. Name(s)   : *Cucurbita foetidissima*
Geog. Reg(s)   : America-Central, America-North (U.S. and Canada), America-South
End-Use(s)   : Forage, Pasture & Feed Grain, Gum & Starch, Medicinal, Miscellaneous, Oil
Domesticated   : Y
Ref. Code(s)   : 169, 220
Summary   : The buffalo gourd is a vine that shows great potential as an oil crop. It also displays the unique attribute of being a source of an edible oil and protein, starch, and forage, a combination no other known crop possesses. The carotenoid level is similar to cottonseed oil's. The vine produces an abundance of fruits, the seeds of which contain about 30-40% edible oil. The whole fruit is used as a soap substitute. An extract from the ground roots is used as a laxative in Mexico. The seed meal is a good source of protein. The roots of the buffalo gourd contain 16% starch. The vines have potential as a forage crop but are frost-sensitive.

This vine originated in the Chihuahuan plateaus of northern Mexico. It grows on the Great Plains, in New Mexico, Arizona, the coastal mountains of southern California, and in semiarid regions.

Bungu, False Sesame

Sci. Name(s)   : *Ceratotheca sesamoides*
Geog. Reg(s) : Africa, Subtropical, Tropical
End-Use(s)   : Oil, Spice & Flavoring
Domesticated : Y
Ref. Code(s) : 170
Summary   : Bungu is a wild annual or perennial herb of tropical Africa occasionally culti-
vated for its edible seeds. Bungu, also known as false sesame, is sometimes used as
an adulterant of true sesame. Pressed seeds yield 37% oil. Plants are usually grown
in the tropics and subtropics.

Candlenut, Candleberry, Indian Walnut

Sci. Name(s)   : *Aleurites moluccana*
Geog. Reg(s) : Asia-Southeast, Tropical
End-Use(s)   : Nut, Oil
Domesticated : Y
Ref. Code(s) : 36, 170
Summary   : The seeds of the candlenut tree produce a quick-drying oil inferior to tung-oil
and linseed-oil. Like linseed-oil, this oil consists of mixtures of glycerides of linoleic,
linolenic and oleic acids. About 60% of the oil is found in the seed kernel, which is
edible and eaten roasted. Candlenut trees are native to Malaysia and widely dis-
tributed throughout the tropics.

Cantaloupe, Melon, Muskmelon, Honeydew, Casaba
(See Fruit)

Cashew
(See Nut)

Castorbean

Sci. Name(s)   : *Ricinus communis*
Geog. Reg(s) : Africa, America-North (U.S. and Canada), America-South, Asia-India
(subcontinent), Subtropical, Tropical
End-Use(s)   : Dye & Tannin, Fat & Wax, Fiber, Forage, Pasture & Feed Grain, Oil,
Ornamental & Lawn Grass
Domesticated : Y
Ref. Code(s) : 126, 170
Summary   : The castorbean plant is a short-lived perennial that is grown for its seeds,
which contain a valuable industrial oil. The plant has naturalized in many tropical and
subtropical countries and is grown commercially in the United States. The plant is
native to Africa. Improved varieties have been developed that increase seed yields.

The oil has many industrial uses. It is water-resistant and can be used for protective coatings, high-grade lubricants, soap and printing ink, textile dyeing, and
leather preservation. The dehydrated oil is a drying agent used in paints and varnishes and is comparable to tung-oil. The hydrogenated oil is used in wax, polishes,
carbon paper, candles, and crayons. It is also used for plastics, ointments, and cosmetics. The remaining oilseed cake, or castor pomace, is used as fertilizer.

Detoxified meal is fed to livestock. Plant stems are made into paper and wallboard. Castorbean plants are grown as ornamentals.

Plants are most successful in warm, frost-free climates. In the United States, crops are irrigated, while in India they are grown as dryland crops. Thorough soil preparation is essential for optimum growth. Harvest begins when seeds are dry and continues while other seed clusters ripen. Fertilization increases yields.

World production of castorbeans in 1984 was about 1 million MT from an area of 16 million ha. Castorbean seeds have been priced at about $160/MT, and the oil has been priced at about $290/MT. Castor pomace has been priced at $40/MT.

Ceara-Rubber
(See Isoprenoid Resin & Rubber)

Chaulmoogra
(See Medicinal)

Chinese Tallow-Tree, Tallow-Tree

Sci. Name(s) : *Sapium sebiferum, Triadica sebifera*
Geog. Reg(s) : Asia, Asia-China, Tropical
End-Use(s) : Dye & Tannin, Fat & Wax, Forage, Pasture & Feed Grain, Oil, Ornamental & Lawn Grass, Timber
Domesticated : Y
Ref. Code(s) : 16, 17, 36, 59, 154, 223
Summary : The Chinese tallow-tree reaches heights of 7.6-9.0 m. It is widely cultivated in parts of China, Japan, and throughout the tropics as an ornamental or shade tree. Trees produce seed pods about 7 mm in length that contain an important drying oil. Seeds are coated with fatty cells, which are used in candles and soaps. The waxy substance was at one time imported to the United States, but has been largely replaced by mineral waxes. Expressed oilseed cakes are used as manure. The leaves provide a black dye. The timber is white and moderately hard and is used for printing blocks. Trees adapt to most soils.

Chufa, Ground Almond, Tigernut, Yellow Nutsedge
(See Forage, Pasture & Feed Grain)

Coconut

Sci. Name(s) : *Cocos nucifera*
Geog. Reg(s) : Tropical
End-Use(s) : Beverage, Fiber, Forage, Pasture & Feed Grain, Fruit, Oil, Timber
Domesticated : Y
Ref. Code(s) : 132, 220
Summary : The coconut palm is a commercially important oil crop that also provides marketable fruit in tropical countries. The hard-shelled fruit, when split open, provides a soft flesh that is eaten fresh or is processed and sold as shredded or dehydrated coconut. The flesh, or copra, is often used in confectionery and cooking.

The flesh yields 60-65% oil through a steam heating process. Coconut oil contains a high percentage of lauric acid, making it useful in quick-lather soaps. It is also used extensively in the manufacture of margarine, cosmetics, and synthetic rubber.

A fiber, coir, is extracted from the outer husk and is used as an inexpensive stuffing material. Copra and coir harvest takes place when the nuts are fully ripened, usually about 11 months before the palm will bloom.

Nuts also have a juice that is collected for drinking 7 months prior to blooming. The wood has been used for building construction. Expressed oil cakes are used as food for livestock.

Coconut palms reach heights of 25-30 m. The lifespan is usually 80-100 years. Palms begin bearing nuts after 7 years growth and reach maximum production in about 12 years. A single tree may produce 400 fruits a year, yielding about 90 kg of flesh and 37 liters of oil.

The United States is a major importer of coconut products. In 1983, the United States imported 29,100 MT of coconut, 43,454 MT of desicated coconut, and 452,280 MT of coconut oil.

Palms are most successful in hot, tropical lowland areas at altitudes below 300 m. They will not tolerate frost. Annual rainfall should be well-distributed, averaging 1,500 mm. Palms thrive in most moist but well-drained soils.

Cohune Nut Palm, Cohune Palm

Sci. Name(s)	: *Orbignya cohune, Attalea cohune*
Geog. Reg(s)	: America-Central
End-Use(s)	: Oil, Ornamental & Lawn Grass, Timber, Vegetable
Domesticated	: N
Ref. Code(s)	: 73, 154, 171
Summary	: The cohune nut palm is a monoecious palm with a trunk reaching 15-18 m in height. Each palm can produce 1,000-2,000 fruits/year, or 1,500 kg/ha of fruit. The fruit seed kernel yields about 50% of a high-quality oil that, when refined, can be used in the manufacture of margarine and candles, as machine oil, and as a substitute for coconut oil. Young leaf buds can be cooked and eaten as cabbage. Mature palm leaves have been used in thatch work, and the trunks have been used for building purposes. Cohune nut palms are popular ornamentals for lining streets.

Palms grow in lowland areas to 550 m in altitude. They are, for the most part, wild, with an estimated average density of 15 palms/ha. Cohune nut palms are native to, and grow mostly in, Central America, especially Mexico. Fruits take approximately 1 year to mature.

Crambe, Colewort

Sci. Name(s)	: *Crambe abyssinica*
Geog. Reg(s)	: America-North (U.S. and Canada)
End-Use(s)	: Forage, Pasture & Feed Grain, Oil
Domesticated	: Y
Ref. Code(s)	: 190
Summary	: Crambe is an experimental oilseed crop, high in erucic acid, presently undergoing research in parts of the United States. Seed production/ha has varied in several states. Processing costs of crambe seed is low enough to compete with another high-

erucic acid oilseed, rapeseed. The residual from crambe seed oil extraction has potential as cattle meal.

Although crambe oil has excellent possibilities, further research on disease resistance, yields, and cultural practices are needed to ensure this crop's success.

Crookneck Pumpkin, Pumpkin, Winter Squash, Squash
(See Vegetable)

Cucumber
(See Vegetable)

Cuddapah Almond, Chirauli Nut

Sci. Name(s)	: *Buchanania latifolia*
Geog. Reg(s)	: Asia-India (subcontinent), Asia-Southeast, Australia, Subtropical
End-Use(s)	: Forage, Pasture & Feed Grain, Gum & Starch, Nut, Oil
Domesticated	: N
Ref. Code(s)	: 195, 220
Summary	: The cuddapah almond is a tree whose fruit seeds provide an edible oil. The seeds are eaten or used in confectionery in Indo-Malaysia, tropical Australia, eastern India and Burma. A gum (chironji-ki-gond) is obtained from the seeds and sold in India as glue. The leaves are considered fairly good fodder. Trees are useful for foresting dry hills.

The trees are moderate-sized and sensitive to frost and drought. They grow in a variety of soils but do best in clay. Shallow, gravelly soils result in stunted growth. Most trees are found in subtropical environments with hot, dry summers and mild winters. Maximum shade temperatures vary from 40 to 46 degrees Celsius, with minimum temperatures from -1 to 13 degrees Celsius. Normal annual rainfall is 750-2,120 mm.

Desert Date
(See Fruit)

Douglas's Meadowfoam

Sci. Name(s)	: *Limnanthes douglasii*
Geog. Reg(s)	: America-North (U.S. and Canada)
End-Use(s)	: Fat & Wax, Oil, Ornamental & Lawn Grass
Domesticated	: Y
Ref. Code(s)	: 70, 93, 106, 231, 234
Summary	: Douglas's meadowfoam (*Limnanthes douglasii*) was originally considered an ornamental plant. Today it is becoming more and more important as a potential oil seed crop. Its oil has the potential for high-quality industrial products such as plastics and lubricants and as a substitute for sperm whale oil. This plant may have a future as livestock feed.

Plantings have been grown experimentally in Maryland, Oregon, California, and Alaska. Douglas's meadowfoam is not as highly favored as other members of the Limnanthes species because of its tendency to shatter. It is still of interest in domestication research and has been highly intercrossed with other species.

The seed oil of meadowfoam is the source of 3 new fatty acids. More than 95% of all fatty acids from limnanthes oil have longer carbon chains than the ones from vegetable oils. Some of the oil derivatives are waxy solids similar to the waxes used in polishes, phonograph records, pharmaceuticals, and cosmetics. A liquid wax similar to jojoba oil is made from limnanthes oil. It has a hardness similar to carnauba and candelilla waxes.

In its natural habitat, Douglas's meadowfoam thrives in wet meadowlands and clay loams. It is native to the Pacific Coast region of the United States, particularly California.

Dragon's Head

Sci. Name(s)	: *Lallemantia iberica*
Geog. Reg(s)	: Africa-North, Asia, Eurasia, Mediterranean
End-Use(s)	: Oil, Vegetable
Domesticated	: Y
Ref. Code(s)	: 7, 220
Summary	: Dragon's head is a plant grown for its fruit seed, which contains 30% of a drying oil used in cooking and lighting in the dry regions of Asia Minor and the Middle East. In Iran, its leaves are eaten as vegetables. It is cultivated to a limited extent in Russia and elsewhere. Plants grow wild throughout the Caucasus and Turkistan.

English Walnut, Persian Walnut
(See Nut)

Ethiopian Mustard

Sci. Name(s)	: *Brassica carinata*
Geog. Reg(s)	: Africa, America-North (U.S. and Canada), Asia-China, Europe
End-Use(s)	: Oil
Domesticated	: Y
Ref. Code(s)	: 169
Summary	: The Ethiopian mustard is grown primarily in Ethiopia as an oil seed crop. It has been introduced to America, Europe, and China. Of the genus Brassica, this mustard shows the greatest yield potential. Its seed pods are shatter-resistant. This plant is more productive if irrigated, and it possesses high levels of erucic acid.

European Beech
(See Ornamental & Lawn Grass)

European Hazel, European Filbert
(See Nut)

Fendler Bladderpod, Fendler Pod

Sci. Name(s)	: *Lesquerella fendleri*
Geog. Reg(s)	: America-North (U.S. and Canada)
End-Use(s)	: Oil
Domesticated	: Y
Ref. Code(s)	: 134, 156, 181

Summary : Fendler bladderpod is an herbaceous plant with potential as an oilseed crop with hydroxy acids. There are limited data for this particular member of the Lesquerella genus. It tolerates a variety of soils and is found throughout the United States, particularly Kansas, southeastern Colorado, and into New Mexico.

Fenugreek
(See Spice & Flavoring)

Figleaf Gourd, Malabar Gourd
Sci. Name(s) : *Cucurbita ficifolia*
Geog. Reg(s) : America-Central, America-South
End-Use(s) : Beverage, Oil, Vegetable
Domesticated : Y
Ref. Code(s) : 132, 170
Summary : The figleaf gourd is a perennial vine that produces fruit 15-20 cm in length. Parts of the fruit are very nutritious and are often eaten as a squash. The flesh is sometimes candied or fermented and made into an alcoholic beverage. The seeds yield an oil and are considered a potential oil and protein source.

Plants have a long history of cultivation in the highlands of Mexico and Central and South America. Crops can be produced at higher altitudes in the tropics and will survive in other areas with cooler climates. They must be grown in areas with adequate rainfall and on well-drained, rich loam soil. Plants flower throughout the year. Fruit can be harvested after 6 weeks to 4 months.

Flax
(See Fiber)

Fox Grape
(See Fruit)

Gordon Bladderpod
Sci. Name(s) : *Lesquerella gordonii, Vesicaria gordonii, Alyssum gordonii*
Geog. Reg(s) : America-North (U.S. and Canada)
End-Use(s) : Oil
Domesticated : N
Ref. Code(s) : 134, 156, 181
Summary : Gordon bladderpod is an herbaceous plant with potential as an oilseed crop. There is limited information for this particular species. Plants are found growing in sunny, open areas of the United States, usually in Kansas, Oklahoma, Utah, Texas, New Mexico, and throughout southern and central Arizona.

Harmala Shrub, African-Rue, Harmel Piganum
(See Medicinal)

Indian-Almond, Country Almond, Tropical-Almond, Myrobalan, Almendro
Sci. Name(s) : *Terminalia catappa*
Geog. Reg(s) : Asia, Tropical

End-Use(s) : Dye & Tannin, Miscellaneous, Oil
Domesticated : Y
Ref. Code(s) : 220
Summary : The Indian-almond is a tree that has been cultivated for its edible oily seeds. The fruit, bark, and roots can be used for tanning. Tree leaves are often fed to silkworms. Indian-almonds are native to tropical Asia and have been introduced and cultivated throughout other tropical areas.

Java-Almond, Kanari
(See Nut)

Jojoba, Goat-Nut
(See Fat & Wax)

Kapok, Silk-Cotton-Tree, Ceiba
(See Fiber)

Kenaf, Bimli, Bimlipatum Jute, Deccan Hemp
(See Fiber)

Lac-tree, Ceylon Oak, Kussum Tree, Malay Lac-tree
(See Fat & Wax)

Levant Cotton, Arabian Cotton, Maltese Cotton, Syrian Cotton
(See Fiber)

Lukrabao, Chaulmogra Tree
(See Medicinal)

Ma-ha-wa-soo
Sci. Name(s) : *Vaupesia cataractarum*
Geog. Reg(s) : America-South
End-Use(s) : Cereal, Oil
Domesticated : N
Ref. Code(s) : 170, 220
Summary : Ma-ha-wa-soo is a tree whose seeds are rich in oil. The seeds contain cyanic poisons, which must be removed by boiling. Once boiled, the seeds are edible. They were eaten by the northwestern Amazon Indians in times of famine and for ceremonial purposes.

Mahua
Sci. Name(s) : *Madhuca longifolia*
Geog. Reg(s) : Asia-India (subcontinent)
End-Use(s) : Beverage, Medicinal, Oil, Vegetable
Domesticated : Y
Ref. Code(s) : 170, 220

Summary   : Mahua is an evergreen tree whose seeds contain 50% of a yellow oil used as a butter substitute. This oil is used for cooking purposes, to manufacture soap and candles or to treat skin diseases. The press cake is used as a fertilizer. The corollas are eaten raw or cooked or used in liqueurs. India is the main oil producer, with most oil produced for domestic consumption. Indian oil mills have crushed 15,000-30,000 MT of seeds/year.

Maize, Corn, Indian Corn
(See Cereal)

Marijuana, Hemp, Marihuana
(See Fiber)

Marking-Nut Tree
(See Dye & Tannin)

Marrow, Pumpkin, Squash, Vegetable Marrow
(See Vegetable)

Mastic
(See Natural Resin)

Mixta Squash, Pumpkin, Winter Squash, Squash
(See Vegetable)

Molasses Grass
(See Insecticide)

Monkey-Pod
(See Timber)

Morula
(See Fruit)

Mu-Oil-Tree, Tung-Mu-Tree
Sci. Name(s)   : *Aleurites montana*
Geog. Reg(s)   : Africa, Asia-China, Asia-India (subcontinent), Asia-Southeast, Eurasia, Subtropical
End-Use(s)   : Oil
Domesticated   : Y
Ref. Code(s)   : 148, 170
Summary   : The mu-oil-tree grows in cool, subtropical regions and is a source of a tung-oil different in chemical composition from true tung-oil but of equal value. Mu-oil is often sold under the same name as tung-oil. China is the major tung-oil producer. Other producing countries are Vietnam, Cambodia, Laos, Zaire, the countries of eastern Africa, Malagasay Republic, South Africa, India, and the USSR.

Mu-oil-trees are native to subtropical China. Trees reach heights up to 18.3 m and do not survive the lowland tropics. They tolerate shallow, stony soils but thrive on slightly acidic soils.

Niger-Seed

Sci. Name(s) : *Guizotia abyssinica*
Geog. Reg(s) : Africa, Asia-India (subcontinent), Temperate, Tropical
End-Use(s) : Forage, Pasture & Feed Grain, Nut, Oil
Domesticated : Y
Ref. Code(s) : 73, 170
Summary : Niger-seed is an annual herb 0.5-1.5 m in height cultivated as an oil seed crop. The seeds contain 30-50% oil and 20% protein. The oil contains 9-17% saturated acids, 31-39% oleic acid, and 51-55% linoleic acid. The highest grades of oil are used for culinary purposes, with lower-quality grades for soap and paint industries. The press cake is rich in albuminoids and is fed to livestock or used as fertilizer. The seeds are fried and eaten. Plants can be cut at flowering and used as green fodder.

Crops are cultivated in Ethiopia and parts of India. They are usually grown as rain-fed crops in areas with up to 180 cm annual rainfall. Crops are most successful in tropical and temperate areas receiving 100-130 cm/year of rain. Plants will produce on most soils but thrive on light loamy ones.

Indian production rates range from 160,000-250,000 MT/year of seeds, of which approximately 23,600 MT are exported. Ethiopian strains of niger-seed contain more linoleic acid and are therefore superior to Indian strains. Because niger-seed is a highly cross-fertilized crop, there is considerable potential for research on maintaining the purity of improved strains and discovering easier methods of developing hybrids.

Oenocarpus bacaba

Sci. Name(s) : *Oenocarpus bacaba*
Geog. Reg(s) : America-South
End-Use(s) : Oil, Vegetable
Domesticated : N
Ref. Code(s) : 107, 132, 169
Summary : Oenocarpus bacaba is a palm whose fruit produces an edible oil similar to olive oil. Palm hearts are eaten fresh and canned and are high in vitamin C and calcium. Fruit yields can be increased if this palm is grown with the *Jessenia bataua*. Palms are found in the valleys of Brazil. In 1983, hearts of palm imported from Costa Rica, Brazil, and other countries were 1,090 MT, valued at $2.4 million.

Oil-Bean Tree, Owala Oil

Sci. Name(s) : *Pentaclethra macrophylla*
Geog. Reg(s) : Africa-West, Tropical
End-Use(s) : Cereal, Forage, Pasture & Feed Grain, Oil, Spice & Flavoring, Timber
Domesticated : Y
Ref. Code(s) : 36, 56
Summary : The oil-bean tree is grown as an oilseed crop. The beans yield 30-36% of a semisolid oil. Seed kernels yield 40-45% oil. These oils are used mainly for culinary purposes, for lubrication, and candle and soap making. The kernels are rich in pro-

teins and low in starch but are used as condiments for flavoring sauces. Expressed oilseed cakes are used for green manure. Seeds can be ground to flour but contain a poisonous alkaloid, pancine. Trees have hard, reddish-yellow timber used for various construction purposes in parts of Africa.

Trees reach up to 40 m in height and are found mainly in tropical West Africa. Oil-bean trees produce beans in their 10th year and will continue to bear regularly. The oil does not enter international trade but is an important product in Africa.

Oiticica Oil, Oiticica

Sci. Name(s)	: *Licania rigida*
Geog. Reg(s)	: America-South
End-Use(s)	: Dye & Tannin, Oil
Domesticated	: N
Ref. Code(s)	: 46, 220, 227
Summary	: The oiticica oil plant provides an oil similar to tung-oil extracted from its fruit seed. This oil contains a conjugated reto-fatty acid and is extremely reactive. It has 15% water-soluble tannins of possible use in the leather industry. About 60% of this oil is used in varnishes and paints.

This tree bears in 10-12 years, and lives to be 100 years. Yields vary yearly, with an average of about 130 kg of fruit/tree. Trees are grown mainly in Brazil.

Okra, Lady's Finger, Gumbo
(See Vegetable)

Olive

Sci. Name(s)	: *Olea europaea*
Geog. Reg(s)	: Africa-North, America-North (U.S. and Canada), Europe, Hawaii, Mediterranean
End-Use(s)	: Forage, Pasture & Feed Grain, Fruit, Miscellaneous, Oil, Timber
Domesticated	: Y
Ref. Code(s)	: 17, 116, 154, 192
Summary	: The olive is a small tree, reaching 7.6 m in height, that produces commercially important, edible fruit. Ripe fruits are pressed for a rich oil with twice the energy value of sugar. Major producers are Italy, Spain, Greece, Turkey, and Tunisia. Olive oil is used as a cooking and salad oil and in food preservation. It is also important for the manufacturing of quality toilet preparations and cosmetics and is used in pharmaceuticals. Residual oil cakes are fed to livestock or used as fertilizer. Olive seeds are an important source of furfural, used in the manufacture of molded products and plastics. Ripe fruits are eaten after being preserved in vinegar or salty liquids. The hard, beautifully-grained wood is made into canes, brushes, and other small domestic articles.

The majority of world olive production is for oil, and the remainder is sold whole. In 1983, the United States produced 50,792 MT of olives, valued at over $22 million. From this crop, 907 MT were crushed for oil, 43,355 MT canned, and 6,077 MT used for other olive products. The United States also imported an additional 33,039 MT of olive oil, valued at $47.4 million.

The olive is native to the Mediterranean and southern Europe. It grows in Hawaii but does not usually flower. In California, olives are grown as commercial fruit trees. They are most successful in semiarid areas with an annual rainfall of 60-75 cm. They require light, deep soils--preferably calcareous, sandy loams. Trees fruit after 6 years and remain productive for 50 years. They tend to produce irregularly, alternating from light to heavy fruit sets.

Opium Poppy
(See Medicinal)

Ouricury Palm, Licuri, Nicuri
(See Fat & Wax)

Paperbark
(See Timber)

Pataua, Seje Ungurahuay

Sci. Name(s)	: *Jessinia bataua*
Geog. Reg(s)	: America-North (U.S. and Canada), America-South, Asia-Southeast
End-Use(s)	: Beverage, Forage, Pasture & Feed Grain, Oil
Domesticated	: N
Ref. Code(s)	: 169
Summary	: Pataua is a palm whose fruits are the source of an edible oil and beverage. The oil is extracted from the mesocarp of the fruits and is then boiled, filtered, and bottled. The bottled oil is good for about a year before going rancid. The beverage (chicha de seje) is prepared by boiling the fruit and macerating the mesocarp. It has a taste similar to chocolate. The pulp by-product is used as animal feed.

The most common palm in the Amazonian valleys, pataua grows in dry or swampy areas. The palm has two panicles from which about 30 kg of fruit is obtained annually. Each fruit weighs about 15 kg, and 8-10% of its weight consists of oil. The palm oil of commerce is obtained from the oil palm (*Elaeis guineensis*), and all palm oil trade and production figures are included in that discussion.

Peanut, Groundnut, Goober, Mani

Sci. Name(s)	: *Arachis hypogaea*
Geog. Reg(s)	: Africa, America-Central, America-North (U.S. and Canada), America-South, Asia, Subtropical, Tropical, West Indies
End-Use(s)	: Fiber, Forage, Pasture & Feed Grain, Nut, Oil
Domesticated	: Y
Ref. Code(s)	: 126, 148, 170
Summary	: The peanut or groundnut is a shrub cultivated for its economically important nuts high in vitamin B complex, protein, and minerals. They enter commerce as raw or roasted nuts or peanut butter. A nondrying oil pressed from the nut seed is used as a cooking oil and in margarine, soap, and lubricants. This high-quality oil is also used in pharmaceutical products. A synthetic textile fiber is obtained from the protein. The expressed oil cakes are fed to livestock. Harvested vines fed to livestock have a nutritional value similar to that of alfalfa.

Peanuts are most productive in tropical and subtropical climates. They do not withstand low temperatures, and they require warm weather and moderate rainfall or irrigation during growing seasons and need hot, dry weather for seed ripening. Crops must be kept weeded and should be rotated every 3 years.

Plants are grown commercially in Asia, Africa, South America, Central America and the Caribbean, and the southern United States. In 1983, the United States produced 1.5 million MT of peanuts, valued at approximately $793 million. The United States exported 223,922 MT of shelled peanuts, valued at $175.5 million. Shrubs are native to South America.

Pecan
(See Nut)

Perilla

Sci. Name(s)	: *Perilla frutescens, Perilla ocymoides, Perilla arguta*
Geog. Reg(s)	: Africa-South, America-North (U.S. and Canada), Asia, Asia-China, Asia-India (subcontinent), Mediterranean
End-Use(s)	: Forage, Pasture & Feed Grain, Oil, Spice & Flavoring
Domesticated	: Y
Ref. Code(s)	: 17, 36, 73, 154
Summary	: Perilla is an annual herb 60-120 cm in height cultivated for its fruit and leaves.

Fruit seeds yield an oil used for domestic culinary purposes. In the United States, this oil is substituted for linseed oil and tung-oil in paints, varnishes, linoleum, oilcloths, printing inks, and similar items. Perilla foliage has a high yield of citral oil with a better gloss, exterior durability, and water resistance than linseed foliage oil. The commercial extraction rate of perilla oil is approximately 37%. Expressed oilseed cakes provide a protein and fiber-rich livestock feed. Perilla leaves are dried and used to flavor food.

Plants are grown in parts of eastern Asia and the Himalayas as oilseed crops. Because fruits do not ripen simultaneously and fall from the plant very quickly, large-scale harvesting is difficult. Perilla is native to the mountainous areas of India and China, where they grow up to 1,200 m in altitude. Commercial plantings of perilla in the United States, Cyprus, and South Africa have met with little economic success. Plants are naturalized in the eastern United States as weeds.

Pili Nut, Elemi
(See Nut)

Pulasan
(See Fruit)

Purging Croton
(See Medicinal)

Radish
(See Vegetable)

Rape, Colza

Sci. Name(s) : *Brassica napus*
Geog. Reg(s) : Asia, Europe
End-Use(s) : Forage, Pasture & Feed Grain, Oil
Domesticated : Y
Ref. Code(s) : 170, 194
Summary : Rape is an herb 1 m in height grown in Europe for fodder and in Japan and Europe for the edible oil (colza oil) extracted from its seeds. Rape is rarely grown in the tropics.

Colza oil is used in baking, as a lubricant, as an illuminant, and in soap manufacture. The expressed seed cake is fed to livestock.

Plants thrive in areas with temperatures of 5-7 degrees Celsius and an annual rainfall of 0.3-4.2 m. Best results are in clay or clay loam soils. Information regarding production and trade is found under bird rape.

Red Silk-Cotton, Silk Cotton
(See Fiber)

Roquette, Garden Rocket, Rocket Salad, Roka

Sci. Name(s) : *Eruca vesicaria*
Geog. Reg(s) : America-North (U.S. and Canada), Asia, Asia-India (subcontinent), Australia, Europe, Mediterranean
End-Use(s) : Essential Oil, Medicinal, Oil, Vegetable
Domesticated : Y
Ref. Code(s) : 205, 220, 238
Summary : Roquette is a plant cultivated in India for its seeds, which yield about 32% of a semidrying oil (jamba oil) used for culinary purposes, as a lubricant, and for burning. This oil must age for 6 months before its acidic taste disappears. Its essential oil, heterosides, consists of a mild stimulant. The young plant leaves are used in salads and have rubefacient properties. Roquette is native to Central Europe, the Mediterranean and Asia, and has been introduced to the United States and Australia.

Roselle
(See Fiber)

Rubber-Tree, Para Rubber
(See Isoprenoid Resin & Rubber)

Safflower

Sci. Name(s) : *Carthamus tinctorius*
Geog. Reg(s) : Africa, Africa-North, America-North (U.S. and Canada), Asia-India (subcontinent), Mediterranean
End-Use(s) : Dye & Tannin, Oil
Domesticated : Y
Ref. Code(s) : 170
Summary : The safflower is an oilseed crop that is gaining importance as a commercial source of oil. The seeds contain 20-38% oil, with thinner-hulled seeds having the

most oil. This oil has a low linolenic acid content, making it useful in paints and varnishes. It retains its color well and has no yellowing effects. In India, the oil is used in cooking, illumination, and soap manufacture. In the Middle East and India, 0.3-0.6% of a scarlet red dye (safflower carthamin) is obtained from the dried florets but is rarely used, having been replaced by aniline dyes. This dye has been used for ceremonial cloth, cakes, bisquits, and rouge. The dye is insoluble in water. It should not be confused with the dye saffron, taken from the stigmas of saffron crocus (*Crocus sativus*).

Principal commercial safflower production is in Afghanistan, the Nile Valley, and Ethiopa. In 1983, the United States produced 100,000 MT of safflower seeds from an area of 105,000 ha. Crops are not generally successful in hot tropical lowlands.

Safflowers are annuals that grow 0.5-1.5 m in height. In their early stages, growth is fairly slow.

Scarlet Poppy, Oriental Poppy
(See Medicinal)

Sea Island Cotton, American Egyptian Cotton, Extra Long Staple Cotton
(See Fiber)

Seaving, Souari, Achotillo

Sci. Name(s)	: *Caryocar amygdaliferum*
Geog. Reg(s)	: America-South
End-Use(s)	: Nut, Oil
Domesticated	: N
Ref. Code(s)	: 140, 220
Summary	: The seaving tree produces edible seeds that yield an oil used as a cooking oil in South America. These seeds have a taste similar to almonds'. The tree grows in the humid evergreen forests of South America, particularly in Ecuador.

Sesame, Simsim, Beniseed, Gingelly, Til

Sci. Name(s)	: *Sesamum indicum, Sesamum orientale*
Geog. Reg(s)	: Africa-West, America-Central, Asia, Asia-China, Asia-India (subcontinent), Asia-Southeast, Tropical
End-Use(s)	: Energy, Essential Oil, Forage, Pasture & Feed Grain, Medicinal, Oil, Spice & Flavoring
Domesticated	: Y
Ref. Code(s)	: 56, 73, 154, 170
Summary	: Sesame is an annual oilseed crop that provides a significant proportion of the world's edible oils. Sesame seeds contain 45-55% of a semidrying oil and 24-29% protein. Sesame oil is used mainly as a salad and cooking oil, and less often, as a substitute for olive oil. The oil is employed in the manufacture of margarine and compound cooking fats, in soaps and paints, and as a lubricant and illuminant. Sesame oil is also used as a vehicle for perfumes, unspecified pharmaceutical products, and as a synergist for pyrethrum in aerosol sprays. Sesamol has been used as an antioxidant in

lard and vegetable oils. The residue from oil extraction is high in vitamin B, calcium, and phosphorous.

Cultural uses of sesame vary. In India, the oil is used as a ghee substitute and for anointing the hair and body, a practice considered healthful. The expressed cake is a protein-rich livestock food, and in India and Java, the cake is eaten in soups or mixed with sugar and eaten as a sweetmeat. West Africans use young plant leaves as a soup vegetable and in certain medicines. Plant stems are burned as fuel.

Major producing countries are India, Mexico, Burma, China, and the Democratic Republic of the Sudan. Important importing countries are Japan, the United States, the USSR, Italy, the United Arab Republic, Egypt, and Belgium, together with Luxembourg.

Sesame thrives in hot, dry, tropical areas under irrigation, unless there is at least 380 mm/annum of rainfall. Plants survive short periods of drought but are susceptible to waterlogging. They grow in most soils but are most productive in sandy loams at elevations approaching 1,220 m. Sesame matures at different rates depending on environments, but the short, single-stemmed varieties mature the earliest. Crops are hand-harvested, which increases production costs and makes competition with other oilseed prices very difficult. Improved, less shatter-prone varieties of sesame are being developed that should help improve its competitiveness.

In 1983, world production was 2.1 million MT, from an area of 6.7 million ha, which yielded 309 kg/ha. The United States imported 42,787 MT, valued at $40 million.

Shea-Butter, Butterseed
(See Fat & Wax)

Smooth Loofah, Sponge Gourd, Dishcloth Gourd, Vegetable Sponge, Luffa, Loofah
(See Miscellaneous)

Sorghum, Great Millet, Guinea Corn, Kaffir Corn, Milo, Sorgo, Kaoling, Durra, Mtama, Jola, Jawa, Cholam Grains, Sweets, Broomcorn, Shattercane, Grain Sorghums, Sweet Sorghums
(See Cereal)

Sour Cherry, Pie Cherry
(See Fruit)

Soybean

Sci. Name(s)	:	*Glycine max*
Geog. Reg(s)	:	America-North (U.S. and Canada), America-South, Asia, Asia-Southeast, Europe, Subtropical, Tropical
End-Use(s)	:	Beverage, Cereal, Forage, Pasture & Feed Grain, Oil, Spice & Flavoring, Vegetable
Domesticated	:	Y
Ref. Code(s)	:	170

Summary   : The soybean is a bushy annual that grows 20-180 cm in height. The bean pods and seeds are a worldwide source of oil and protein. Fermented pods are used to make soya sauce, which is one of the major flavorings and sauces used in eastern Asia. The pods are also an important ingredient in Worcestershire and other western sauces. The seeds are a good source of vitamin B and are dried to produce soya milk, an important protein supplement in infant diets. From the whole bean pod, a nutritious soya flour is prepared that is a popular ingredient in health foods.

A semidrying oil extracted from the seeds contains lecithin and is used as an edible oil in margarine, shortening, and salad oils and as a wetting and stabilizing agent in food, cosmetics, and pharmaceutical preparations. The oil also has an industrial use in paints, linoleums, oilcloths, printing inks, soaps, insecticides, and disinfectants. After oil extraction, soya meal remains, which is fed to livestock or used for manufacturing synthetic fiber, adhesives, textiles, and waterproofing agents. Soybean plants provide good pasturage, fodder, hay, and silage. The plants can also be grown as cover crops.

Soybean plants thrive in subtropical areas and have been introduced into most tropical countries. In the United States, soybean production is predominant in the corn belt area. The United States produces a wide range of soybean products.

In 1984, world production of soybeans was approximately 90 million MT from an area of 52 million ha. In the same year, the United States imported 10,947 MT of soybeans, valued at $3.2 million, and exported 19.5 million MT of soybeans, valued at $543 million.

Day length determines the maturation rate of soybeans. Before growing crops extensively, it is important to test for their adaptation to local conditions. Generally, soybeans mature in 75-200 days. In the United States, yields are 1,700-2,700 kg/ha of dried seeds, 8.2-9.1 MT/ha of green matter, and 0.91-4.5 MT/ha of hay.

Spotted Gum
(See Timber)

Sunflower

Sci. Name(s)   : *Helianthus annuus*
Geog. Reg(s)   : Africa-North, America-North (U.S. and Canada), America-South, Eurasia, Mediterranean, Subtropical, Temperate, Tropical
End-Use(s)   : Dye & Tannin, Fiber, Forage, Pasture & Feed Grain, Oil, Ornamental & Lawn Grass
Domesticated   : Y
Ref. Code(s)   : 126, 154, 170
Summary   : The sunflower is a tall annual herb 1.5-3 m in height and is considered the second most important oilseed crop in the world. The oilseeds are produced in large flower heads 10-30 cm in diameter. There are 2 classes of sunflowers: oilseeds, which produce small, dark seeds with higher oil content and lower hull content; and, the confectionery or garden oilseed, which produces larger, light-colored seeds with thicker hulls that are easily separated from the kernel.

Seeds of the oilseed type contain 48% of a semidrying oil, high in oleic-linolenic acid content. Refined oil is used in the salad oil industries and for lighting. The expressed oil cake is high in protein and is a valuable livestock feed. The cake and meal from dehulled seeds is also an excellent protein source in human diets. It is considered superior to most vegetable proteins and is more closely balanced in essential amino acids. The protein value is as high as that of standard egg protein.

There is a sizable market for the edible seeds of the garden variety. The seed kernels provide quality protein, vitamins, and minerals and are used in commercial bird seed mixtures. Plant leaves are used as fodder. A fiber is obtained from the stem. The sunflower plant is strikingly tall and bears a large, vividly colored flower, making it a popular ornamental. The flowers yield a yellow dye.

Sunflowers are native to the western United States and have been introduced to most temperate and tropical countries. Plants thrive in direct sunlight in areas with adequate rainfall and low to moderate humidity. They will tolerate most soils. In temperate areas, the largest producers of sunflower oil are Argentina, Bulgaria, Romania, Yugoslavia, the USSR, the United States, and Uruguay. In the tropics and subtropics, producing countries are Ethiopia, Morocco, Tanzania, and Turkey.

In 1983, the United States produced 1.4 million MT of sunflower seeds, valued at $398.8 million. In the same year, the United States exported 791,809 MT of sunflower seed, valued at $222.6 million, and 281,409 MT of sunflower oil, valued at $152.1 million. Commercial processing of 1 MT of seed for oil yields 400 kg of oil, 350 kg of meal, and 200 kg of hulls.

Tacoutta, Wing Sesame

Sci. Name(s)	: *Sesamum alatum*
Geog. Reg(s)	: Africa-West, Tropical
End-Use(s)	: Cereal, Oil, Spice & Flavoring
Domesticated	: N
Ref. Code(s)	: 8, 119, 170
Summary	: Tacoutta (*Sesamum alatum*) is an erect herb 0.6-0.9 m in height that produces edible seeds rich in an oil used in soups, cakes, and other food. Tacoutta seeds are also used as adulterants of sesame (*Sesamum indicum*). Seeds are ground to a flour. Plants grow in parts of tropical Africa.

Tarwi, Tarhui, Chocho, Pearl Lupine, Tarin Altramuz, Muti, Ullus

Sci. Name(s)	: *Lupinus mutabilis*
Geog. Reg(s)	: America-South
End-Use(s)	: Forage, Pasture & Feed Grain, Medicinal, Oil, Ornamental & Lawn Grass
Domesticated	: Y
Ref. Code(s)	: 56, 169
Summary	: Tarwi is a plant whose seeds have high oil and protein content and is a cooking oil rich in unsaturated fatty acids, including linoleic acid. This oil and protein constitute 50% of the seed weight but is deficient in the amino acid methionine, with adequate amounts of lysine and cystine. It requires processing prior to eating. Tarwi is an excellent manure crop. It is grown as an ornamental and used as a poison in folk medicines.

This plant is a native of the Andean plains but is domesticated and grows in the high mountain valleys and basins at altitudes of 1,800-4,000 m. It is tolerant to acid, sandy soils, animal predators, drought, and frost. It is shatter-resistant but not self-pollinating. Experimental yields show 50 MT/ha wet matter with 1,750 kg of protein.

Tea
(See Beverage)

Teosinte
(See Forage, Pasture & Feed Grain)

Tomato
(See Fruit)

Tree Cotton
(See Fiber)

Tung-Oil-Tree

Sci. Name(s)	: *Aleurites fordii*
Geog. Reg(s)	: America-North (U.S. and Canada), America-South, Asia-China, Asia-Southeast, Australia, Eurasia
End-Use(s)	: Oil
Domesticated	: Y
Ref. Code(s)	: 148, 170
Summary	: Tung-oil-trees reach heights of 12 m and are the source of a high quality drying oil extracted from their seeds (*Aleurites fordii*). Tung-oil is a quick-drying oil with good waterproofing qualities, used in varnish and paint manufacture and in making linoleum, oilcloth, and insulating compounds. The expressed oil cake is poisonous but is a good fertilizer. Seeds yield 15-20% tung-oil.

Trees are native to the cooler parts of west and Central China. They are most successfully grown in warm temperate areas and are suited to the lowland tropics. They can be grown at higher tropical elevations. China is the main producer of tung-oil, with lesser production in the southeastern United States, Argentina, Brazil, Paraguay, Australia, Burma, Vietnam, Cambodia, Laos, and the USSR.

Upland Cotton, Cotton
(See Fiber)

Vernonia galamensis

Sci. Name(s)	: *Vernonia galamensis*
Geog. Reg(s)	: Africa-West, Tropical
End-Use(s)	: Oil
Domesticated	: Y
Ref. Code(s)	: 169
Summary	: Vernonia galamensis is an annual herb with seeds that produce an oil rich in natural epoxy acid, vernolic acid, and triglyceride. It has potential in the prepared baked film or coatings industry. Tests reveal its excellent flexibility, resistance to chipping, and adherence to substrates. It is also resistant to mineral acid, alkali, detergents, and solvents. This herb is found in eastern Zimbabwe and throughout tropical West Africa.

Watermelon
(See Fruit)

Waxgourd, White Gourd
(See Vegetable)

White Melilot, White Sweetclover
(See Forage, Pasture & Feed Grain)

White Mustard
(See Forage, Pasture & Feed Grain)

Wild Sesame

Sci. Name(s)	: *Sesamum radiatum*
Geog. Reg(s)	: Africa-West
End-Use(s)	: Cereal, Oil, Spice & Flavoring
Domesticated	: Y
Ref. Code(s)	: 8, 170
Summary	: Wild sesame (*Sesamum radiatum*) is an herb reaching up to 1.2 m in height and occasionally cultivated for its edible seeds in West Africa. The seeds yield an edible oil used in soups, cakes, and flour. They are used as an adulterant of true sesame.

Wine Grape, Grape, European Grape, California Grape
(See Beverage)

Wine Palm, Coco de Chile, Coquito Palm, Honey Palm, Wine Palm of Chile, Chile Coco Palm
(See Beverage)

Winged Bean, Goa Bean, Asparagus Pea, Four-Angled Bean, Manilla Bean, Princess Pea
(See Vegetable)

Ornamental & Lawn Grass

Aconite, Monkshood, Friar's-Cap
(See Medicinal)

African Bermudagrass, Masindi Grass

Sci. Name(s)	:	*Cynodon transvaalensis*
Geog. Reg(s)	:	Africa-East, Africa-South
End-Use(s)	:	Erosion Control, Ornamental & Lawn Grass
Domesticated	:	Y
Ref. Code(s)	:	171
Summary	:	African bermudagrass is more common than Bermuda grass as a lawn grass

because it is less coarse. In Uganda, this grass is used on golf greens. It provides good erosion control. African bermudagrass is native to South Africa.

Annatto
(See Dye & Tannin)

Annual Bluegrass, Low Speargrass, Dwarf Meadow Gold, Six-Weeks Grass, Plains Bluegrass
(See Forage, Pasture & Feed Grain)

Annual Ryegrass, Italian Ryegrass, Australian Ryegrass
(See Forage, Pasture & Feed Grain)

Australian Blackwood
(See Timber)

Australian Tea-Tree
(See Erosion Control)

Babassu, Orbignya speciosa
(See Oil)

Bahia Grass, Bahiagrass
(See Forage, Pasture & Feed Grain)

Bamboo, Phyllostachys
(See Timber)

Barbados Aloe, Mediterranean Aloe, True Aloe
(See Medicinal)

Barbados Cherry, Acerola, West Indian-Cherry, Barbados-Cherry, West Indian Cherry
(See Fruit)

Barwood, Camwood
(See Dye & Tannin)

Beach, She Oak
(See Timber)

Beach Plum, Shore Plum
(See Erosion Control)

Bermudagrass, Star Grass, Bahama Grass, Devilgrass
(See Erosion Control)

Betel Palm, Betel Nut, Areca Palm

Sci. Name(s)	: *Areca catechu, Areca cathecu*
Geog. Reg(s)	: Asia-India (subcontinent), Asia-Southeast
End-Use(s)	: Medicinal, Nut, Ornamental & Lawn Grass
Domesticated	: Y
Ref. Code(s)	: 148, 171
Summary	: The betel palm provides seeds that are chewed as a masticatory. The seeds, called betel nuts, are wrapped in a betel pepper leaf with lime and chewed. In the Far East, the seeds and other parts of the palm have various medicinal qualities, mainly as a vermifuge. In Florida and other tropical areas, the palm is grown as an ornamental.

Betel palms are often grown with other cultivated trees. They require tropical conditions and should be grown in shady areas on soils with good drainage. Lime in the soil may lessen nut yields. Palms stand 12-30 m in height. They bear seeds in 7-8 years, reach full bearing in 10-15 years, and continue producing for about 40 more years. Growth and production of the palm is primarily in southeastern Asia, particularly India, eastern Pakistan, Sri Lanka, Malaysia, and Indonesia.

Big Bluegrass

Sci. Name(s)	: *Poa ampla*
Geog. Reg(s)	: America-North (U.S. and Canada)
End-Use(s)	: Forage, Pasture & Feed Grain, Ornamental & Lawn Grass
Domesticated	: Y
Ref. Code(s)	: 17
Summary	: Big bluegrass is native to North America and is an important pasture and range grass sometimes cultivated for lawns. This grass provides a nutritious and palatable forage. There is very little information concerning specific big bluegrass yields or general characteristics.

Black Locust, False Acacia
(See Timber)

Blue Clitoria, Butterfly Pea, Asian Pigeon-Wings, Butterfly Bean, Kordofan Pea
(See Forage, Pasture & Feed Grain)

Blue Wildrye

Sci. Name(s)	: *Elymus glaucus*
Geog. Reg(s)	: America-North (U.S. and Canada), Asia-Central
End-Use(s)	: Erosion Control, Forage, Pasture & Feed Grain, Ornamental & Lawn Grass
Domesticated	: Y
Ref. Code(s)	: 14, 96, 214
Summary	: Blue wildrye grass is a grass often grown as an ornamental. It is also valued as a soil and sand binder, a secondary range grass and a good source of hay. It grows in a variety of soils and locations at low to medium altitudes in North America from British Columbia to California and from the midwest to the southwest. It also grows in parts of Turkistan.

Bolo-Bolo
(See Fiber)

Borage

Sci. Name(s)	: *Borago officinalis*
Geog. Reg(s)	: Mediterranean
End-Use(s)	: Beverage, Medicinal, Miscellaneous, Ornamental & Lawn Grass, Spice & Flavoring, Vegetable
Domesticated	: Y
Ref. Code(s)	: 194, 220
Summary	: Borage is a Mediterranean herb reaching heights of about 1 m. It is widely cultivated for its leaves and flowers. The leaves are used in salads and to flavor liqueurs. Dried leaves and flowers are brewed as a refreshing diuretic drink. Borage plants are usually cultivated as ornamentals because of their attractive blue to purple flowers that bloom throughout the growing season. Flowers are said to have a taste similar to cucumbers. The flowers attract bees and are encouraged by apiculturists. Blossoms may be candied and used in confections. Medicinally, preparations of the plant are used in the treatment of jaundice, coughs, fever, dermatitis, and kidney ailments.

Plants will grow in most soils and can adapt to different environmental conditions. Best climates are in areas with temperatures of 5-21 degrees Celsius and an annual rainfall of 0.3-1.3 m.

Bowstring Hemp, Snake-Plant
(See Fiber)

Breadfruit
(See Vegetable)

Broad-Leaved Lavender
(See Essential Oil)

Buffelgrass
(See Forage, Pasture & Feed Grain)

360

Cainito, Star-Apple
(See Fruit)

Calendula, Pot-Marigold
Sci. Name(s) : *Calendula officinalis*
Geog. Reg(s) : Europe
End-Use(s) : Medicinal, Ornamental & Lawn Grass, Spice & Flavoring
Domesticated : Y
Ref. Code(s) : 16, 170, 225
Summary : Calendula is a cultivated ornamental and drug plant. It is native to southern Europe. Since its flowers bloom throughout the season, it makes an attractive ornamental and long-lasting garden plant. The flower heads can be used for flavoring soups and stews, and the florets are used in various home remedies. It is an annual herb reaching heights of 30-60 cm. Plants are easily grown in most warm, loose soils.

Candytuft, Rocket Candytuft
Sci. Name(s) : *Iberis amara*
Geog. Reg(s) : Europe
End-Use(s) : Medicinal, Ornamental & Lawn Grass, Spice & Flavoring, Weed
Domesticated : Y
Ref. Code(s) : 16, 220
Summary : Candytuft, or rocket candytuft, is an annual often grown as ornamental edging or in flower beds. Florists raise candytufts to sell as cut flowers. The plants are relatively easy to grow because they are successful in most rich garden soils in full sunlight. Plants are weeds of cultivated land in Britain and Central and southern Europe.

Formerly, candytuft leaves were used medicinally as an antiscorbutic and in treatment of rheumatism. Today, the seeds are a component of many home remedies. They are used as substitutes for mustard seed.

Carpetgrass
(See Forage, Pasture & Feed Grain)

Cassia
(See Timber)

Castorbean
(See Oil)

Centipede Grass
Sci. Name(s) : *Eremochloa ophiuroides*
Geog. Reg(s) : America-North (U.S. and Canada), Asia-Southeast
End-Use(s) : Erosion Control, Ornamental & Lawn Grass
Domesticated : Y
Ref. Code(s) : 96, 154, 220
Summary : Centipede grass provides rapid, dense turf and is a good lawn grass and erosion control agent. This grass was introduced to parts of North America and is native to Southeast Asia. It thrives in shade.

Ceriman, Monstera

Sci. Name(s)	: *Monstera deliciosa*
Geog. Reg(s)	: America-Central, America-North (U.S. and Canada), Tropical
End-Use(s)	: Fruit, Ornamental & Lawn Grass
Domesticated	: Y
Ref. Code(s)	: 65, 171
Summary	: Ceriman is a creeping plant popular as a tropical ornamental. It has fruit edible when fully ripe. In Mexico, this fruit is eaten in mixed fruit salads. In the United States, vines are grown in Florida under half shade in a manner similar to pineapple culture.

Plants require moist, warm climates for fruit production and fruit better if grown on the ground. Fruit matures 9-12 months after flowering and is harvested in 14-18 months. Cerimans are native to Mexico.

Champac

Sci. Name(s)	: *Michelia champaca*
Geog. Reg(s)	: Asia-Central
End-Use(s)	: Medicinal, Ornamental & Lawn Grass, Spice & Flavoring, Timber
Domesticated	: Y
Ref. Code(s)	: 154
Summary	: Champac is a fragrant Himalayan tree whose blossoms serve ornamental purposes. The tree blooms throughout the year. The decorative flowers are worn in the hair, made into necklaces, or manufactured into perfume. The tree wood is fine-grained and used for buildings and furniture. The bark, seeds, and leaves are sometimes used in medicines. The bitter bark is used to adulterate the spice cinnamon.

Cherry Laurel, English Laurel
(See Erosion Control)

Cherry Plum, Myrobalan Plum

Sci. Name(s)	: *Prunus cerasifera, Prunus myrobalan*
Geog. Reg(s)	: America-North (U.S. and Canada), Asia-Central, Europe, Subtropical, Temperate
End-Use(s)	: Fruit, Miscellaneous, Ornamental & Lawn Grass
Domesticated	: Y
Ref. Code(s)	: 17, 192
Summary	: The cherry plum is a deciduous tree that reaches 7.6 m in height. It is grown as an ornamental hedge and widely used as a rootstock for other plums. Its fruit is similar to, but larger than, cherries.

For optimum growth, trees are most successful in light, well-drained soils in the temperate and subtropical climates of Central Asia, Europe, and North America. Trees are usually propagated by cuttings or from seed and are fairly easy to grow.

Chewings Fescue
(See Forage, Pasture & Feed Grain)

Chinese Lacquer Tree, Varnish Tree, Lacquer Tree, Japanese Lacquer Tree, Urushi
(See Natural Resin)

Chinese Rhubarb, Medicinal Rhubarb, Canton Rhubarb
(See Medicinal)

Chinese Tallow-Tree, Tallow-Tree
(See Oil)

Chinese Yam, Cinnamon-Vine
(See Tuber)

Chives
(See Vegetable)

Clary, Clary Wort, Clary Sage
(See Essential Oil)

Cohune Nut Palm, Cohune Palm
(See Oil)

Colchicum, Meadow Saffron, Autumn Crocus
(See Medicinal)

Colonial Bentgrass, Rhode Island Bentgrass, Browntop, Common Bentgrass
(See Forage, Pasture & Feed Grain)

Common Fig, Adriatic Fig
(See Fruit)

Common Lavender, English Lavender
(See Essential Oil)

Common Mignonette, Sweet Reseda
(See Essential Oil)

Common Persimmon, Persimmon, American Persimmon
(See Fruit)

Common Rue, Garden Rue
(See Medicinal)

Common Thyme
(See Spice & Flavoring)

Cooba
(See Timber)

Corn Salad, Lamb's Lettuce, European Cornsalad
(See Vegetable)

Costa Rican Guava, Wild Guava
(See Fruit)

Cowberry, Red Whortleberry, Mountain Cranberry, Lingon Berry, Rock Cranberry, Lingen
(See Fruit)

Cowslip, Marsh-Marigold
(See Vegetable)

Creeping Bent-Grass, Fiorm, Red Top
(See Forage, Pasture & Feed Grain)

Crowfoot Grass
(See Weed)

Crownvetch, Trailing Crownvetch
(See Forage, Pasture & Feed Grain)

Curuba, Carua, Casabanana

Sci. Name(s)	: *Sicana odorifera*
Geog. Reg(s)	: America-North (U.S. and Canada), America-South
End-Use(s)	: Fruit, Ornamental & Lawn Grass, Vegetable
Domesticated	: Y
Ref. Code(s)	: 16, 170, 220
Summary	: Curuba is a fast-growing vine climbing to heights of 9.1-15.2 m. It is cultivated in South America as an ornamental. Its fruits are occasionally eaten as domestic vegetables and in preserves. In the United States, curuba is found along the Gulf Coast.

Dichondra

Sci. Name(s)	: *Dichondra repens, Dichondra micantha, Dichondra carolinensis*
Geog. Reg(s)	: America-North (U.S. and Canada), Asia, Eurasia, Europe, Subtropical, Tropical
End-Use(s)	: Erosion Control, Ornamental & Lawn Grass
Domesticated	: Y
Ref. Code(s)	: 153, 197, 218
Summary	: Dichondra is a creeping perennial herb, with stems reaching up to 50 cm in height, that provides a hardy lawn grass. It roots easily and crowds out other grasses. Dichondra thrives on sandy soil below 457 m in the United States, western Europe, and eastern Asia. It is commonly cultivated in California as a ground cover and widely distributed throughout the tropics and subtropics.

Douglas-Fir
(See Timber)

Douglas's Meadowfoam
(See Oil)

Dyer's-Greenwood, Dyer's-Greenweed
(See Dye & Tannin)

Eggplant, Brinjal, Melongene, Aubergine
(See Vegetable)

Egyptian Carissa
(See Fruit)

Elephant Bush, Spekboom

Sci. Name(s) : *Portulacaria afra*
Geog. Reg(s) : Africa-South, America-North (U.S. and Canada), Hawaii
End-Use(s) : Dye & Tannin, Forage, Pasture & Feed Grain, Miscellaneous, Ornamental & Lawn Grass, Vegetable
Domesticated : Y
Ref. Code(s) : 15, 17, 119, 154, 226
Summary : The elephant bush is a small, succulent shrub or tree that grows up to 3.6 m in height and is used for fodder. Its leaves are eaten raw. Shrubs produce fragrant flowers that attract bees. There are sweet and sour leaf varieties, with the sour ones containing large quantities of tannin.

Shrubs are native to the arid regions of South Africa and are grown in the continental United States and Hawaii as ornamentals and are known as jade or purslane trees.

Emblic, Myrobalan, Emblica
(See Fruit)

English Walnut, Persian Walnut
(See Nut)

Eucalyptus
(See Timber)

European Beech

Sci. Name(s) : *Fagus sylvatica*
Geog. Reg(s) : Asia, Asia-Central, Europe
End-Use(s) : Energy, Medicinal, Nut, Oil, Ornamental & Lawn Grass, Timber
Domesticated : Y
Ref. Code(s) : 220
Summary : European beech trees grow from Europe to Asia Minor and Central Asia. The water-resistant wood is used in shipbuilding and furniture making. It also provides

excellent fuel. Creosote, which is a wood preservative and has medicinal properties, is obtained from the wood through a process of destructive distillation. Wood charcoal, carbo ligni depuratus, is used as an absorbent in alkali and phosphorus poisoning. Tree nuts are edible and are fed to livestock and poultry. Beech seeds yield a nondrying oil used in the manufacture of soap and for lighting. The edible oil may also be used for culinary purposes. Trees are often grown as hedges or ornamentals.

European Pennyroyal
(See Essential Oil)

Fennel, Florence Fennel, Finocchio
(See Spice & Flavoring)

Field Horsetail
(See Medicinal)

Fox Grape
(See Fruit)

French Lavender
(See Essential Oil)

Galanga, Kent Joer

Sci. Name(s)	: *Kaempferia galanga*
Geog. Reg(s)	: Asia-India (subcontinent), Asia-Southeast
End-Use(s)	: Essential Oil, Medicinal, Ornamental & Lawn Grass, Spice & Flavoring
Domesticated	: Y
Ref. Code(s)	: 77, 154, 171, 180
Summary	: Galanga is a stemless herb from India grown mainly as ornamentals. In India, its aromatic roots produce 2.4-3.9% of a volatile oil used in perfumery. In Southeast Asia, the roots are used as a spice to flavor curries and for unspecified medicinal uses.

Giant Reed
(See Miscellaneous)

Giri Yam Pea, Yam Bean, Akitereku, African Yam Bean, Haricot Igname
(See Tuber)

Governor's-Plum, Ramontchi, Rukam

Sci. Name(s)	: *Flacourtia indica*
Geog. Reg(s)	: Africa-West, Asia, Asia-Southeast, Tropical
End-Use(s)	: Fruit, Medicinal, Ornamental & Lawn Grass
Domesticated	: Y
Ref. Code(s)	: 36, 214, 226
Summary	: Governor's-plum is a shrubby plant with juicy, edible fruit that is eaten raw and has an agreeable flavor. It is important to grow several plants together, as single specimens do not fruit. In Madagascar, the bark is used with oil as an antirheumatic

liniment, root ash is used as a kidney remedy, and the leaf for asthma and gynecological remedies. Plants are often grown as ornamental hedges. Their timber is too small and crooked for use. Governor's-plums thrive in dry, open places and are native to tropical Africa and southern Asia.

Guaiac, Guaiacum
(See Natural Resin)

Gutta-Percha
(See Isoprenoid Resin & Rubber)

Heartnut, Cordate Walnut, Siebold Walnut
(See Nut)

Henna

Sci. Name(s)	:	*Lawsonia inermis*
Geog. Reg(s)	:	Africa-North, America-North (U.S. and Canada), America-South, Asia-Central, Asia-India (subcontinent), Asia-Southeast, Europe, Mediterranean
End-Use(s)	:	Dye & Tannin, Essential Oil, Medicinal, Ornamental & Lawn Grass, Timber
Domesticated	:	Y
Ref. Code(s)	:	154, 194
Summary	:	Henna is a perennial shrub most often grown as an ornamental hedge, but it

can be used as a dye source. The leaves and young branches produce a reddish-orange dye used in Europe and North America in hair shampoos, dyes, conditioners, and rinses. In Asia, henna dye is used to color hair, skin, nails, teeth, silk, wool, and leather. The dye can also be mixed with indigo for a more complex color range. Extracts are used as wood stain and fabric dye. An essential oil is extracted from the flowers and used in perfumery. Leaves, bark, and seeds are used for certain medicinal purposes in the East. Henna wood is hard and can be used for making small objects.

Plants thrive in areas with temperatures 19-27 degrees Celsius and annual rainfall of 0.2-4.2 m. Henna is native to North Africa and southern Asia and has naturalized in tropical America, Egypt, India, and parts of the Middle East.

Highbush Blueberry, Swamp Blueberry
(See Fruit)

Inca-Wheat, Quihuicha, Quinoa, Love-Lies-Bleeding
(See Cereal)

India Rubber Fig, Rubber-Plant
(See Isoprenoid Resin & Rubber)

Indian-Tobacco

Sci. Name(s)	:	*Lobelia inflata*
Geog. Reg(s)	:	America-North (U.S. and Canada), Subtropical, Tropical

End-Use(s) : Medicinal, Ornamental & Lawn Grass
Domesticated : Y
Ref. Code(s) : 154, 205
Summary : The Indian-tobacco plant is grown for ornamental purposes in many tropical and subtropical areas. Traditionally, it was used by North American Indians to treat asthma and similar conditions. This plant has been administered for treating chronic bronchitis and whooping cough. It can be applied externally to relieve pain caused by rheumatism, bruises, bites, and certain skin conditions. It is fatal in large doses.

Itabo, Izote, Palmita, Ozote, Spanish Bayonnette
(See Vegetable)

Japan Wax
(See Fat & Wax)

Japanese Chestnut
Sci. Name(s) : *Castanea crenata*
Geog. Reg(s) : Asia
End-Use(s) : Nut, Ornamental & Lawn Grass, Timber
Domesticated : Y
Ref. Code(s) : 140, 154
Summary : The Japanese chestnut tree has hard, durable wood valued in construction. These trees are often grown as ornamentals and yield edible nuts similar but inferior to the American chestnut. Japanese chestnuts are eaten boiled or roasted. Trees are native to Japan.

Japanese Lawngrass, Zoysia, Korean Grass
Sci. Name(s) : *Zoysia japonica*
Geog. Reg(s) : America-North (U.S. and Canada), Asia, Asia-China
End-Use(s) : Ornamental & Lawn Grass, Vegetable
Domesticated : Y
Ref. Code(s) : 17, 96, 213, 214
Summary : Japanese lawngrass is a tough, wiry grass that requires minimal mowing. Its seeds and stem bases are edible, with limited commercial value. It is suited to areas with long, hot summers and deficient rainfall and grows well in China, Japan, and the southeastern United States.

Jasmine, Poet's Jessamine
(See Essential Oil)

Java-Almond, Kanari
(See Nut)

Jojoba, Goat-Nut
(See Fat & Wax)

368

Juniper, Common Juniper
(See Essential Oil)

Karanda

Sci. Name(s) : *Carissa carandas*
Geog. Reg(s) : America-North (U.S. and Canada), Asia-India (subcontinent)
End-Use(s) : Fruit, Ornamental & Lawn Grass
Domesticated : Y
Ref. Code(s) : 167, 170, 225
Summary : Karanda is a shrub or tree with edible fruits. In India, this fruit is used for pickles and preserves. Plants are introduced throughout the United States but are rarely cultivated. Shrubs are heavily branched and often used as hedges. Karanda is native to India.

Kentucky Bluegrass, June Grass

Sci. Name(s) : *Poa pratensis*
Geog. Reg(s) : America-North (U.S. and Canada), Europe, Hawaii, Temperate
End-Use(s) : Forage, Pasture & Feed Grain, Ornamental & Lawn Grass
Domesticated : Y
Ref. Code(s) : 17, 20, 154
Summary : Kentucky bluegrass is a perennial with creeping rootstocks, smooth culms that reach up to 62 cm in height, and narrow leaves. It is commonly cultivated for lawns and nutritious pasture in humid and temperate regions with limey soils. Its spreading root system creates a close mat of turf and provides durable lawn or pasture.

The grass needs at least 3 years to established itself. It starts quickly in the spring and is fairly cold-hardy and drought-resistant. For maximum richness and quality, plants are cut for hay prior to or during the first bloom. This European grass is naturalized in the mainland United States and grows at higher altitudes in Hawaii.

Kikuyu Grass
(See Forage, Pasture & Feed Grain)

Kinka Oil Ironweed
(See Medicinal)

Kiwi, Yangtao, Chinese-Gooseberry
(See Fruit)

Kudzu Vine, Kudzu
(See Tuber)

Liberica Coffee, Liberian Coffee
(See Beverage)

Lily Turf, Dwarf Lily Turf, Mondo Grass, Snake's Beard
Sci. Name(s) : *Ophiopogon japonicus, Mondo japonicum, Liriope japonica*
Geog. Reg(s) : Asia-Southeast, Hawaii

End-Use(s)   : Erosion Control, Ornamental & Lawn Grass, Tuber
Domesticated   : Y
Ref. Code(s)   : 119, 154
Summary   : Lily turf is an excellent turf-forming, grass-leaved lily used mainly for edges in gardens and as a substitute grass. Lily turf is common in Hawaii and grows in nearly any hot, dry, sandy area where regular grass is at a disadvantage. This grass is most successful in mountainous areas 700-1,200 m in elevation.

This grass is native to eastern Asia. Its dark green leaves are 10-30 cm long and form erect clusters. Its tuberous roots are eaten after some preparation.

Limeberry
(See Fruit)

Longan
(See Fruit)

Loquat
(See Fruit)

Mandrake, European Mandrake, Mandragora
(See Medicinal)

Manila Grass, Zoysia, Japanese Carpet Grass
Sci. Name(s)   : *Zoysia matrella*
Geog. Reg(s)   : Africa, Asia, Asia-Southeast, Australia, Hawaii, Pacific Islands, Tropical
End-Use(s)   : Erosion Control, Forage, Pasture & Feed Grain, Ornamental & Lawn Grass
Domesticated   : Y
Ref. Code(s)   : 17, 23, 122, 214
Summary   : Manila grass is considered poor fodder because of its tough foliage and low yields. It forms durable, dense mats and is cultivated as a lawn grass or soil binder. This grass is shade-tolerant and grows in open, sandy, saline soils in tropical areas. It is common in the Philippines, Japan, Hawaii, Australia, Asia, and Africa. Manila grass grows up to approximately 400 m in elevation.

Mauritius Hemp
(See Fiber)

Mayflower, Apamate, Roble
(See Timber)

Myrtle, Indian Buchu
(See Spice & Flavoring)

Natal-Palm
(See Fruit)

Negro-Coffee, Coffee Senna, Senna Coffee
(See Beverage)

Opium Poppy
(See Medicinal)

Ouabain, Cream Fruit
(See Medicinal)

Peach, Nectarine
(See Fruit)

Pepper Tree, Brazilian Pepper Tree, California Peppertree
(See Gum & Starch)

Perennial Ryegrass, English Ryegrass
(See Forage, Pasture & Feed Grain)

Periwinkle, Madagascar Periwinkle, Bright-Eyes
(See Medicinal)

Persian Clover, Shaftal, Birdseye Clover, Reversed Clover
(See Forage, Pasture & Feed Grain)

Peru Balsam, Balsam-of-Peru
(See Natural Resin)

Pimentchien, Cherry Capsicum, Aji
(See Spice & Flavoring)

Pomegranate
(See Fruit)

Pomerac, Malay Apple
(See Fruit)

Pot Majoram
(See Spice & Flavoring)

Purslane, Pusley, Wild Portulaca, Akulikuli-Kula

Sci. Name(s)	: *Portulaca oleracea*
Geog. Reg(s)	: Africa-North, America-Central, America-South, Asia-China, Europe, Hawaii, Mediterranean, Pacific Islands
End-Use(s)	: Forage, Pasture & Feed Grain, Ornamental & Lawn Grass, Spice & Flavoring, Vegetable, Weed
Domesticated	: Y
Ref. Code(s)	: 17, 81, 86, 154, 237

Summary   : Purslane is an herb with thick stems containing vitamins B and C used as a flavoring and vegetable in cooking. Plants are collected wild in Mexico, Europe, China, and the Middle East, while cultivated forms exist in England, Holland, and France. Purslane is grown as a garden ornamental in Hawaii. Plants are palatable to livestock and poultry. It is a weed in cultivated areas and lawns. The cultivars most commonly grown as ornamentals are 'Grandiflora' from Brazil and 'Lutea' from the Society Islands.

Quaking Grass

Sci. Name(s)   : *Briza humilis*
Geog. Reg(s)   : America-North (U.S. and Canada), Europe
End-Use(s)   : Ornamental & Lawn Grass
Domesticated   : Y
Ref. Code(s)   : 146, 164
Summary   : Quaking grass is an annual grass occasionally grown as an ornamental. It is native to Europe and introduced to the United States.

Red Fescue
(See Forage, Pasture & Feed Grain)

Red Ironbark
(See Timber)

Red Mulberry

Sci. Name(s)   : *Morus rubra*
Geog. Reg(s)   : America-North (U.S. and Canada)
End-Use(s)   : Forage, Pasture & Feed Grain, Fruit, Ornamental & Lawn Grass, Timber
Domesticated   : Y
Ref. Code(s)   : 17, 65, 220
Summary   : The red mulberry tree (*Morus rubra*) is grown in the southern United States as an ornamental and shade tree because of its height of 21 m. Two varieties of red mulberry, the 'Hicks' and 'Stubbs', bear sweet fruits usually fed to poultry and pigs. The fruit is extremely soft, easily injured and has limited export potential. Trees have light, coarse-grained wood used in parts of North America for shipbuilding, fences and barrelmaking. Trees thrive in rich, lowland soils.

Ringworm Bush
(See Medicinal)

Roman Chamomile, Chamomile
(See Essential Oil)

Rosemary
(See Spice & Flavoring)

Sand Lovegrass
Sci. Name(s)   : *Eragrostis trichodes*

Geog. Reg(s) : America-North (U.S. and Canada)
End-Use(s) : Forage, Pasture & Feed Grain, Ornamental & Lawn Grass
Domesticated : Y
Ref. Code(s) : 34, 212
Summary : Sand lovegrass is an easily cultivated, palatable grass popular as an ornamental in the United States. It thrives in sandy and mixed soils.

Sapucaja, Sapucaia Nut

Sci. Name(s) : *Lecythis minor, Lecythis elliptica*
Geog. Reg(s) : America-Central, America-South, Tropical, West Indies
End-Use(s) : Fruit, Ornamental & Lawn Grass
Domesticated : Y
Ref. Code(s) : 132, 140, 148
Summary : Sapucaja is a tree reaching up to 8 m in height grown for its edible fruit, which is a good source of fat, protein, carbohydrates, and minerals. There is minor cultivation of this tree in Trinidad, Central America, and parts of South America. Sapucaja is grown as an ornamental. Trees favor hot, tropical, lowland conditions and are not frost-tolerant.

Scarlet Poppy, Oriental Poppy
(See Medicinal)

Scarlet Runner Bean

Sci. Name(s) : *Phaseolus coccineus*
Geog. Reg(s) : America-Central, Temperate, Tropical
End-Use(s) : Ornamental & Lawn Grass, Vegetable
Domesticated : Y
Ref. Code(s) : 132, 170
Summary : The scarlet runner bean is a perennial vine cultivated for its pods, seeds, and fleshy tap roots. Pods are cooked, and seeds are usually dried. Young pods and dry seeds are processed as frozen or canned beans. There is potential for development of this plant in limited agricultural areas of the highland tropics. Vines are often planted as ornamentals.

Plants occur wild at altitudes of 1,828 m in the uplands of Central America. They grow well in temperate climates and are found in the tropics at altitudes above 1,000 m. Vines are most successful in slightly acidic, moist, well-drained loams. They tolerate heavy rains but not drought. Green pods and seeds are collected as early as 2.5 months after planting. The average yield for dry seeds is 1 MT/ha.

Sea Island Cotton, American Egyptian Cotton, Extra Long Staple Cotton
(See Fiber)

Sea Rush, Sparto

Sci. Name(s) : *Juncus maritimus*
Geog. Reg(s) : Temperate, Tropical
End-Use(s) : Erosion Control, Fiber, Ornamental & Lawn Grass
Domesticated : Y

Ref. Code(s)   : 154
Summary   : Sea rush, or sparto, grows in damp, temperate areas and in the mountains of the tropics. Its rushes are used to make mats. The plant is used as a ground cover or ornamental. There is little information concerning other uses for this plant.

Sheep Fescue

Sci. Name(s)   : *Festuca ovina*
Geog. Reg(s)   : America-North (U.S. and Canada), Europe
End-Use(s)   : Erosion Control, Forage, Pasture & Feed Grain, Ornamental & Lawn Grass
Domesticated   : Y
Ref. Code(s)   : 96, 212, 214, 220
Summary   : Sheep fescue provides valuable pasture in Europe and North America. The hardy grass grows on poor, well-drained soil and survives most drought conditions. It produces attractive, durable lawns and is used for sand stabilization. Sheep fescue is native to Europe.

Shingle Tree
(See Timber)

Showy Crotalaria
(See Erosion Control)

Siberian Crabapple
(See Fruit)

Sierra Plum, Pacific Plum, Klamath Plum
(See Fruit)

Signal Grass, Palisade Grass
(See Forage, Pasture & Feed Grain)

Silky Oak
(See Timber)

Siris, East Indian Walnut, Koko
(See Timber)

Smilograss
(See Forage, Pasture & Feed Grain)

Southern Red Cedar, Eastern Red-Cedar
(See Essential Oil)

Southernwood, Lad's Love, Southern Wormwood
Sci. Name(s)   : *Artemisia abrotanum*
Geog. Reg(s)   : Asia, Europe
End-Use(s)   : Beverage, Medicinal, Ornamental & Lawn Grass

374

Domesticated : Y
Ref. Code(s) : 194, 220
Summary : Southernwood is an erect, perennial, shrubby garden plant most often grown for ornamental purposes. Its aromatic leaves are made into a stimulating beverage, used as an antiseptic, and used as a detergent. Former uses of the leaves were to control menstruation and to combat intestinal worms.

The aromatic plant grows to approximately 2 m in height. Plants grow in temperate Asia and southern Europe and thrive in good soil with full sun.

Spanish Broom, Weaver's Broom
(See Fiber)

Spanish Sainfoin
(See Forage, Pasture & Feed Grain)

Springer Asparagus, Jessop
Sci. Name(s) : *Asparagus densiflorus cv. Sprengeri*
Geog. Reg(s) : Africa-South, Subtropical, Tropical
End-Use(s) : Erosion Control, Ornamental & Lawn Grass
Domesticated : Y
Ref. Code(s) : 137
Summary : Springer asparagus is a cultivar of asparagus fern (*Asparagus densiflorus*) which is grown as ground cover in the tropics and subtropics. It is also grown as an ornamental indoor plant.

Asparagus ferns require adequate lighting and thorough watering. Soil should not remain waterlogged, but consistently dry soil results in leafless plants. Temperatures can drop as low as 7 degrees Celsius with no plant damage, but best results are from temperatures of 21-27 degrees Celsius. *Asparagus densiflorus* is native to South Africa.

St. Augustine Grass, Buffalo Grass of Australia, Pimento Grass of Jamaica
Sci. Name(s) : *Stenotaphrum secundatum*
Geog. Reg(s) : America-Central, America-North (U.S. and Canada), America-South, Australia, West Indies
End-Use(s) : Forage, Pasture & Feed Grain, Ornamental & Lawn Grass
Domesticated : Y
Ref. Code(s) : 20, 36
Summary : St. Augustine grass is a perennial reaching 30 cm in height, often grown as a lawn grass. It provides important pasture on porous soils. St. Augustine grass is found along the southern coast of the United States and in the West Indies, Mexico, Argentina, and Australia. It is most successful on moist, sandy soils.

Strawberry Clover
(See Forage, Pasture & Feed Grain)

Strawberry Guava, Cattley Guava, Waiawi-'ula'ula, Wild Guava, Purple Guava
(See Fruit)

Sugar Gum
(See Timber)

Sugi
(See Timber)

Sunflower
(See Oil)

Surinam-Cherry, Pitanga Cherry
(See Fruit)

Swamp Cypress
(See Timber)

Sweet Birch, Black Birch, Cherry Birch, Mountain Mahogany
(See Essential Oil)

Sweet Marjoram, Knotted Marjoram
(See Spice & Flavoring)

Switchgrass
(See Forage, Pasture & Feed Grain)

Tall Fescue

Sci. Name(s)	:	*Festuca arundinacea*
Geog. Reg(s)	:	America-North (U.S. and Canada), Asia, Europe, Temperate
End-Use(s)	:	Forage, Pasture & Feed Grain, Ornamental & Lawn Grass
Domesticated	:	Y
Ref. Code(s)	:	96, 183, 212, 238
Summary	:	Tall fescue is a tall grass grown in the northern United States as an ornamental border for roadsides and meadows. It is native to Europe and temperate Asia and widely introduced as a meadow and pasture grass in the northern United States. This grass provides fodder and hay. Yields tend to increase after the 1st crop. Tall fescue thrives in wet meadows.

Tamarind
(See Fruit)

Tarwi, Tarhui, Chocho, Pearl Lupine, Tarin Altramuz, Muti, Ullus
(See Oil)

Teff
(See Cereal)

Teosinte
(See Forage, Pasture & Feed Grain)

Texas Bluegrass

Sci. Name(s)	:	*Poa arachnifera*
Geog. Reg(s)	:	America-North (U.S. and Canada), Hawaii
End-Use(s)	:	Forage, Pasture & Feed Grain, Ornamental & Lawn Grass
Domesticated	:	Y
Ref. Code(s)	:	17, 20, 184
Summary	:	Texas bluegrass (*Poa arachnifera*) has creeping rootstocks and long, slender leaves. It is sometimes cultivated for winter pasture and as a lawn grass in many parts of the United States and grows to a limited extent at higher elevations in Hawaii. This grass has potential as pasture in southern climates. It grows faster and is hardier than Kentucky bluegrass (*Poa pratensis*) and produces 3-4 times as much pasture or hay.

Tolu Balsam, Balsam-of-Tolu
(See Natural Resin)

Trifoliate Orange

Sci. Name(s)	:	*Poncirus trifoliata*
Geog. Reg(s)	:	Asia, Asia-China
End-Use(s)	:	Miscellaneous, Ornamental & Lawn Grass
Domesticated	:	Y
Ref. Code(s)	:	94, 95, 170
Summary	:	The trifoliate orange is a medium-sized shrub grown as an ornamental. It produces an inedible fruit. It is widely cultivated in Japan and introduced elsewhere.

Shrubs are hardy and cold-resistant. In Japan, they are used as rootstocks for the Satsuma orange. Bigeneric hybrids are produced from this plant and other cultivated species of the Citrus and Fortunella genera. The trifoliate orange is native to northern China.

Tropical Carpetgrass, Savanna Grass
(See Forage, Pasture & Feed Grain)

Tuberose
(See Essential Oil)

Turkish Hazel, Turkish Filbert
(See Nut)

Urucu Timbo, Fish Poison
(See Insecticide)

Velvet Bentgrass, Brown Bent-Grass

Sci. Name(s) : *Agrostis canina*
Geog. Reg(s) : America-North (U.S. and Canada), Asia, Europe
End-Use(s) : Forage, Pasture & Feed Grain, Ornamental & Lawn Grass
Domesticated : Y
Ref. Code(s) : 15, 42, 164
Summary : Velvet bentgrass is cultivated as a lawn, golf green, and pasture grass. It provides good pasture in the acidic soils of Britain.

This grass is native to Britain, Europe, and Asia from the Caucasus to the Himalayas and northward. Velvet bentgrass has been introduced to the United States.

Vetiver, Khus-Khus
(See Essential Oil)

Vogel Tephrosia
(See Erosion Control)

Western Elderberry, Blue Elderberry, Blueberry Elder
(See Fruit)

White Tephrosia
(See Erosion Control)

Wild Rice, Southern Wildrice, Indian Rice, Tuscarora Rice
(See Cereal)

Wild Thyme, Creeping Thyme

Sci. Name(s) : *Thymus serpyllum*
Geog. Reg(s) : Asia, Europe
End-Use(s) : Beverage, Essential Oil, Medicinal, Ornamental & Lawn Grass, Spice & Flavoring
Domesticated : Y
Ref. Code(s) : 194, 205
Summary : Wild thyme (*Thymus serpyllum*) is a small, creeping, perennial herb that is most often grown as an ornamental. Its flowers can be dried and have been used medicinally as an antiseptic and expectorant. An oil can be extracted from the leaves that has been used for certain pharmaceutical and cosmetic products. Dried leaves can be brewed as a tea or used as flavoring in some culinary dishes.

Herbs grow best on well-drained sandy soils. They are native to Europe and northwestern Asia. For United States import statistics on thyme as an essential oil and thyme as a spice, see common thyme.

Yellow Flame, Yellow Poinciana, Soga

Sci. Name(s) : *Peltophorum pterocarpum, Peltophorum inerme*
Geog. Reg(s) : Asia-Southeast, Tropical
End-Use(s) : Dye & Tannin, Forage, Pasture & Feed Grain, Medicinal, Ornamental & Lawn Grass, Timber

Domesticated : Y
Ref. Code(s) : 17, 170
Summary : Yellow flame is an ornamental tree, native to Malaysia. It is cultivated in the tropics as an ornamental and shade tree. It is good for lining streets because its roots spread deep and do not buckle the pavement. It is windfirm and is not attacked by boring beetles.

The timber is strong but not durable. It is used for making planks and other miscellaneous items. The bark is used for dyeing cotton and has been used in the Javanese batik industries. It also has astringent properties and has been sold in Javanese drug stores as "Kayu timor," a preparation used for treating dysentery, and in lotions for sprains, muscular aches, and ulcers. Preparations are also used for soothing eye irritations, for gargles, and for tooth powders. Leaves provide fodder.

Yellow Lupine, European Yellow Lupine
(See Forage, Pasture & Feed Grain)

Spice & Flavoring

Aji, Capsicum chinense
Sci. Name(s) : *Capsicum chinense*
Geog. Reg(s) : America-South, West Indies
End-Use(s) : Spice & Flavoring
Domesticated : Y
Ref. Code(s) : 170, 172
Summary : Aji is a plant reaching 45-75 cm in height with pungent fruit similar to tabasco peppers. This fruit is dried and used as a culinary seasoning. Plants are cultivated in northern South America and the West Indies and are most successful at lower altitudes.

American Licorice, Wild Licorice
Sci. Name(s) : *Glycyrrhiza lepidota*
Geog. Reg(s) : America-Central, America-North (U.S. and Canada)
End-Use(s) : Erosion Control, Medicinal, Spice & Flavoring, Sweetener, Weed
Domesticated : Y
Ref. Code(s) : 56
Summary : The American licorice is a wild, perennial herb 6-9 m in height. Its fleshy, sweet roots were used by Native Americans as a spice and masticatory and are now used like licorice (*Glycyrrhiza glabra*) in confectionery and flavoring. It does not have the commercial importance of licorice. Roots have also been used in folk medicines to treat coughs and colds. The plant provides substantial erosion control but can become a noxious weed.
  Plants grow in North America and Mexico. They are common in areas with alluvial and sandy soils, and they displace other weeds.

Ammi
(See Medicinal)

Anise
(See Essential Oil)

Applemint
Sci. Name(s) : *Mentha rotundifolia*
Geog. Reg(s) : Europe
End-Use(s) : Spice & Flavoring
Domesticated : Y
Ref. Code(s) : 205, 220
Summary : Applemint is an aromatic, perennial herb to 40 cm in height, cultivated for its apple-mint scented leaves. The leaves are used as flavoring in cooking and were used in the confection industries. Herbs are native to Europe, where they thrive in damp soils. There is no commercial cultivation of the plant, but it is a common horticultural item.

Argus Pheasant-Tree
(See Fruit)

Avens

Sci. Name(s)	: *Geum urbanum*
Geog. Reg(s)	: Europe
End-Use(s)	: Insecticide, Medicinal, Spice & Flavoring
Domesticated	: Y
Ref. Code(s)	: 205
Summary	: Avens is a perennial herb that grows on rhizomes 3-7 cm long. This rhizome smells like cloves and is often used as a potherb in broths and soups and as a flavoring in ale. The rhizomes may also be hung with clothes to repel moths. The dried rhizomes and fresh flowering plants are used as an astringent or an appetite-stimulating tonic.

The herb is native to Europe. It grows in soils that are damp and rich in nitrogen.

Balsam-Pear, Bitter Cucumber, Bitter Gourd, Balsam Apple
(See Vegetable)

Basil, Sweet Basil
(See Essential Oil)

Bay, Bay-Rum-Tree
(See Essential Oil)

Benoil Tree, Horseradish-Tree
(See Oil)

Bergamot
(See Essential Oil)

Bitter Almond
(See Essential Oil)

Bitter Orange
(See Essential Oil)

Black Cumin

Sci. Name(s)	: *Nigella sativa*
Geog. Reg(s)	: Africa-North, Asia, Europe
End-Use(s)	: Medicinal, Spice & Flavoring
Domesticated	: N
Ref. Code(s)	: 36, 220
Summary	: Black cumin is an annual that is occasionally cultivated for its seeds, which are used for seasoning food and are sometimes mixed with breads. The plant grows from

Central Europe to North Africa and West Asia. At one time, the seeds were used to treat intestinal worms and jaundice.

Black Mustard

Sci. Name(s)	: *Brassica nigra*
Geog. Reg(s)	: Europe
End-Use(s)	: Essential Oil, Oil, Spice & Flavoring
Domesticated	: Y
Ref. Code(s)	: 170, 194
Summary	: The seeds of black mustard (*Brassica nigra*) were the first of the Brassica species to provide table mustard. Table mustard is now composed of processed black mustard seed flour mixed with the less pungent seed flour from white and brown mustard. Black mustard seeds yield 28% of a fixed oil used in medicines and soaps. They also contain about 1% of a volatile oil used greatly diluted as a counterirritant. The oil is valued in the food industry for its distinctive aroma.

Black mustard is an annual plant reaching heights of 2 m. Plants are grown in areas with temperatures of 5-27 degrees Celsius and an annual rainfall of 0.3-4.2 m. Best results are in sandy loams. Plants are seldom grown in the tropics.

Borage
(See Ornamental & Lawn Grass)

Boston Marrow, Pumpkin, Winter Squash, Squash, Marrow
(See Vegetable)

Brown Mustard, Indian Mustard, Mustard Greens
(See Oil)

Bungu, False Sesame
(See Oil)

Burnet, Garden Burnet, Small Burnet
(See Weed)

Burweed
(See Fiber)

Cabbage Rose
(See Essential Oil)

Cacao Blanco, Nicaraguan Cacao, Patashiti, Bacao

Sci. Name(s)	: *Theobroma bicolor*
Geog. Reg(s)	: America-Central, America-South, Tropical
End-Use(s)	: Fruit, Spice & Flavoring
Domesticated	: N
Ref. Code(s)	: 170, 220

Summary : Cacao blanco is a short tree cultivated for its large fruit, which contains an edible pulp. Fruit seeds have a high cacao butter content and are used like cocoa beans. Trees thrive in the rain forests of tropical America from sea level to altitudes approaching 990 m. They range in habitat from Mexico to Brazil.

Calendula, Pot-Marigold
(See Ornamental & Lawn Grass)

Candytuft, Rocket Candytuft
(See Ornamental & Lawn Grass)

Caper

Sci. Name(s) : *Capparis spinosa*
Geog. Reg(s) : Europe, Mediterranean
End-Use(s) : Medicinal, Spice & Flavoring
Domesticated : Y
Ref. Code(s) : 194, 220
Summary : The caper bush is a perennial shrub with stems reaching 1.5 m in height, cultivated for its flower buds, which are pickled and used as a pungent condiment in salads and sauces. Extracts from the plant are effective treatment for enlarged capillaries and dry skin.

There are 2 varieties, one spiny and one smooth. Plants are most successful in areas with temperatures 13-27 degrees Celsius and an annual rainfall of 0.3-2.6 m. They thrive in dry, well-drained soils and are found throughout the Mediterranean and Europe.

Caraway

Sci. Name(s) : *Carum carvi*
Geog. Reg(s) : Africa-North, America-North (U.S. and Canada), Europe, Mediterranean, Temperate
End-Use(s) : Essential Oil, Spice & Flavoring
Domesticated : Y
Ref. Code(s) : 170, 194
Summary : Caraway is an annual or biennial European herb 30-80 cm in height, cultivated in temperate regions for its seeds. The seeds are used to flavor baked goods, cheeses, sausages, and pickles. The fruit contains 3-6% of an essential oil rich in carvone (50-60%), used for flavoring liqueurs and medicines. An inferior oil (caraway chaff oil) from the husks and stalks is used for scenting soaps. Caraway seed is produced in eastern and northern Europe, Egypt, the United States, and Morocco.

Caraway thrives in areas with temperatures 6-19 degrees Celsius and an annual rainfall of 0.4-1.3 m. Plants mature in 15 months and are harvested when seeds turn brown. They are grown in a variety of soils but prefer upland, well-tilled soils. Herbs are susceptible to frost, overcrowding, and mechanical damage. In 1984, the United States imported 5.7 MT of caraway as an essential oil, valued at $227,000.

Cardamom, Cardamon

Sci. Name(s) : *Elettaria cardamomum*

Geog. Reg(s) : Africa-North, Asia-India (subcontinent), Mediterranean
End-Use(s) : Essential Oil, Medicinal, Miscellaneous, Spice & Flavoring
Domesticated : Y
Ref. Code(s) : 55, 171
Summary : The cardamom is a perennial herb with stems reaching 1.8-2.7 m tall and whose fruit provides the cardamom spice of commerce. Southern India is the largest producer of cardamom. The estimated area in 1966 was 70,000 ha, but production has declined.

The dried fruits are used as a spice, masticatory, and in medicine. In the Middle East, the fruit is used for flavoring coffee. In the United States, cardamom is used as an aromatic stimulant, carminative, and flavoring agent. The plant yields a volatile oil used in perfumery, for flavoring liqueurs and bitters, and in preparation of tinctures.

The cardamom is propagated vegetatively or by seeds. Plants begin to bear 3 years after planting. Fruits are picked just before ripening at intervals of 30-40 days. The first crop is usually small. The economic lifespan of the plant is from 10-15 years. The average yield of the dried cardamom fruit is usually 112-280 kg/ha/annum. Trees are most successful if grown at elevations of 914-1,523 m.

Carrot, Queen Anne's Lace
(See Vegetable)

Cassava, Manioc, Tapioca-Plant, Yuca, Mandioca, Guacomole
(See Energy)

Cassia
(See Essential Oil)

Celery
(See Vegetable)

Champac
(See Ornamental & Lawn Grass)

Cherry Laurel, English Laurel
(See Erosion Control)

Chervil
Sci. Name(s) : *Anthriscus cerefolium, Chaerophyllum sativum*
Geog. Reg(s) : America-North (U.S. and Canada), Europe
End-Use(s) : Essential Oil, Medicinal, Spice & Flavoring
Domesticated : Y
Ref. Code(s) : 42, 194, 220
Summary : Chervil is an annual herb whose leaves yield an essential oil used in flavoring. The oil has been used medicinally as a diuretic, expectorant, and stimulant. Fresh leaves have a flavor similar to parsley with an aroma like anise. They are used to complement other herbs in cooking. Chervil is most often used in combination with basil, chives, and tarragon.

Commercial production is in France and the western United States. The herb grows wild in Britain. Plants thrive in areas with temperatures 7-21 degrees Celsius and an annual rainfall of 0.5-1.3 m. Soils should be well-drained, and plants must have cool, shady or partially shaded surroundings. Chervil leaves are harvested 8-12 weeks after seeding.

Chia, Ghia
(See Essential Oil)

Chili Pepper, Tabasco, Bird Pepper, Tabasco Pepper

Sci. Name(s)	: *Capsicum frutescens*
Geog. Reg(s)	: Subtropical, Tropical
End-Use(s)	: Medicinal, Spice & Flavoring
Domesticated	: Y
Ref. Code(s)	: 170, 194
Summary	: The chili or tabasco pepper plant (*Capsicum frutescens*) produces a fruit that is dried and powdered to produce a pungent red or cayenne pepper used in seasonings. It is used in sauces and occasionally in beverages. Dried fruit is used as an internal stimulant and external counterirritant.

Plants are short-lived, shrubby perennials growing to 2 m in height. They grow in tropical and subtropical areas from sea level to 1,830 m or more. Best results are obtained from areas with temperatures of 7-29 degrees Celsius with an average annual rainfall of 0.3-4.6 m. Plants thrive in well-drained, sandy or silt loam soils. They are sensitive to cold.

Chinese Chives, Chinese Onion, Oriental Garlic
(See Vegetable)

Chives
(See Vegetable)

Cimaru, Tonka Bean

Sci. Name(s)	: *Dipteryx odorata*
Geog. Reg(s)	: America-South, West Indies
End-Use(s)	: Essential Oil, Gum & Starch, Spice & Flavoring
Domesticated	: Y
Ref. Code(s)	: 56, 170
Summary	: Cimaru is a wild tree about 25-30 m tall that is native to tropical America. Its cured beans are a commercial source of the flavoring coumarin. Coumarin has a characteristic flavor and odor and is used commercially to flavor and scent tobacco. It is also used in perfumery, confectionery, liqueurs, and soaps. A gum, kino, is exuded from incised bark.

In Venezuela, Colombia, and Brazil, beans are collected mainly from wild trees, but some cultivation work has been done in Venezuela, Trinidad, and Dominica.

Cimaru grows in shady, damp places on fertile, humus-rich soils. With little or no attention, trees bear beans in 7-10 years. If carefully cultivated, trees produce in their 2nd year. Dried seed yields are 0.45-0.91 kg/tree.

Cinnamon

Sci. Name(s)	: *Cinnamomum verum, Cinnamomum zeylanicum*
Geog. Reg(s)	: Africa-South, Asia-India (subcontinent)
End-Use(s)	: Essential Oil, Spice & Flavoring
Domesticated	: Y
Ref. Code(s)	: 172, 225
Summary	: Cinnamon (*Cinnamomum verum*) is the economic species of Cinnamomum.

Its synonym is *Cinnamomum zeylanicum*. Cinnamon is cultivated for its bark, used as flavoring, for a distilled essential oil, and for extracted oleoresins, also used as flavoring. Cinnamon leaves yield a cinnamon leaf oil used in flavoring and perfumery. Cinnamon oil is a source of eugenol.

The bark or quills are exported primarily to the United Kingdom for use in food processing industries or for oil or oleoresin production. Rough, unscraped bark or Seychelles cinnamon is mixed with other spices and has a low essential oil content. Cinnamon spice is used in baked goods, candy, table sauces, and pickles.

Cinnamon leaf oil is exported from Sri Lanka, the Seychelles, and the Malagasay Republic. It is used in perfumes and flavorings and in the synthesis of vanillin. In 1984, the United States imported 33 MT of cinnamon as an essential oil, valued at $252,000, and 2790 MT of cinnamon ground or unground, valued at $4.9 million. The United Kingdom is the largest importer of the oleoresins.

Cinnamon trees are native to Sri Lanka. They reach heights of 9 m and are most successful in areas with warm, wet climates and temperatures of 27 degrees Celsius. Soil should be well-drained and receive an annual rainfall of 2,000-2,540 mm. Trees produce well below 500 m in altitude.

Citron
(See Fruit)

Clary, Clary Wort, Clary Sage
(See Essential Oil)

Clove

Sci. Name(s)	: *Syzygium aromaticum, Eugenia caryophyllus, Caryophyllus aromaticus, Eugenia aromatica, Eugenia caryophyllata*
Geog. Reg(s)	: Africa-West, Asia-Southeast, Tropical
End-Use(s)	: Essential Oil, Medicinal, Spice & Flavoring
Domesticated	: Y
Ref. Code(s)	: 170
Summary	: Clove trees are evergreen trees 12-15 m in height whose unopened, dried

flower buds are used as an important commercial spice. In the East, cloves are used as a table spice, in the making of curry powder, and to flavor betel quids. In the West, cloves are an ingredient in mixed spices and are used in sauces, meat seasonings, and as flavoring for certain pastries. Medicinally, cloves serve as a stimulant, antispasmodic, and carminative.

The clove can be mixed with tobacco to make clove cigarettes. The dried fruits, called mother of cloves, can also be used as a spice. Clove oil is from the cloves, stems, and leaves and is used as a flavoring agent and in perfumes, soaps, bath salts, medicine, and dentistry. It can also be used as a clearing agent in microscopy.

Clove trees are indigenous to the Spice Islands of the Moluccas. They do well in the maritime climates of the lowland tropics and grow best in deep, well-drained sandy, acid loams. Trees flower at 6 years and reach full bearing in about 20 years. Healthy trees may produce up to 100 years. In 1984, the United States imported 814 MT of clove as an essential oil, valued at $3.7 million.

Cocoa, Cacao
(See Beverage)

Common Lavender, English Lavender
(See Essential Oil)

Common Licorice

Sci. Name(s)	: *Glycyrrhiza glabra*
Geog. Reg(s)	: Africa-North, Asia, Asia-China, Eurasia, Europe, Mediterranean, Temperate
End-Use(s)	: Forage, Pasture & Feed Grain, Miscellaneous, Spice & Flavoring, Sweetener
Domesticated	: Y
Ref. Code(s)	: 56, 194
Summary	: The common licorice is a perennial herb 1-1.5 m in height whose dried rhizomes and roots are used to flavor candy and tobacco. The roots contain glycyrrhizin, which is 50 times sweeter than cane sugar. The tobacco industry is the largest single user of an ammoniated form--glycyrrhizic acid. It is used to enhance the flavor of tobacco, chocolate, and maple. The manufactured excess is used as fire extinguishing agents, insulation for fiberboards, or compost for mushrooms. Licorice paste, extracts, and mafeo syrup are other by-products. Plants are sometimes fed to livestock. Licorice roots are often chewed with betel quids in India.

Licorice is native to Eurasia, North Africa, and western Asia. Plants are grown in temperate regions. The oriental variety is grown in the Near East, Syria, Iraq, Russia, Iran, and China. The Spanish variety grows in Spain, Italy, and Greece. Licorice plantations have shown economic potential. Plants are most successful in deep, sandy soils in warm areas.

Plants are harvested in 3-5 years. Harvest continues once every 3-5 years thereafter. About 2,000-4,000 kg/ha of unprocessed roots are produced each year. In 1984, the United States imported 17,196 MT of licorice root and extract, valued at $2.15 million.

Common Rue, Garden Rue
(See Medicinal)

Common Thyme

Sci. Name(s)	: *Thymus vulgaris*
Geog. Reg(s)	: America-North (U.S. and Canada), Europe, Mediterranean
End-Use(s)	: Essential Oil, Medicinal, Miscellaneous, Ornamental & Lawn Grass, Spice & Flavoring, Sweetener
Domesticated	: Y

Ref. Code(s) : 170, 194
Summary : Thyme (*Thymus vulgaris*) is a small, perennial herb cultivated as a commercial flavoring. Leaves and flower tops are used fresh or dried as seasoning in stuffings and other culinary preparations. An essential oil from the leaves is used in perfumery. A derivative of the oil, thymol, has antiseptic properties and is used as flavoring in toothpastes and pharmaceutical products. Herbs are grown as ornamentals. Flowers attract bees and are the source of thyme honey.

Plants are cultivated in France, Spain, Portugal, Greece, and the western United States. They are native to the Mediterranean. Plants grow well in areas with temperatures 7-25 degrees Celsius, an annual rainfall of 0.4-2.8 m and on well-drained, dry, calcareous soils. Harvest occurs when plants are in bloom. They should be renewed every 3-4 years. In 1984, the United States imported 722 MT of crude and manufactured thyme, valued at $1.2 million, and 8.5 MT of thyme oil as an essential oil, valued at $189,000.

Coriander

Sci. Name(s) : *Coriandrum sativum*
Geog. Reg(s) : Africa-North, America-Central, America-North (U.S. and Canada), America-South, Asia-China, Asia-India (subcontinent), Eurasia, Mediterranean
End-Use(s) : Essential Oil, Medicinal, Miscellaneous, Spice & Flavoring
Domesticated : Y
Ref. Code(s) : 154, 170, 180, 194
Summary : Coriander is an annual herb cultivated for its fruit, which is dried and used in curry powders or pickled when immature. Fruit contains 0.5-1% of an essential oil used as flavoring in certain condiments, bakery products, and alcoholic beverages. It is also used to improve the flavor of medicinal preparations. In traditional medicine, coriander is an antispasmodic, carminative, stimulant, and stomachic. Fresh leaves are used to flavor soups and stews. Its flowers attract bees.

The herb reaches heights of 1 m and requires an annual rainfall of 0.3-2.6 m for best results. It thrives on deep, fertile loam soils.

Coriander is cultivated in North Africa, Central and South America, China, North America, the Middle East, India, Indonesia, and the USSR. Plants may fail to set seed in lowland tropics.

Corn Mint, Japanese Mint, Field Mint
(See Essential Oil)

Costmary, Mint Geranium, Alecost, Bible-Leaf Mace

Sci. Name(s) : *Chrysanthemum balsamita, Pyrethrum balsamita*
Geog. Reg(s) : America-North (U.S. and Canada), Asia, Europe
End-Use(s) : Essential Oil, Medicinal, Spice & Flavoring
Domesticated : Y
Ref. Code(s) : 205
Summary : Costmary is an herb reaching 1 m in height, cultivated for its scented foliage. Its leaves are used to flavor salads, soups, cakes, poultry dishes, home-made beers, and ales. They contain a volatile oil used in stomachics and soothing ointments for burns and stings but seldom in traditional medicines. Costmary is native to western

Asia and naturalized in North America and Europe. This herb is considered wild and grows on most soils in full sun. It does not flower in the shade.

Cowslip, Marsh-Marigold
(See Vegetable)

Cumin

Sci. Name(s) : *Cuminum cyminum*
Geog. Reg(s) : Africa-North, Asia-India (subcontinent), Europe, Mediterranean
End-Use(s) : Essential Oil, Spice & Flavoring
Domesticated : Y
Ref. Code(s) : 170, 194
Summary : Cumin is an annual herb 25 cm in height, cultivated in Europe, the Middle East, and North Africa for its fruit, called cumin seeds. These seeds are valued as condiments and for an essential oil used in perfumery and as a culinary flavoring.

Plants are most successful in areas with temperatures of 9-26 degrees Celsius and an annual rainfall of 0.3-2.7 m. They require mild temperatures during their growing season and do not survive long, dry, heat spells. Cumin bears fruit 60-90 days after sowing. It is native to the Mediterranean.

Curry-Leaf-Tree

Sci. Name(s) : *Murraya koenigii*
Geog. Reg(s) : Asia-India (subcontinent)
End-Use(s) : Essential Oil, Medicinal, Spice & Flavoring, Timber
Domesticated : Y
Ref. Code(s) : 36, 55, 170, 208, 220
Summary : Curry-leaf-tree is grown in India for its leaves, which are dried and used as flavoring in curries and chutneys. The bark, leaves, and roots are made into tonics used in local medicines. An essential oil (limbolee oil) is extracted from the seeds. Curry-leaf-tree wood is hard and durable and used for various domestic tools and carved items. Trees are grown in gardens and the leaves collected for sale in domestic markets.

Cut-Egg Plant, Mock Tomato
(See Vegetable)

Damask Rose
(See Essential Oil)

Dandelion, Common Dandelion
(See Beverage)

Dill

Sci. Name(s) : *Anethum graveolens*
Geog. Reg(s) : America-Central, America-North (U.S. and Canada), Asia-India (subcontinent), Eurasia, Europe, Mediterranean, Pacific Islands, Tropical
End-Use(s) : Essential Oil, Medicinal, Spice & Flavoring

Domesticated : Y
Ref. Code(s) : 170, 194
Summary : Dill is an annual or biennial Eurasian herb up to 1 m tall whose aromatic seeds are used extensively for culinary flavoring.

In the United States, dill seeds and dried leaves are used ground or whole as flavoring in cooking. The seeds are used as substitutes for caraway seeds. They are also made into dill water for medicinal purposes. Dill water is especially useful for reducing problems of flatulence in babies. Dill seeds yield about 3% of an essential oil, used in cosmetics and perfume industries. In 1983, 597 MT of dill seed were imported, with a value of $533,000.

The herb grows in Europe, the United States, India, and to a limited extent in the tropics. Dill is produced commercially in India and Pakistan, with less production in Egypt, Fiji, Mexico, the United States, and Europe.

Dill thrives in areas with temperatures of 6-26 degrees Celsius with an annual rainfall of 0.5-1.7 m. Plants are hardy but need long, cool days and deep, fertile loams for best results.

Dyer's-Greenwood, Dyer's-Greenweed
(See Dye & Tannin)

East Indian Lemongrass, Malabar Grass
(See Essential Oil)

Eastern Elderberry, American Elderberry, American Elder
(See Fruit)

Elephant Garlic, Great-Headed Garlic, Kurrat, Leek
Sci. Name(s) : *Allium ampeloprasum*
Geog. Reg(s) : America-North (U.S. and Canada), Asia-Southeast, Europe, Mediterranean
End-Use(s) : Spice & Flavoring, Vegetable
Domesticated : Y
Ref. Code(s) : 171, 194
Summary : Elephant garlic is a home-garden plant grown for its bulbs used as flavoring. The plant resembles garlic but is distinguishable by its size and lack of bulbils. Elephant garlic is grown in the Mediterranean, Europe and the United States and is native to Southeast Asia.

European Pennyroyal
(See Essential Oil)

Fennel, Florence Fennel, Finocchio
Sci. Name(s) : *Foeniculum vulgare*
Geog. Reg(s) : America-South, Asia-China, Asia-India (subcontinent), Asia-Southeast, Mediterranean, Temperate, Tropical
End-Use(s) : Essential Oil, Medicinal, Ornamental & Lawn Grass, Spice & Flavoring, Vegetable
Domesticated : Y

Ref. Code(s) : 170, 194, 205
Summary : Fennel is an aromatic herb whose leaves are used as garnish or potherbs. Leaf stalks are eaten in salads or blanched as cooked vegetables. The fruit contains about 60% of a high-quality essential oil used in flavorings. Seeds are used as an antispasmodic. There are several forms of fennel, which differ in leaf color and morphology and are grown as ornamentals and fresh vegetables.

Plants are native to the Mediterranean, are naturalized in most temperate countries, and have been introduced into several tropical countries. They thrive on well-drained, loamy soil. Their 1st year yield is low, and it takes several harvests to reach maximum yield.

Fenugreek

Sci. Name(s) : *Trigonella foenum-Graecum*
Geog. Reg(s) : Africa-North, Asia, Asia-India (subcontinent), Europe
End-Use(s) : Energy, Essential Oil, Forage, Pasture & Feed Grain, Gum & Starch, Medicinal, Oil, Spice & Flavoring
Domesticated : Y
Ref. Code(s) : 56, 170
Summary : Fenugreek is an annual herb reaching 60 cm in height cultivated for its edible seeds, important as condiments and as flavoring in artificial maple syrup, cheese, and curries. About 5% of a celery-scented oil is extracted from the seeds and used to flavor butterscotch, cheese, licorice, pickles, rum, syrup, and vanilla. This oil has potential in the perfume and cosmetic industries. Seeds contain 0.4-1.2% drug diospenin, used in the synthesis of hormones. Seed husks are used as an industrial source of mucilage, and the residue is made into oil, sapogenin, and protein-rich fractions. Organic residue is used for biomass fuel or manures, while inorganic residue is used for "inorganic" fertilizers. In India, plants are grown for forage.

Herbs are native to southern Europe and Asia. They thrive in temperate areas with low to moderate rainfall on deep, well-drained, loamy soils. Seeds ripen 3-5 months after planting. In England, experimental plantings showed seed yields of 3,700 kg/ha; in Morocco, seed yields are recorded at 1,000 kg/ha; and in India, 9-10 MT/ha of green forage is produced.

Fitweed, False Coriander

Sci. Name(s) : *Eryngium foetidum*
Geog. Reg(s) : Asia-China, Asia-India (subcontinent), Asia-Southeast, Europe
End-Use(s) : Essential Oil, Medicinal, Spice & Flavoring
Domesticated : Y
Ref. Code(s) : 36, 220
Summary : Fitweed is an herb with roots ground as a substitute for coriander. This herb is cultivated in Europe, China, India, and the Malaysian Pennisula. It contains a volatile oil with medicinal qualities that have not been thoroughly investigated.

Galanga, Kent Joer
(See Ornamental & Lawn Grass)

Garden Myrrh, Sweet Scented Myrrh
(See Medicinal)

Garlic

Sci. Name(s)	: *Allium sativum*
Geog. Reg(s)	: America-North (U.S. and Canada), America-South, Asia, Asia-India (subcontinent), Europe, Mediterranean, Worldwide
End-Use(s)	: Essential Oil, Medicinal, Spice & Flavoring, Vegetable
Domesticated	: Y
Ref. Code(s)	: 171, 194
Summary	: Garlic is a hardy perennial member of the onion group (Allium) whose cloves are 2nd only to bulb onions in use. The cloves are used for flavoring foods, and fresh cloves are sold as market items but are less popular for industrial and domestic use. Dehydrated garlic in powdered or granulated form is replacing fresh cloves. Oil and oleoresins contain alliin, which provides the characteristic flavor and smell of garlic. The oil and oleoresin are used as culinary flavoring. Broken-down alliin enzymes produce allicin, which has antibacterial properties. Garlic has a wide range of medicinal uses in both traditional and folk medicines.

Crops are grown worldwide but most extensively in countries bordering the Mediterranean, and specifically in India, Egypt, France, Mexico, and Brazil. Garlic is cultivated commercially in the western United States, Egypt, Italy, Bulgaria, Hungary, and Taiwan. Plants may fail to produce bulbs in areas with cool temperatures and short days. In the tropics, best results are at higher altitudes. Plants thrive on irrigated, fertile, well-drained clay loams. Bulbs mature 4-6 months after planting. Clove yields are 4,500-11,000 kg/ha. In 1983, the United States produced 84,000 MT of garlic from an area of 6,000 ha. In addition, the United States imported 12,638 MT, valued at $10.5 million.

Giant Granadilla, Barbardine
(See Fruit)

Gilo
(See Fruit)

Ginger

Sci. Name(s)	: *Zingiber officinale*
Geog. Reg(s)	: Africa-West, Asia-India (subcontinent), Asia-Southeast, Tropical, West Indies
End-Use(s)	: Beverage, Essential Oil, Medicinal, Spice & Flavoring, Vegetable
Domesticated	: Y
Ref. Code(s)	: 17, 171, 220
Summary	: Ginger is an herb whose underground stem is an important spice known for its flavor, pungency, and aroma. Ginger is cultivated in tropical Asia and most other tropical areas from sea level to 1,500 m. It continues to thrive under adverse environmental conditions but requires fertile soil, partial shade, and warm temperatures. The highest quality ginger is produced in the central mountains of Jamaica. Other ginger exporting countries are India and Nigeria, and principal importers are the

United Kingdom, South Yemen, and the United States. Yields vary, but with improved cultural practices, 20-30 MT/ha of ginger is possible.

Herbs are used for culinary purposes, especially in Chinese cooking, and for ginger beer, ginger ale, and ginger wine. Preserved and crystallized ginger is also popular, and an essential oil is used in flavoring essences and perfumery. Medicinally, ginger relieves upset stomachs. Green ginger is produced mainly in China.

Goat Chili, Rocoto

Sci. Name(s)	: *Capsicum pubescens*
Geog. Reg(s)	: America-Central, America-North (U.S. and Canada), America-South
End-Use(s)	: Spice & Flavoring
Domesticated	: Y
Ref. Code(s)	: 170, 172
Summary	: Goat chili is a pepper plant grown for its pungent leaves and fruit dried and used in seasonings. These peppers are grown at 1,500-3,300 m in elevation in the Andes of Colombia, Ecuador, Bolivia and Peru. They have been introduced to the United States but have little horticultural value. Goat chili cannot be crossed with any other member of the Capsicum species.

Greater Galanga

Sci. Name(s)	: *Alpinia galanga*
Geog. Reg(s)	: Asia-Southeast
End-Use(s)	: Spice & Flavoring
Domesticated	: Y
Ref. Code(s)	: 92, 171
Summary	: Greater galanga is an herb reaching about 2 m in height, with rhizomes that provide a mild spice and flavoring in various foods. These rhizomes are popular in Malaysia and Indonesia as a spice in curries and as flavoring in medicines. Plants are cultivated extensively throughout Southeast Asia and Indonesia. They thrive in rich, well-cultivated soil in partial shade.

Guarana
(See Beverage)

Holy Basil
(See Medicinal)

Horsegram
(See Cereal)

Horseradish

Sci. Name(s)	: *Armoracia rusticana*
Geog. Reg(s)	: America-North (U.S. and Canada), Europe, Tropical
End-Use(s)	: Spice & Flavoring
Domesticated	: Y
Ref. Code(s)	: 170, 194

Summary : Horseradish is an herb growing to 1 m in height whose roots are grated as a pungent condiment for meat dishes. It is produced primarily in the United States and Europe.

This herb is native to southeastern Europe and is occasionally grown in the higher-altitude tropics. For optimum growth, plants need temperatures of 5-19 degrees Celsius with an annual rainfall of 0.5-1.7 m and deep, moist, rich soils in semishade.

Huon-Pine
(See Timber)

Hyssop
(See Essential Oil)

Icecream Bean
(See Fruit)

Inca-Wheat, Quihuicha, Quinoa, Love-Lies-Bleeding
(See Cereal)

Indian Borage
Sci. Name(s) : *Coleus amboinicus*
Geog. Reg(s) : Asia-Southeast, West Indies
End-Use(s) : Medicinal, Spice & Flavoring
Domesticated : Y
Ref. Code(s) : 170, 220
Summary : Indian borage is a small, succulent herb 50 cm in height, grown in Southeast Asia and the West Indies for its aromatic leaves. The leaves are used to season stuffings and meats and as a substitute for sage seasoning and true borage. In Malaysia, preparations of the leaves have been used for washing clothes and hair. Medicinally, the leaves are made into poultices that soothe scorpion bites, headaches, and other maladies. Indian borage is native to Indonesia.

Indian Long Pepper
Sci. Name(s) : *Piper longum*
Geog. Reg(s) : Asia-India (subcontinent)
End-Use(s) : Medicinal, Spice & Flavoring
Domesticated : Y
Ref. Code(s) : 170, 172
Summary : Indian long pepper plants are slender climbers whose wild habitat ranges from the foothills of the central Himalayas to the hills of southern India. Plants are grown in India and Sri Lanka for their spikes, which are dried and used as a spice. Pepper spikes are picked just prior to ripening. The spice is used in pickles, preserves, and curries. The dried roots and thickened stem parts have been used in certain Indian medicines. Indian long pepper plants do not enter international trade.

Indian-Bark, Indian Cassia

Sci. Name(s) : *Cinnamomum tamala*
Geog. Reg(s) : Asia-India (subcontinent)
End-Use(s) : Medicinal, Spice & Flavoring
Domesticated : Y
Ref. Code(s) : 36, 67, 170
Summary : Indian-bark is an economically important tree of northern India cultivated for its bark and leaves. The bark is prepared like cinnamon, while the leaves have commercial value as a spice for flavoring curries. The leaves have unspecified medicinal uses.

   Trees are moderate-sized evergreens, but their wood has no commercial value as timber.

Jasmine, Poet's Jessamine
(See Essential Oil)

Javanese Long Pepper

Sci. Name(s) : *Piper retrofractum, Piper officinarum*
Geog. Reg(s) : Asia-Southeast
End-Use(s) : Medicinal, Spice & Flavoring
Domesticated : Y
Ref. Code(s) : 170, 172
Summary : The Javanese long pepper (*Piper retrofractum*) is a climbing vine that resembles the Indian long pepper (*Piper longum*) and grows wild throughout Malaysia. It is grown for a spice used for seasoning pickles, preserves, and curries. This spice may have undocumented medicinal uses.

Juniper, Common Juniper
(See Essential Oil)

Kenaf, Bimli, Bimlipatum Jute, Deccan Hemp
(See Fiber)

Lemon
(See Fruit)

Lesser Galanga

Sci. Name(s) : *Alpinia officinarum, Languas officinarum*
Geog. Reg(s) : Asia-China, Asia-Southeast
End-Use(s) : Medicinal, Spice & Flavoring
Domesticated : Y
Ref. Code(s) : 171
Summary : The rhizomes of lesser galanga provide a pungent spice popular in Malaysia and Indonesia as food flavoring and for medicinal purposes. The fruit is sometimes used as a substitute for cardamom. The herb is usually less than 1 m high and is native to China.

Lime
(See Fruit)

Lovage, Common Lovage

Sci. Name(s) : *Levisticum officinale*
Geog. Reg(s) : Europe
End-Use(s) : Essential Oil, Medicinal, Spice & Flavoring, Vegetable
Domesticated : Y
Ref. Code(s) : 61, 220
Summary : Lovage is an herb of southern European origin. Its blanched leaves are eaten as vegetables, and its fruit is used as flavoring for foods and confectionery. The roots and other parts of the plant have a celerylike flavor and are sometimes added to salads. They have been used as a stimulant, digestive aid, stomachic, diuretic, and menstrual control. An essential oil (oil of lovage) is also distilled from them.

Plants thrive on deep, fertile, moist soils in sunny locations. They are perennials and will last for several years, once established.

Mango
(See Fruit)

Maritime Wormwood, Sea Wormwood
(See Medicinal)

Mastic
(See Natural Resin)

Melegueta-Pepper, Grains of Paradise, Guinea Grains

Sci. Name(s) : *Aframomum melegueta*
Geog. Reg(s) : Africa-West
End-Use(s) : Medicinal, Spice & Flavoring
Domesticated : Y
Ref. Code(s) : 171
Summary : The melegueta-pepper is a West African plant 1 m tall whose cardamom-flavored seeds are used as a spice and carminative. They were used in Britain to spice wine and strengthen beer. In West Africa, the seeds are used in local medicines and for flavoring food. The fruit pulp is chewed as a refreshing stimulant.

Mugwort

Sci. Name(s) : *Artemisia vulgaris*
Geog. Reg(s) : Asia, Temperate
End-Use(s) : Spice & Flavoring, Vegetable
Domesticated : Y
Ref. Code(s) : 15, 220
Summary : Mugwort is a perennial, aromatic herb with leaves and shoots used as flavoring or tobacco substitutes. In Japan, young leaves are boiled and eaten as vegetables. Mugwort grows in most temperate areas of the northern hemisphere.

Myrtle, Indian Buchu

Sci. Name(s) : *Myrtus communis*
Geog. Reg(s) : America-North (U.S. and Canada), Asia, Europe, Mediterranean
End-Use(s) : Essential Oil, Ornamental & Lawn Grass, Spice & Flavoring
Domesticated : Y
Ref. Code(s) : 56, 154, 205
Summary : Myrtle is a strong-scented evergreen shrub 1-3 m in height whose berries and flower buds are dried and used to flavor wine and food. Distilled flowers and leaves yield an essential oil used in perfume. This oil was once used in the making of toilet water (eau d'ange). Myrtle is now grown primarily for ornamental purposes.

Shrubs are native to the Mediterranean and western Asia and are cultivated to a limited extent in Great Britain and the southern United States. Shrubs are most productive in full sunlight in well-drained, fairly fertile soil. Their growth is affected in areas with cooler climates. They are usually wild but make good garden plants. The flowers and leaves are harvested as needed, and there is no yield data available.

Nardus Grass, Citronella Grass
(See Essential Oil)

Native Eggplant
(See Vegetable)

Nutmeg, Mace

Sci. Name(s) : *Myristica fragrans*
Geog. Reg(s) : Asia-Southeast, Tropical, West Indies
End-Use(s) : Essential Oil, Medicinal, Spice & Flavoring
Domesticated : Y
Ref. Code(s) : 148, 170
Summary : *Myristica fragrans* is a tree 10-18 m in height that produces 2 commercially important spices--nutmeg from the dried seed and mace from the dried aril. The tree is native to the Moluccas. Today, it is grown primarily in Indonesia and Grenada.

As a commercial product, nutmeg spice is grated and marketed as a flavoring for milk dishes, cakes, punch, and possets. Mace is manufactured as flavoring for savory dishes, ketchup, and pickles. Local uses of the spices vary. In Malaysia, sweet meats and jellies are made from dried seed husks.

By-products of the broken seeds are nutmeg butter and mace, used in ointments and perfumery. Distillation of nutmeg oil produces a volatile essential oil also used in perfumery and as external medicine. The butter and oil contain myristicin, a poisonous narcotic, 4-5 g of which will produce signs of poisoning in humans. In 1984, the United States imported 106 MT of nutmeg oil as essential oil, valued at $1,926,000. For medicinal purposes, nutmeg is the more important of the 2 spices, having stimulative, carminative, astringent, and aphrodisiac properties.

Trees grow well in hot, humid tropical areas in light soils. They reach a fruiting peak at 15-20 years and continue to bear for at least 30-40 more years. Trees usually fruit twice a year, each tree producing 1,500-2,000 fruits/year. Yields vary from 226.5-453 kg/year of nutmeg and up to 90.6 kg/year of mace.

Oil-Bean Tree, Owala Oil
(See Oil)

Okra, Lady's Finger, Gumbo
(See Vegetable)

Onion, Common Onion
(See Vegetable)

Opium Poppy
(See Medicinal)

Oregano, Pot Marjoram, Wild Marjoram

Sci. Name(s)	: *Origanum vulgare, Origanum hirtum, Origanum virens*
Geog. Reg(s)	: America-North (U.S. and Canada), Europe
End-Use(s)	: Essential Oil, Spice & Flavoring
Domesticated	: Y
Ref. Code(s)	: 17, 194, 205
Summary	: Oregano is a hardy, sweet, perennial herb whose leaves are collected as a commercial flavoring. They are dried for flavoring or distilled for an essential oil used to flavor meats and stuffings. The oil and dried leaves are used in cosmetics. Most of the commercial products are obtained from wild plants.

The herb is native to Europe and has naturalized in North America. Oregano produces the superior spice compared to Spanish oregano, which also produces the spice. It thrives in dry, rocky, calcareous soils but is easily grown in most warm garden soils. Leaves are harvested 2-6 times a year. Mature plants reach heights of 1 m. In 1984, the United States imported 4,350 MT of origanum (oregano) in crude and manufactured form, valued at $5.6 million.

Orris Root, Florentine Iris
(See Essential Oil)

Padang-Cassia

Sci. Name(s)	: *Cinnamomum burmannii*
Geog. Reg(s)	: Asia-Southeast
End-Use(s)	: Essential Oil, Spice & Flavoring
Domesticated	: Y
Ref. Code(s)	: 36, 172
Summary	: Padang-cassia (*Cinnamomum burmannii*) is an Indonesian tree whose bark and leaves yield essential oils. Cassia oil is used in flavoring and perfumery. Dried bark provides a spice of commerce known by many names, among them Indonesian cassia and cassia vera. Extracted oleoresins are popular in the flavor industries of western Europe and North America. The United States is the largest importer of cassia, and there is a considerable market for it in the East.

Trees are cultivated in western Sumatra, Java, and east to Timor at elevations from sea level to 2,000 m. They are most successful on light, rich, sandy loams. Bark and leaf harvest begins in about 5 years. Subsequent cuttings are taken for about 15 more years. For 1984 United States import statistics on cassia as an essential oil, see cassia.

Paroquet Bur, Pulut-Pulut
(See Fiber)

Parsley
Sci. Name(s) : *Petroselinum crispum, Petroselinum hortense, Petroselinum sativum*
Geog. Reg(s) : America-North (U.S. and Canada), Europe, Temperate, Tropical
End-Use(s) : Essential Oil, Spice & Flavoring
Domesticated : Y
Ref. Code(s) : 170, 194
Summary : Parsley is an erect, biennial herb whose leaves and seeds are used as a garnish and flavoring. The leaves are an excellent source of vitamin C. Leaves and seeds yield volatile oils that contain apiol. Parsley seed oil is used as a fragrance in perfumes, soaps, and creams.

Herbs are cultivated in the western United States and in France, Hungary, and several other European countries. They are most successful on rich, moist, well-drained soils. Parsley is native to southern Europe. It has naturalized in most temperate countries and is fairly successful in the tropics, although it dies out near the equator.

Parsnip, Wild Parsnip
(See Tuber)

Passion Fruit, Purple Granadilla
(See Beverage)

Patchouli
(See Essential Oil)

Pawpaw
(See Fruit)

Pepper, Black Pepper
Sci. Name(s) : *Piper nigrum*
Geog. Reg(s) : Africa, America-South, Asia-India (subcontinent), Tropical
End-Use(s) : Essential Oil, Medicinal, Spice & Flavoring
Domesticated : Y
Ref. Code(s) : 148, 170
Summary : The pepper plant is a climbing, woody, perennial vine that provides one of the most important and widely used spices. The dried fruit spikes produce black pepper, and white pepper is from fruit that has been thoroughly soaked and the mesocarp removed. Pepper has extensive culinary uses, serving as a stimulative on the digestive organs, producing an increased flow of saliva and gastric juices. White pepper has higher market prices than does black pepper. An essential oil is distilled from the fruit and used in perfumery.

Commercial pepper plants are often grown as 2nd crops in stands of tree crops or as climbers on specially planted shade trees. Best results are from vines grown on posts under adequate care. Properly cultivated vines yield 4.5-9.1 kg/vine of spikes. First harvest is made 2.5-3 years after planting, and production continues for 12-15 more years.

Annual world exports are 45,350-68,025 MT. India is the largest exporter. About a third of the world's pepper supply is imported by the United States. Pepper consumption in the United States is approximately 100 g/caput/annum. Brazil and the Malagasay Republic have limited production of pepper.

Plants thrive in the hot, wet tropics at low altitudes. They are sensitive to waterlogging and do best in well-drained, alluvium-rich soils.

Pepper Tree, Brazilian Pepper Tree, California Peppertree
(See Gum & Starch)

Peppergrass, Virginia Pepperweed, Pepper Grass
(See Cereal)

Peppermint
(See Essential Oil)

Perejil
(See Vegetable)

Perennial Soybean, Pempo
(See Forage, Pasture & Feed Grain)

Perilla
(See Oil)

Peru Balsam, Balsam-of-Peru
(See Natural Resin)

Pimentchien, Cherry Capsicum, Aji

Sci. Name(s) : *Capsicum baccatum* var. *pendulum*
Geog. Reg(s) : America-South
End-Use(s) : Ornamental & Lawn Grass, Spice & Flavoring
Domesticated : Y
Ref. Code(s) : 170, 172
Summary : Pimentchien is a pepper plant grown as an ornamental and for its fruit dried as a seasoning. Pimentchien fruits are usually smaller than most other Capsicums and are often marketed as capsicums or chillies. It is considered more pungent than paprika.

This plant grows at lower altitudes in South America and is found both wild and cultivated. Varieties of pimentchien are introduced to the United States. The plants mature early and have high yields but are rarely used in breeding because of sterility.

Pimento, Allspice

Sci. Name(s) : *Pimenta dioica, Pimenta officinalis*
Geog. Reg(s) : America-Central, West Indies
End-Use(s) : Beverage, Essential Oil, Spice & Flavoring, Timber
Domesticated : Y
Ref. Code(s) : 170, 172
Summary : The pimento is a small, evergreen tree 7-10 m in height that is indigenous to the West Indies and Central America. The tree's dried, unripe fruits are used in a culinary spice of commerce, allspice. Allspice is used in pickles, ketchups, and sausages and in curing meats. Pimento berry oil is extracted from the dried fruit and can yield 3-4.5% of a volatile oil called allspice oil. In Jamaica, pimento leaf oil is extracted for use as a volatile oil. Pimento oil is used in flavoring essences and perfumes and as a source of eugenol and vanillin. In Jamaica, a rumlike drink is made from the ripe berries. Tree saplings have been made into umbrella handles and walking sticks, which were at one time exported from Jamaica.

Jamaica is the only large producer of pimento products, followed in decreasing order of importance by Mexico, Guatemala, Honduras, Belize, and Grenada. Trees have a long life span and are usually in full bearing at 20-25 years, giving excellent yields every 3 years. Young trees of 10-15 years produce 22.7 kg/year of green berries. West Germany is one of the foremost importers, followed by the United States and the United Kingdom. The United States has been the main market for pimento leaf oil, but the USSR has increased its levels of imports as well.

Pistachio Nut
(See Nut)

Pot Majoram

Sci. Name(s) : *Majorana onites, Origanum onites*
Geog. Reg(s) : Asia, Europe
End-Use(s) : Ornamental & Lawn Grass, Spice & Flavoring
Domesticated : Y
Ref. Code(s) : 17, 205, 220
Summary : Pot majoram is an herb grown in southern Europe and Asia Minor and cultivated for its leaves, used as a substitute for sweet majoram. It is also grown as an ornamental. Pot majoram withstands a variety of conditions but thrives in full sun on well-drained soils. It is not used medicinally.

Purslane, Pusley, Wild Portulaca, Akulikuli-Kula
(See Ornamental & Lawn Grass)

Red Pepper, Bell Pepper, Green Pepper, Sweet Pepper, Paprika Pepper, Chili Pepper, Mango Pepper

Sci. Name(s) : *Capsicum annuum* var. *annuum*
Geog. Reg(s) : Asia-India (subcontinent), Subtropical, Tropical
End-Use(s) : Spice & Flavoring
Domesticated : Y
Ref. Code(s) : 170, 194

Summary : The red pepper plant (*Capsicum annuum*) is an herbaceous annual reaching 1 m in height grown for its mild-flavored fruits which are dried and used as culinary seasoning. Two other varieties of red pepper are the sweet pepper plant and paprika plant. The sweet pepper, also known as the green or bell pepper, is pitted, cooked, or used in salads. Paprika fruits are preserved and used in cheese preparations or ground and used in condiments and as a spice in dishes such as Hungarian goulash.

Red pepper plants grow in tropical and subtropical areas from sea level to 1,830 m. The 1st fruit is picked and dried after a month. About 45 kg of fresh, ripe fruit produces 11-14 kg of the dried product. Plants thrive in well-drained, sandy or silt loam soil and are frost-sensitive. India is a major exporter of dried red peppers. Annual exports have reached over 10,157 MT.

Rose Geranium, Geranium
(See Essential Oil)

Roselle
(See Fiber)

Rosemary
Sci. Name(s) : *Rosmarinus officinalis*
Geog. Reg(s) : America-North (U.S. and Canada), Europe, Mediterranean
End-Use(s) : Essential Oil, Ornamental & Lawn Grass, Spice & Flavoring
Domesticated : Y
Ref. Code(s) : 17, 64, 94, 180, 194
Summary : Rosemary is a evergreen shrub reaching 1.8 m in height that is cultivated for its fresh, flowering tops, leaves, and twigs, which are distilled to obtain essential oils used in perfumery, soaps, and certain technical preparations. This oil contains cineole, camphor, and borneol. Dried rosemary leaves make a good seasoning for certain foods, such as poultry and stews. The plant makes an attractive hedge and is used as an ornamental.

Shrubs are native to the Mediterranean, Portugal, and northwestern Spain. They do well in the United States but need winter protection when grown in the northern states. Plants thrive on rocky or sandy soils that have lime. Major producing areas are the Mediterranean countries, the United States, and England. Plants are harvested once or twice a year. In 1984, the United States imported 81.5 MT of rosemary as an essential oil, valued at $707,000, and 317.2 MT of rosemary crude and manufactured, valued at $196,000.

Rosewood
(See Essential Oil)

Rosha Grass, Palmarosa
(See Essential Oil)

Saffron, Saffron Crocus
(See Dye & Tannin)

Sage, Garden Sage, Common Sage

Sci. Name(s)	: *Salvia officinalis*
Geog. Reg(s)	: America-North (U.S. and Canada), Europe, Mediterranean
End-Use(s)	: Beverage, Essential Oil, Insecticide, Spice & Flavoring
Domesticated	: Y
Ref. Code(s)	: 31, 154, 170, 194, 220
Summary	: Sage is a shrubby herb reaching 60 cm in height that provides a spice whose dried or fresh leaves can be used whole, chopped, or crushed to flavor poultry, meats, sausages, and other foods. Fresh leaves are used to flavor pickles, salads, and cheeses. About 1.2-2.5 % of an essential oil can be extracted from the dried leaves and is used to lengthen the shelf life of fats and meats. The oil has also been used in some perfumes and cosmetics and as a natural insect repellent. Sage leaves may also be brewed as tea.

Plants are cultivated in Yugoslavia (Dalmatia), Italy, Greece, Turkey, the USSR, the United States, and western Europe. Sage is native to the Mediterranean region. Herbs do not produce well when grown in the lowland tropics. They are most successful on nitrogen-rich, clay loam soils exposed to full sunlight.

Sage can be propagated by seed, plant division, layering, or cuttings. Plants usually last 2-6 years. Leaves may be harvested in the 1st year. About 2 or 3 harvests can be made from 1 plant. Sage can be grown as an ornamental.

Saigon Cinnamon, Cassia

Sci. Name(s)	: *Cinnamomum aromaticum, Cinnamomum loureirii*
Geog. Reg(s)	: Asia, Asia-China, Asia-Southeast
End-Use(s)	: Essential Oil, Medicinal, Spice & Flavoring
Domesticated	: Y
Ref. Code(s)	: 172, 220
Summary	: Saigon cinnamon (*Cinnamomum loureirii*) is an evergreen tree cultivated for its bark, which is a spice of commerce in Indochina. The bark, leaves, and roots also yield essential oils and oleoresins that are economically important, but do not compare in the spice market with the essential oils of true cinnamon (*Cinnamomum verum*) and cassia (*Cinnamomum cassia*). The bark and root essential oils contain cinnamaldehyde and the leaf and twig oils are made up of mostly citral with less eugenol.

In the United States, spice millers buy Saigon cinnamon in rough bark form and market it as "whole cinnamon." It is considered to have the best flavor of all the cinnamons and is used widely in baked goods and processed foods and sold as the ground domestic spice cassia cinnamon. The bark has also been used medicinally as an astringent and digestive stimulant.

Trees are grown primarily in China, Japan, and Java. Bark is harvested when trees reach 10-12 years of age. Saigon cinnamon grows on most well-drained soils, usually along the sides of valleys. It is also grown on flat land, either on plantations or mixed with other crops. An average annual rainfall of 2,500-3,000 mm gives the best production. For 1984 United States imports statistics for cassia as an essential oil, see cassia, and for cinnamon as an essential oil or spice, see cinnamon.

Scotch Spearmint, Red Mint
(See Essential Oil)

Sesame, Simsim, Beniseed, Gingelly, Til
(See Oil)

Siam Benzoin, Siam Balsam
(See Medicinal)

Sour Orange, Seville Orange
(See Essential Oil)

Soybean
(See Oil)

Spanish Oregano, Origanum
(See Essential Oil)

Spanish Thyme, False Thyme

Sci. Name(s)	: *Lippia micromera*
Geog. Reg(s)	: America-South
End-Use(s)	: Spice & Flavoring
Domesticated	: Y
Ref. Code(s)	: 154, 170
Summary	: Spanish thyme is a South American home garden herb often substituted for true thyme in soups, meatloafs, and salads. For United States import statistics on thyme as an essential oil and thyme as a spice, see common thyme.

Spearmint
(See Essential Oil)

Star-Anise Tree, Star Anise
(See Essential Oil)

Summer Savory

Sci. Name(s)	: *Satureja hortensis*
Geog. Reg(s)	: Europe, Temperate
End-Use(s)	: Essential Oil, Medicinal, Spice & Flavoring
Domesticated	: Y
Ref. Code(s)	: 31, 103, 194, 210, 233
Summary	: Summer savory is an herbaceous annual reaching 30 cm in height with leaves that provide the commercially important spice savory. Savory leaves and stems are used fresh or dried as flavorings in seasonings, stews, meat and poultry dishes, and vegetables. Leaves also yield an essential oil that is used primarily in the food industries and to a lesser extent in the perfume industries. Savory leaf oil has also demonstrated antimicrobial and antidiarrheic activity.

Plants are cultivated in France, Yugoslavia, and Spain, and the spice is exported mainly to the United States. Summer savory does best in warm temperate areas on humus-rich soils in full sun with an annual rainfall of 0.3-1.3. Leaves are harvested during the first flowering stage, and harvests can be made twice in the following years.

Sweet Balm, Lemon Balm
(See Essential Oil)

Sweet Bay, Bay, Laurel, Grecian Laurel

Sci. Name(s)	: *Laurus nobilis*
Geog. Reg(s)	: Africa-North, America-Central, America-North (U.S. and Canada), Asia, Europe, Mediterranean
End-Use(s)	: Essential Oil, Spice & Flavoring
Domesticated	: Y
Ref. Code(s)	: 154, 170, 194
Summary	: Sweet bay (*Laurus nobilis*) is an evergreen shrub or small tree cultivated for its scented leaves dried as a flavoring used in soups, fish, meats, stews, puddings, vinegars, and certain beverages. An essential oil obtained from the leaves is replacing the dried, whole leaves as flavoring. Plants are cultivated commercially in parts of Europe, North Africa, the Mediterranean, and Central and North America. They thrive in areas with temperatures of 8-25 degrees Celsius and an annual rainfall of 0.3-2.2 m. They tolerate light frost. Sweet bay is native to the Mediterranean and Asia Minor.

Sweet Flag, Calamus, Flagroot
(See Essential Oil)

Sweet Marjoram, Knotted Marjoram

Sci. Name(s)	: *Marjorana hortensis, Origanum majorana*
Geog. Reg(s)	: Europe, Mediterranean
End-Use(s)	: Essential Oil, Medicinal, Ornamental & Lawn Grass, Spice & Flavoring
Domesticated	: Y
Ref. Code(s)	: 194, 205, 220
Summary	: Sweet marjoram is an herb native to the Mediterranean region and southern Europe. It is cultivated in both the Old and New World for its leaves, which are used to flavor meat dishes. Oil can be obtained from the leaves by distillation and used as a flavoring. The dried leaves have been used to relieve upset stomachs. Dried marjoram contains tyhnol and carvacrol. The essential oils are composed of terpineol, borneol, mucilage, and tannic acid. It also has uses in cosmetic and scented articles.

Sweet marjoram is sometimes grown as an ornamental. It grows best in well-drained, fertile soil in sunny places.

Sweet Woodruff

Sci. Name(s)	: *Asperula odorata, Galium odoratum*
Geog. Reg(s)	: Africa-North, Europe
End-Use(s)	: Essential Oil, Medicinal, Spice & Flavoring
Domesticated	: Y
Ref. Code(s)	: 194

Summary   : Sweet woodruff (*Asperula odorata*) is a wild, perennial herb from which coumarin, tannin asperuloside, fatty and essential oils, and a bitter principle are obtained. The fresh leaves of the plant are used to flavor nonalcoholic and alcoholic beverages. The dried leaves are used in sachets and perfumery. Medicinal uses of the plant have been as an antispasmodic, diaphoretic, diuretic, and stomachic. The essential oil is used as a carminative and mild expectorant.

Herbs are generally collected from moist wooded locations. They have been successfully cultivated in parts of Europe and North Africa. Sweet woodruff is now classified as *Galium odoratum*.

Tacoutta, Wing Sesame
(See Oil)

Tamarind
(See Fruit)

Tarragon, French Tarragon
Sci. Name(s)   : *Artemisia dracunculus*
Geog. Reg(s)   : America-North (U.S. and Canada), Europe
End-Use(s)   : Essential Oil, Spice & Flavoring
Domesticated   : Y
Ref. Code(s)   : 194, 205
Summary   : Tarragon is a perennial herb that reaches 90 cm in height and is cultivated for its pungent, aromatic leaves used as flavoring for salad and meats. The leaves also yield an essential oil (estragon oil) used in perfumes and liqueurs.

Plantings usually last about 3 years and are then reestablished. Harvests are midseason and in early autumn. During winter, plant roots should be mulched for protection. Tarragon thrives in climates with temperatures of 7-17 degrees Celsius with an average annual rainfall of 0.3-1.3 m. Plants should be grown in sunny areas on dry, well-drained soils. The herb is cultivated commercially in Europe and the United States. In 1983, United States crude tarragon imports were 46 MT, valued at $295,000.

Thyme
(See Essential Oil)

Tiberato, Indian Nightshade
(See Vegetable)

Tolu Balsam, Balsam-of-Tolu
(See Natural Resin)

Tonka Bean
Sci. Name(s)   : *Dipteryx oppositifolia*
Geog. Reg(s)   : America, America-South, Tropical
End-Use(s)   : Spice & Flavoring
Domesticated   : Y

Ref. Code(s) : 94, 119, 170
Summary : The tonka bean (*Dipteryx oppositifolia*) is a wild tropical American tree whose beans are cured and used as the commercial source of the flavoring coumarin. The coumarin, like that obtained from the cimaru tree (*Dipteryx odorata*), is used mainly to flavor tobacco, perfumes, soaps, liqueurs, and confectionery.

Tonka bean production is mainly in South America, and harvests vary depending on the care of the trees. Wild trees take several years to produce, and cultivated trees start producing within 2 years. Yields average about 0.454 to 0.908 kg/tree of dried beans, but yields of 23 kg/tree have been recorded.

Turmeric

Sci. Name(s) : *Curcuma domestica*
Geog. Reg(s) : America-South, Asia-India (subcontinent), Asia-Southeast, Europe, Subtropical, Tropical, West Indies
End-Use(s) : Dye & Tannin, Essential Oil, Medicinal, Spice & Flavoring
Domesticated : Y
Ref. Code(s) : 171, 172
Summary : Turmeric is a perennial herb growing up to 1 m high whose cured and dried rhizomes produce an aromatic yellow-orange powder that is used as a spice. It is also used as a coloring agent and for the preparation of solvent-extracted oleoresin. The rhizomes yield an essential oil with little or no market value.

The spice has widespread culinary uses. It is used on rice, in curry powders, and in processed foods and sauces. The oleoresins are used in processed foods and in certain unspecified medicinal preparations. The powder is also used as a textile dye and as a coloring agent in pharmacies, confectionery, and food industries.

Major producing countries are India, Asia, Bangladesh, Taiwan, and China. Small exporters are Haiti, Jamaica, and Peru. The oleoresins are produced by the North American and European food industries. The United States and the United Kingdom import the spice in polished fingers, because they have the highest contents of pigment and are the lowest in bitter principle content.

Turmeric plants are cultivated in most tropical and subtropical countries receiving adequate rainfall or irrigation. In the Himalayan foothills, plants thrive up to 1,220 m in elevation. Plants should be grown in loam or alluvial soils that are well-drained. They can be rotated with finger millet, rice, and sugarcane. They can also be grown in mixed plantings with castor, maize, finger millet, onions, brinjal, and tomatoes.

Vanilla
(See Essential Oil)

Water Dropwort
(See Vegetable)

Watercress
(See Vegetable)

West Indian Lemongrass, Sere Grass, Lemon Grass
(See Essential Oil)

White Cottage Rose, Rose
(See Essential Oil)

White Mustard
(See Forage, Pasture & Feed Grain)

Wild Sesame
(See Oil)

Wild Thyme, Creeping Thyme
(See Ornamental & Lawn Grass)

Winter Savory

Sci. Name(s)	:	*Satureja montana*
Geog. Reg(s)	:	Africa-North, Mediterranean
End-Use(s)	:	Essential Oil, Spice & Flavoring
Domesticated	:	Y
Ref. Code(s)	:	31, 91, 194, 210
Summary	:	Winter savory is a small, woody, perennial shrub whose leaves and stems are

used for flavoring. Fresh and dried leaves are used primarily as flavoring in season-ings, meat dishes, and vegetables. An essential oil is distilled from the leaves for use in the perfume industries. Oil contains phenols, carvacrol, and thymol.

Leaves are harvested during the 1st flowering season and can be harvested twice in subsequent years if plants are grown as perennials. Winter savory is a hardy shrub that can withstand temperatures as low as 7 degrees Celsius. Plants will thrive in any sandy, well-drained soil with average moisture, in areas receiving an annual rainfall of 0.7-1.7 m. Plants are often grown as edgings for herb gardens or garden borders. Winter savory is native to the Mediterranean region and North Africa.

Winter's Grass, Citronella Grass
(See Essential Oil)

Winter's-Bark
(See Medicinal)

Yard-Long Bean, Asparagus Bean
(See Vegetable)

Yellow Gentian
(See Beverage)

Yellow Melilot, Yellow Sweetclover
(See Forage, Pasture & Feed Grain)

Sweetener

African Oil Palm, Oil Palm
(See Oil)

American Licorice, Wild Licorice
(See Spice & Flavoring)

Benoil Tree, Horseradish-Tree
(See Oil)

Butternut
(See Nut)

Carob, St. John's-Bread
(See Gum & Starch)

Chicory
(See Beverage)

Common Licorice
(See Spice & Flavoring)

Common Thyme
(See Spice & Flavoring)

Date Palm
(See Fruit)

English Walnut, Persian Walnut
(See Nut)

Fishtail Palm, Toddy Palm

Sci. Name(s) : *Caryota urens*
Geog. Reg(s) : Asia-India (subcontinent), Asia-Southeast
End-Use(s) : Cereal, Fiber, Gum & Starch, Miscellaneous, Sweetener
Domesticated : Y
Ref. Code(s) : 63, 171
Summary : The fishtail palm is tapped for a sugary sap, which contains 76.6-83.5% sucrose, 0.76-0.9% reducing sugar, 1.79-2.27% protein, and 6.6-8.34% pectin gums. The sap is drunk with tea or coffee. Sap yields from a single palm inflorescence are 7-14 liters/day. If several inflorescences are tapped, 20-27 liters may be produced daily.

A fine, pliable fiber obtained from the leaf sheaths is made into brooms. The fiber is occasionally exported from Sri Lanka. The pith of the trunk is used for the production of sago, a starchy preparation used in foods and as a textile stiffener.

Fishtail palms flower at 15 years and continue for several more years. They reach heights of 10-20 m. Fishtail palms grow in hot and moist parts of India, Sri Lanka, Burma, and Thailand.

Garden Myrrh, Sweet Scented Myrrh
(See Medicinal)

Indianfig, Spineless Cactus, Prickly Pear, Indianfig Pricklypear
(See Fruit)

Jerusalem-Artichoke, Girasole
(See Tuber)

Kaa He'e
Sci. Name(s)   : *Stevia rebaudiana*
Geog. Reg(s)   : America-South
End-Use(s)   : Sweetener
Domesticated : Y
Ref. Code(s)   : 220
Summary   : Kaa he'e is a plant that has been used as a substitute for sugar. It contains es-tevin, which is 150 times sweeter than sugar. There is very little information about this plant and little or no available information concerning its potential as a natural sweetener. Plants are native to South America and are grown primarily in Paraguay.

Kangaroo Apple
(See Fruit)

Miracle-Fruit, Katempe
Sci. Name(s)   : *Thaumatococcus daniellii*
Geog. Reg(s)   : Africa
End-Use(s)   : Sweetener
Domesticated : N
Ref. Code(s)   : 36
Summary   : The miracle-fruit is a bush that produces very sweet berries. The berries can cause a kind of paralysis of the taste buds, leaving a long-lasting sweet taste. Plants grow freely along the coast of Africa.

Palmyra Palm, African Fan Palm
Sci. Name(s)   : *Borassus flabellifer*
Geog. Reg(s)   : Asia, Asia-India (subcontinent)
End-Use(s)   : Fiber, Miscellaneous, Sweetener
Domesticated : N
Ref. Code(s)   : 67, 171
Summary   : The Palmyra palm produces a sap that contains 12% sucrose. In India, the sap is an important source of jaggery (sugar). Fiber is obtained by beating the base of the leaf stalks. A single palm tree has produced 120,000 liters of sap during its tapping life.

Palmyra palms form trunks at 10 years and will flower at 20 years. Tapping begins when palms are 25-30 years old and may continue for another 30 years. Palms are erect, growing 12-15 m. Some palms have been known to reach 30 m in height. Stem diameters are 0.5-0.6 m. Palms have crowns of large fan-shaped leaves.

Palms grow throughout the drier parts of tropical Asia, and India is the main producer. Palms will not produce satisfactorily in humid tropical areas.

Pineapple
(See Fruit)

Serendipity-Berry, Diel's Fruit
Sci. Name(s)	: *Dioscoreophyllum cumminsii*
Geog. Reg(s)	: Africa-West, Tropical
End-Use(s)	: Fruit, Sweetener, Tuber
Domesticated	: Y
Ref. Code(s)	: 97, 119, 209
Summary	: The serendipity-berry is an herbaceous perennial grown for its sweet berries,

which are used as natural noncarbohydrate sweetener substitutes. The substance, a low-molecular-weight protein called monellin, has potential use in combating tooth decay and in low calorie diets for diabetics and dieters. The fruit pulp is eaten raw and the tubers are edible.

The berry is native to West tropical Africa and grows wild in tropical forests. Very little experimental work has been done on the cultivation of this plant. It can be propagated from its tubers, by cuttings, or by seed. Through better germination techniques, the serendipity-berry may have potential for large-scale commercial berry cultivation.

Sorghum, Great Millet, Guinea Corn, Kaffir Corn, Milo, Sorgo, Kaoling, Durra, Mtama, Jola, Jawa, Cholam Grains, Sweets, Broomcorn, Shattercane, Grain Sorghums, Sweet Sorghums
(See Cereal)

Sugar Maple, Rock Maple, Hard Maple
(See Timber)

Sugar Palm, Gomuti Palm
(See Gum & Starch)

Sugarcane
Sci. Name(s)	: *Saccharum officinarum*
Geog. Reg(s)	: Asia-Southeast, Hawaii, Pacific Islands, West Indies
End-Use(s)	: Beverage, Energy, Fat & Wax, Fiber, Forage, Pasture & Feed Grain, Miscellaneous, Sweetener
Domesticated	: Y
Ref. Code(s)	: 19, 154, 220
Summary	: Sugarcane is a large, perennial grass that is one of the most important sources

of raw sugar. Raw sugar is a major export and import product throughout the world.

Important by-products are bagasse, molasses, filter mud cakes, and cane wax. Bagasse is residue used as fuel for sugar factories, in livestock feed, and for the production of compressed fiberboard, paper pulp, plastics, furfural, and cellulose. Molasses is a heavy liquid used for edible and industrial purposes. It contains 50% sugar by weight and is used in confectionery and as an additive in livestock feeds. It is also the source of industrial alcohol and such potable alcohols as rum and gin. A protein-rich yeast is also obtained. Filter mud cakes are used as fertilizer. Cane wax has potential use in the manufacture of furniture, shoe and leather polishes, electrical insulating materials, and waxed paper.

Major sugar producing areas are the West Indies, Hawaii, certain Pacific islands, Java, the Philippines, Africa, Madagascar, the Democratic Republic of the Sudan, and Sri Lanka. In the United States, some sugar plantations are found in the Florida Everglades area, in the Mississippi River alluvial flats of Lousiana, and in Hawaii.

Sugarcane is successful on iron-rich soils, black soils, brown clay loams, alluvial soils, and organic soils. It is fairly tolerant of saline soils. Plants require at least 2-2.2 m of rainfall or adequate irrigation. In most tropical areas, cane occupies land for a period of 12-18 months. Harvest season is in the dry, cool part of the year. Hawaii is considered a prime environment for flowering and for large-scale production of viable seed.

World production in 1984-85 was 97.5 million MT of sugar, or 2% more than world production in 1983-84. World consumption during 1984-85 is estimated at 96.1 million MT, a figure up 0.5% from the previous year. Estimated world molasses output for 1984-85 was 33.4 million MT, or 2% above the lower 1983-84 output. The largest increases in molasses production were in the European Economic Community, Africa, and Asia. The United States cane sugar output was down in 1984-85; less area being harvested. Area and production for the United States mainland are 246,000 ha of cane from which 15.4 million MT of sugar was obtained. In Hawaii, 34,000 ha of cane produced 7.8 million MT.

Sweet Berry, Miraculous-Berry, Miraculous Fruit

Sci. Name(s) : *Synsepalum dulcificum*
Geog. Reg(s) : Africa-West
End-Use(s) : Sweetener
Domesticated : N
Ref. Code(s) : 17
Summary : Sweet berry is a plant with sweet, acidic fruits native to West Africa. Information concerning its potential as a natural sweetener is limited.

Teosinte
(See Forage, Pasture & Feed Grain)

Wine Palm, Coco de Chile, Coquito Palm, Honey Palm, Wine Palm of Chile, Chile Coco Palm
(See Beverage)

Timber

Abarco, Colombian Mahogany

Sci. Name(s) : *Cariniaria pyriformis*
Geog. Reg(s) : America-South
End-Use(s) : Timber
Domesticated : Y
Ref. Code(s) : 228
Summary : Abarco is an evergreen tree that grows in northern Colombia and Venezuela. It can reach heights of 40-50 m. The timber is used in heavy and light construction as well as for veneer and plywood. Annual production is 10-20 cu.m/ha.

This tree grows at altitudes up to 600 m, and needs an annual rainfall of 2,000-4,000 mm and an average temperature range of 22-30 degrees Celsius. Abarcos grow best in medium- to heavy-textured soils and have a moderate tolerance for shade.

Afara, Limba

Sci. Name(s) : *Terminalia superba*
Geog. Reg(s) : Africa-West
End-Use(s) : Miscellaneous, Timber
Domesticated : Y
Ref. Code(s) : 228
Summary : Afara is a buttressed, wide-crowned, semievergreen tree found in West Africa from Sierra Leone to Zaire. It is used as an agricultural shade and is a source of highly valued decorative timber. It is easy to saw and preserve but lacks durability. Its saw timber is used in light construction, furniture, box making, and boat building, while its roundwood is used for shortfiber pulp, veneer, and plywood. Annual timber yields are 10-14 cu.m/ha.

This tree reaches heights of 40-60 m and is found at an altitudinal range from sea level to 500 m with an annual rainfall of 1,300-1,900 mm and temperatures of 24-27 degrees Celsius. It withstands dry seasons of 1-3 months, thrives on medium-textured, deep soils and adapts to most soils in full sun.

African Oak, Muvule, African Teak, Iroko

Sci. Name(s) : *Chlorophora excelsa*
Geog. Reg(s) : Africa, Tropical
End-Use(s) : Timber
Domesticated : Y
Ref. Code(s) : 170, 216
Summary : The African oak is a tropical African tree grown for its timber, which is valued in furniture making, structural work, ship building, cabinet work, and as a substitute for true teak. It is marketed as African teak but the two are not botanically related. The wood grain is interlocked or wavy and can be given various finishing treatments. It is somewhat resistant to fungal diseases but is susceptible to wood borers.

African Star-Apple
(See Fruit)

Aleppo Pine

Sci. Name(s) : *Pinus halepensis*
Geog. Reg(s) : Europe, Mediterranean
End-Use(s) : Erosion Control, Natural Resin, Timber
Domesticated : Y
Ref. Code(s) : 228
Summary : The aleppo pine is an evergreen tree found throughout the Mediterranean, from Spain to Turkey. It is most often used for protection rather than timber production. It is moderately fire-resistant, termite-resistant, and frost-resistant and is used for shade, shelter, windbreak, dune fixation, and erosion control. Its saw timber is used for heavy and light construction and its roundwood serves as building poles and fence posts. A resin is obtained from the bark. Timber production is 5-11 cu.m/ha/annum.

This tree grows to heights of 15-18 m. It adapts to most soil conditions and occurs naturally at altitudes of 1,500-2,500 m with 400-800 mm of rain and temperatures of 15-20 degrees Celsius. It requires considerable light.

Alpine Ash

Sci. Name(s) : *Eucalyptus delegatensis, Eucalyptus gigantea*
Geog. Reg(s) : Australia, Pacific Islands
End-Use(s) : Timber
Domesticated : Y
Ref. Code(s) : 228
Summary : Alpine ash is an evergreen tree reaching heights of 50-60 m and grown in southeastern New South Wales, Victoria, and Tasmania. Its timber is used in light construction and its roundwood for transmission poles, veneer, and plywood. Yearly timber production is 10-25 cu.m/ha.

This tree grows at altitudes of 2,000-3,000 m. A dry season may last up to 2 months. Needed annual precipitation is 1,000-2,000 mm. The alpine ash thrives on medium-textured soils in full sun for maximum growth.

American Beech

Sci. Name(s) : *Fagus grandifolia, Fagus americana, Fagus ferruginea*
Geog. Reg(s) : America-North (U.S. and Canada)
End-Use(s) : Medicinal, Miscellaneous, Nut, Timber
Domesticated : Y
Ref. Code(s) : 220
Summary : American beech is a tree of North America. Its hard, heavy wood is used commercially for veneers, barrel making, and boxes. Through a distillation process, beech wood produces creosote, which is used as an antiseptic, expectorant, and antipyretic. Wood charcoal is used in art. American beech nuts are edible.

American Black Walnut, Black Walnut, Eastern Black Walnut

Sci. Name(s) : *Juglans nigra*
Geog. Reg(s) : America-North (U.S. and Canada)
End-Use(s) : Nut, Timber
Domesticated : Y

Ref. Code(s)   : 131, 182, 220

Summary   : The American black walnut is one of the largest members of the walnut family, with trees reaching heights of 45 m. This tree is a source of durable hardwood considered one of the most valuable hardwoods in the United States and exported to Europe. It is used for interior work, gunstocks, and other fine construction work. American black walnuts are edible, slightly pungent, and used in confectionery.

Trees are native to the deciduous forests of the eastern United States and Canada. They thrive in most deep, fertile, well-drained soils but do not tolerate waterlogging or the presence of alkali.

American Chestnut

Sci. Name(s)   : *Castanea dentata*
Geog. Reg(s)   : America-North (U.S. and Canada)
End-Use(s)   : Dye & Tannin, Nut, Timber
Domesticated   : Y
Ref. Code(s)   : 59, 216
Summary   : The American chestnut is a fast-growing tree whose commercial potential as a timber and nut source has been seriously affected by chestnut blight, a fungal disease. Trees usually reach about 4-6 m in height before the blight visibly affects their growth. They rarely survive to produce substantial yields although they have the potential to yield large sweet nuts.

The American chestnut timber is light, soft, and resistant to decay. It is fairly durable when exposed to the weather and has been used to make gates, furniture, and telephone poles. It is also used for paneling. The bark and wood of the chestnut are rich in tannins. Trees are found in the eastern part of North America.

Annatto
(See Dye & Tannin)

Apple
(See Fruit)

Arbor-Vitae, Northern White-Cedar
(See Essential Oil)

Argus Pheasant-Tree
(See Fruit)

Arizona Cypress

Sci. Name(s)   : *Cupressus arizonica, Cupressus glabra*
Geog. Reg(s)   : America-Central, America-North (U.S. and Canada)
End-Use(s)   : Energy, Erosion Control, Timber
Domesticated   : Y
Ref. Code(s)   : 228
Summary   : Arizona cypress is an open-crowned evergreen tree. It reaches 10-20 m in height and is found in southern Arizona, New Mexico, and northern Mexico. The timber is used for furniture and boxes. The roundwood is used as fence posts, fuel,

and charcoal. This tree also serves as a windbreak and helps hinder erosion. Annual timber production is 3-5 cu.m/ha.

This tree is shade tolerant and frost-resistant. It grows at altitudes of 1,500-2,800 m. Dry seasons may last 4-7 months. Annual rainfall is 250-750 mm. Average temperatures are 15-18 degrees Celsius. Trees grow best on light- to medium-textured soils.

Arundinaria

Sci. Name(s)	: *Arundinaria* spp.
Geog. Reg(s)	: Africa-East, America-North (U.S. and Canada), Asia-China, Europe
End-Use(s)	: Timber
Domesticated	: Y
Ref. Code(s)	: 171
Summary	: The Arundinaria species is a family of bamboos that has been significant in

world trade. The main commercial species was *Arundinaria amabilis*. This is still grown commercially in the Kwangtung Province of China and is used for making split and glued fishing rods in Britain and America because of its particularly strong and resilient wood. Another type, *Arundinaria alpina*, is a mountain bamboo grown mainly in the mountainous areas of East Africa.

Most Arundinarias grow in open clumps in mountainous areas at elevations of 2,330-3,300 m. The average culm height is 12 m with single culm diameters of 5 cm.

Australian Blackwood

Sci. Name(s)	: *Acacia melanoxylon*
Geog. Reg(s)	: Africa-East, Australia, Pacific Islands
End-Use(s)	: Energy, Erosion Control, Ornamental & Lawn Grass, Timber
Domesticated	: Y
Ref. Code(s)	: 228
Summary	: The Australian blackwood is an evergreen tree found in southeastern Australia

and Tasmania and is grown for its timber, which is easily sawn but difficult to preserve and is used for fine furniture and light construction. The roundwood is used for fence posts, fuel, charcoal, and veneer. The tree itself serves as a shade, windbreak, and ornamental. Annual timber production is 5-12 cu.m/ha.

Trees are of medium height and reach 18-30 m. They grow at altitudes of 1,500-2,500 m, in areas with an annual rainfall of 900-2,700 mm and temperatures of 12-18 degrees Celsius. Australian blackwoods are termite-resistant and grow best in deep, light- to medium-textured soils. Young trees are shade tolerant.

Babul
(See Gum & Starch)

Baccaurea ramiflora

Sci. Name(s)	: *Baccaurea ramiflora, Baccaurea parriflora, Baccaurea sapida*
Geog. Reg(s)	: Asia-Southeast
End-Use(s)	: Timber
Domesticated	: N
Ref. Code(s)	: 36

Summary   : Baccaurea ramiflora is a tree that provides hardy, durable wood used to make walking sticks. This tree is not commercially utilized but may have potential as a substitute for boxwood. Baccaurea ramiflora is native to Southeast Asia.

Bacury
(See Fruit)

Balata
(See Isoprenoid Resin & Rubber)

Balsa
Sci. Name(s)   : *Ochroma lagopus, Ochroma pyramidalis, Ochroma grandiflora*
Geog. Reg(s)   : America-Central, America-South, West Indies
End-Use(s)   : Timber
Domesticated   : Y
Ref. Code(s)   : 228
Summary   : The balsa is a short-lived evergreen tree reaching heights of 15-20 m. Its timber is not naturally durable but the saw timber is used for insulation and the roundwood for shortfiber pulp. Balsa trees are found in Central America, the West Indies, and South America. Annual timber production is 17-30 cu.m/ha. Timber stands need ideal growth conditions of deep, moist, fertile soils in full sun if commercial density requirements are to be met. Trees grow from sea level to 1,000 m in altitude in areas receiving 1,500-3,000 mm with annual temperatures of 22-28 degrees Celsius.

Balsam Fir, Canada Balsam, American Silver Fir
(See Essential Oil)

Bamboo, Phyllostachys
Sci. Name(s)   : *Phyllostachys* spp.
Geog. Reg(s)   : Asia, Temperate
End-Use(s)   : Erosion Control, Ornamental & Lawn Grass, Timber, Vegetable
Domesticated   : Y
Ref. Code(s)   : 17, 132
Summary   : Phyllostachys bamboo is the second major but less well-known genus of the tall, woody, evergreen grasses known as bamboo. Bamboo provides a strong, durable timber and can be made into various domestic items such as garden stakes, poles, or fence material. Plants are also popular as ornamental hedges or potted plants. Young shoots of bamboo are edible and used mostly in cooking, although they have no nutritional value. This fast-growing grass also provides erosion control.

Bamboo is Asian in origin. It is primarily a temperate region plant that does not thrive in tropical areas. Plants grow on most well-drained, moist, fertile soils and need adequate time and space for productive growth. The 1st harvest takes place after 5 years with a possible yield of 1 MT/ha of shoots.

Bambusa, Bamboo
Sci. Name(s)   : *Bambusa* spp.
Geog. Reg(s)   : America-Central, Asia, Tropical

End-Use(s) : Fiber, Timber, Vegetable
Domesticated : Y
Ref. Code(s) : 171
Summary : The most widely distributed species of bamboo is *Bambusa vulgaris* and is one of the few bamboos known in cultivation. Its strong culm is used to make a fine quality paper pulp. *Bambusa vulgaris* is susceptible to attack from the bamboo borer. This plant is reportedly native to Asia.

In the tropics, bamboo culms are used for house construction, huts, rafts, bridges, and scaffolding. Split and flattened culms are used for floorboards and other purposes. In Central America, the culms are considered the best banana tree props. Young shoots of the bamboo are edible. Bambusa culms are 10-20 m tall.

Barwood
(See Dye & Tannin)

Barwood, Camwood
(See Dye & Tannin)

Batai
Sci. Name(s) : *Albizia falcataria, Albizia moluccana, Albizia falcata*
Geog. Reg(s) : Asia, Asia-Southeast
End-Use(s) : Erosion Control, Timber
Domesticated : Y
Ref. Code(s) : 228
Summary : Batai is an open-crowned deciduous tree that originated in the north Moluccas and Indonesia and has naturalized in the Far East. It is 25-35 m in height and is grown mainly for timber. Its saw timber is used in light construction and its roundwood is used for veneer or plywood. Batai also aids in soil improvement and is used as an agricultural shade. Annual timber production is 20-40 cu.m/ha

Trees thrive at altitudes up to 1,200 m. They require 2,000-4,000 mm of rainfall annually, and temperatures of 22-29 degrees Celsius. Batai is adaptable and grows on light- to heavy-textured soils. It requires direct sunlight for optimum growth.

Beach, She Oak
Sci. Name(s) : *Casuarina equisetifolia*
Geog. Reg(s) : Asia-Southeast, Australia
End-Use(s) : Energy, Erosion Control, Ornamental & Lawn Grass, Timber
Domesticated : Y
Ref. Code(s) : 228
Summary : Beach is a light-crowned evergreen tree found on the coastal dunes of Southeast Asia and Australia. Its hardwood is used for heavy construction and boat building. The roundwood is used for building poles, transmission poles, fence posts, fuel, and charcoal. The tree's nitrogen-fixing properties aid in soil improvement and its resistance to salt winds makes it a good windbreak. It is also grown for ornamental purposes. Six to 18 cu.m/ha/annum of timber are produced.

This tree reaches heights of 20-40 m and can withstand saline soils but grows best in soils with light texture, in full sun. It is found at altitudes from sea level to

1,400 m with 750-1,100 mm of rain annually, and temperatures of 18-26 degrees Celsius. Beaches are termite-resistant.

Bhutan Cypress

Sci. Name(s)	: *Cupressus torulosa*
Geog. Reg(s)	: Asia-India (subcontinent)
End-Use(s)	: Erosion Control, Timber
Domesticated	: Y
Ref. Code(s)	: 228
Summary	: Bhutan cypress is an evergreen tree standing 30-40 m tall. It grows from the western Himalayas to Bhutan, where its timber is used for both heavy and light construction. The roundwood is used for building poles and fence posts and the tree is used for shade and as a windbreak. It must be pruned to prevent extensive knotting. Twelve to 17 cu.m/ha/annum of timber are obtained.

This tree adapts to most soils and grows at altitudes of 1,500-2,800 m with an annual rainfall of 650-1,100 mm and a mean annual temperature of 12-19 degrees Celsius. Bhutan cypresses are frost-resistant and require moderate light for optimum growth.

Bignay, Chinese Laurel, Salamander Tree
(See Fruit)

Binuang

Sci. Name(s)	: *Octomeles sumatrana*
Geog. Reg(s)	: Asia-Southeast, Pacific Islands
End-Use(s)	: Timber
Domesticated	: Y
Ref. Code(s)	: 228
Summary	: Binuang is a buttressed evergreen tree that grows naturally in Indonesia, the Philippines, New Guinea, and the Solomon Islands. Its timber is susceptible to stain and borers but is used to make boxes, veneer, and plywood. This tree is often attacked by defoliators. Annual timber yields are 25-40 cu.m/ha.

This tree requires deep soils of light to medium texture, altitudes up to 500 m, a dry season of 1 month, and 2,000-5,000 mm of rain annually. Mean annual temperatures should be 24-30 degrees Celsius with a lot of light.

Black Box

Sci. Name(s)	: *Eucalyptus largiflorens, Eucalyptus bicolor*
Geog. Reg(s)	: Australia
End-Use(s)	: Energy, Timber
Domesticated	: Y
Ref. Code(s)	: 228
Summary	: Black box is an evergreen tree that grows in the semiarid areas of Central Australia. It grows to heights of 10-18 m and its roundwood is used for building poles, fence posts, fuel, and charcoal. The tree also serves as a shade and windbreak. Annual timber is 7-9 cu.m/ha.

This frost-resistant tree grows at altitudes up to 1,200 m with an annual precipitation of 250-500 mm and a mean annual temperature of 19-25 degrees Celsius. It grows best on heavy-textured soils in full sunlight.

Black Locust, False Acacia

Sci. Name(s) : *Robinia pseudoacacia*
Geog. Reg(s) : America-North (U.S. and Canada)
End-Use(s) : Energy, Erosion Control, Forage, Pasture & Feed Grain, Ornamental & Lawn Grass, Timber
Domesticated : Y
Ref. Code(s) : 228
Summary : Black locust is a spiny deciduous tree found in the Appalachian and Ozark regions of the United States. It reaches heights of 20-25 m. Its timber is difficult to preserve, has good natural durability, is easy to saw, and has good bending properties. The roundwood is used for fence posts, fuel, and charcoal. Foliage provides fodder. It is used for shade, shelter, dune fixing, erosion control, and as an ornamental. Its annual timber yield is 4-5 cu.m/ha.

The necessary climatic conditions are altitudes of 1,500-2,500 m, with a mean annual rainfall of 500-700 mm, and temperatures of 10-18 degrees Celsius. It prefers light- to medium-textured soils and for maximum growth needs considerable light. It is frost-resistant.

Black Sapote, Zapte Negro
(See Fruit)

Black Wattle
(See Dye & Tannin)

Blue Gum

Sci. Name(s) : *Eucalyptus globulus*
Geog. Reg(s) : Australia, Pacific Islands
End-Use(s) : Energy, Timber
Domesticated : Y
Ref. Code(s) : 228
Summary : Blue gum is an evergreen tree reaching heights of 40-50 m. It is grown in southern Victoria and Tasmania, where its timber is used in both heavy and light construction. The roundwood is used for poles, posts, fuel, charcoal, veneer, and plywood. Annual timber production is 10-40 cu.m/ha.

This tree grows best at altitudes of 1,500-3,000 m. Mean annual rainfall should be approximately 900-1,800 mm. Temperatures should go no lower than 4 degrees Celsius and no higher than 30 degrees Celsius. The tree is sensitive to frost, requires considerable light for optimum growth, and does best on medium- to heavy-textured soils.

Blue Gum

Sci. Name(s) : *Eucalyptus saligna*
Geog. Reg(s) : Australia

End-Use(s) : Energy, Timber
Domesticated : Y
Ref. Code(s) : 228
Summary : Blue gum is an evergreen tree grown in New South Wales and extreme south-east Queensland. This tree of 35-45 m in height is used in heavy and light construction. Its timber is difficult to saw and is often used for building and transmission poles, veneer, plywood, fuel, and charcoal.

Light- to medium-textured soils are best, and trees require a lot of light. They grow at altitudes of 500-2,100 m. The mean annual rainfall is 1,000-4,000 mm with a dry season of 0-2 months. The tree's mean minimum temperature is 2-12 degrees Celsius, while its mean maximum temperature is 20-35 degrees Celsius.

Brown Barrel

Sci. Name(s) : *Eucalyptus fastigata*
Geog. Reg(s) : Africa-South, Australia
End-Use(s) : Natural Resin, Timber
Domesticated : Y
Ref. Code(s) : 228
Summary : Brown barrel (*Eucalyptus fastigata*) is an evergreen tree reaching 30-40 m in height, which grows in the highlands of eastern New South Wales. It is used primarily in the manufacture of veneer, plywood, and boxes and is used in light construction. The wood contains resin pockets but in general this timber is easy to preserve, season, and saw. Annual timber yields are 21-28 cu.m/ha.

This tree grows at altitudes of 1,600-2,500 m. Established trees tolerate dry seasons of 2 months. They are found in areas where the average annual rainfall is 750-1,100 mm and temperatures are 4-26 degrees Celsius. They are moderately resistant to frost and grow most often on medium-textured soil.

Burma Cedar, Toona

Sci. Name(s) : *Toona ciliata* var. *australis, Cedrela toona*
Geog. Reg(s) : Asia-India (subcontinent), Asia-Southeast
End-Use(s) : Timber
Domesticated : Y
Ref. Code(s) : 228
Summary : Burma cedar is a deciduous tree that reaches heights of 30-35 m and is found in Southeast Asia from India to Thailand. Its timber is easy to saw and season and has decorative features. Its saw timber is used for light construction, furniture and boxes, and its roundwood for veneer and plywood. Annual timber production is 7-18 cu.m/ha.

It is found at altitudes from sea level to 1,200 m in areas with an annual rainfall of 850-1,800 mm and temperatures of 22-28 degrees Celsius. It withstands a dry season of 2-6 months and prefers soils light to medium in texture in full sun. The shoot borer Hypsipyla causes severe damage to the tree.

Butternut
(See Nut)

Cajeput
(See Medicinal)

Canary Island Pine

Sci. Name(s)	:	*Pinus canariensis*
Geog. Reg(s)	:	Africa-North
End-Use(s)	:	Natural Resin, Timber
Domesticated	:	Y
Ref. Code(s)	:	228
Summary	:	The Canary Island pine is an evergreen tree reaching 20-30 m in height with durable and easily preserved timber. The saw timber is used in heavy and light construction and its roundwood for building and transmission poles. The wood produces useful resins. Trees provide good shade and shelter for other crops. Annual timber production is 8-18 cu.m/ha.

The pine coppices are fairly fire-resistant and frost-hardy. Trees grow in the Canary Islands at altitudes of 1,500-2,500 m in areas with 600-1,750 mm of rain and temperatures of 14-19 degrees Celsius. The best growth is from deep soils, light to medium in texture, in full sun.

Caribbean Pine

Sci. Name(s)	:	*Pinus caribaea* var. *bahamensis*
Geog. Reg(s)	:	West Indies
End-Use(s)	:	Natural Resin, Timber
Domesticated	:	Y
Ref. Code(s)	:	228
Summary	:	The Caribbean pine (var. *bahamensis*) is a fairly windfirm evergreen tree reaching 15-20 m in height and grown for its timber in the Bahamas. Its saw timber is used for boat building and, light and heavy construction, and its roundwood for poles and fence posts. A resin is obtained from the tree trunk. Timber production is 10-28 cu.m/ha/annum.

In the Bahamas, this tree is found at altitudes up to 1,000 m in areas with an annual rainfall of 1,000-1,500 mm and temperatures of 22-26 degrees Celsius. It is best suited to light-textured soils and tolerates shallow soils in full sun. This tree is sensitive to frost.

Caribbean Pine

Sci. Name(s)	:	*Pinus caribaea* var. *hondurensis*
Geog. Reg(s)	:	America-Central
End-Use(s)	:	Natural Resin, Timber
Domesticated	:	Y
Ref. Code(s)	:	228
Summary	:	The Caribbean pine (var. *hondurensis*) reaches 35-40 m in height and grows along the Atlantic coast of Central America. It is grown for its timber, used in boat building and light and heavy construction, and its roundwood, used for transmission poles and fence posts. A resin is extracted from the trunk of this tree. Timber production is 10-40 cu.m/ha/annum.

This tree grows at an altitude of 1,000 m and is found in areas with 660-4,000 mm of rain and temperatures of 21-27 degrees Celsius. It needs fertile soil of light to medium texture and full sun. This tree is sensitive to frost, susceptible to needle blight, and attacked by the Dendroctonus beetle. It is moderately fire-resistant.

Caribbean Pine

Sci. Name(s)   : *Pinus caribaea* var. *caribaea*
Geog. Reg(s)   : Pacific Islands, West Indies
End-Use(s)   : Natural Resin, Timber
Domesticated   : Y
Ref. Code(s)   : 228
Summary   : The Caribbean pine (var. *caribaea*) is a frost-sensitive evergreen tree that reaches 20-27 m in height and grows in western Cuba and the Isle of Pines from sea level to 500 m in elevation. Its saw timber is used in heavy and light construction and boat building and its roundwood as transmission poles, fence posts, and longfiber pulp. The wood yields a useful resin. Annual timber production is 10-20 cu.m/ha.

This pine requires an annual rainfall of 1,050-1,800 mm and annual temperatures of 24-26 degrees Celsius. It grows best in light- to medium-textured soils.

Carob, St. John's-Bread
(See Gum & Starch)

Carolina Poplar, Eastern Cottonwood

Sci. Name(s)   : *Populus deltoides* var. *deltoides, Populus deltoides* var. *missouriensis*
Geog. Reg(s)   : America-North (U.S. and Canada)
End-Use(s)   : Timber
Domesticated   : Y
Ref. Code(s)   : 228
Summary   : The Carolina poplar is a deciduous tree of Missouri and Mississippi in the United States. It reaches heights of 25-30 m. Its timber is soft and difficult to preserve and is used primarily for making matches. The roundwood is used for short-fiber pulp, veneer, and plywood. Annual timber production is 20-40 cu.m/ha.

This tree is found at altitudes of 2,000-3,000 m with a mean annual rainfall of 1,200-3,000 mm and a mean annual temperature of 12-16 degrees Celsius. It prefers deep, medium-textured, fertile soil and wide spacing. For maximum growth it requires considerable light. There is wide origin variation. It is susceptible to defoliators, leaf rusts, and borers.

Cassia

Sci. Name(s)   : *Cassia siamea, Cassia florida*
Geog. Reg(s)   : Asia-India (subcontinent), Asia-Southeast
End-Use(s)   : Energy, Erosion Control, Forage, Pasture & Feed Grain, Ornamental & Lawn Grass, Timber
Domesticated   : Y
Ref. Code(s)   : 228
Summary   : Cassia is an evergreen tree reaching 6-7 m in height. It is found in Southeast Asia, including India, Sri Lanka, and Malaya. Its durable timber is used to make fine

furniture. The roundwood is used for building poles, fuel, and charcoal. The leaves are used as fodder. The tree itself serves as a windbreak and ornamental. Cassia is susceptible to a serious disease called *Phaeolus manihotis*, which kills the roots. Timber production is 11-30 cu.m/ha/annum.

This tree has an altitudinal range from sea level to 1,000 m. It grows best in areas with a dry season of 4-6 months, an annual rainfall of 650-950 mm, and temperatures of 22-28 degrees Celsius. The tree prefers deep, light- to medium-textured soil and direct sunlight.

Cedar, Cedro

Sci. Name(s) : *Cedrella odorata, Cedrella mexicana*
Geog. Reg(s) : America-Central, America-South
End-Use(s) : Timber
Domesticated : Y
Ref. Code(s) : 228
Summary : Cedar is a deciduous tree reaching 30-40 m in height and found growing in Central and South America. It is used for light construction, veneer, and plywood. Annual timber production is 11-22 cu.m/ha.

This cedar grows at altitudes up to 1,200 m with 1,600-2,500 mm of rain and temperatures of 22-32 degrees Celsius. It grows best on fertile soils, light to heavy in texture, and needs direct sunlight. It is attacked by root borers. Damage can be prevented by mixing cedars with other species to maintain optimum vigor.

Champac
(See Ornamental & Lawn Grass)

Chicle, Sapodilla, Naseberry, Nispero, Chicle Tree
(See Gum & Starch)

Chinese Chestnut
(See Nut)

Chinese Tallow-Tree, Tallow-Tree
(See Oil)

Chinquapin, Allegheny Chinkapin
(See Nut)

Chir Pine

Sci. Name(s) : *Pinus roxburghii, Pinus longifolia*
Geog. Reg(s) : Asia-Central, Asia-India (subcontinent)
End-Use(s) : Natural Resin, Timber
Domesticated : Y
Ref. Code(s) : 228
Summary : Chir pine is an open-crowned evergreen tree reaching heights of 30-35 m, grown in the Himalayas from Afghanistan to Bhutan, and valued for its spiral-grained timber. Its saw timber is used for light construction and boxes and the roundwood for

building poles, fence posts, and longfiber pulp. It also produces resins. Annual timber production is 7-14 cu.m/ha.

This tree grows at altitudes of 1,200-2,500 m with an annual rainfall of 750-1,100 mm and annual temperatures of 12-20 degrees Celsius. It needs a dry season of 2-4 months and tolerates shallow soils but prefers light- to heavy-textured ones in full sun. The chir pine is frost- and fire-resistant.

Clausena dentata
(See Fruit)

Coconut
(See Oil)

Cohune Nut Palm, Cohune Palm
(See Oil)

Cooba
Sci. Name(s) : *Acacia salicina*
Geog. Reg(s) : Australia
End-Use(s) : Energy, Forage, Pasture & Feed Grain, Ornamental & Lawn Grass, Timber
Domesticated : Y
Ref. Code(s) : 228
Summary : Cooba is an open-crowned evergreen tree grown in the semiarid regions of New South Wales. The roundwood is used for fence posts, fuel, and charcoal. The tree is sometimes used as a shade, windbreak, and ornamental. Tree leaves and shoots are used as fodder.

Trees grow at altitudes up to 2,000 m, in areas with temperatures of 16-26 degrees Celsius and an annual rainfall of 300-700 mm. The trees reach heights of 10-16 m. They grow best in medium to heavy soils and will tolerate saline soils. They are moderately frost-resistant. Annual timber production is 3-5 cu.m/ha.

Coolabah
Sci. Name(s) : *Eucalyptus microtheca*
Geog. Reg(s) : Australia
End-Use(s) : Energy, Miscellaneous, Timber
Domesticated : Y
Ref. Code(s) : 228
Summary : Coolabah is an evergreen tree found in the arid and semiarid regions of north Australia. Its timber has a natural durability and is resistant to termites. The roundwood is used for fence posts, fuel, and charcoal. The tree itself can be used for shade and shelter. Five to 10 cu.m/ha of timber are collected annually.

This evergreen tree stands 10-15 m in height and is found at altitudes up to 1,000 m. For optimum growth it requires considerable light and temperatures of 21-27 degrees Celsius with a mean temperature range of 4-40 degrees Celsius. The mean annual rainfall needed is 250-500 mm. It prefers a heavy-textured soil.

426

Cork Oak
(See Miscellaneous)

Curry-Leaf-Tree
(See Spice & Flavoring)

Cypress Pine

Sci. Name(s)	: *Callitris glauca, Callitris hugelii*
Geog. Reg(s)	: Australia
End-Use(s)	: Timber
Domesticated	: Y
Ref. Code(s)	: 228
Summary	: The cypress pine is an evergreen tree of inland Queensland and New South Wales. This tree produces timber of good natural durability used for light construction and fence posts. Trees make good windbreaks. Two to 10 cu.m/ha of timber are produced annually.

This light-crowned tree grows 18-22 m in height at altitudes of 500-1,500 m. It requires 450-800 mm of rain annually and temperatures of 18-24 degrees Celsius. Cypress pines need considerable radiation and prefer light- to medium-textured soils.

Dafo, Swamp Oak

Sci. Name(s)	: *Terminalia brassii*
Geog. Reg(s)	: Pacific Islands
End-Use(s)	: Timber
Domesticated	: Y
Ref. Code(s)	: 228
Summary	: Dafo is an evergreen tree that reaches heights of 30-35 m and is grown in the Solomon Islands, Bougainville, and New Ireland. Its saw timber is used for light construction and boxes and the roundwood for shortfiber pulp, veneer, and plywood. The saw timber is susceptible to stain and borers. Annual timber production is 25-35 cu.m/ha.

Trees grow from sea level to 500 m in elevation in areas with an annual rainfall of 2,000-5,000 mm and temperatures of 23-28 degrees Celsius. They prefer deep soils, light to medium in texture, and full sun for maximum growth.

Damar

Sci. Name(s)	: *Agathis dammara, Agathis loranthifolia*
Geog. Reg(s)	: Asia-Southeast, Europe, Pacific Islands
End-Use(s)	: Natural Resin, Timber
Domesticated	: Y
Ref. Code(s)	: 228
Summary	: Damar is an evergreen tree reaching heights of 45-60 m and found in Malaysia, Indonesia, the Philippines, New Guinea, and New Britain. It is grown for its timber, used for light construction and boat building, and its roundwood, used for veneer or plywood. The timber exudes a gum. Annual timber production is 20-30 cu.m/ha.

This sturdy tree grows at altitudes of 100-1,600 m in areas with 2,000-4,000 mm of rain and temperatures of 19-28 degrees Celsius. Trees are most productive in soils light to heavy in texture.

Damas

Sci. Name(s)	: *Conocarpus lancifolius*
Geog. Reg(s)	: Africa
End-Use(s)	: Energy, Forage, Pasture & Feed Grain, Timber
Domesticated	: Y
Ref. Code(s)	: 228
Summary	: Damas is an evergreen tree found in Somalia. It grows 15-18 m in height and produces a timber of interlocked grain and good durability. The saw timber is used for light construction and boat building, and the roundwood is used for building poles, fence posts, fuel, and charcoal. The leaves are used for fodder. The tree itself is a shade and shelter. Annual timber production is 5-10 cu.m/ha.

Damas grows at altitudes up to 800 m with 250-600 mm of rain yearly and temperatures of 24-30 degrees Celsius. It needs light-textured soil and can tolerate moderately saline soils. Damas demands considerable light.

Dendrocalamus, Bamboo

Sci. Name(s)	: *Dendrocalamus* spp.
Geog. Reg(s)	: Asia-China, Asia-India (subcontinent), Asia-Southeast
End-Use(s)	: Fiber, Miscellaneous, Timber, Vegetable
Domesticated	: Y
Ref. Code(s)	: 10, 68, 170
Summary	: The Dendrocalamus species includes the tallest and biggest bamboo. This bamboo has erect culms with deciduous and variable-sized leaves. Average bamboo culms are over 30 m tall and reach up to 25 cm in diameter. *Dendrocalamus strictus* is the most common and is India's principal source of acceptable quality paper pulp. It is also used for building and making furniture, baskets, sticks, and other products.

Another member of the species, *Dendrocalamus asper*, is commonly cultivated by Chinese market gardeners in western Malaysia and Java for its large, edible bamboo shoots. The giant bamboo of Burma is *Dendrocalamus giganteus* and reaches heights of 30 m with culm diameters of 25 cm.

Divi-Divi
(See Dye & Tannin)

Douglas-Fir

Sci. Name(s)	: *Pseudotsuga menziesii*
Geog. Reg(s)	: America-North (U.S. and Canada), Temperate
End-Use(s)	: Beverage, Dye & Tannin, Energy, Fat & Wax, Ornamental & Lawn Grass, Timber
Domesticated	: Y
Ref. Code(s)	: 17, 75, 139, 220
Summary	: The Douglas-fir is a tree reaching 60 m in temperate climates. In the areas where it grows well, it is considered a major timber tree. Its wood is light yellow to

red in color and has variable hardness. It is used for general construction work, fuel, and plywood. The bark is used for tanning and the leaves are brewed for tea. It is hardy and adaptable as an ornamental and is sold for Christmas trees.

Dundas Mahogany

Sci. Name(s)	: *Eucalyptus brockwayi*
Geog. Reg(s)	: Australia
End-Use(s)	: Dye & Tannin, Energy, Erosion Control, Timber
Domesticated	: Y
Ref. Code(s)	: 228
Summary	: The dundas mahogany is an evergreen tree that is found growing in a restricted

area in western Australia. Its timber is tough and is most often used for building poles, fuel, and charcoal. It is also the source of a tannin. The tree is used as a shade and windbreak. Four to 5 cu.m/ha of timber are produced annually.

This tree can be grown at altitudes up to 1,500 m with annual rainfall of 250-400 mm and temperatures of 19-25 degrees Celsius. For optimum growth, considerable light is required. Dundas mahogany can tolerate moderately saline soils.

East Indian Sandalwood
(See Essential Oil)

English Walnut, Persian Walnut
(See Nut)

Eucalyptus

Sci. Name(s)	: *Eucalyptus* spp.
Geog. Reg(s)	: America, Australia, Tropical
End-Use(s)	: Beverage, Dye & Tannin, Energy, Erosion Control, Essential Oil, Medicinal, Miscellaneous, Ornamental & Lawn Grass, Timber
Domesticated	: Y
Ref. Code(s)	: 170, 214, 238
Summary	: The eucalyptus consists of a large number of species, and approximately 20 of

these are commercially important. It is a tall tree used as a windbreak and for timber. Tree leaves yield an essential oil used medicinally as an inhalant, soap, gargle spray, or lozenge. It provides rutin, which is used for the treatment of hypertension. The aromatic oil is used in perfumery. The leaves are also brewed as tea. A tannin is extracted from the tree bark. Eucalyptus trees grow in most tropical countries. They can withstand climatic changes and various soil conditions.

Trees are a good source of firewood and are also used in the manufacture of newsprint, fiberboard, and rayon. Trees are often grown as ornamentals and sometimes as bee trees. They are used to stop erosion along river banks. Eucalyptus can tolerate saline and swampy soil conditions. In 1984, the United States imported 384 MT of eucalyptus as an essential oil, valued at $2 million.

Eucalyptus, Broad-Leaved Peppermint Tree
(See Essential Oil)

Eucalyptus urophylla

Sci. Name(s) : *Eucalyptus urophylla*
Geog. Reg(s) : Asia, Asia-Southeast
End-Use(s) : Energy, Miscellaneous, Timber
Domesticated : Y
Ref. Code(s) : 228
Summary : Eucalyptus urophylla is an evergreen tree previously included in *Eucalyptus alba* and *Eucalyptus decaisneana*. It is found in the Western Sunda Islands of Indonesia and stands 35-45 m in height. Its easy-sawing timber is used for building poles, fence posts, fuel, and charcoal. It can also be used for dissolving pulp. This tree is susceptible to termite attack, but 20-30 cu.m/ha of timber are produced anually. This light-demanding tree is found at altitudes of 200-1,500 m with a mean annual temperature of 18-28 degrees Celsius. Rainfall averages about 1,100-1,950 mm yearly. It requires a medium- to heavy-textured soil and is frost-tender.

Eurabbie

Sci. Name(s) : *Eucalyptus St. Johnii, Eucalyptus bicostata*
Geog. Reg(s) : Australia
End-Use(s) : Energy, Timber
Domesticated : Y
Ref. Code(s) : 228
Summary : Eurabbie is an evergreen tree found in Victoria and southeast New South Wales. It reaches 30-40 m in height. Its saw timber is important in heavy construction and can be used for poles, posts, fuel, and charcoal. The tree serves as a shade, shelter, and windbreak. The annual timber yield is 8-14 cu.m/ha.

This tree grows at altitudes of 2,000 to 3,500. Rainfall of 750-2,000 mm and temperatures of 9-16 degrees Celsius are necessary. Medium- to heavy-textured soils are preferred. For optimum growth, considerable radiation is required. Eurabbie is moderately frost-resistant.

European Beech
(See Ornamental & Lawn Grass)

European Chestnut, Spanish Chestnut
(See Nut)

European Hazel, European Filbert
(See Nut)

Flat-Topped Yate

Sci. Name(s) : *Eucalyptus occidentalis*
Geog. Reg(s) : Australia
End-Use(s) : Dye & Tannin, Energy, Erosion Control, Timber
Domesticated : Y
Ref. Code(s) : 228
Summary : Flat-topped yate is an open-crowned evergreen tree. It reaches 20-25 m in height and is grown in the semiarid areas of southwestern Western Australia. The

tree is used for erosion control and shade. Its timber is very strong and rarely sawn. It is used for heavy construction, building poles, fence posts, fuel, and charcoal. It is also the source of a tannin. Annual timber production is 3-8 cu.m/ha.

This moderately frost-resistant tree is found at altitudes up to 1,400 m. Best temperatures are 20-25 degrees Celsius. The mean annual rainfall is 350-550 mm. This tree prefers medium- to heavy-textured soils and will tolerate moderately saline soils. For optimum growth, it needs considerable radiation.

Flooded Gum, Rose Gum

Sci. Name(s)	: *Eucalyptus grandis*
Geog. Reg(s)	: Australia
End-Use(s)	: Energy, Timber
Domesticated	: Y
Ref. Code(s)	: 228
Summary	: Flooded gum is an evergreen tree standing 40-55 m in height. It is grown in

coastal Queensland and New South Wales and is one of the most productive plantation timber crops, with an annual production of 24-70 cu.m/ha. The saw timber is used for heavy and light construction and the roundwood for building and transmission poles, fence posts, veneer, plywood, fuel, and charcoal.

This tree is found at altitudes up to 2,100 m in areas with a mean annual rainfall of 1,000-4,000 mm and temperatures of 17-26 degrees Celsius. Light- to medium-textured soils are best, and for optimum growth considerable sunlight is needed.

Forest Red Gum, Mysore "Hybrid", Izabl, Eucalyptus "C"

Sci. Name(s)	: *Eucalyptus tereticornis, Eucalyptus umbellata*
Geog. Reg(s)	: Australia, Pacific Islands
End-Use(s)	: Energy, Timber
Domesticated	: Y
Ref. Code(s)	: 228
Summary	: Forest red gum is an evergreen tree reaching 35-45 m in height and is found in

Queensland and Papua New Guinea. Its timber of interlocking grain has a good natural durability and is easy to saw. It is used for heavy and light construction, boat building, poles, posts, fuel, and charcoal. Timber production is 12-25 cu.m/ha/annum.

This tree is subject to termite attacks when young. Variations occur within differing provinces. It is often grown at altitudes up to 1,800 m. Precipitation should be 500-1,000 mm annually. The temperature range is 17-27 degrees Celsius. It requires considerable sunlight and will withstand a slight frost.

Gaboon, Mahogany, Okoume

Sci. Name(s)	: *Aucoumea klaineana*
Geog. Reg(s)	: Africa
End-Use(s)	: Timber
Domesticated	: Y
Ref. Code(s)	: 228
Summary	: Gaboon is a buttressed tree reaching heights of 30-40 m and found in the

African rain forests of Gabon and the Congo. Its saw timber is used for light construction and the roundwood for veneer or plywood, but both are of poor quality and are difficult to preserve. Annual timber production is 15-30 cu.m/ha.

This tree adapts to soils ranging from light to heavy in texture and requires full sun. It thrives at altitudes up to 500 m with 1,600-3,000 mm of rain annually and temperatures of 25-33 degrees Celsius.

Green Wattle

Sci. Name(s) : *Acacia decurrens, Acacia decurrens* var. *normalis*
Geog. Reg(s) : Australia
End-Use(s) : Dye & Tannin, Energy, Erosion Control, Timber
Domesticated : Y
Ref. Code(s) : 228
Summary : Green wattle is a tree that grows in Australia and can reach heights of 6-12 m. This tree is found in Victoria, New South Wales, and Queensland and is a timber crop. The roundwood is used for building poles, fence posts, and as a fuel and charcoal source. The timber is also a source of a tannin. The tree is sometimes grown for shade, shelter, and as a windbreak. Yearly timber production is 6-16 cu.m/ha.

Trees will continue to grow at altitudes of 1,500-2,500 m. Annual rainfall requirements are 900-1,600 mm, with temperatures of 12-18 degrees Celsius. Green wattles prefer deep, light to medium soils and will tolerate shade.

Grey Ironbark

Sci. Name(s) : *Eucalyptus paniculata*
Geog. Reg(s) : Australia
End-Use(s) : Energy, Timber
Domesticated : Y
Ref. Code(s) : 228
Summary : Grey ironbark is an evergreen tree 25-30 m in height. It is found in central and southern New South Wales. It has tough timber with interlocking grain and natural durability but is difficult to saw. It is used mainly for heavy construction and sleepers. The roundwood is used for building and transmission poles, fence posts, fuel, and charcoal. Production of timber is 9-18 cu.m/ha/annum.

This moderately frost-resistant tree is found at altitudes of 500-1,500 m. It requires considerable radiation for optimum growth, and temperatures should be 18-23 degrees Celsius. Annual precipitation is approximately 750-1,300 mm. A soil of light to medium texture is preferred. It is tolerant of shallow soils.

Guaiac, Guaiacum
(See Natural Resin)

Gum-Barked Coolibah

Sci. Name(s) : *Eucalyptus intertexta*
Geog. Reg(s) : Australia
End-Use(s) : Energy, Timber
Domesticated : Y
Ref. Code(s) : 228

Summary : Gum-barked coolibah is an evergreen tree that grows in central and southeastern Australia. It reaches heights of 15-20 m. Its roundwood is tough, with an interlocking grain, and is used for fence posts, fuel, and charcoal. Production of timber is 4-5 cu.m/ha/annum.

This tree needs considerable radiation for optimum growth. It is found at altitudes up to 1,250 m with 250-400 mm of annual rainfall and a mean annual temperature of 18-25 degrees Celsius. It tolerates moderately saline soils of light to medium texture.

Gympie, Messmate

Sci. Name(s) : *Eucalyptus cloeziana*
Geog. Reg(s) : Australia
End-Use(s) : Timber
Domesticated : Y
Ref. Code(s) : 228
Summary : Gympie is an evergreen tree reaching heights of 35-45 m and grown in certain parts of Queensland for its timber, suitable for transmission poles and heavy construction. Annual timber production is 15-34 cu.m/ha.

This tree grows best at altitudes up to 1,500 m with rainfall of 900-1,650 mm yearly and annual temperatures of 18-26 degrees Celsius. It is shade-tolerant and requires medium-textured soil.

Heartnut, Cordate Walnut, Siebold Walnut
(See Nut)

Henna
(See Ornamental & Lawn Grass)

Hickory Nut
(See Nut)

Hoop Pine

Sci. Name(s) : *Araucaria cunninghamii*
Geog. Reg(s) : Australia, Pacific Islands
End-Use(s) : Timber
Domesticated : Y
Ref. Code(s) : 228
Summary : The hoop pine is a windfirm evergreen tree reaching heights of 35-45 m and is grown in Papua New Guinea and tropical Australia. Its timber is used in heavy and light construction, as transmission poles, fence posts, veneer, and plywood. Annual timber production is 10-18 cu.m/ha.

This tree grows at altitudes up to 2,000 m with 1,000-1,800 mm of rain annually and temperatures of 16-26 degrees Celsius. This pine grows best on deep, fertile, medium- to heavy-textured soils in full sun.

Horsebean
Sci. Name(s) : *Parkinsonia aculeata*

Geog. Reg(s) : America-North (U.S. and Canada), America-South
End-Use(s) : Energy, Erosion Control, Timber
Domesticated : Y
Ref. Code(s) : 228
Summary : The horsebean is a spiny, short-lived evergreen tree whose roundwood is used for fuel and charcoal. Trees serve as windbreaks and erosion control. They grow 4-5 m in height at altitudes up to 1,400 m and are found from Texas to Peru. Trees need a mean annual rainfall of 250-400 mm with a dry season of 6-8 months and a mean annual temperature of 20-28 degrees Celsius. They prefer light- to medium-textured soils and tolerate moderately saline soils. Horsebeans need considerable light.

Huisache, Cassie Flower, Kolu, Sweet Acacia, Cassie
(See Essential Oil)

Huon-Pine
Sci. Name(s) : *Dacrydium franklinii*
Geog. Reg(s) : Asia-Southeast, Australia, Pacific Islands
End-Use(s) : Essential Oil, Spice & Flavoring, Timber
Domesticated : Y
Ref. Code(s) : 220
Summary : The huon-pine is an evergreen tree found in Southeast Asia, New Caledonia, Australia, and New Zealand. It is a commercial source of timber in New Zealand and Tasmania. Its wood is fairly hard, straight-grained, and strong with a pleasant smell and is used for light construction work. An oil (Huon-pine oil) containing methyl eugenol is distilled from the wood and used in toilet waters, medicinal soaps, and as a source of vanillin.

Idigbo, Framire
Sci. Name(s) : *Terminalia ivorensis*
Geog. Reg(s) : Africa-West
End-Use(s) : Miscellaneous, Timber
Domesticated : Y
Ref. Code(s) : 228
Summary : Idigbo is a buttressed, open-crowned, deciduous tree found in West Africa. Its saw timber is used in heavy and light construction and furniture making, and its roundwood is used for shortfiber pulp, veneer, and plywood. This timber is difficult to preserve but has a valuable decorative quality to it. Trees are grown as an agricultural shade. Annual timber production is 8-17 cu.m/ha.

This tree is found at altitudes from sea level to 700 m in areas with an annual rainfall of 1,300-3,000 mm, a dry season of 2 months, and annual temperatures of 24-26 degrees Celsius. It prefers light to medium soils and full sun for maximum growth. Young trees are susceptible to termites.

Imbe
(See Beverage)

Indian Bael, Bael Fruit, Bengal Quince, Bilva, Siniphal, Bael Tree
(See Fruit)

Jackfruit, Jack
(See Fruit)

Japanese Chestnut
(See Ornamental & Lawn Grass)

Kadam

Sci. Name(s)	:	*Anthocephalus chinensis, Anthocephalus indica, Anthocephalus rich, Anthocephalus cadamba*
Geog. Reg(s)	:	Asia-India (subcontinent), Asia-Southeast, Pacific Islands
End-Use(s)	:	Timber
Domesticated	:	Y
Ref. Code(s)	:	228
Summary	:	Kadam is a deciduous tree reaching heights of 20-30 m and grown in Assam, Bengal, Burma, Sri Lanka, the Philippines, Indonesia, and New Guinea. Its valuable timber is used in light construction and for veneer or plywood.

This tree is shade-tolerant and thrives in light- to medium-textured soils from sea level to 1,000 m in elevation. The needed rainfall is 1,300-4,000 mm annually with an average temperature of 20-32 degrees Celsius.

Kapok, Silk-Cotton-Tree, Ceiba
(See Fiber)

Kariis, Karii
(See Fruit)

Khasya Pine, Khasi Pine, Benguet Pine

Sci. Name(s)	:	*Pinus kesiya, Pinus khasya royle, Pinus insularis*
Geog. Reg(s)	:	Africa, Asia-India (subcontinent), Asia-Southeast
End-Use(s)	:	Energy, Natural Resin, Timber
Domesticated	:	Y
Ref. Code(s)	:	228
Summary	:	Khasya pine is an evergreen tree that reaches 30-35 m. It grows in the Luzon province of the Philippines, and in India, Thailand, Burma, and Africa, where it produces a good quality timber. Saw timber is used for heavy and light construction and boxes. Roundwood is used for building poles, veneer, plywood, fuel, and charcoal. It produces a natural resin. Annual production of timber is 10-30 cu.m/ha.

This tree is found at altitudes of 1,000-2,000 m in areas with 700-1,800 mm of annual precipitation and a mean annual temperature of 17-22 degrees Celsius. It grows on soils of light to heavy texture and is termite-resistant and susceptible to frost. For optimum growth it requires considerable radiation. This tree shows wide variation in origin. It is susceptible to damping off and to attack by *Dothistomiella pini* and shoot borers.

Kiri

Sci. Name(s) : *Paulownia tomentosa*
Geog. Reg(s) : Asia
End-Use(s) : Timber
Domesticated : Y
Ref. Code(s) : 228
Summary : Kiri is a deciduous, windfirm tree reaching 12-16 m in height and is found in Japan. Its timber is used for furniture making. Pruning is necessary for this tree to meet high market specifications. Kiri is subject to canker and defoliator attacks. Annual timber production is 25-35 cu.m/ha.

This tree requires a lot of sunlight and deep, fertile soils of medium texture. It is only moderately frost-resistant. It is found at altitudes of 1,300-1,800 m in areas with temperatures of 20-24 degrees Celsius.

Klinki Pine

Sci. Name(s) : *Araucaria hunsteinii, Araucaria klinki*
Geog. Reg(s) : Pacific Islands
End-Use(s) : Timber
Domesticated : Y
Ref. Code(s) : 228
Summary : The klinki pine is an evergreen tree reaching 40-80 m in height, grown in New Guinea, and used for heavy and light construction. The roundwood serves as transmission poles, fence posts, veneer, and plywood. About 20-30 cu.m/ha/annum of timber are produced.

This tree is found at altitudes of 200-1,400 m in areas receiving 1,600-4,000 mm of rain with temperatures of 20-27 degrees Celsius. The klinki pine prefers deep, fertile soils of medium to heavy texture, needs direct sunlight, and is frost-sensitive.

Lac-tree, Ceylon Oak, Kussum Tree, Malay Lac-tree
(See Fat & Wax)

Laurel, Salmwood

Sci. Name(s) : *Cordia alliodora, Cordia gerecanthus*
Geog. Reg(s) : America-Central, America-South, West Indies
End-Use(s) : Timber
Domesticated : Y
Ref. Code(s) : 228
Summary : Laurel is a deciduous tree reaching 25-30 m in height and grown in Central America, the West Indies, and South America for its timber. The wood shows some resistance to termites and is used for light construction, furniture, veneer, and plywood. Trees are grown as an agricultural shade. About 10-20 cu.m/ha of timber are produced annually.

This tree is found at altitudes from sea level to 1,500 m with 1,000-4,000 mm of rain annually and temperatures of 20-27 degrees Celsius. It prefers medium- to heavy-textured soils and needs full sun for optimum growth.

Leucaena-Hawaiian Type

Sci. Name(s) : *Leucaena leucocephala, Leucaena glauca*
Geog. Reg(s) : America-Central, Asia, Asia-Southeast, Hawaii
End-Use(s) : Energy, Erosion Control, Forage, Pasture & Feed Grain, Timber
Domesticated : Y
Ref. Code(s) : 228
Summary : Leucaena (Hawaiian type) is a small, open-crowned evergreen tree 3-5 m in height. It is grown in Central America, from Mexico to Salvador and is naturalized in the Philippines, Hawaii, and parts of Asia. Its timber is used for fuel and fence posts, its foliage provides fodder, and the tree itself serves as a windbreak and agricultural shade. It is also used for erosion control and soil improvement. Annual production of timber is 20-25 cu.m/ha.

Leucaena is able to withstand shade when young but later demands a lot of light and is only moderately frost-resistant. It tolerates salt winds and has shown a high resistance to pests and disease. This tree is found at an altitudinal range up to 800 m with 600-1,000 mm of rainfall and a mean annual temperature of 20-26 degrees Celsius. It grows in shallow soils of light to heavy texture.

Leucaena-Salvadorian Type

Sci. Name(s) : *Leucaena leucocephala, Leucaena glauca*
Geog. Reg(s) : America-Central
End-Use(s) : Energy, Erosion Control, Forage, Pasture & Feed Grain, Timber
Domesticated : Y
Ref. Code(s) : 228
Summary : Leucaena (Salvadorian type) is an evergreen tree reaching heights of 15-20 m. It occurs naturally in southwestern Mexico and central Guatemala. It is used in light construction and for building and transmission poles, fence posts, fuel, and charcoal. Its leaves are used as fodder. This tree serves as an agricultural shade and aids in soil improvement. Timber production is 30-40 cu.m/ha/annum.

This tree requires considerable light and is only moderately frost-resistant. It has an altitudinal range up to 800 m. It is best suited to a climate with 600-1,000 mm of rain annually and temperatures of 20-28 degrees Celsius. Soils of light to heavy texture are best.

Loblolly Pine

Sci. Name(s) : *Pinus taeda*
Geog. Reg(s) : America-North (U.S. and Canada)
End-Use(s) : Natural Resin, Timber
Domesticated : Y
Ref. Code(s) : 228
Summary : The loblolly pine is an evergreen reaching 40-50 m in height and found in the eastern and southeastern United States. Its saw timber is used in heavy and light construction and for making boxes, and its roundwood is used for transmission poles, longfiber pulp, veneer, and plywood. It produces resins. Annual timber production is 12-30 cu.m/ha.

This pine is found at altitudes of 1,300-2,400 m in areas with an annual rainfall of 900-2,200 mm and temperatures of 13-19 degrees Celsius. For maximum growth it requires light- to medium-textured soil and full sun. It has wide variation in origin. While it is resistant to *Diplodia pini*, it is subject to a canker-forming rust, *Cronartum fusiform*, found in the United States.

Longleaf Pine
(See Essential Oil)

Lukrabao, Chaulmogra Tree
(See Medicinal)

Mahogany, Acajou, Caoba

Sci. Name(s)	: *Swietenia macrophylla*
Geog. Reg(s)	: America-Central, America-South
End-Use(s)	: Timber
Domesticated	: Y
Ref. Code(s)	: 228
Summary	: Mahogany is a deciduous tree reaching 30-40 m in height and found in Central and South America. Its timber is known for its decorative quality and is moderately durable and easily sawn and seasoned. Its saw timber is used in light construction and furniture and boat building, and its roundwood for veneer and plywood. Timber yields are approximately 7-11 cu.m/ha/annum.

This tree is moderately windfirm and is found at altitudes of 50-1,400 m in areas with an annual rainfall of 1,600-4,000 mm, a dry season of 4 months, and annual temperatures of 23-28 degrees Celsius. It prefers medium- to heavy-textured soils and full sun for maximum growth. Seedlings are vulnerable to the Hypsipyla shoot borer.

Maidens Gum

Sci. Name(s)	: *Eucalyptus maidenii*
Geog. Reg(s)	: Australia
End-Use(s)	: Energy, Timber
Domesticated	: Y
Ref. Code(s)	: 228
Summary	: Maidens gum is an evergreen tree that grows in southeast New South Wales and Victoria. It is 35-45 m in height and is most often grown for its roundwood, although its saw timber is sometimes used in heavy construction. The roundwood is used for building and transmission poles, fence posts, fuel,and charcoal. Twenty to 35 cu.m/ha/annum of timber are produced.

This tree is found at altitudes of 1,000-2,100 m with 760-2,000 mm of precipitation and a mean annual temperature of 16-20 degrees Celsius. For maximum growth it needs considerable light, and it is susceptible to frost. It grows on soils of light to heavy texture.

Mammy-Apple, Mammee-Apple
(See Fruit)

Mango
(See Fruit)

Manna Gum

Sci. Name(s) : *Eucalyptus viminalis*
Geog. Reg(s) : Australia, Pacific Islands
End-Use(s) : Timber
Domesticated : Y
Ref. Code(s) : 228
Summary : Manna gum is an evergreen tree reaching heights of 25-30 m and found in the upland areas of Victoria, Tasmania, and New South Wales. It is a source of a second-class timber used for light construction, transmission poles, veneer, and plywood. This tree is susceptible to attacks by the Gonipterus beetle. Timber production is 10-30 cu.m/ha/annum.

This tree prefers deep soils, light to heavy in texture, and full sun. It is moderately frost-resistant. It grows at an altitudinal range of 2,000-3,000 m in elevation in areas with an annual rainfall of 750-2,500 mm and temperatures of 10-16 degrees Celsius.

Maritime Pine

Sci. Name(s) : *Pinus pinaster, Pinus maritima*
Geog. Reg(s) : Europe
End-Use(s) : Timber
Domesticated : Y
Ref. Code(s) : 228
Summary : The maritime pine is a small, windfirm evergreen tree that reaches heights of 20-30 m and is found in Portugal. Its saw timber is used in heavy and light construction and for making boxes, while its roundwood is used for building and transmission poles, fence posts, and longfiber pulp. Annual timber production is 12-24 cu.m/ha.

This tree is moderately tolerant of salt winds and frost. It tolerates shallow soils and is adaptable to most soil conditions. For maximum growth it needs full sun. It grows at an altitude of 1,400-2,500 m with an annual rainfall of 625-1,300 mm and annual temperatures of 12-18 degrees Celsius.

Mayflower, Apamate, Roble

Sci. Name(s) : *Tabebuia pentaphylla*
Geog. Reg(s) : America-Central
End-Use(s) : Miscellaneous, Ornamental & Lawn Grass, Timber
Domesticated : Y
Ref. Code(s) : 228
Summary : The mayflower is an open-crowned, deciduous tree reaching 25-30 m in height that occurs naturally from southern Mexico to Venezuela and Ecuador. Its timber has good natural durability, is easy to saw and preserve, and has a desirable decorative quality. The saw timber is used in light construction and furniture and the roundwood for veneer and plywood. The tree is important as an agricultural shade and ornamental. Annual timber production is 10-20 cu.m/ha.

This tree is found at altitudes of 100-1,000 m in areas with an annual rainfall of 1,250-2,500 mm and temperatures of 22-27 degrees Celsius and a dry season of

3 months. It prefers light- to medium-textured soils but adapts to most soil conditions. For maximum growth it needs considerable sunlight.

Mesquite, Algorrobo

Sci. Name(s) : *Prosopis juliflora*
Geog. Reg(s) : America, America-Central, America-North (U.S. and Canada)
End-Use(s) : Energy, Forage, Pasture & Feed Grain, Timber
Domesticated : Y
Ref. Code(s) : 228
Summary : Mesquite is a spiny, deciduous tree found from the southwestern United States through Central America to Ecuador. It reaches heights of 5-10 m. Although durable, its timber is of poor quality. Its roundwood is used for fence posts, fuel, and charcoal. The seed pods are used for fodder. Annual timber yields are 3-5 cu.m/ha.

This tree is found in areas from sea level up to 2,000 m with a mean annual temperature of 200-600 mm and a temperature of 16-28 degrees Celsius. It grows on soils of light to heavy texture and tolerates saline soils. For maximum growth it needs considerable radiation.

Messmate

Sci. Name(s) : *Eucalyptus obliqua*
Geog. Reg(s) : Australia, Pacific Islands
End-Use(s) : Energy, Timber
Domesticated : Y
Ref. Code(s) : 228
Summary : Messmate is an evergreen tree that reaches heights of 50-60 m and is found in New South Wales, Victoria, and Tasmania. The tree is sometimes used as a shade and shelter but is most often used for light construction and furniture. The roundwood is used for poles, posts, veneer, plywood, fuel, and charcoal.

This frost-resistant tree adapts to a variety of soils, although a medium- to heavy-textured soil is preferred. It grows at altitudes of 2,000-3,000 m with an annual rainfall of 1,000-3,000 mm and a mean annual temperature of 9-16 degrees Celsius. It is moderately fire-resistant. For optimum growth, messmate requires considerable light.

Mexican Cypress, Portuguese Cedar, Kenya Cypress

Sci. Name(s) : *Cupressus lusitanica, Cupressus lindleyi, Cupressus glauca*
Geog. Reg(s) : America-Central
End-Use(s) : Timber
Domesticated : Y
Ref. Code(s) : 228
Summary : The Mexican cypress is an evergreen tree reaching heights of 25-30 m. It grows in the mountains of Mexico and Guatemala and its timber is used for light and heavy construction. The fine, soft wood is difficult to preserve but is used for joinery and furniture. The roundwood is used for building poles and plywood. This tree serves as a shade, shelter, and windbreak. Timber production is 15-40 cu.m/ha/annum.

This tree is found at altitudes of 1,300-3,300 m with an annual precipitation of 1,000-1,500 mm and mean temperatures of 10-17 degrees Celsius. It thrives in deep, medium-textured soils. It is shade-tolerant and frost-resistant.

Mexican White Pine

Sci. Name(s) : *Pinus ayacahuite*
Geog. Reg(s) : America-Central
End-Use(s) : Natural Resin, Timber
Domesticated : Y
Ref. Code(s) : 228
Summary : The Mexican white pine is an evergreen tree reaching 30-35 m in height and grown for its timber and resin. The saw timber is used in light construction, furniture, and boxes, and its roundwood for longfiber pulp and as veneer and plywood. Annual timber production is 8-15 cu.m/ha.

Trees grow in Central America from southern Mexico to Guatemala at altitudinal ranges of 1,800-3,100 m and need an annual rainfall of 1,200-2,500 mm and temperatures of 13-17 degrees Celsius. For optimum growth, trees require deep, fertile soils that are light to medium textured. They are generally resistant to pests and diseases.

Michoacan Pine

Sci. Name(s) : *Pinus michoacana*
Geog. Reg(s) : America-Central
End-Use(s) : Natural Resin, Timber
Domesticated : Y
Ref. Code(s) : 228
Summary : The Michoacan pine is an evergreen tree reaching heights of 20-25 m and grown for its timber in central and southern Mexico. Its timber is easily sawn and used for light and heavy construction and boxes; its roundwood is used for transmission poles, veneer, plywood, and longfiber pulp. A resin is also produced. Timber production is 6-12 cu.m/ha/annum.

This tree prefers deep, medium- to heavy-textured soils. It is commonly found at altitudes of 1,000-2,300 m with an annual rainfall of 1,000-1,700 mm and annual temperatures of 14-21 degrees Celsius. It is somewhat fire- and frost-resistant and requires considerable sun for maximum growth.

Mindanao Gum

Sci. Name(s) : *Eucalyptus deglupta, Eucalyptus naudiniana*
Geog. Reg(s) : Asia-Southeast, Pacific Islands
End-Use(s) : Energy, Timber
Domesticated : Y
Ref. Code(s) : 228
Summary : Mindanao gum is an evergreen tree that reaches heights of 50-60 m. It grows in the Philippines, Sulawesi, Papua New Guinea, and New Britain. Its timber is lighter than most other eucalypti. It is used for heavy and light construction, furniture, boat building, fuel, charcoal, veneer, and plywood. Timber production is 14-50 cu.m/ha/annum.

This tree is somewhat termite-resistant. It needs considerable light for maximum growth and deep, fertile soils of light to medium texture. It can grow at altitudes up to 1,800 m with a mean annual rainfall of 2,000-5,000 mm and a mean annual temperature of 20-32 degrees Celsius.

Monkey-Pod

Sci. Name(s)	: *Lecythis ollaria*
Geog. Reg(s)	: Asia-Southeast, Hawaii
End-Use(s)	: Nut, Oil, Timber
Domesticated	: Y
Ref. Code(s)	: 170, 220
Summary	: Monkey-pods are tall, wide-spreading trees grown for their beautiful red-gray wood valued for carving, particularly in Hawaii and the Philippines. This wood is very resistant to marine borers and often used for piers and other aquatic structures. Monkey-pod seeds are edible and contain sapucaja oil, used for lighting and soap manufacture.

Monterey Cypress

Sci. Name(s)	: *Cupressus macrocarpa, Cupressus hartwegii*
Geog. Reg(s)	: America-North (U.S. and Canada)
End-Use(s)	: Energy, Erosion Control, Timber
Domesticated	: Y
Ref. Code(s)	: 228
Summary	: The Monterey cypress is an evergeen tree found in the United States in Monterey along the California coast. It reaches 15-25 m in height and provides timber used in light construction and for building poles, fence posts, veneer, plywood, fuel, and charcoal. The tree must be pruned to prevent branching and knotty timber. It is frost-resistant and withstands salt winds. It is often used as a windbreak and for erosion control. Annual timber production is 11-25 cu.m/ha.

This evergreen tree is found at altitudes of 1,200-3,500 m with a mean annual rainfall of 700-1,600 mm and a mean annual temperature of 14-20 degrees Celsius. It grows on light- to medium-textured soils and is tolerant of shade and moderately saline soils.

Monterey Pine, Radiata

Sci. Name(s)	: *Pinus radiata, Pinus insignis*
Geog. Reg(s)	: America-North (U.S. and Canada), Australia, Europe, Pacific Islands
End-Use(s)	: Timber
Domesticated	: Y
Ref. Code(s)	: 228
Summary	: The Monterey pine is an evergreen found in isolated areas along the California coast. Its saw timber is used for light and heavy construction and boxes. The roundwood is used for building and transmission poles, fence posts, longfiber pulp, veneer, and plywood. This tree serves as a windbreak. Annual timber production is 12-30 cu.m/ha.

This tree is suited to altitudes of 1,500-3,000 m with an annual rainfall of 650-1,600 mm, a dry season of 2-3 months, and annual temperatures of 11-18 degrees

442

Celsius. For optimum growth, it requires moderate sun and light to medium soils. It is susceptible to frost and tolerant of salt winds but can be damaged by *Diplodia pini*.

Montezuma Pine

Sci. Name(s) : *Pinus montezumae*
Geog. Reg(s) : America-Central
End-Use(s) : Energy, Natural Resin, Timber
Domesticated : Y
Ref. Code(s) : 228
Summary : Montezuma pine is an evergreen tree reaching heights of 25-30 m. Trees must be pruned to prevent the timber from becoming knotty. Its saw timber is used for light and heavy construction and boxes. Its roundwood is used for fence posts, fuel, charcoal, veneer, plywood, and longfiber pulp. Other products include resins. Timber production is 6-12 cu.m/ha/annum.

This tree is found in the highlands of Mexico and Guatemala at altitudes of 1,400-3,000 m. Required climatic conditions include a mean annual rainfall of 900-1,600 mm and a mean annual temperature of 11-18 degrees Celsius. It prefers deep, fertile soils of light to medium texture. For optimum growth, considerable light is required but it tolerates shade when young. It is moderately fire- and frost-resistant.

Mountain Ash

Sci. Name(s) : *Eucalyptus regnans*
Geog. Reg(s) : Australia, Pacific Islands
End-Use(s) : Timber
Domesticated : Y
Ref. Code(s) : 228
Summary : Mountain ash is an evergreen tree reaching heights of 60-100 m and grown in Victoria and Tasmania. Its white, odor-free timber is used in light construction and furniture and its roundwood for transmission poles, veneer, and plywood. The annual timber yield is 11-15 cu.m/ha.

This tree is frost-resistant, needs full sun, and thrives on deep soils medium to heavy in texture. It is grown at altitudes of 2,000-3,200 m in areas with an annual rainfall of 1,000-2,000 mm, a dry season of 2 months, and annual temperatures of 10-16 degrees Celsius. The mountain ash is susceptible to the Gonipterus beetle.

Mountain Gum

Sci. Name(s) : *Eucalyptus dalrympleana*
Geog. Reg(s) : Australia, Pacific Islands
End-Use(s) : Timber
Domesticated : Y
Ref. Code(s) : 228
Summary : Mountain gum is an evergreen tree that reaches heights of 25-35 m and is found in southern New South Wales, Victoria, and Tasmania. It is grown for its timber, used in light construction and boxes. Annual timber production is 8-10 cu.m/ha.

This tree grows at altitudes of 2,000-3,500 m. The dry season usually lasts for 2 months. Annual rainfall is 750-1,500 mm in areas with temperatures of 10-14 degrees Celsius.

Musizi

Sci. Name(s) : *Maesopsis eminii*
Geog. Reg(s) : Africa-Central, Tropical
End-Use(s) : Timber
Domesticated : Y
Ref. Code(s) : 228
Summary : The musizi is a deciduous, short-lived tree reaching 30-40 m in height with timber that has several economic uses. The saw wood is used in light construction, furniture, and boxes, and the roundwood as building poles, shortfiber pulp, veneer, and plywood. This timber is not naturally durable. Trees grow in tropical Central Africa. Annual timber production is 8-20 cu.m/ha.

It grows at altitudes of 100-700 m in areas receiving 1,200-3,000 mm of rain. Trees reach optimum growth under annual temperatures of 22-27 degrees Celsius. Musizis like deep, fertile soils, light to medium in texture, and full sun.

Narrow-Leaved Ironbark

Sci. Name(s) : *Eucalyptus crebra, Eucalyptus racemosa*
Geog. Reg(s) : Australia, Tropical
End-Use(s) : Energy, Timber
Domesticated : Y
Ref. Code(s) : 228
Summary : Narrow-leaved ironbark is an evergreen tree that grows in inland tropical New South Wales and Queensland. It reaches 20-25 m in height. Its timber is rarely sawn, but is used for heavy construction, building poles, fuel, and charcoal. Timber yields are 3-8 cu.m/ha/annum.

This tree is moderately frost-resistant and requires considerable light. It grows at altitudes up to 1,400 m with about 500-700 mm of rainfall and temperatures of 18-26 degrees Celsius. Light- to medium-textured soils are most successful for optimum growth.

Neem

Sci. Name(s) : *Azadirachta indica, Melia azadirachta*
Geog. Reg(s) : Asia-India (subcontinent), Asia-Southeast
End-Use(s) : Energy, Forage, Pasture & Feed Grain, Timber
Domesticated : Y
Ref. Code(s) : 228
Summary : Neem is an evergreen tree that grows in the drier areas of India, Burma, Thailand, and Cambodia. It reaches about 20-25 m in height and has several uses. The tree itself is used for shade and shelter, its timber for light construction, building poles, fence posts, fuel, and charcoal, and its leaves for fodder. Annual timber production is 5-18 cu.m/ha.

This tree grows best in deep, fertile soils. It has a moderate demand for light and is frost-tender. Neem grows best at altitudes up to 500 m with 450-1,000 mm of rain and a dry season of 5-7 months. Temperatures are 24-33 degrees Celsius.

Obeche, Samba, Wawa, Ayous

Sci. Name(s)	: *Triplochiton scleroxylon*
Geog. Reg(s)	: Africa-West
End-Use(s)	: Timber
Domesticated	: Y
Ref. Code(s)	: 228
Summary	: The obeche is a deciduous, buttressed tree reaching heights of 40-50 m and found in West Africa. Its timber is difficult to preserve but easy to saw and season. It is considered a premier quality, white, odor-free hardwood. Its saw timber is used for light construction, boxes, and furniture and its roundwood for shortfiber pulp, veneer, and plywood. Annual timber production is 6-18 cu.m/ha.

This tree is found at altitudes from sea level to 500 m in areas with an annual rainfall of 1,600-3,000 mm, temperatures of 24-29 degrees Celsius, and a dry season of 2 months. It prefers soils light to medium in texture and full sun. Obeche is susceptible to attack by defoliators, borers, and stain.

Ocotea, Carela sassafraz
(See Essential Oil)

Oil-Bean Tree, Owala Oil
(See Oil)

Olive
(See Oil)

Opepe, Bilinga

Sci. Name(s)	: *Nauclea diderichii*
Geog. Reg(s)	: Africa-Central, Africa-West
End-Use(s)	: Energy, Timber
Domesticated	: Y
Ref. Code(s)	: 228
Summary	: Opepe is an evergreen tree of West and Central tropical Africa that attains heights of 30-40 m. Its timber is decorative with an interlocking grain and natural durability. It is used for heavy construction and boat building. Its roundwood is used for transmission poles, fuel, and charcoal. This tree is susceptible to attack by a variety of borers. Annual production of timber is 3-10 cu.m/ha.

This light-demanding tree is found at altitudes up to 500 m. Optimum rainfall is 2,000-2,500 mm annually with a mean temperature range of 24-38 degrees Celsius. A soil of light to medium texture is preferred.

Paperbark

Sci. Name(s)	: *Malaleuca leucadendron, Malaleuca leucadendra, Malaleuca quinquenervia*
Geog. Reg(s)	: Asia-Southeast, Australia

End-Use(s)   : Energy, Erosion Control, Oil, Timber
Domesticated : Y
Ref. Code(s)   : 228
Summary   : The paperbark is an evergreen tree whose timber is naturally durable and easily preserved. The saw timber is used in heavy construction and the roundwood serves as building and transmission poles, shortfiber pulp, fuel, and charcoal. A useful oil is also produced from tree leaves. Trees provide shade and shelter and are often used for erosion control. Paperbarks grow in Southeast Asia from Burma to Indonesia, in the Philippines, and in tropical Australia. Annual timber production is 10-16 cu.m/ha.

         Paperbark coppices demand considerable light for good growth. They withstand saline soils of light to heavy texture. Trees grow from sea level to altitudes of 800 m in areas with 800-1,600 mm of rainfall with a mean annual temperature of 22-28 degrees Celsius.

Parana Pine

Sci. Name(s)   : *Araucaria angustifolia, Araucaria brasiliana*
Geog. Reg(s)   : America-South
End-Use(s)   : Timber
Domesticated : Y
Ref. Code(s)   : 228
Summary   : The Parana pine is an evergreen tree reaching heights of 25-30 m and found in southern Brazil, mainly Parana, and is grown for its timber, used for heavy and light construction work, furniture making, transmission poles, and fence posts. Annual timber production is 10-23 cu.m/ha.

         This pine tree grows at altitudes of 1,500-2,200 m with a needed annual rainfall of 1,250-2,200 mm. Temperatures are 12-18 degrees Celsius. This tree thrives in deep, fertile soils of medium texture and does not tolerate shade.

Patula Pine

Sci. Name(s)   : *Pinus patula*
Geog. Reg(s)   : America-Central
End-Use(s)   : Timber
Domesticated : Y
Ref. Code(s)   : 228
Summary   : The Patula pine is an evergreen tree of southern central Mexico that reaches heights of 20-30 m and is grown for its strong, light timber. The saw timber is used for light construction and boxes, while the roundwood is used for transmission poles, fence posts, and longfiber pulp. Annual timber production is 15-40 cu.m/ha.

         This tree prefers deep soils of light to medium texture and, for optimum growth, requires full sun. It is found at altitudes of 1,400-3,200 m in areas with an annual rainfall of 750-2,000 mm and annual temperatures of 12-18 degrees Celsius. It is a moderately frost-resistant tree but is subject to attack by *Diplodia pini*, which causes cankers and dieback.

Pear, Common Pear
(See Fruit)

Pecan
(See Nut)

Peru Balsam, Balsam-of-Peru
(See Natural Resin)

Phalsa
(See Beverage)

Pimento, Allspice
(See Spice & Flavoring)

Pino Blanco

Sci. Name(s)	: *Pinus pseudostrobus*
Geog. Reg(s)	: America-Central
End-Use(s)	: Energy, Natural Resin, Timber
Domesticated	: Y
Ref. Code(s)	: 228
Summary	: The pino blanco is a small evergreen tree that reaches heights of 25-35 m. This tree grows in the highlands of Mexico, Guatemala, and Honduras. Its timber becomes knotty if not pruned. Its saw timber is used for light construction and boxes and its roundwood for fence posts, fuel, charcoal, and longfiber pulp. It also contains resins. Annual timber production is 15-30 cu.m/ha.

Trees grow at altitudes of 1,300-2,800 m with a dry season lasting 3 months, a mean annual temperature of 13-18 degrees Celsius, and a mean annual rainfall of 1,000-1,500 mm. It prefers soils of medium to heavy texture and requires a moderate amount of light for maximum growth. It is moderately frost-sensitive and is widely subject to origin variation. It is susceptible to attack by *Diplodia pini*.

Pinus brutia

Sci. Name(s)	: *Pinus brutia, Pinus halepensis* var. *brutia*
Geog. Reg(s)	: Mediterranean
End-Use(s)	: Natural Resin, Timber
Domesticated	: Y
Ref. Code(s)	: 228
Summary	: Pinus brutia is an evergreen tree that is moderately fire- and termite-resistant and extremely frost-hardy. It is grown in the northeast Mediterranean, Cyprus, and Turkey, and its main products are timber and a resin. Its saw timber is used in heavy and light construction and for furniture. Its roundwood serves as transmission poles and fence posts. The caterpillar *Thaumetopoea wilkinsoni* is a major pest in the Mediterranean. Two to 6 cu.m/ha/annum of timber are produced.

Climatic requirements for Pinus brutia are altitudes of 1,500-2,500 m, a mean annual rainfall of 400-900 mm, and temperatures of 15-20 degrees Celsius. It is adaptable to a variety of soils and will tolerate shallow soils.

Pinus greggii

Sci. Name(s)	: *Pinus greggii, Pinus pseudopatula*

Geog. Reg(s)   : America-Central
End-Use(s)   : Timber
Domesticated : Y
Ref. Code(s)   : 228
Summary   : The Pinus greggii is an evergreen tree whose saw timber is used in light construction and whose roundwood is used as fence posts and longfiber pulp. The timber is soft and weak but is easily preserved. The frost-resistant tree grows in the mountains of Mexico at altitudes of 1,700-3,100 m. Annual timber production is 5-13 cu.m/ha.

This tree grows to heights of 15-18 m and needs an average rainfall of 650-800 mm and a mean annual temperature of 10-17 degrees Celsius for optimum growth. It prefers deep soils of medium to heavy texture but is adaptable to a range of soils. It needs a lot of light.

Pinus merkusii

Sci. Name(s)   : *Pinus merkusii*
Geog. Reg(s)   : Asia-Southeast
End-Use(s)   : Natural Resin, Timber
Domesticated : Y
Ref. Code(s)   : 228
Summary   : Pinus merkusii is a tall evergreen tree reaching heights of 50-60 m at maturity. Its primary product is a saw timber, which is used for heavy and light construction and making boxes. Its roundwood is used for transmission poles, longfiber pulp, veneer, and plywood. This tree is also the source of a resin. Annual production of timber is 12-27 cu.m/ha. It requires altitudes of 800-1,600 m and temperatures of 19-23 degrees Celsius for optimum growth. Annual rainfall should be about 2,000-3,000 mm, with a dry season lasting not longer than 2 months. This tree is termite-resistant and for maximum growth needs considerable radiation. While it prefers light- to medium-textured soils, it can adapt to most soil conditions. It is subject to attack by Looper caterpillars in Indonesia and also to damping-off in the nursery.

Pinus oocarpa

Sci. Name(s)   : *Pinus oocarpa*
Geog. Reg(s)   : America-Central
End-Use(s)   : Natural Resin, Timber
Domesticated : Y
Ref. Code(s)   : 228
Summary   : Pinus oocarpa is a Central American evergreen tree. This light-crowned tree stands 20-30 m in height at maturity and is found from central Mexico to Nicaragua. Its saw timber is used for light construction and boxes. Its roundwood is used for transmission poles, fence posts, and longfiber pulp. A resin can also be obtained from this tree. Timber production is 10-40 cu.m/ha/annum, depending on the province in which the tree is grown.

This tree is found at altitudes of 1,000-2,400 m with a mean annual temperature of 13-21 degrees Celsius and a mean annual rainfall of 750-1,500 mm. For optimum growth, considerable light is required and soils of light to heavy texture. It can tolerate shallow soils. The tree is termite-resistant and shows wide origin variation.

Pinus pumilio, Swiss Mountain Pine, Dwarf Pine
(See Essential Oil)

Pinus strobus var. chiapensis

Sci. Name(s)	: *Pinus strobus* var. *chiapensis*
Geog. Reg(s)	: America-Central
End-Use(s)	: Natural Resin, Timber
Domesticated	: Y
Ref. Code(s)	: 228
Summary	: Pinus strobus var. chiapensis is a light, feathery-crowned evergreen tree. It reaches heights of 25-30 m. It is grown in southern Mexico and Guatemala. Its saw timber is used for light construction and furniture, and the roundwood is used for longfiber pulp. Resins are also produced. Annual timber production is 10-30 cu.m/ha.

It prefers soils that are light textured as well as deep and fertile for maximum growth. This tree grows at altitudes of 600-1,800 m with a mean annual rainfall of 1,000-1,600 mm, a dry season of 2-3 months, and a mean annual temperature of 17-23 degrees Celsius. For maximum growth, it needs considerable radiation.

Plum, Common Plum, European Plum, Garden Plum, Prune Plum
(See Fruit)

Pochote

Sci. Name(s)	: *Bombacopsis quinatum, Bombacopsis sepium*
Geog. Reg(s)	: America-Central, America-South
End-Use(s)	: Timber
Domesticated	: Y
Ref. Code(s)	: 228
Summary	: Pochote is a spiny deciduous tree reaching 30-40 m in height, found from Nicaragua to Venezuela, and used for light construction and furniture making. This tree grows at altitudes from sea level to 800 m in areas with an average rainfall of 800-1,200 mm annually, a dry season of 3-5 months, and temperatures of 20-27 degrees Celsius. It does best in full sun. This tree is prone to attacks by beetles and weevils.

Ponderosa Pine

Sci. Name(s)	: *Pinus ponderosa* var. *arizonica, Pinus arizonica*
Geog. Reg(s)	: America-Central, America-North (U.S. and Canada)
End-Use(s)	: Energy, Erosion Control, Natural Resin, Timber
Domesticated	: Y
Ref. Code(s)	: 228
Summary	: The Ponderosa pine is an evergreen tree that grows in New Mexico, Arizona, and northern Mexico. It reaches heights of 25-30 m. It is used for its timber, as a windbreak, and for erosion control. Its saw timber is used in heavy and light construction and for making boxes, and its roundwood for transmission poles, fence posts, fuel, charcoal, longfiber pulp, veneer, and plywood. In addition, it contains resins.

Trees prefer light to medium soils and are tolerant of shallow soils. Young trees tolerate shade, but for optimum growth they need considerable radiation. They are found at altitudes of 2,000-3,000 m with a mean annual rainfall of 650-900 mm, a dry season of 1-4 months, and a mean annual temperature of 8-15 degrees Celsius. They are fire- and frost-resistant and windfirm.

Primavera

Sci. Name(s)	: *Cybistax donnellsmithii, Tabebuia donnellsmithii*
Geog. Reg(s)	: America-Central
End-Use(s)	: Timber
Domesticated	: Y
Ref. Code(s)	: 228
Summary	: Primavera is a deciduous tree reaching 25-33 m in height and found in southern Mexico and along the Pacific coast of Guatemala. Its timber has poor natural durability but is easy to preserve and is used primarily for light construction, furniture, veneer, and plywood. Timber production is 20-30 cu.m/ha/annum.

This tree thrives in deep, light- to medium-textured soils in full sun. It grows from sea level to 600 m in altitude in areas receiving 1,000-3,000 mm/annum of rainfall and temperatures of 23-28 degrees Celsius.

Pulasan
(See Fruit)

Raintree, Saman

Sci. Name(s)	: *Samanea saman, Pithecolobium saman*
Geog. Reg(s)	: America-South
End-Use(s)	: Forage, Pasture & Feed Grain, Miscellaneous, Timber
Domesticated	: Y
Ref. Code(s)	: 228
Summary	: The raintree is a small, open-crowned tree reaching heights of 15 m. It is found in Ecuador, Colombia, and Venezuela. Its timber is of moderate natural durability, is easy to saw but difficult to preserve, and has an interlocked grain. Its saw timber is used in heavy and light construction, furniture, and boxes, and its roundwood for fence posts, veneer, and plywood. This tree also produces fodder and is an agricultural shade. Annual timber production yields are 25-35 cu.m/ha.

This tree grows at altitudes from sea level to 700 m in areas with a mean annual rainfall of 760-3,000 mm and temperatures of 22-28 degrees Celsius. It needs a dry season of 2-4 months. Trees grow on light- to heavy-textured soils and need considerable light for maximum growth.

Ramon, Breadnut
(See Fruit)

Rattan

Sci. Name(s)	: *Calamus rotang*
Geog. Reg(s)	: Asia-China, Asia-Southeast
End-Use(s)	: Fiber, Miscellaneous, Timber

Domesticated : Y
Ref. Code(s) : 16, 171
Summary : Rattan is an easily grown tree whose light, flexible canes are valued for furniture, walking sticks, and other products. Most canes are taken from trees in Malaysian forests. Some trees are cultivated and yield usable canes in 6 years. Full production occurs in 15 years. For basketry and other woven products, the outer cane is stripped into slender lengths. Dried canes are exported to Europe, the United States, and the Far East.

Young plants thrive in rooting mediums rich in leaf mold. Older trees need more substantial soil with ground bone, charcoal nutrients, and plenty of water.

Red Ironbark

Sci. Name(s) : *Eucalyptus sideroxylon*
Geog. Reg(s) : Australia
End-Use(s) : Dye & Tannin, Energy, Ornamental & Lawn Grass, Timber
Domesticated : Y
Ref. Code(s) : 228
Summary : Red ironbark is an ornamental evergreen tree that occurs naturally in inland Victoria. Its timber is rarely sawn, but is used for heavy construction, fence posts, fuel, and charcoal. It also produces a tannin. Timber production is 4-9 cu.m/ha/annum.

This tree is susceptible to attacks by the Gonipterus beetle and termites. It adjusts to shallow and moderately saline soils of light to heavy texture. It demands considerable light and is most often found up to 2,000 m in altitude with a mean annual rainfall of 420-750 mm and mean annual temperatures of 19-24 degrees Celsius.

Red Mahogany

Sci. Name(s) : *Eucalyptus resinifera, Eucalyptus hemilampia*
Geog. Reg(s) : Australia
End-Use(s) : Energy, Timber
Domesticated : Y
Ref. Code(s) : 228
Summary : Red mahogany is an evergreen tree reaching heights of 30-40 m. It grows in coastal Queensland and New South Wales. It has naturally durable timber of tough interlocking grain, which is easy to saw and is used for heavy construction, boat building, poles, posts, fuel, and charcoal. Young trees are susceptible to termite attacks. Timber production is 5-11 cu.m/ha/annum.

Trees are frost-sensitive. They are grown at altitudes of 900-2,000 m with a mean annual rainfall of 1,000-3,000 mm and a mean annual temperature of 15-21 degrees Celsius. They prefer soils of light to medium texture and are shade-tolerant in their early stages.

Red Mulberry
(See Ornamental & Lawn Grass)

Red Quebracho
(See Dye & Tannin)

Red River Gum

Sci. Name(s)	: *Eucalyptus camaldulensis, Eucalyptus rostrata*
Geog. Reg(s)	: Australia
End-Use(s)	: Energy, Timber
Domesticated	: Y
Ref. Code(s)	: 228
Summary	: The red river gum is an evergreen tree used mainly for its timber. It is 30-40 m in height and grows in Australia, north of latitude 32 degrees south. This crop, and the one immediately following, have identical common and scientific names. Growing requirements vary however. Its tough timber with interlocking grain is used for heavy construction, and its roundwood is used for building and transmission poles, fuel, and charcoal. Timber production is 15-25 cu.m/ha/annum.

This tree is found at altitudes up to 1,400 m with a rainfall averaging 250-1,250 mm annually and temperatures of 19-26 degrees Celsius. Trees need direct sunlight for optimum growth and are frost-sensitive. Red river gums tolerate moderately saline soils.

Red River Gum

Sci. Name(s)	: *Eucalyptus camaldulensis, Eucalyptus rostrata*
Geog. Reg(s)	: Australia
End-Use(s)	: Energy, Timber
Domesticated	: Y
Ref. Code(s)	: 228
Summary	: The red river gum is another evergreen tree that grows in Australia but not in Tasmania. It grows south of latitude 32 degrees south. This crop, and the one immediately preceding, have identical common and scientific names. Growing requirements vary however. It reaches heights of 35-45 m and its timber is used for heavy construction. Roundwood is used for building and transmission poles, fuel, and charcoal. Annual timber production is 10-22 cu.m/ha.

Trees grow at altitudes of 500-2,000 m with a mean annual precipitation of 400-1,000 mm and temperatures of 16-24 degrees Celsius. They tolerate shallow soils and are moderately frost-resistant. For optimum growth, they need considerable light.

Red Sanderswood, Red Sandalwood, Calialur Wood
(See Dye & Tannin)

Red Silk-Cotton, Silk Cotton
(See Fiber)

Russian Olive

Sci. Name(s)	: *Elaeagnus angustifolia*
Geog. Reg(s)	: Africa-North, Asia, Asia-Central, Europe, Mediterranean
End-Use(s)	: Energy, Erosion Control, Timber
Domesticated	: Y
Ref. Code(s)	: 228
Summary	: The Russian olive is a spiny deciduous tree found growing in southern Europe, the Middle East, and from Central Asia to the Himalayas. It is 4-8 m in height. Its

452

timber is used for fuel and charcoal, and trees are used for windbreaks, dune fixers, and erosion control. Annual timber production is 3-6 cu.m/ha.

Altitudinal range for this tree is 1,600-2,500 m. Annual rainfall is approximately 250-600 mm, and mean temperatures are normally 8-14 degrees Celsius. It tolerates saline soils but requires considerable light for optimum growth. Trees are frost-resistant.

Salmon Gum

Sci. Name(s) : *Eucalyptus salmonophloia*
Geog. Reg(s) : Australia
End-Use(s) : Energy, Timber
Domesticated : Y
Ref. Code(s) : 228
Summary : Salmon gum is an evergreen tree capable of growing to heights of 18-25 m. This tree originates inland, southwest of western Australia. Its strong timber is naturally durable and is used for heavy construction, poles, posts, fuel, and charcoal. Timber production figures are unavailable.

This tree is frost-sensitive and requires considerable light. It grows at altitudes up to 1,800 m with 250-500 mm of rainfall annually and mean annual temperatures of 17-25 degrees Celsius. It adapts to soils ranging in texture from light to heavy.

Sappanwood, Japan Wood
(See Dye & Tannin)

Seagrape
(See Fruit)

Senegal Rosewood, Barwood, West African Kino, Red Barwood
(See Dye & Tannin)

Shingle Tree

Sci. Name(s) : *Acrocarpus fraxinifolius*
Geog. Reg(s) : Asia-India (subcontinent), Asia-Southeast
End-Use(s) : Energy, Ornamental & Lawn Grass, Timber
Domesticated : Y
Ref. Code(s) : 228
Summary : The shingle tree is a deciduous tree found growing in western India, Assam, and Burma. This tree reaches heights of 30-50 m and is used for light construction, furniture, and boxes. The roundwood is used as fuel and charcoal. The tree is occasionally grown for shade or ornamental purposes. Up to 10 cu.m/ha/annum of timber are obtained.

This tree grows at altitudes up to 1,500 m. Rainfall requirements are 1,100-1,600 mm with temperatures of 19-28 degrees Celsius. Shingle trees grow best in deep soils of medium texture. They demand little care but should be well-spaced and protected from frost.

Shining Gum

Sci. Name(s) : *Eucalyptus nitens*
Geog. Reg(s) : Australia
End-Use(s) : Timber
Domesticated : Y
Ref. Code(s) : 228
Summary : Shining gum is an evergreen tree that stands 40-50 m in height and is commonly found in southeastern New South Wales and Victoria. Its saw timber is used for light construction and its roundwood for building and transmission poles. Timber production is 12-28 cu.m/ha/annum.

This tree requires an annual rainfall of 750-1,500 mm with a dry season of 2 months. Its altitude range is 2,000-3,500 m. The necessary annual temperature is 9-15 degrees Celsius. For optimum growth, full sun and a medium- to heavy-textured soil are needed. It is moderately frost-resistant.

Silky Oak

Sci. Name(s) : *Grevillea robusta, Grevillea umbricata*
Geog. Reg(s) : Australia
End-Use(s) : Energy, Miscellaneous, Ornamental & Lawn Grass, Timber
Domesticated : Y
Ref. Code(s) : 228
Summary : Silky oak is an evergreen tree occurring naturally in Queensland and New South Wales. It is 25-35 m in height and has sawn timber that resembles oak. This is used for light construction, furniture, joinery, veneer, plywood, fuel, and charcoal. Trees are ornamental and also used for agricultural shade. Timber production averages 5-10 cu.m/ha/annum.

This tree demands considerable light and is susceptible to frost. Altitudes of 800-2,100 m with an annual rainfall of 700-1,200 mm and mean annual temperatures of 13-21 degrees Celsius are most suitable. It adapts to most soils of light or medium texture.

Siris, East Indian Walnut, Koko

Sci. Name(s) : *Albizia lebbek*
Geog. Reg(s) : Asia-India (subcontinent), Asia-Southeast
End-Use(s) : Energy, Erosion Control, Forage, Pasture & Feed Grain, Gum & Starch, Ornamental & Lawn Grass, Timber
Domesticated : Y
Ref. Code(s) : 228
Summary : Siris is an open-crowned deciduous tree that reaches heights of 25-30 m. It grows in Burma, India, and the Andaman Islands at altitudes up to 1,400 m. It needs 500-1,000 mm of rainfall annually and temperatures of 20-28 degrees Celsius. It has several uses. The saw timber is used for light construction and furniture. The roundwood is used for fuel and charcoal. It is also used as a fodder and is the source of a gum. The tree itself serves as a shade, windbreak, and ornamental. Annual production is 18-28 cu.m/ha. It is adaptable, can grow on a variety of soils, and is termite-resistant.

Slash Pine

Sci. Name(s) : *Pinus elliottii* var. *elliottii*
Geog. Reg(s) : America-North (U.S. and Canada)
End-Use(s) : Natural Resin, Timber
Domesticated : Y
Ref. Code(s) : 228
Summary : The slash pine is light-crowned evergreen tree that reaches 20-30 m in height. Its saw timber is used in light and heavy construction, boats, and boxes, and its roundwood serves as building and transmission poles, fence posts, and longfiber pulp. This moderately durable timber produces useful resins. Annual timber production is 10-20 cu.m/ha.

This salt-tolerant, frost-hardy tree grows on the coastal plains of the southeastern United States at altitudes of 500-2,500 m in areas with 650-2,500 mm of rain and annual temperatures of 15-24 degrees Celsius. Trees thrive in soils of light to heavy texture, as well as shallow soils. The slash pine is the most resistant to a common pine disease called *Diplodea pini*.

Sour Cherry, Pie Cherry
(See Fruit)

Southern Mahogany

Sci. Name(s) : *Eucalyptus botryoides*
Geog. Reg(s) : Australia
End-Use(s) : Energy, Erosion Control, Timber
Domesticated : Y
Ref. Code(s) : 228
Summary : Southern mahogany is an evergreen tree found in coastal Victoria and New South Wales. It reaches heights of 20-25 m. Its timber is used for light and heavy construction, transmission poles, fuel, and charcoal. This tree tolerates salt winds and is often used as a windbreak. Annual timber production is 15-35 cu.m/ha.

This tree grows at altitudes of 800-1,800 m with a mean annual rainfall of 650-1,000 mm and temperatures of 16-22 degrees Celsius. Soil texture should be medium to heavy. Moderately saline soils are tolerated, and for optimum growth considerable radiation is required.

Southern Red Cedar, Eastern Red-Cedar
(See Essential Oil)

Spotted Gum

Sci. Name(s) : *Eucalyptus citriodora, Eucalyptus maculata* var. *citriodora*
Geog. Reg(s) : Australia
End-Use(s) : Energy, Oil, Timber
Domesticated : Y
Ref. Code(s) : 228
Summary : Spotted gum is an evergreen tree 30-40 m in height. It is grown in central and northern coastal Queensland. Its timber is used for heavy and light construction,

building and transmission poles, fuel, and charcoal. This tree is also the source of an oil. Annual timber production is 10-21 cu.m/ha.

This tree grows at altitudes up to 1,800 m with a mean annual precipitation of 650-1,600 mm and temperatures of 17-24 degrees Celsius. It requires deep soils of light to medium texture and needs considerable light for optimum growth.

Spotted Gum

Sci. Name(s)   : *Eucalyptus maculata*
Geog. Reg(s)   : Australia
End-Use(s)   : Energy, Timber
Domesticated   : Y
Ref. Code(s)   : 228
Summary   : Spotted gum is an evergreen tree of Queensland and New South Wales. It is fairly fire-resistant and windfirm and can reach heights of 30-45 m. Its timber is used for heavy and light construction. The roundwood is often used for poles, posts, veneer, plywood, fuel, and charcoal. Its annual timber yields are 21-35 cu.m/ha.

This tree grows at altitudes of 1,000-2,000 m. Temperatures should be no lower than 2 degrees Celsius and no higher than 32 degrees Celsius. A mean annual rainfall of 620-1,250 mm is necessary. It prefers deep soils of medium to heavy texture and needs considerable light for maximum growth.

Sugar Gum

Sci. Name(s)   : *Eucalyptus cladocalyx*
Geog. Reg(s)   : Australia
End-Use(s)   : Energy, Erosion Control, Ornamental & Lawn Grass, Timber
Domesticated   : Y
Ref. Code(s)   : 228
Summary   : Sugar gum is an evergreen tree grown in parts of southern Australia. It reaches 25-30 m in height. Its timber is used for heavy construction, building and transmission poles, fuel, and charcoal. The tree is used as a shade, windbreak, and ornamental. Thirteen to 22 cu.m/ha/annum of timber are produced.

This tree grows at altitudes up to 2,000 m with annual rainfall of 400-650 mm and temperatures of 8-18 degrees Celsius during the coldest months and 22-32 degrees Celsius during the hottest months. It prefers shallow soils of light to medium texture. For optimum growth it needs considerable light.

Sugar Maple, Rock Maple, Hard Maple

Sci. Name(s)   : *Acer saccharum*
Geog. Reg(s)   : America-North (U.S. and Canada)
End-Use(s)   : Miscellaneous, Sweetener, Timber
Domesticated   : Y
Ref. Code(s)   : 14, 216
Summary   : The sugar maple is a North American tree, reaching heights of 30 m, which is a commercial source of a durable, shock-resistant, light-colored timber used in furniture and cabinetry. In the United States, maple wood is also used for paneling, veneers, plywood, heavy-duty flooring, and musical instruments. Trees yield 2-6% of a syrupy sap that provides a commercial sugar maple syrup. Trees are grown in

Canada, Newfoundland, and the northern United States in lowland areas at elevations up to 1,600 m.

Sugi

Sci. Name(s)	: *Cryptomeria japonica*
Geog. Reg(s)	: Asia, Asia-China
End-Use(s)	: Erosion Control, Ornamental & Lawn Grass, Timber
Domesticated	: Y
Ref. Code(s)	: 228
Summary	: Sugi is an evergreen tree that grows 35-50 m in height. It is grown in China and Japan, where its soft timber is used for light construction, veneer, and plywood. Trees withstand prolonged shade, salt winds, and frost and are often used as windbreaks and agricultural shade. Ten to 33 cu.m/ha of timber are produced annually.

Trees grow at altitudes of 1,500-2,400 m with 1,500-2,500 mm of rain annually and temperatures of 10-18 degrees Celsius. They prefer deep, fertile soils.

Swamp Cypress

Sci. Name(s)	: *Taxodium distichum*
Geog. Reg(s)	: America-North (U.S. and Canada)
End-Use(s)	: Energy, Ornamental & Lawn Grass, Timber
Domesticated	: Y
Ref. Code(s)	: 228
Summary	: The swamp cypress occurs in the southeastern United States. It is a deciduous tree reaching heights of 30-40 m. Its timber has good natural durability and is easy to saw and season. Its saw timber is used for light construction and furniture and its roundwood for fence posts, fuel, and charcoal. It is valued as an ornamental. Annual timber production is 4-8 cu.m/ha.

While this tree is adaptable to most soil conditions, it prefers light to heavy soils, and the young seedlings are tolerant of shade. It is not frost-resistant.

Swamp Mahogany

Sci. Name(s)	: *Eucalyptus robusta, Eucalyptus multiflora*
Geog. Reg(s)	: Australia
End-Use(s)	: Energy, Timber
Domesticated	: Y
Ref. Code(s)	: 228
Summary	: Swamp mahogany is a small evergreen tree 25-30 m in height. It is found on the coasts of south Queensland and New South Wales. Its coarse-grained, red wood is used in both heavy and light construction and its roundwood is used for poles, posts, fuel, and charcoal. Young trees are often attacked by the Gonipterus beetle and termites.

This frost-sensitive tree grows best on deep soils of medium to heavy texture in full sun. Optimum climatic conditions include an altitudinal range up to 1,500 m, a mean annual rainfall of 1,000-3,000 mm, and a mean annual temperature of 12-28 degrees Celsius.

Sweet Birch, Black Birch, Cherry Birch, Mountain Mahogany
(See Essential Oil)

Tallow Wood

Sci. Name(s) : *Eucalyptus microcorys*
Geog. Reg(s) : Australia
End-Use(s) : Timber
Domesticated : Y
Ref. Code(s) : 228
Summary : Tallow wood is an evergreen tree reaching 25-30 m in height and grown in northeastern New South Wales and southeastern Queensland for its tough, naturally durable timber used for heavy and light construction. The roundwood is used for building and transmission poles and fence posts.

This frost-sensitive tree tolerates moderate shade, prefers fertile, medium-textured soils, and grows at altitudes of 500-2,000 m in areas with a dry season of 2 months, an annual rainfall of 900-1,500 mm, and temperatures of 17-23 degrees Celsius.

Tamarisk

Sci. Name(s) : *Tamarix articulata, Tamarix aphylla*
Geog. Reg(s) : Asia-Central
End-Use(s) : Energy, Erosion Control, Timber
Domesticated : Y
Ref. Code(s) : 228
Summary : Tamarisk is a tree reaching heights of 10-15 m and is found in Central Asia from Arabia to Afghanistan. Its timber has good natural durability. Its saw timber is used for furniture and its roundwood for fuel and charcoal. The tree is planted for erosion control, dune fixation, and as a windbreak.

It is found at altitudes from sea level to 1,400 m in areas with a mean annual rainfall of 200-500 mm and temperatures of 18-28 degrees Celsius. It withstands a dry season of 6-8 months. For maximum growth it needs light- to medium-textured soils and considerable radiation. It tolerates very saline soils and is moderately frost-resistant.

Tamarugo, Tamarugal

Sci. Name(s) : *Prosopis tamarugo*
Geog. Reg(s) : America-South
End-Use(s) : Energy, Miscellaneous, Timber
Domesticated : Y
Ref. Code(s) : 228
Summary : Tamarugo is a deciduous tree reaching heights of 8-12 m. Its tough timber has good natural durability. The roundwood is used for fence posts, fuel, and charcoal. The tree is used for shade, shelter, and windbreaks. Annual timber yields are 2-4 cu.m/ha.

An altitudinal range of 1,500-2,500 m with an annual rainfall of 200-400 mm and a temperature of 10-16 degrees Celsius are required. It tolerates saline and light- to heavy-textured soils. For maximum growth, considerable radiation is necessary.

Tan Wattle

Sci. Name(s) : *Acacia auriculiformis, Acacia auriculaeformis*
Geog. Reg(s) : Australia, Pacific Islands
End-Use(s) : Dye & Tannin, Energy, Erosion Control, Timber
Domesticated : Y
Ref. Code(s) : 228
Summary : Tan wattle is an evergreen tree that reaches heights of 15-25 m. It is grown for timber in the coastal parts of Queensland, Papua New Guinea, and the Solomon Islands. The roundwood is used for building poles, fence posts, charcoal, and fuel, and the saw timber for small furniture. The timber is a source of tannin. The tree is used for shade, erosion control, and soil improvement. Annual timber production is 10-20 cu.m/ha.

This tree grows at altitudes ranging from sea level to 500 m in areas with a mean annual rainfall of 1,300-1,700 mm. Average annual temperatures are 24-29 degrees Celsius. Trees adapt to most soil conditions.

Teak, Tec, Teca

Sci. Name(s) : *Tectona grandis*
Geog. Reg(s) : Africa, Asia-India (subcontinent), Asia-Southeast
End-Use(s) : Energy, Timber
Domesticated : Y
Ref. Code(s) : 228
Summary : Teak is a deciduous tree that grows naturally on the Indian subcontinent and in Southeast Asia. It reaches heights of 30-40 m. Its strong timber has good natural durability and is easy to preserve. It has special decorative qualities and, because of silica in the wood, is a fine hardwood. Its saw timber is used for heavy and light construction, furniture, boxes, and boat building. Its roundwood is used for building and transmission poles, fence posts, fuel, charcoal, veneer, and plywood. Timber yields are 6-18 cu.m/ha/annum.

Trees are found at altitudes from sea level to 900 m in areas with a mean annual rainfall of 1,250-2,500 mm and mean annual temperatures of 22-26 degrees Celsius. It survives a dry season of 3-5 months. Teak prefers medium to heavy, deep, fertile soils. In pure stands, there is often a problem with soil erosion. Teak is moderately fire-resistant and demonstrates wide variation in origin. For maximum growth, it needs considerable radiation.

Trees are subject to attacks by Atta ants, which cause defoliation during the first year. In Africa there is a problem with root rot, and in Asia there are problems with leaf skeletonizers.

Tenasserim Pine

Sci. Name(s) : *Pinus merkusiana, Pinus merkusii*
Geog. Reg(s) : Asia-India (subcontinent), Asia-Southeast
End-Use(s) : Natural Resin, Timber
Domesticated : Y
Ref. Code(s) : 228
Summary : Tenasserim pine is an evergreen tree that reaches 30-40 m in height and is found in Southeast Asia from northwestern India to Cambodia. Its timber is

moderately durable. Its saw timber is used in heavy and light construction and for making boxes, and its roundwood for transmission poles, longfiber pulp, veneer, and plywood. A resin is obtained from its trunk. Annual timber production is 8-18 cu.m/ha.

This tree is fire-resistant, windfirm, and termite-resistant. It adapts to most soil conditions but prefers light- to heavy-textured soils. Optimum climatic conditions include an altitudinal range from sea level to 900 m, an annual rainfall of 1,000-2,800 mm, and annual temperatures of 21-28 degrees Celsius. For maximum growth, it needs considerable sun. This tree is less susceptible to shoot borer attack than *Pinus kesiya*.

Terentang, Ketekete

Sci. Name(s)	: *Campnosperma brevipetiolata*
Geog. Reg(s)	: Pacific Islands
End-Use(s)	: Timber
Domesticated	: Y
Ref. Code(s)	: 228
Summary	: Terentang is a buttressed evergreen tree reaching 30-50 m in height and found growing in the Solomon Islands, the Moluccas, and New Guinea. Its timber is used to make boxes, veneer, and plywood, and the tree is susceptible to insect and fungal attack. Timber production is 30-40 cu.m/ha/annum.

This tree thrives in deep soils in full sun at altitudes up to 1,000 m with 2,000-5,000 mm of rainfall and a mean annual temperature of 23-28 degrees Celsius.

Tuart

Sci. Name(s)	: *Eucalyptus gomphocephala*
Geog. Reg(s)	: Australia
End-Use(s)	: Dye & Tannin, Energy, Erosion Control, Timber
Domesticated	: Y
Ref. Code(s)	: 228
Summary	: Tuart is an evergreen tree found only on the southwestern coast of western Australia. It is 20-30 m in height with tough timber used for fence posts, fuel, and charcoal. It is also a source of a tannin. Trees are used in erosion control. Annual timber production is 8-15 cu.m/ha.

This species of Eucalyptus grows at altitudes of 500-2,000 m. It needs 500-1,000 mm of annual rainfall and temperatures of 16-22 degrees Celsius for optimum growth. It prefers light- to medium-textured soils and will tolerate moderately saline soils. It is able to withstand salt winds and is sensitive to frost. It demands considerable light for best production.

Udjung Atup
(See Essential Oil)

Umbrella Pine, Stone Pine

Sci. Name(s)	: *Pinus pinea*
Geog. Reg(s)	: Mediterranean
End-Use(s)	: Energy, Timber

Domesticated : Y
Ref. Code(s) : 228
Summary : The umbrella pine is an open-crowned evergreen tree 15-25 m in height. It is found in the northern and eastern parts of the Mediterranean. Its saw timber is used for light construction and boxes and its roundwood for fence posts, fuel, and charcoal. Annual timber production is 3-5 cu.m/ha.

It is found at an altitudinal range of 1,500-2,500 m with a mean annual rainfall of 400-800 mm and a mean temperature range of 14-18 degrees Celsius. It needs a dry season of 4-6 months. For best results, this tree prefers light- to medium-textured shallow soils and considerable radiation. It is fire- and termite-resistant and moderately frost-resistant. Trees should be well-spaced. They are attacked by the processionary caterpillar, *Thaumetopoea wilkinsoni*, which is found in the Mediterranean area.

Umbrella Tree, Parasolier

Sci. Name(s) : *Musanga cecropioides, Musanga smithii*
Geog. Reg(s) : Africa-Central, Africa-West, Tropical
End-Use(s) : Timber
Domesticated : Y
Ref. Code(s) : 228
Summary : The umbrella tree occurs naturally in West and Central tropical Africa. It is 12-15 m in height and its only product is a shortfiber pulp from the roundwood. The tree has some soil-improvement abilities. Annual timber production is 30-35 cu.m/ha.

This tree is located in areas up to 200 m in altitude with an annual precipitation of 2,000-5,000 mm, a dry season lasting at least 2 months, and temperatures of 25-30 degrees Celsius. It is best suited to soils of light to medium texture.

West Indian Pine

Sci. Name(s) : *Pinus occidentalis, Pinus cubensis*
Geog. Reg(s) : West Indies
End-Use(s) : Natural Resin, Timber
Domesticated : Y
Ref. Code(s) : 228
Summary : The West Indian pine is a light-crowned evergreen tree reaching 25-35 m in height and grown on the island of Hispaniola and in eastern Cuba for its timber and resin. Annual timber production is 5-10 cu.m/ha. The saw timber is used for light and heavy construction, and boxes and the roundwood for transmission poles, fence posts, and longfiber pulp.

This tree requires 1,300-1,500 mm rain and temperatures of 18-24 degrees Celsius for optimum growth. It occurs naturally at altitudes up to 1,500 m, readily adapts to most soil conditions, but prefers light to heavy soils in full sun. It is termite-resistant.

White Mulberry
(See Miscellaneous)

Winter's-Bark
(See Medicinal)

Yellow Box

Sci. Name(s)	: *Eucalyptus melliodora*
Geog. Reg(s)	: Australia
End-Use(s)	: Energy, Timber
Domesticated	: Y
Ref. Code(s)	: 228
Summary	: Yellow box is an inland-growing evergreen tree of New South Wales and Victoria. It reaches 20-25 m in height and is grown mainly for its roundwood, which is used for fence posts, fuel, and charcoal. This moderately frost-resistant tree is used for shade, shelter, and windbreaks. Timber production is 2-6 cu.m/ha/annum.

It is most often found at altitudes of 1,000-2,000 m. The necessary rainfall is from 625-900 mm annually. Temperatures should be 16-21 degrees Celsius with a mean maximum temperature during the hottest month of 27-33 degrees Celsius and during the coldest month of 5-12 degrees Celsius. A medium- to heavy-textured soil is best, and considerable direct radiation is essential for optimum growth.

Yellow Flame, Yellow Poinciana, Soga
(See Ornamental & Lawn Grass)

Yemane, Gmelina

Sci. Name(s)	: *Gmelina arborea*
Geog. Reg(s)	: Asia-China, Asia-India (subcontinent), Asia-Southeast
End-Use(s)	: Energy, Timber
Domesticated	: Y
Ref. Code(s)	: 228
Summary	: Yemane is a short-lived deciduous tree standing 20-30 m in height. It grows in Southeast Asia, from Pakistan to Cambodia and southern China. Its white, odorless timber is used for light construction, building poles, veneer, plywood, fuel, and charcoal. Annual timber production is 18-32 cu.m/ha.

This tree occurs naturally at altitudes up to 800 m with a mean annual rainfall of 1,000-2,500 mm and temperatures of 21-28 degrees Celsius. It is best suited to fertile, light- to heavy-textured soils. It requires considerable light. Yemane is resistant to termites.

Tuber

Airpotato, Aerial Yam, Potato Yam, Bulbil-Bearing Yam

Sci. Name(s) : *Dioscorea bulbifera*
Geog. Reg(s) : Africa, America-North (U.S. and Canada), America-South, Asia, Tropical, West Indies
End-Use(s) : Tuber
Domesticated : Y
Ref. Code(s) : 48, 128, 171
Summary : The airpotato (*Dioscorea bulbifera*) is cultivated as a tuber crop in the tropics of Asia and Africa, the southern United States, the Caribbean and tropical South America. Its tubers are hard, bitter, and unpalatable, while its large bulbils are edible, succulent, and eaten raw or prepared like yams. Although airpotatoes are popular, they do not compete with yams. There are 2 forms of airpotatoes, the Asian and the African.

Ajipo, Tuberosus Yam Bean, Yam Bean, Potato Bean
(See Insecticide)

Anu, Capucine
(See Vegetable)

Cassava, Manioc, Tapioca-Plant, Yuca, Mandioca, Guacomole
(See Energy)

Chayote
(See Vegetable)

Chinese Artichoke, Japanese Artichoke, Artichoke Betony

Sci. Name(s) : *Stachys sieboldii, Stachys affinis, Stachys tuberifera*
Geog. Reg(s) : Asia, Asia-China, Europe
End-Use(s) : Tuber
Domesticated : Y
Ref. Code(s) : 220, 237
Summary : The Chinese artichoke is a perennial plant 30-45 cm in height and cultivated for its edible tubers, which contain tetrasaccharide stachyose in place of starch. These tubers are eaten boiled or fried. Chinese artichokes are native to China and Japan. Principal areas of cultivation are France and Belgium.

Chinese Waterchestnut, Waternut

Sci. Name(s) : *Eleocharis dulcis*
Geog. Reg(s) : Asia, Asia-China
End-Use(s) : Gum & Starch, Tuber
Domesticated : Y
Ref. Code(s) : 13, 36, 143
Summary : The Chinese waterchestnut is a perennial leafless sedge that thrives in moist places at low altitudes, usually below 50 m but sometimes up to 1,000 m in altitude.

They are cultivated in open wetlands or shallow, swampy areas, and in salt, brackish, or fresh water. Plants produce tiny edible tubers that are a good source of starch. There is a limited market for these in Japan.

Chinese Yam, Cinnamon-Vine

Sci. Name(s) : *Dioscorea oppositifolia, Dioscorea opposita, Dioscorea batatas*
Geog. Reg(s) : Asia, Asia-China
End-Use(s) : Ornamental & Lawn Grass, Tuber, Weed
Domesticated : Y
Ref. Code(s) : 48, 62, 128, 171
Summary : The Chinese yam (*Dioscorea oppositifolia*) is a woody plant with stems climbing to a height of 3 m. It produces edible, underground potatolike tubers 1 m in length, and aerial tubers, or bulbils, which could be used for propagation except for their small size and slow maturation rate. The underground tubers are nutritious and can be cooked like potatoes.

There is extensive tuber cultivation in China, Korea, and Japan. Because of the tuber length and thinness, harvesting can be difficult. The plants are often cultivated as ornamentals but they can escape easily and become weeds in waste areas. Chinese yams will tolerate a colder climate than any other economic species of the Dioscorea genus.

Chufa, Ground Almond, Tigernut, Yellow Nutsedge
(See Forage, Pasture & Feed Grain)

Composite Yam, Wild Yam
(See Medicinal)

Convolvulacea Yam

Sci. Name(s) : *Dioscorea convolvulacea*
Geog. Reg(s) : America-Central, Tropical
End-Use(s) : Gum & Starch, Tuber
Domesticated : Y
Ref. Code(s) : 48, 119
Summary : The convolvulacea yam is grown throughout the wetter areas of the tropics for its starchy tubers. The tubers are eaten raw or cooked. There is little or no trade, as it is grown mainly for domestic consumption. This yam is native to Mexico.

Cornroot, Topee Tambu, Leren, Sweet Corm-Root, Allouya

Sci. Name(s) : *Calathea allouia*
Geog. Reg(s) : America-South, Tropical, West Indies
End-Use(s) : Tuber
Domesticated : Y
Ref. Code(s) : 16, 171
Summary : Cornroot is a perennial herb reaching 1 m in height and cultivated in the West Indies and northern South America for its small, edible, potatolike tubers. Crops are produced in 10 months. Yields reach approximately 10 MT/ha of tubers. Plants

thrive in moist, tropical climates on fertile loams in areas with temperatures of 18-35 degrees Celsius.

Dasheen, Elephant's Ear, Taro, Eddoe

Sci. Name(s)	: *Colocasia esculenta, Colocasia esculentum, Colocasia antiquorum*
Geog. Reg(s)	: Africa-West, Asia-Southeast, Hawaii, Mediterranean, Subtropical, Tropical, West Indies
End-Use(s)	: Cereal, Gum & Starch, Miscellaneous, Tuber, Vegetable
Domesticated	: Y
Ref. Code(s)	: 126, 132, 220
Summary	: Dasheen, or taro, is a perennial herb. It is cultivated throughout the tropics for

its tubers, which contain large quantities of small starch grains. The tubers, sometimes reaching up to 15 cm in diameter, take about 7 months to reach maturity. They contain more protein, calcium, and phosphorus than potatoes. The tubers are eaten mainly as energy food because they are low in fats and proteins.

The tubers and leaves are eaten boiled. They are grated and fermented to make poi or fried as chips. Flour is made from the dried corms. The tubers are a good source of calories. There is not much potential for expansion of dasheen production unless better methods of preparation are developed.

The plant can reach heights of 1 m if grown under good conditions. Dasheen is most successful in rich soils from low to medium elevations, usually up to 1,800 m. It is grown in the moist parts of the tropics and frost-free subtropics. Dasheen is cultivated in the West Indies and Hawaii and there are commercial plantings in Egypt. West Africa is the largest producing region. In 1984, all tuber production on a worldwide basis was estimated as exceeding 3 million MT.

Eboe Yam, White Guinea Yam, White Yam, Guinea Yam

Sci. Name(s)	: *Dioscorea rotundata*
Geog. Reg(s)	: Africa-West
End-Use(s)	: Tuber
Domesticated	: Y
Ref. Code(s)	: 48, 66, 105, 171, 222
Summary	: The eboe yam is an annual vine grown for its edible tubers, popular in West

Africa where they are prepared as a food called fufu. These tubers are hard and unpalatable if harvested prematurely. Most tuber cultivation is along the West African coast. Crops are produced in 8-10 months.

Giant Granadilla, Barbardine
(See Fruit)

Ginseng, American Ginseng
(See Medicinal)

Giri Yam Pea, Yam Bean, Akitereku, African Yam Bean, Haricot Igname

Sci. Name(s)	: *Sphenostylis stenocarpa, Vigna ornata, Sphenostylis ornata*
Geog. Reg(s)	: Africa-West, Europe, Tropical
End-Use(s)	: Ornamental & Lawn Grass, Tuber

Domesticated : Y
Ref. Code(s) : 111, 220
Summary : The giri yam pea is a vigorous vine 1.5-2 m in height, grown for its edible, starchy tubers cooked similarly to the Irish potato. Its seeds are edible. In Europe, the flowers are used as ornamentals.

Vines are found wild and cultivated in tropical West Africa. They thrive from sea level to altitudes approaching 1,800 m. Harvests are made 8 months after planting. In Africa, tuber yields are recorded at 1,800 kg/ha with dry seed yields of 300-500 kg/ha.

Hausa Potato, Kaffir Potato

Sci. Name(s) : *Plectranthus esculentus*
Geog. Reg(s) : Africa-West
End-Use(s) : Tuber
Domesticated : Y
Ref. Code(s) : 92, 170
Summary : The hausa potato is an herbaceous perennial reaching heights of 1 m. Grown for its aromatic tubers, it is often substituted for the Irish potato in West Africa. Plants produce crops in about 6 months with tuber yields of 4.5-6 MT/ha. Plants require an annual rainfall of 100 cm for satisfactory production.

Jerusalem-Artichoke, Girasole

Sci. Name(s) : *Helianthus tuberosus*
Geog. Reg(s) : America-North (U.S. and Canada), Europe, Temperate, Tropical
End-Use(s) : Beverage, Forage, Pasture & Feed Grain, Gum & Starch, Miscellaneous, Sweetener, Tuber, Vegetable, Weed
Domesticated : Y
Ref. Code(s) : 154, 170, 200
Summary : The Jerusalem-artichoke is a tuberous perennial herb 1-3 m in height, grown for its edible tubers, which contain approximately 8-18% carbohydrates, mainly inulin and related inulides with fructose and glucose. Inulin starch is one of the few types of carbohydrates tolerated by diabetics. These carbohydrates are used in industrial alcohol and in the preparation of a beerlike beverage. The tubers are eaten as vegetables, used for livestock feed, as a commercial source of fructose, and for the preparation of 5-hydroxy-methyl-furfural. Average fructose yields are about 6% from fresh tubers.

Tubers are perishable, harvesting is labor intensive, and plants easily become weeds--all of which create drawbacks to extensive commercial production.

Plants grow in most temperate and tropical countries on sandy or loamy soil. In France, they grow on soil too dry or infertile for beet or potato production. Irrigation and fertilization produce the best and most consistent results. Plants should be grown in areas with short days to aid tuber formation. Tubers are ready for harvest in 4-6 months. Plants grown on sandy soils in Europe produced 30 MT/ha of tubers.

Kudzu Vine, Kudzu

Sci. Name(s) : *Pueraria lobata, Pueraria thunbergiana, Pueraria hirsuta, Pueraria triloba*
Geog. Reg(s) : Asia, Subtropical, Temperate

End-Use(s) : Erosion Control, Forage, Pasture & Feed Grain, Ornamental & Lawn Grass, Tuber
Domesticated : Y
Ref. Code(s) : 17, 24, 36, 56, 154
Summary : Kudzu is a vine that produces tough-skinned, starchy tubers that are often used in oriental cooking. Vines are grown as ornamentals, ground cover, or green manure. As a fodder, kudzu leaves have a nutritional value like that of clover or alfalfa and provide fairly nutritious feed. Kudzu forage yields are about 5 MT/ha. About 25,000 plants or crowns can be harvested from a plantation of 1 ha.

Plants thrive on well-drained loam soils in areas with an annual rainfall of 9.7-21.4 dm. Once established, plants grow slowly until their 2nd year, when they begin spreading rapidly. Plants are utilized in their 2nd year. Seed production is considered fairly low; kudzu produces about 200 kg/ha of seed. Kudzu is native to the Far East and is now grown throughout most subtropical areas and in some warm temperate areas.

Lily Turf, Dwarf Lily Turf, Mondo Grass, Snake's Beard
(See Ornamental & Lawn Grass)

Nutgrass, Purple Nutsedge, Nutsedge
(See Weed)

Oca
Sci. Name(s) : *Oxalis tuberosa*
Geog. Reg(s) : America-South
End-Use(s) : Tuber
Domesticated : Y
Ref. Code(s) : 92, 111, 170, 237
Summary : Oca is a high-altitude, carbohydrate-rich root crop cultivated in the Andes Mountains from Colombia to Bolivia. Plants are also grown commercially in southern New Zealand. These tubers are eaten baked, boiled, or dehydrated. Bitter forms of tubers contain calcium oxalate crystals, an important source of calcium and iron, and must be processed before eating. Tubers are generally watery and contain 1.1% protein and 13% carbohydrates.

Tubers can be harvested after 8 months and are most successfully grown at elevations of 2,742-4,265 m in areas with cool temperatures and light frost. Tuber yields are 5-7 MT/ha.

Parsnip, Wild Parsnip
Sci. Name(s) : *Pastinaca sativa*
Geog. Reg(s) : Europe, Temperate
End-Use(s) : Medicinal, Spice & Flavoring, Tuber, Vegetable
Domesticated : Y
Ref. Code(s) : 170, 220
Summary : Parsnip is a biennial European herb cultivated commercially in many temperate countries for its thick, edible taproot and leafy plant tops eaten cooked as vegetables. Roots and fresh leaves have diuretic, sedative, and appetite-stimulative

properties. Root extracts are used to flavor schnapps. Roots contain a starch that hydrolyzes on exposure to cold, creating a sweet root flavor.

Plants do not do well in tropical lowlands but are successfully grown at higher altitudes. Parsnips grow wild in meadows with deep nitrogenous or calcareous soils.

Peruvian-Carrot, Arracacha

Sci. Name(s)	: *Arracacia xanthorrhiza*
Geog. Reg(s)	: America-South
End-Use(s)	: Gum & Starch, Tuber
Domesticated	: Y
Ref. Code(s)	: 170, 237
Summary	: The Peruvian-carrot is a temperate herb, cultivated for its large, starchy roots. The crop is an important starch source in the mountainous regions of Bolivia, Peru, Ecuador, and Colombia.

In southern and central Brazil, there is sizable commercial production of the Peruvian-carrot. Root harvest occurs 9-10 months after planting in Brazil, and at 10-14 months in the Andes.

Potato, European Potato, Irish Potato, White Potato

Sci. Name(s)	: *Solanum tuberosum, Solanum phureja*
Geog. Reg(s)	: Worldwide
End-Use(s)	: Cereal, Forage, Pasture & Feed Grain, Gum & Starch, Tuber, Vegetable
Domesticated	: Y
Ref. Code(s)	: 31, 132, 170, 220
Summary	: The potato is an annual plant 30-100 cm high whose tubers are an extremely important vegetable throughout the world. The tubers are also an important source of flour and ethyl alcohol. Potatoes can be cooked in a variety of ways or can be used as stock food. Tubers contain 8-28% starch and 1-4% protein.

Plants are grown throughout the world but are not successful in the lowland tropics. They are usually grown in mountainous regions in areas with temperatures dropping to 15.5 degrees Celsius. They thrive on light, sandy, well-drained soils in areas with short days and cool evening temperatures. With early maturing cultivars, harvest occurs 100 days from planting. With later maturing cultivars, harvest occurs 200 days from planting.

In 1983, world production was 286.5 million MT from an area of 20.2 million ha. Major producing countries were the USSR, China, and Poland. The USSR produced 29% of the world's potatoes, China, 17%, and Poland, 12%. The United States represented 5% of the world's production with 14.8 million MT from an area of 501,000 ha. The Netherlands was the leading potato exporter and the Federal Republic of Germany was the leading importer.

Rampion
(See Vegetable)

Ratala, Country-Potato

Sci. Name(s)	: *Coleus parviflorus*
Geog. Reg(s)	: Asia-Southeast

End-Use(s) : Tuber
Domesticated : Y
Ref. Code(s) : 170
Summary : Ratala is a small herb cultivated in Southeast Asia for its aromatic tubers, substituted for potatoes. This plant produces tubers in 6 months and yields are approximately 3,363-6,726 kg/ha.

Rutabaga, Swede
(See Vegetable)

Sacred Lotus, Lotus, East Indian Lotus
Sci. Name(s) : *Nelumbo nucifera*
Geog. Reg(s) : America-North (U.S. and Canada), Asia, Asia-China, Asia-India (subcontinent), Hawaii
End-Use(s) : Fiber, Gum & Starch, Medicinal, Tuber
Domesticated : Y
Ref. Code(s) : 36, 111, 237
Summary : The sacred lotus is a perennial pond herb whose roots are used for culinary purposes and as a source of starch. Lotus roots are 60-120 cm in size and are considered a delicacy in Oriental cooking. Roots are rarely processed for starch because of the high labor requirement involved in harvest. In China, the seed pods are brewed as a medicinal tea. The starchy seeds may also be cooked as food. Sap from the tuber has been used medicinally, while leaf fibers provide lamp wicks. Roots are mature in 6-9 months.

Plants tend to crowd out most other aquatic plants. They are grown mainly in Japan, China, Hawaii, and India, and to a limited degree in California. Commercial production in Hawaii has declined since 1967.

Salsify, Oyster-Plant, Vegetable Oyster
Sci. Name(s) : *Tragopogon porrifolius*
Geog. Reg(s) : America-North (U.S. and Canada), Eurasia
End-Use(s) : Gum & Starch, Tuber, Weed
Domesticated : Y
Ref. Code(s) : 17, 170, 220
Summary : Salsify is a plant that is cultivated for its edible taproot. Roots also contain a latex that has been used for chewing gum by the Indians of British Columbia. Plants are native to Eurasia and have naturalized in the United States as a weed. When cultivated, seeds are sown in the spring and the roots left underground during the winter.

Serendipity-Berry, Diel's Fruit
(See Sweetener)

Sudan Potato, Hausa Potato, Coleus-Potato
Sci. Name(s) : *Solenostemon rotundifolius*
Geog. Reg(s) : Africa, Asia-India (subcontinent), Asia-Southeast, Tropical
End-Use(s) : Tuber
Domesticated : Y

470

Ref. Code(s) : 111
Summary : The Sudan potato is an herbaceous annual 15-30 cm in height, which is cultivated for its tubers, used as a potato substitute, and prepared similarly. Sudan potatoes have been replaced by other starchy foods, specifically the cassava and Irish
potato.

Plants are cultivated on a small-scale basis in parts of Southeast Asia, Sri Lanka,
Malaysia, Indonesia, and tropical Africa. In India, they are grown as monsoon crops.
They thrive in areas with high rainfall, cool nights, and well-drained, sandy loams.
Crops mature in 5-8 months. Recorded tuber yields are 7.5-15 MT/ha.

Sweet Potato

Sci. Name(s) : *Ipomoea batatas*
Geog. Reg(s) : Africa, America-North (U.S. and Canada), America-South, Asia, Asia-China,
Asia-Southeast, Hawaii, Pacific Islands
End-Use(s) : Forage, Pasture & Feed Grain, Gum & Starch, Tuber
Domesticated : Y
Ref. Code(s) : 126, 132, 148, 170
Summary : The sweet potato is a perennial vine that is cultivated for its edible tubers.
Sweet potatoes are important crops in the Caribbean and the United States. The tu-
bers contain high concentrations of carbohydrates in the form of sugar and starch.
They are eaten boiled, baked, candied with syrup, or pureed. They are processed like
potatoes into fried chips, starch, and flour, or are canned and dehydrated. Sweet
potatoes are a prime source of starch, glucose, syrup, and alcohol. Plant tops and
leaves are used as potherbs and the vines as fodder.

China, Africa, Japan, the United States, Indonesia, Brazil, Vietnam, and New
Zealand are major producers of sweet potatoes. In general, yields vary with the cultivar, the environmental conditions, and the cultural practices. The range is 6-35
MT/ha/crop. Several varieties have been developed that produce well in the tropics
and are fairly disease-resistant: 'Nemagold' withstands nematodes and is considered
good for canning; 'Goldrush' is favored in baking; and 'Centennial' produces high
yields in Puerto Rico.

Sweet potatoes thrive in areas with an average temperature of 24 degrees Celsius
and 76-127 cm of rain/year. Plants tolerate conditions from drought to high rainfall,
as long as soils are well-drained. Tuber formation favors short days and dry weather,
and leaf and stem growth favors long days. Crops grow best in slightly acidic, sandy
loams. Plants mature in 3-6 months. Sweet potatoes are native to tropical America.

Turnip
(See Vegetable)

Turnip Chervil, Turnip-Rooted Chervil
Sci. Name(s) : *Chaerophyllum bulbosum*
Geog. Reg(s) : Europe, Temperate
End-Use(s) : Tuber
Domesticated : Y
Ref. Code(s) : 220

Summary : The turnip chervil is a biennial herb, cultivated for its fleshy, edible roots, which grows in the northern temperate areas, especially in Europe.

Ullucu

Sci. Name(s) : *Ullucus tuberosus*
Geog. Reg(s) : America-South
End-Use(s) : Tuber
Domesticated : Y
Ref. Code(s) : 170, 220, 237
Summary : Ullucu is a frost-hardy herb grown for its tubers eaten like potatoes. It is an important food source in the Andes Mountains of South America. Yields are 5-9 MT/ha. Fresh tubers contain 80-85% water, 1% protein, and 14% carbohydrates.

Vigna vexillata

Sci. Name(s) : *Vigna vexillata*
Geog. Reg(s) : Africa, Tropical
End-Use(s) : Tuber
Domesticated : Y
Ref. Code(s) : 24, 84, 141, 170
Summary : Vigna vexillata is a twining herb grown for its tuberous roots eaten like sweet potatoes. It withstands salt winds and drought and is both wild and cultivated across tropical Africa, the Democratic Republic of the Sudan, and Ethiopia. Tubers have excellent nutritive value.

Wild Yam, Floribunda Yam
(See Medicinal)

Winged Yam, Greater Yam, Water Yam, Ten Months Yam

Sci. Name(s) : *Dioscorea alata*
Geog. Reg(s) : Africa-West, West Indies
End-Use(s) : Tuber, Vegetable
Domesticated : Y
Ref. Code(s) : 31, 128, 159, 171, 221, 237
Summary : The winged yam is the highest yielding root crop of the cultivated yam genus (Dioscoreas), popular worldwide, particularly in West Africa, as a source of carbohydrates. The roots are baked, boiled, roasted, fried, or used raw as a salad vegetable. The tubers are high in starch but lack protein. Normal growing period is 8-10 months. In the West Indies, the 'White Lisbon' is a high-yielding crop.

Global yam production has reached 20 million MT/annum. Plants are most successful in areas with temperatures of 25-30 degrees Celsius and heavy rains of 1-3 m/year. They should be grown at altitudes of 15-914 m. Winged yams, grown under good conditions, can yield of 7-16 MT/ha.

Yam Bean, Erosus Yam Bean, Jicama

Sci. Name(s) : *Pachyrhizus erosus, Pachyrhizus angulatus, Dolichus erosus*
Geog. Reg(s) : America-Central, Asia-China, Asia-India (subcontinent), Asia-Southeast, Hawaii, Tropical

End-Use(s) : Gum & Starch, Insecticide, Tuber, Vegetable
Domesticated : Y
Ref. Code(s) : 56, 154, 170
Summary : The yam bean, or jicama, is a perennial vine that produces an edible watery tuber. The tubers contain 10% starch and are usually marketed like potatoes, fresh or in processed form. Market-sized tubers weigh about 0.2-1 kg each. Young pods are eaten like French beans. Ripe pods are harvested for their large proportion of the toxin rotenone, which is used as a fish and insect poison.

Yam beans are grown from seed in well-tilled, loose, sandy soils, either on hills or staked. They are quite successful in the hot, wet tropics. If the plant is grown for tubers, the inflorescences should be removed, preventing pod and seed formation. The tubers will become too fibrous for eating 4-8 months after planting. Plants provide good green manure.

Vines once grew wild in Mexico and northern Central America. They now grow in southeastern Asia, India, and Hawaii, and have naturalized in southern China and Thailand.

Yautia, Tannia, Tannier, Cocoyam, Ocumo, Malanga, Macabo

Sci. Name(s) : *Xanthosoma sagittifolium*
Geog. Reg(s) : Africa-West, America, Pacific Islands, Tropical
End-Use(s) : Tuber
Domesticated : Y
Ref. Code(s) : 126, 132, 154, 171
Summary : Yautia is a root crop grown for its starchy lateral tubers, eaten roasted or boiled, which are richer in carbohydrates, calcium, phosphorus, and iron than the Irish potato. While the main root is inedible, it has potential as animal feed. These roots are substituted for yams in the West African dish, fufu.

Plants are native to the tropical rain forests of Africa and require high rainfall and moist soils for adequate growth. They prefer warm, sunny conditions but tolerate light shade.

Crops are harvested after 9 months and tubers can be stored for more than 18 months. Yautia yields are 3-30 MT/ha of tubers. Plants are grown throughout the Pacific, the Americas, West Africa, and other tropical areas.

Yellow Melilot, Yellow Sweetclover
(See Forage, Pasture & Feed Grain)

Vegetable

African Locust Bean
(See Oil)

Angled Luffa, Angled Loofah, Singkwa Towelgourd, Seequa, Dishcloth Gourd
(See Fruit)

Anu, Capucine

Sci. Name(s)	: *Tropaeolum tuberosum*
Geog. Reg(s)	: America-South
End-Use(s)	: Tuber, Vegetable
Domesticated	: Y
Ref. Code(s)	: 170
Summary	: Anu is a twining herb grown for its edible, tuberous roots eaten as vegetables. It is cultivated at high altitudes in the Andes Mountains from Colombia to Bolivia. Anu can be confused with oca and potatoes.

Archucha, Wild Cucumber
(See Fruit)

Asparagus, Garden Asparagus

Sci. Name(s)	: *Asparagus officinalis*
Geog. Reg(s)	: Africa-North, America-North (U.S. and Canada), Asia, Europe
End-Use(s)	: Beverage, Vegetable
Domesticated	: Y
Ref. Code(s)	: 42, 148, 220
Summary	: The asparagus is a glabrous, perennial herb that is cultivated in Europe, Asia, North Africa, and North America for its edible young shoots. These are eaten as vegetables, either green or blanched. Asparagus seeds have been used as a coffee substitute.

Herbs should be grown in neutral or slightly acidic, loose soil. They should be planted at sufficient depth to promote good shoot development. In the tropics, commercial plantings of asparagus are hampered because of the dormancy period required to produce shoots of marketable size. For this reason asparagus may not be an economical tropical vegetable.

Asparagus is harvested once a year, starting after 2 full seasons of growth. Shoots are cut over a span of 2-8 weeks, depending on the strength of individual plants.

Asparagus Broccoli, Sprouting Broccoli

Sci. Name(s)	: *Brassica oleracea* var. *italica*
Geog. Reg(s)	: America-North (U.S. and Canada), Europe, Tropical
End-Use(s)	: Vegetable
Domesticated	: Y
Ref. Code(s)	: 170, 200

Summary   : The asparagus broccoli, also called cauliflower, is a vegetable cultivated for its terminal and auxiliary buds. It is harvested when buds reach 15 cm in diameter.

Although cauliflower tolerates the hot and dry climates of the tropics better than its relative the broccoli (*Brassica oleracea* var. *botrytis*), too much heat keeps the head from forming. Insects, lack of water, and excessive cold also prevent head development. Plants are cultivated extensively in Europe and North America.

Asparagus Pea, Winged Pea

Sci. Name(s)   : *Tetragonolobus purpureus, Lotus tetragonolobus*
Geog. Reg(s)   : Mediterranean
End-Use(s)   : Vegetable
Domesticated   : Y
Ref. Code(s)   : 170
Summary   : The asparagus pea is an herb that grows wild in the Mediterranean region but is occasionally cultivated for its young pods eaten as vegetables.

Azuki Bean, Adzuki Bean

Sci. Name(s)   : *Vigna angularis*
Geog. Reg(s)   : Asia-China, Subtropical, Temperate, Tropical
End-Use(s)   : Forage, Pasture & Feed Grain, Medicinal, Vegetable
Domesticated   : Y
Ref. Code(s)   : 5, 24, 115, 126, 132, 220
Summary   : The azuki bean is an important vegetable protein in eastern Asia. Its adaptation and culture are similar to those of the soybean but it lacks the oil content of the soybean. The beans are important as a protein supplement. They are generally eaten boiled, candied, or in paste form. Plants are grown for grain, forage, and green manure. After harvest, they will become self-pollinators, as well as cross-pollinators. Beans contain saponins and sapogenol, which are glucosides used in medicine for breaking down red blood cells. Azuki beans are native to China. They thrive in most temperate and subtropical areas and can be grown in the tropics.

Balsam-Pear, Bitter Cucumber, Bitter Gourd, Balsam Apple

Sci. Name(s)   : *Momordica charantia*
Geog. Reg(s)   : Asia-India (subcontinent), Asia-Southeast, Tropical
End-Use(s)   : Spice & Flavoring, Vegetable
Domesticated   : Y
Ref. Code(s)   : 132, 170
Summary   : The balsam-pear (*Momordica charantia*) is an annual climbing vine that produces bitter fruits eaten as cooked vegetables, pickled, and used in curries. This fruit is an excellent source of vitamin C and loses most of its bitterness if soaked in salt water. Seeds provide a condiment in India, and shoots and leaves are eaten like spinach. Balsam pears have limited market potential.

Vines flower 30-35 days after planting and fruit is picked 15-20 days later. Plants require trellising and are most successful in hot to moderate temperatures in tropical lowlands on most soils, except sand. Yield data is limited, but vines generally produce 10 MT/ha of fruit. Vines are cultivated in India, Indonesia, Malaysia, and Singapore.

Bambarra Groundnut, Voandzou, Ground Pea

Sci. Name(s) : *Voandzeia subterranea*
Geog. Reg(s) : Africa-West, America-South, Asia, Tropical
End-Use(s) : Cereal, Vegetable
Domesticated : Y
Ref. Code(s) : 84, 132, 154, 170
Summary : Bambarra groundnut is a heavily branched herb grown for its edible seeds used as a nutritious pulse. These seeds are eaten fresh or boiled and have a taste similar to the garden pea. Dried seeds can be roasted and ground into flour. Young pods are eaten raw and shelled seeds are canned in parts of Africa. The bambarra groundnut is an excellent source of protein, iron, and B vitamins.

This plant grows on poor soils in hot climates where other pulse crops fail. It also gives higher yields than other pulses. Crops grow in tropical lowlands and interior valleys at medium altitudes on well-drained soils. Once established, they withstand droughts. Bambarra groundnuts are cultivated mainly for domestic consumption throughout tropical Africa, Brazil, Surinam, and the Philippines. The most extensive seed production is in Zambia. Crops mature 4 months after sowing with average yields of 560 kg/ha of dried seed.

Bamboo, Phyllostachys
(See Timber)

Bambusa, Bamboo
(See Timber)

Baobab, Monkey-Bread

Sci. Name(s) : *Adansonia digitata*
Geog. Reg(s) : Africa
End-Use(s) : Beverage, Fiber, Oil, Vegetable
Domesticated : N
Ref. Code(s) : 170
Summary : Baobab trees have a variety of uses. In tropical Africa, the young leaves are eaten as vegetables, a fiber from the inner bark is made into ropes, and the bark is beaten into cloth. The fruit pulp contains tarranic acid and is made into a beverage or used as food flavoring. The seed kernels are edible and contain 12-15% oil. Baobabs have a remarkably long life span, with some trees living as long as 5,000 years.

Barnyardgrass, Barnyard Millet
(See Forage, Pasture & Feed Grain)

Beet, Beetroot, Garden Beet

Sci. Name(s) : *Beta vulgaris*
Geog. Reg(s) : America-North (U.S. and Canada), Asia, Asia-Southeast, Europe, Tropical
End-Use(s) : Vegetable
Domesticated : Y
Ref. Code(s) : 148, 170

Summary : The beet (*Beta vulgaris*) is an herb producing roots eaten boiled as a vegetable in salads and pickled and canned. The vulgaris species includes chad, spinach beets, sugar beets, and mangel.

In parts of Europe, Asia, and the East Indies, beets grow wild. In the tropics, they do well but high temperatures hamper root development. Best results in the tropics are from medium to high elevations.

In the United States, domestic beets are grown as cultivated garden herbs and sold as domestic market produce. Best varieties for growing roots and leafy tops are the 'Long Season' and 'Early Wonder Tall Top'.

Benoil Tree, Horseradish-Tree
(See Oil)

Black Gram, Urd, Wooly Pyrol

Sci. Name(s) : *Vigna mungo, Phaseolus mungo*
Geog. Reg(s) : America, America-North (U.S. and Canada), Asia-India (subcontinent), Asia-Southeast, Tropical, West Indies
End-Use(s) : Cereal, Erosion Control, Forage, Pasture & Feed Grain, Vegetable
Domesticated : Y
Ref. Code(s) : 56, 126, 132, 170, 237
Summary : The black gram is an important pulse crop in India. The beans are boiled and eaten whole or in paste form. They can also be parched and ground into flour. The green pods are eaten as vegetables. Black gram plants are used as a green manure and cover crop and as short-term forage. The hulls and straw are used as cattle feed but are considered inferior to other cattle fodders because of the hairy stems. Seed flour is considered an excellent soap substitute.

Plants have been introduced into the tropics and grow in India from sea level to 1,828 m. Black grams are drought-hardy and can grow in areas with limited rainfall. Plants will not grow well in the wet tropics but do well on heavy clay soils with adequate drainage. They are grown in parts of Southeast Asia, the West Indies, the United States, and drier areas of the tropics.

Black gram matures in about 80-120 days and yield up to 0.907 MT/ha. The seeds contain 19-24% protein and 1-2% oil.

Blue Clitoria, Butterfly Pea, Asian Pigeon-Wings, Butterfly Bean, Kordofan Pea
(See Forage, Pasture & Feed Grain)

Bondue, Livid Amaranth, Wild Blite

Sci. Name(s) : *Amaranthus lividus, Amaranthus thunbergii*
Geog. Reg(s) : Africa, Europe, Tropical
End-Use(s) : Vegetable, Weed
Domesticated : N
Ref. Code(s) : 3, 220
Summary : Bondue is an erect, annual tropical African herb grown for its fruits eaten as vegetables. This plant occurs most often as a weed in cultivated areas, particularly on the Cape Peninsula of Africa. Bondue is native to Europe.

Borage
(See Ornamental & Lawn Grass)

Boston Marrow, Pumpkin, Winter Squash, Squash, Marrow
Sci. Name(s)   : *Cucurbita maxima*
Geog. Reg(s)   : Africa, America-North (U.S. and Canada), America-South, Asia, Europe
End-Use(s)   : Fruit, Oil, Spice & Flavoring, Vegetable
Domesticated   : Y
Ref. Code(s)   : 132, 170
Summary   : Boston marrows are annual vines that produce nutritious, edible fruit. In the United States and Europe, this fruit is grown commercially for canning. There is a small market for the seeds as a snack food and they are a source of protein and an edible oil. Immature fruits are eaten as vegetables while mature fruits are used for baking and jams. In Africa and Asia, leaves and flowers are cooked and used as potherbs.

Immature fruit is picked at 7-8 weeks, while mature fruit ripens in 3-4 months. Boston marrows grow throughout the world at low to high altitudes in areas with hot temperatures and adequate rainfall. They are probably native to South America. In the United States, there are various breeding programs for the improvement of this crop.

Breadfruit
Sci. Name(s)   : *Artocarpus altilis*
Geog. Reg(s)   : Pacific Islands, Tropical
End-Use(s)   : Fiber, Forage, Pasture & Feed Grain, Fruit, Ornamental & Lawn Grass, Vegetable
Domesticated   : Y
Ref. Code(s)   : 148, 170
Summary   : The breadfruit tree bears an edible fruit used as an important staple food in the Polynesian islands. Its fruit is considered a vegetable. There are both seedless and seeded varieties of breadfruit. The seeded type is commonly referred to as the breadnut. These nuts are eaten boiled or roasted.

The mild-flavored fruit pulp is a fair source of vitamins A and B, is high in starch, rich in calcium, and eaten cooked, boiled, baked, roasted, or fried. For long-term storage, it is cooked and dried. The bark provides a useful fiber and an exuded latex, used for caulking boats. Breadfruit leaves are fed to livestock. Trees provide shade and are grown as ornamentals.

Breadfruits are tropical trees and thrive best in humid, lowland areas. Annual rainfall should be 1.5-2.5 m with temperatures of 21-32 degrees Celsius. Young plants require shade, while mature plants need increasing sunlight. Trees grow in most deep, well-drained soils, reach heights of 20 m, and are usually planted 9-12 m apart. They bear when 3-6 years old. Mature trees yield up to 700 fruits/year. Fresh fruit cannot be kept for more than a few days.

Broadbean, Horsebean, Field Bean, Tick Bean, Windsor Bean
Sci. Name(s)   : *Vicia faba, Faba vulgaris*

Geog. Reg(s)	: Africa-North, America-Central, America-South, Europe, Mediterranean, Temperate, Tropical
End-Use(s)	: Forage, Pasture & Feed Grain, Miscellaneous, Nut, Vegetable
Domesticated	: Y
Ref. Code(s)	: 56, 126, 170, 198
Summary	: The broadbean is grown as a garden and field crop and is used for fodder and hay. Garden crops are cultivated for the green shell beans and field crops for the dried beans. The roasted seed is eaten like peanuts. They are rich in minerals, calcium, and phosphorus as well as vitamins. Seeds are considered a meat extender and substitute for protein and skim milk. Broadbeans are one of the most important winter crops for human consumption in the Middle East and Latin America.

Plants are native to the Mediterranean area. They thrive in temperate climates but can be grown in the tropics as a winter crop. They do not produce pods in the low, hot tropics and are particularly sensitive to hot weather during the blooming season. For best development, beans need a cool season. They are not drought-resistant but are more tolerant of acidic soils than most legumes. Plants depend on fertile soil with lots of lime and an adequate and sustained water supply. Ingestion of the pollen can cause anemia or collapse.

In England, dried broadbean yields reach 2,018-2,690 kg/ha and about 1-2 MT of straw. Tropical countries report yields of 134-170 kg/ha of beans.

Broccoli, Cauliflower

Sci. Name(s)	: *Brassica oleracea* var. *botrytis*
Geog. Reg(s)	: America-North (U.S. and Canada), Europe, Temperate, Tropical
End-Use(s)	: Vegetable
Domesticated	: Y
Ref. Code(s)	: 148, 170, 200
Summary	: Broccoli (*Brassica oleracea* var. *botrytis*) is a hardy, temperate crop reaching 1-1.5 m in height and cultivated as a vegetable. It is high in vitamin C and other vitamins and minerals. Flower heads and stalks are eaten fresh or cooked in soups and pickled. Broccoli is also important as a quick-frozen vegetable.

This plant is a valuable leafy green vegetable crop in tropical areas and thrives at medium elevations. The Indian cultivar 'Early Patna' survives best in the lowland tropics. Broccoli has adapted to most parts of the United States and Europe and is frost-hardy.

Broccoli reaches maturity later than other Brassica species and needs more care and richer soil than cabbage. Flower heads and stalks are harvested continually for 4 months.

Brown Mustard, Indian Mustard, Mustard Greens
(See Oil)

Brussels-Sprouts

Sci. Name(s)	: *Brassica oleracea* var. *gemmifera*
Geog. Reg(s)	: Europe, Mediterranean
End-Use(s)	: Vegetable
Domesticated	: Y

Ref. Code(s) : 170, 200
Summary : Brussels-sprouts (*Brassica oleracea* var. *gemmifera*) are a frost-tolerant, cool-season crop whose compact heads and sprouts are eaten as vegetables. It is a form of nonheading cabbage. Crops are generally not successful in the lowland tropical areas. Brussels-sprouts need fertile, moist soil for best growth results. Brussel-sprouts originated in the Mediterranean and southwestern Europe.

Burdock, Great Burdock, Edible Burdock
(See Medicinal)

Burnet, Garden Burnet, Small Burnet
(See Weed)

Butterfly Pea
(See Erosion Control)

Cabbage, Savoy Cabbage
Sci. Name(s) : *Brassica oleracea* var. *capitata*
Geog. Reg(s) : America-North (U.S. and Canada), Europe, Tropical
End-Use(s) : Forage, Pasture & Feed Grain, Vegetable
Domesticated : Y
Ref. Code(s) : 170, 200
Summary : Cabbage is a vegetable eaten cooked or raw. The red variety is often pickled. Sauerkraut is made from sliced, fermented cabbage. A large cultivar, 'Drumhead', is fed to livestock.

This vegetable is primarily a cool-climate crop but there have been successful plantings in the lowland tropics. The best tropical cultivars are the 'Jersey Wakefield' and the 'Charleston Wakefield'. Most types tolerate frost but not heat. In the southern United States, crops are grown every season except summer, but along the Pacific coast, cabbage is grown year-round.

There are many varieties of cabbage that range in color from green to purple and in leaf style from smooth to ruffled with head shapes flat to pointed. The most common cabbage is the round-headed green one.

In 1983, the United States produced 1.5 million MT of cabbage from an area of 74,000 ha, which yielded 20,809 kg/ha. In addition, the United States imported 14,002 MT of fresh, chilled, or frozen cabbage, 447 MT of sauerkraut, and 120 MT of prepared or preserved cabbage. The total value of these imports was approximately $2.8 million.

Calabash Gourd, Calabash, Bottle Gourd, White Flowered Gourd
Sci. Name(s) : *Lagenaria siceraria, Lagenaria leucantha, Lagenaria vulgaris*
Geog. Reg(s) : Africa, America-North (U.S. and Canada), Asia, Tropical
End-Use(s) : Miscellaneous, Vegetable
Domesticated : Y
Ref. Code(s) : 170, 237
Summary : The calabash gourd is an annual vine with hard-shelled fruit used in many ways. Mature fruit rinds are made into utensils and containers. In California,

immature fruit flesh is used for culinary purposes and, in the Orient, is sliced into thin strips and air dried for cooking. The fruit is eaten as a vegetable and is similar to summer squash.

Vines are grown in most semiarid regions but are native to the tropical lowland areas of South-Central Africa. About 3-6 months after planting, fruits are harvested. Each plant yields 10-15 fruits.

Cantaloupe, Melon, Muskmelon, Honeydew, Casaba
(See Fruit)

Cardoon
Sci. Name(s) : *Cynara cardunculus*
Geog. Reg(s) : Europe, Mediterranean
End-Use(s) : Vegetable
Domesticated : Y
Ref. Code(s) : 220
Summary : Cardoon is an herb cultivated in Europe for its edible leaf stalks eaten as vegetables. It grows mainly in the Mediterranean region and is a close relative of the artichoke (*Cynara scolymus*) but without its uses.

Carrot, Queen Anne's Lace
Sci. Name(s) : *Daucus carota*
Geog. Reg(s) : America-North (U.S. and Canada)
End-Use(s) : Beverage, Essential Oil, Spice & Flavoring, Vegetable
Domesticated : Y
Ref. Code(s) : 148, 170, 237
Summary : The cultivated carrot has a swollen taproot that is eaten as a vegetable and pickled. The roots are a rich source of carotene, vitamins, and sugar. They are pressed for juice, canned, or dehydrated. Seeds contain an essential oil used for flavoring and in perfumery.

Carrots need moist soil for optimum growth and root formation. The best soils are sandy or organic soils. Roots are harvested in 75-85 days. The best temperatures are 16-24 degrees Celsius; higher temperatures result in toughness. Carrots can be stored for several months before marketing and not experience a loss in quality.

In 1983, 651,172 MT of carrots were produced in the United States for the fresh market and about 359,317 MT were processed. The total value of production was approximately $207 million.

Cassava, Manioc, Tapioca-Plant, Yuca, Mandioca, Guacomole
(See Energy)

Catjang, Jerusalem Pea, Marble Pea, Catjan
Sci. Name(s) : *Vigna unguiculata* subsp. *cylindrica*
Geog. Reg(s) : Africa, Africa-West, America-North (U.S. and Canada), Asia, Asia-India (subcontinent), Tropical
End-Use(s) : Cereal, Forage, Pasture & Feed Grain, Vegetable
Domesticated : Y

Ref. Code(s) : 17, 61, 126, 132, 141, 220
Summary : Catjang is a tropical herb grown for forage. Its pods are eaten as vegetables. Plants are grown for seeds, pods, and leaves in parts of West Africa. They are easily digested and are a nutritious food source rich in lysine and tryptophan, a characteristic different from most from other cereals. This variety is less commonly grown than others. It is susceptible to frost, requires a long-growing season, and is grown additionally in parts of Africa, India, North America, and Asia.

Celeriac, Turnip-Rooted Celery, Knob Celery, Celery Root

Sci. Name(s) : *Apium graveolens* var. *rapaceum*
Geog. Reg(s) : America-North (U.S. and Canada), Asia, Europe
End-Use(s) : Vegetable
Domesticated : Y
Ref. Code(s) : 170
Summary : Celeriac is a variety of celery. This herb has edible, turniplike roots but its leaf stalks do not develop fully. It is grown in parts of Europe, Asia, and North America.

Celery

Sci. Name(s) : *Apium graveolens* var. *dulce*
Geog. Reg(s) : America-North (U.S. and Canada), Asia-China, Asia-India (subcontinent), Europe, Temperate
End-Use(s) : Essential Oil, Spice & Flavoring, Vegetable
Domesticated : Y
Ref. Code(s) : 170, 194, 200
Summary : Celery is a biennial, temperate herb whose tender leaf stalks are a major commercial vegetable. These are eaten fresh, dried, or in a flavoring salt. The pungent seeds are used to flavor salads, soups, and vegetable and meat dishes. Dried ripe fruit yields 2-3% of an essential oil used as a fixative for medicines, in perfumery, liqueurs, and cosmetics. In China, France, India, Italy, Pakistan, the United States, and Great Britain, celery is cultivated mainly as a market vegetable and spice.

Celery is a marsh plant and should be grown in rich, sandy loams with frequent watering. Stalks are harvested 8 months after seeding. In 1982, United States celery production amounted to 816,000 MT and was valued at $189 million.

Chayote

Sci. Name(s) : *Sechium edule*
Geog. Reg(s) : America-Central, Subtropical, Tropical
End-Use(s) : Fruit, Gum & Starch, Tuber, Vegetable
Domesticated : Y
Ref. Code(s) : 81, 132, 170
Summary : Chayote is a vine 9-30 m in length with edible fruits eaten boiled, baked, or combined with other dishes. Young leaves and vine tips are eaten as vegetables. The tuberous roots are considered a valuable starch source.

This perennial is indigenous to southern Mexico and Central America and has a long history of use as a vegetable in those regions. Chayote grows throughout the tropics and thrives above 305 m and in areas receiving moderate rainfall. Fruits are harvested 3-5 months after planting and fruiting continues for several months. Plants are grown as annuals in the tropics and perennials in the subtropics.

Chicory
(See Beverage)

Chinese Amaranth, Tampala

Sci. Name(s) : *Amaranthus tricolor*
Geog. Reg(s) : Africa, America-Central, America-North (U.S. and Canada), America-South, Asia-China, Asia-Southeast, Pacific Islands, Tropical
End-Use(s) : Vegetable
Domesticated : Y
Ref. Code(s) : 15, 170
Summary : Chinese amaranth (*Amaranthus tricolor*) is an annual that reaches approximately 1.5 m in height. The amaranth family has been generally divided into two groups: the grain amaranths, most prominent in Mexico and the new world; and the vegetable amaranths, of which *Amaranthus tricolor* is one. (*A. gangeticus* is a synonym.)

The growth of *A. tricolor* is presently concentrated in the higher-rainfall, low-elevation tropics. It is used as a hot-season, leafy green vegetable in regions of Africa, Indonesia, New Guinea, China, and, to a lesser degree, in Central and South America. It is also grown in the United States, but as in the case of the countries above, only as a home gardening vegetable. In China, the leafy greens are sold in local markets on a limited basis. Apparently marketability is hampered by a relatively short shelf life and damage incurred in handling. Its major advantage lies in the leafy green's excellent quality as a cooked vegetable for hot-season production when most of the cooler-season leafy greens will not grow. No production, consumption, or yield data are available.

Chinese Cabbage, Pe-Tsai

Sci. Name(s) : *Brassica pekinensis, Brassica rapa*
Geog. Reg(s) : Asia, Asia-China
End-Use(s) : Vegetable
Domesticated : Y
Ref. Code(s) : 170, 200
Summary : The Chinese cabbage (*Brassica pekinensis*) is an annual heading variety of Brassica cultivated as a vegetable and similar to the leafy variety, pak-choi (*Brassica chinensis*). It is hardy, tolerates cool climates, and thrives on rich, well-drained, moist soils. Plants are grown throughout Asia, primarily in China and Japan.

Chinese Chives, Chinese Onion, Oriental Garlic

Sci. Name(s) : *Allium tuberosum*
Geog. Reg(s) : Asia, Asia-Central, Asia-India (subcontinent), Asia-Southeast
End-Use(s) : Spice & Flavoring, Vegetable
Domesticated : Y
Ref. Code(s) : 171, 194

Summary : Chinese chives are perennial herbs grown for their garlic-flavored edible leaves and young flower stalks used for seasoning or eaten as vegetables.

Plants grow wild and are cultivated in Southeast Asia. Their habitat extends from Mongolia and Japan to the Philippines and northern India. Chinese chives are rarely successful in tropical areas.

Chives

Sci. Name(s) : *Allium schoenoprasum*
Geog. Reg(s) : Temperate, Tropical, Worldwide
End-Use(s) : Medicinal, Ornamental & Lawn Grass, Spice & Flavoring, Vegetable
Domesticated : Y
Ref. Code(s) : 171, 194
Summary : Chives are perennial herbs widely distributed throughout the world as home garden plants. Their mild-flavored leaves are eaten or used as a garnish in soups, stews, egg dishes, salads, cheese, and cream cheese. In California, chives are harvested and frozen for later consumption. Plants make attractive ornamentals in temperate countries. In traditional medicine, chives are used as a digestive and appetite stimulant.

Plants tolerate cold temperatures and drought. In the tropics, they thrive at high altitudes and need full sun and long daylight hours for bulbset. Leaves are harvested several times throughout the year.

Chlabato, Ghiabato

Sci. Name(s) : *Ipomoea eriocarpa, Ipomoea hispida*
Geog. Reg(s) : Asia-India (subcontinent)
End-Use(s) : Forage, Pasture & Feed Grain, Vegetable
Domesticated : Y
Ref. Code(s) : 170
Summary : Chlabato is a twining annual or perennial herb that is grown in India for use as a vegetable and green fodder. Plant leaves are prepared and eaten like spinach. There is very little information about the economics or production of this crop.

Cohune Nut Palm, Cohune Palm
(See Oil)

Collards, Boreocole, Kale

Sci. Name(s) : *Brassica oleracea* var. *acephala*
Geog. Reg(s) : Europe, Hawaii, Temperate, Tropical, West Indies
End-Use(s) : Forage, Pasture & Feed Grain, Vegetable
Domesticated : Y
Ref. Code(s) : 170, 200
Summary : Collards (*Brassica oleracea* var. *acephala*) are a nonheading form of cabbage cultivated as a vegetable. It is a member of the wild cabbage family. Plants reach 6-12 cm in height. Two varieties are commonly grown, the "thousand-headed" type, which produces a lot of foliage and is fed to livestock, and the "curled" type with curled leaves boiled as green vegetables.

Plants thrive in cool, moist climates and at higher elevations in tropical areas. Crops are harvested 3-4 months after planting. Vegetable breeding and experimentation have received more attention in temperate areas but little work has been done in the tropics, with the exception of Puerto Rico and Hawaii.

Coltsfoot
(See Medicinal)

Common Bean, French Bean, Kidney Bean, Runner Bean, Snap Bean, String Bean, Garden Bean, Green Bean, Haricot Bean

Sci. Name(s)	: *Phaseolus vulgaris*
Geog. Reg(s)	: Africa-West, America-South, Subtropical, Temperate
End-Use(s)	: Cereal, Forage, Pasture & Feed Grain, Vegetable
Domesticated	: Y
Ref. Code(s)	: 132, 170
Summary	: The common bean is an erect bush or twining vine. Its beans are excellent sources of protein, vitamin B, and iron. Bean pods are often sold in processed forms, usually as frozen, canned, or dehydrated beans. Bean seeds are sometimes processed into dry flour or eaten boiled. Bean plants are used as forage.

Plants are grown throughout the middle and high altitudes of tropical America and parts of tropical Africa but are not considered an important food source in the hot, humid, lowland tropics. Common beans are more prevalent in the subtropics and temperate regions in areas with medium rainfall. Plants thrive from sea level to altitudes approaching 2,500 m. They thrive in well-aerated and well-drained moist soils. Excess rainfall promotes disease.

Dry seed harvest begins 3-6 months from planting when plants are mature. Young, tender snap beans are collected before seed formation, while green shell beans are harvested when the young pods develop full-sized seeds.

Common Reed Grass, Carrizo
(See Miscellaneous)

Corn Salad, Lamb's Lettuce, European Cornsalad

Sci. Name(s)	: *Valerianella locusta, Valerianella olitoria*
Geog. Reg(s)	: Africa-North, America-North (U.S. and Canada), Asia, Europe, Temperate
End-Use(s)	: Ornamental & Lawn Grass, Vegetable
Domesticated	: Y
Ref. Code(s)	: 17, 42, 61
Summary	: Corn salad is a temperate shrub that grows easily on arable land. It is often found in grain fields. Plants are grown as potherbs, salad plants, and ornamentals. Corn salad is considered hardy and easy to cultivate. It is common throughout the British Isles, the Madeira Islands, North Africa, western Asia, and North America.

Cowpea, Southern Pea, Black-Eyed Pea, Crowder Pea
(See Forage, Pasture & Feed Grain)

Cowslip, Marsh-Marigold

Sci. Name(s) : *Caltha palustris*
Geog. Reg(s) : America-North (U.S. and Canada), Asia, Europe, Temperate
End-Use(s) : Ornamental & Lawn Grass, Spice & Flavoring, Vegetable
Domesticated : Y
Ref. Code(s) : 16, 220
Summary : Cowslip is a hardy perennial herb whose roots, stems, and leaves are eaten as vegetables in North America, Europe, and temperate Asia. Pickled flower buds are a substitute for capers. Cowslip flowers do well in the cut-flower market. Plants grow on wet ground and thrive in rich soils.

Crookneck Pumpkin, Pumpkin, Winter Squash, Squash

Sci. Name(s) : *Cucurbita moschata*
Geog. Reg(s) : America-Central, Tropical
End-Use(s) : Oil, Vegetable
Domesticated : Y
Ref. Code(s) : 132, 170
Summary : The crookneck pumpkin (*Cucurbita moschata*) is an annual vine grown for its nutritious, soft-shelled fruits. Fruit seeds are used as snack food and have potential as an oil and protein source.
　　　　　　Crookneck pumpkins are heat-tolerant and widely grown in the tropics at low to medium altitudes. They are cultivated in Central America.

Cuculmeca

Sci. Name(s) : *Dioscorea macrostachya*
Geog. Reg(s) : America-Central
End-Use(s) : Vegetable
Domesticated : N
Ref. Code(s) : 12, 119
Summary : Cuculmeca is a member of the Dioscorea species grown for its leaves and shoots used in stews. It is native to Mexico. There is limited information concerning uses or economic potential for this crop.

Cucumber

Sci. Name(s) : *Cucumis sativus*
Geog. Reg(s) : Asia-Southeast, Tropical, Worldwide
End-Use(s) : Oil, Vegetable
Domesticated : Y
Ref. Code(s) : 170
Summary : The cucumber is a trailing herb grown for its edible fruit, used as salad vegetables and pickled. In Indonesia and western Malaysia, plant leaves are eaten in salads or cooked and eaten as spinach. Cucumber seeds are edible and yield an edible oil. In the tropics, cucumbers are raised for domestic markets.
　　　　　　Cucumbers are cultivated throughout most parts of the world in areas with warm climates and low rainfall. They are most successful on rich, moist, well-drained sandy loams. Fruit harvests begin 2 months after sowing. Cucumbers are native to northern India.

Curuba, Carua, Casabanana
(See Ornamental & Lawn Grass)

Cut-Egg Plant, Mock Tomato

Sci. Name(s)	: *Solanum aethiopicum*
Geog. Reg(s)	: Africa, Tropical
End-Use(s)	: Spice & Flavoring, Vegetable
Domesticated	: Y
Ref. Code(s)	: 8, 119, 170
Summary	: The cut-egg plant is a branched herb or subshrub 0.3-0.6 m in height, grown for its leaves and immature fruits. The leaves are used as potherbs and the immature fruits as vegetables and seasoning. This plant is native to parts of Africa and grown to some extent throughout the tropics.

Dandelion, Common Dandelion
(See Beverage)

Dasheen, Elephant's Ear, Taro, Eddoe
(See Tuber)

Dendrocalamus, Bamboo
(See Timber)

Dragon's Head
(See Oil)

Durian
(See Fruit)

Eggplant, Brinjal, Melongene, Aubergine

Sci. Name(s)	: *Solanum melongena*
Geog. Reg(s)	: Africa-North, Asia, Asia-India (subcontinent), Europe, Mediterranean, Temperate, Tropical
End-Use(s)	: Ornamental & Lawn Grass, Vegetable
Domesticated	: Y
Ref. Code(s)	: 116, 132, 148, 170, 220
Summary	: The eggplant is a short-lived perennial herb that provides a useful fruit in many tropical and temperate areas. This fruit is eaten boiled, fried, or stuffed, while unripe fruits are used in curries. They contain 92% moisture, 6% carbohydrates, 1% protein, 0.3% fats, and several minerals, and are fairly good sources of calcium, phosphorus, iron, and vitamin B. Eggplants are sometimes grown as ornamentals.

Plants thrive in areas with moderate to hot temperatures on sandy, well-drained loams. Fruit is harvested after 3 months. Major producing countries are Japan, Turkey, Italy, Egypt, and Iraq. Eggplants were first cultivated in India.

Elephant Bush, Spekboom
(See Ornamental & Lawn Grass)

Elephant Garlic, Great-Headed Garlic, Kurrat, Leek
(See Spice & Flavoring)

Endive

Sci. Name(s)	:	*Cichorium endivia*
Geog. Reg(s)	:	Asia-India (subcontinent), Mediterranean, Tropical
End-Use(s)	:	Vegetable
Domesticated	:	Y
Ref. Code(s)	:	132, 170
Summary	:	Endive is an annual or biennial herb grown for its individual leaves and loose heads, which are blanched and eaten in salads. It resembles but is more heat-tolerant than lettuce. The leaves are a source of vitamin A and iron. It is grown with little difficulty in the tropics. Leaves are harvested after 6 weeks and heads at 8-10 weeks. Yields are about 9-11 MT/ha. Endive is possibly native to India and the Mediterranean area. This herb is grown in rich soils at medium to high altitudes in areas receiving low to moderate rainfall.

Ensete, Abyssinian Banana

Sci. Name(s)	:	*Ensete ventricosum, Ensete edule*
Geog. Reg(s)	:	Africa, Asia
End-Use(s)	:	Fiber, Vegetable
Domesticated	:	N
Ref. Code(s)	:	154, 171
Summary	:	Ensete is an Asian tree whose thickened stems are a staple food source in southern Ethiopia. The stems and swollen underground stems are eaten cooked or fermented. Fiber from the stems is made into cordage and sacking. The fruit is dry and inedible.

Fat Hen, Lamb's-Quarters, White Goosefoot, Lambsquarter
(See Cereal)

Fennel, Florence Fennel, Finocchio
(See Spice & Flavoring)

Field Horsetail
(See Medicinal)

Figleaf Gourd, Malabar Gourd
(See Oil)

Garbanzo, Chickpea, Gram

Sci. Name(s)	:	*Cicer arietinum*
Geog. Reg(s)	:	Africa-North, America-Central, Asia, Europe, Mediterranean
End-Use(s)	:	Cereal, Forage, Pasture & Feed Grain, Medicinal, Vegetable
Domesticated	:	Y
Ref. Code(s)	:	132, 148, 170

Summary : The garbanzo is an annual, cool-season herb, reaching 30-70 cm in height, which is an important pulse crop in India and the Middle East, where it is cultivated for domestic consumption. Other areas of production are in Asia, Spain, and Mexico.

The whole dried seeds are eaten cooked, boiled, or made into dhal. Seeds are ground to flour and used in Indian confectionery. Green pods and young shoots are eaten as vegetables. The rest of the plant is fed to livestock. Pod hairs contain 94% percent malic acid and 6% oxalic acid and are used in certain medicines.

Plants need semiarid climates and mild temperatures for best production. They are harvested 4-6 months after sowing and yield 447-1,789 kg/ha of dried pulse.

Garden Angelica, Angelica
(See Essential Oil)

Garden Cress

Sci. Name(s) : *Lepidium sativum*
Geog. Reg(s) : Europe, Worldwide
End-Use(s) : Vegetable
Domesticated : Y
Ref. Code(s) : 92, 170, 185, 200, 215
Summary : Garden cress is a cool-climate, hardy annual reaching 7.6-30.6 cm in height and grown for its edible leaves and seedling shoots which are eaten as leafy salad vegetables throughout the world.

Plants require cool, rich soil for rapid, satisfactory growth. Leaves and seedlings are harvested 10-14 days after planting. Plants are native to Europe and naturalized throughout most parts of the world.

Garden Myrrh, Sweet Scented Myrrh
(See Medicinal)

Garden Orach, Butler Leaves, Orach, Mountain Spinach, Orache

Sci. Name(s) : *Atriplex hortensis*
Geog. Reg(s) : Asia, Europe
End-Use(s) : Forage, Pasture & Feed Grain, Vegetable
Domesticated : Y
Ref. Code(s) : 42, 220
Summary : Garden orach is an herb cultivated and cooked as a vegetable. It can be fed to livestock. This herb is native to Europe and Asia.

Garden Rhubarb

Sci. Name(s) : *Rheum rhaponticum*
Geog. Reg(s) : Asia-China, Europe, Temperate, Tropical
End-Use(s) : Beverage, Medicinal, Vegetable
Domesticated : Y
Ref. Code(s) : 72, 116, 170, 178, 220
Summary : Garden rhubarb is an economically important herb whose succulent petioles are used in pies and sauces. A refreshing juice is made from it and it can be

fermented into wine. The rhizomes contain a purgative, chrysarobin. The leaves have a high percentage of calcium oxalate and are poisonous.

Rhubarb was once used medicinally in parts of China and England. It has adapted to temperate areas and can be grown in the higher altitudes of the tropics. Garden rhubarb is native to southeastern Russia.

Garland Chrysanthemum, Spring Chrysanthemum

Sci. Name(s)	: *Chrysanthemum coronarium*
Geog. Reg(s)	: Asia, Asia-China
End-Use(s)	: Vegetable
Domesticated	: Y
Ref. Code(s)	: 220, 225
Summary	: The garland chrysanthemum is an herb 1-1.2 m tall cultivated for its edible young leaves and seedlings in China and Japan. This plant grows mainly in Okinawa and the southern Ryukyu Islands of Japan.

Garlic
(See Spice & Flavoring)

Gherkin, West Indian Gherkin

Sci. Name(s)	: *Cucumis anguria*
Geog. Reg(s)	: Africa, America-South, Asia, West Indies
End-Use(s)	: Vegetable
Domesticated	: Y
Ref. Code(s)	: 170, 200
Summary	: Gherkin is a trailing herb common in Brazil and the West Indies and similar to small cucumbers in appearance and use. Its fruit is usually pickled or cooked as a vegetable.

This herb is sensitive to cold, susceptible to fungal diseases, and attacked by aphids and cucumber beetles. For best results, gherkins should be planted in warm, fertile, irrigated soils. Gherkins are native to Asia and Africa.

Giant Granadilla, Barbardine
(See Fruit)

Gilo
(See Fruit)

Ginger
(See Spice & Flavoring)

Globe Artichoke, Artichoke

Sci. Name(s)	: *Cynara scolymus*
Geog. Reg(s)	: Africa-North, Subtropical, Temperate
End-Use(s)	: Medicinal, Vegetable
Domesticated	: Y
Ref. Code(s)	: 148, 170, 205

490

Summary : The globe artichoke is a thistlelike herb grown for its immature flower heads, which are eaten boiled. The plant has been used as a diuretic and a stimulant to liver cells because of the combined action of its constituents. It has medicinal value as a treatment for jaundice, liver insufficiency, anemia, and liver damage caused by poisons. It is a major component in proprietary digestive tonics.

Plants must be grown in areas with cool climates because hot temperatures cause flower heads to open, increasing their amount of fiber and toughness. They are sensitive to frost. Best results are obtained from moist, humus-rich soils in temperate and subtropical areas. Top production is reached in the 3rd year and plants should be replaced by their 5th year. Globe artichokes are native to North Africa.

Goosegrass, Fowl Foot Grass, Yardgrass
(See Weed)

Gow-Kee, Chinese Matrimony Vine, Chinese Wolfberry

Sci. Name(s)	: *Lycium chinense*
Geog. Reg(s)	: Asia
End-Use(s)	: Erosion Control, Vegetable
Domesticated	: Y
Ref. Code(s)	: 154, 214, 220
Summary	: Gow-kee is a hardy, east Asian ornamental shrub of the potato family, grown

for its leaves, which are eaten as vegetables in oriental cooking. It spreads rapidly and provides good ground cover. It thrives in ordinary, well-drained soil and is easy to cultivate.

Grass Pea, Chickling Vetch, Chickling Pea
(See Cereal)

Guar, Clusterbean
(See Gum & Starch)

Hairy Vetch, Winter Vetch, Russian Vetch
(See Forage, Pasture & Feed Grain)

Hops
(See Beverage)

Icecream Bean
(See Fruit)

Inca-Wheat, Quihuicha, Quinoa, Love-Lies-Bleeding
(See Cereal)

India Rubber Fig, Rubber-Plant
(See Isoprenoid Resin & Rubber)

Indian Mulberry
(See Dye & Tannin)

Itabo, Izote, Palmita, Ozote, Spanish Bayonnette

Sci. Name(s)	: *Yucca elephantipes*
Geog. Reg(s)	: America-Central, America-North (U.S. and Canada), America-South, West Indies
End-Use(s)	: Erosion Control, Fiber, Ornamental & Lawn Grass, Vegetable
Domesticated	: Y
Ref. Code(s)	: 17, 132, 154, 220
Summary	: The itabo tree has flowers that are rich in vitamin C and, in Mexico and Central America, are eaten in soups or with eggs. Plant hearts are edible and rich in calcium and have potential as processed gourmet items. Trees are grown as tall hedges, garden ornamentals, or as terracing for coffee plantations. Itabo leaves yield a fiber used for twine, cloth, and baskets. Trees prefer the well-drained, sandy loams found in Central, North, and South America and the West Indies. Itabo needs more trials in order to measure its full potential.

Jackbean, Horsebean, Swordbean

Sci. Name(s)	: *Canavalia ensiformis*
Geog. Reg(s)	: America-Central, Tropical, West Indies
End-Use(s)	: Forage, Pasture & Feed Grain, Medicinal, Vegetable
Domesticated	: Y
Ref. Code(s)	: 126, 170
Summary	: The jackbean is a bushy, erect herb reaching 1-2 m in height and grown for its young pods and green seeds, eaten as vegetables, and as a green manure. It has little value as fodder and is palatable only after drying. An enzyme, urease, which promotes the hydrolyisis of urea, and lectin are extracted for medicinal purposes.

Jackbeans are native to Central America and the West Indies and are introduced throughout the tropics. Plants grow well in subhumid climates in regions with an annual rainfall of 900-1,200 mm. They are fairly drought-resistant and tolerate limited shade and waterlogging. Yields of green matter are 14.5-18 MT/ha, and dried bean yields are 800-1,000 kg/ha.

Jackfruit, Jack
(See Fruit)

Japanese Lawngrass, Zoysia, Korean Grass
(See Ornamental & Lawn Grass)

Jerusalem-Artichoke, Girasole
(See Tuber)

Kamraj

Sci. Name(s)	: *Helminthostachys zeylanica*
Geog. Reg(s)	: Asia, Asia-Southeast, Australia
End-Use(s)	: Fiber, Medicinal, Vegetable

Domesticated : Y
Ref. Code(s) : 36, 220, 225
Summary : Kamraj is a relatively rare fern with a thick, fleshy rhizome. It grows over a broad geographical range, from the Himalayas to Australia. In the Philippines, young fern fronds are eaten as vegetables. The underground stems were once used to treat malaria and have been chewed with betel quids as a tonic. Rhizomes are exported to China. Old plant stems are used in basketry. The fern grows in lightly shaded areas.

Kangaroo Apple
(See Fruit)

Kohlrabi
Sci. Name(s) : *Brassica oleracea* var. *gongylodes*
Geog. Reg(s) : Europe, Tropical
End-Use(s) : Vegetable
Domesticated : Y
Ref. Code(s) : 170, 200
Summary : Kohlrabi (*Brassica oleracea* var. *gongylodes*) is a vegetable cultivated in Europe for its short, thick stems cooked like turnips and its leaves eaten like spinach. Plants do reasonably well in the lowland tropics. Kohlrabi needs fertile, moist soil.

Lablab, Lablab Bean
(See Forage, Pasture & Feed Grain)

Lac-tree, Ceylon Oak, Kussum Tree, Malay Lac-tree
(See Fat & Wax)

Leadtree, Lead Tree
(See Forage, Pasture & Feed Grain)

Lentil
(See Cereal)

Lettuce, Garden Lettuce, Celtuce
Sci. Name(s) : *Lactuca sativa*
Geog. Reg(s) : Africa-North, America-North (U.S. and Canada), Asia-China, Eurasia, Mediterranean, Temperate, Tropical
End-Use(s) : Vegetable
Domesticated : Y
Ref. Code(s) : 132, 170
Summary : Lettuce is an annual or biennial herb with edible leaves used widely as salad vegetables in temperate areas. Lettuce leaves are a fair source of nutrients. Plants are most successful in low to moderate, humid climates and at medium- to high-altitudes in areas with average rainfall. They thrive in rich, loamy soil. Certain varieties can be grown in China and the tropics.

Herbs are native to the Middle East. Full heads can be cut after 12 weeks.
Yields are 10-25 MT/ha of lettuce. In 1983, the United States produced 3 million MT
of lettuce, valued at $800 million, from an area of 91,271 ha.

Lima Bean, Sieva Bean, Butter Bean, Madagascar Bean
Sci. Name(s) : *Phaseolus lunatus, Phaseolus limensis, Phaseolus inamoenus, Phaseolus
turkinensis*
Geog. Reg(s) : Africa, America-Central, America-North (U.S. and Canada), Tropical,
Worldwide
End-Use(s) : Vegetable
Domesticated : Y
Ref. Code(s) : 126, 132, 170
Summary : The lima bean (*Phaseolus lunatus*) is an annual or perennial vine grown for its
edible beans, which comprise a commercialized industry in the United States. Lima
beans are used as dried shelled beans or whole green beans in the canning and freez-
ing industries. Green shelled beans are eaten boiled as vegetables. Bean seeds are
good sources of protein and fair sources of vitamin B complex.

This plant originated in Guatemala and spread throughout Mexico and into the
United States. It is widespread in the lowland tropics and Africa. Lima beans are
most successful planted in neutral or slightly acidic, well-drained soils. Plants thrive in
areas with moderate to high temperatures and low to moderate rainfall. They are
more tolerant of wet seasons than common beans (*Phaseolus vulgaris*). Plants are
grown from sea level to 2,437 m in elevation. Harvest occurs 2-8 months after sowing.
Green seed yields are 2-8 MT/ha, and dried seed yields are 500-600 kg/ha.

Lovage, Common Lovage
(See Spice & Flavoring)

Mahua
(See Oil)

Maize, Corn, Indian Corn
(See Cereal)

Malabar-Spinach, Indian Spinach, Vine-Spinach
Sci. Name(s) : *Basella alba*
Geog. Reg(s) : Tropical
End-Use(s) : Vegetable
Domesticated : Y
Ref. Code(s) : 170
Summary : Malabar-spinach is a coarse, viny herb grown for its stems and leaves, eaten as
spinach. It is grown throughout the lowland tropics. Leaves can be picked 80-90 days
after planting.

Marrow, Pumpkin, Squash, Vegetable Marrow
Sci. Name(s) : *Cucurbita pepo*
Geog. Reg(s) : America-Central, America-North (U.S. and Canada)

End-Use(s) : Nut, Oil, Vegetable
Domesticated : Y
Ref. Code(s) : 132, 170
Summary : Marrow is an annual vine with nutritious hard- or soft-shelled fruit. Hard-shelled fruit flesh is usually canned or frozen. The seeds are eaten as snacks and are a source of oil and protein. Three varieties are most commonly grown: variety pepo--field pumpkin; variety medullosa--vegetable marrow; and variety melopepo--bush squash, pumpkin, or summer squash.

Plants are most successful in hot climates but tolerate cool temperatures. They are grown at low to medium altitudes in areas with adequate rainfall. Fruit is harvested after 6 weeks. Marrow is a common crop in northern Mexico and the southwestern United States.

Mexican Husk Tomato, Tomatillo, Jamberry, Tomatillo Ground Cherry, Husk-Tomato

Sci. Name(s) : *Physalis ixocarpa*
Geog. Reg(s) : America-Central, America-North (U.S. and Canada)
End-Use(s) : Fruit, Vegetable
Domesticated : Y
Ref. Code(s) : 17, 58, 81
Summary : The Mexican husk tomato is an annual herb grown for its edible fruit. In Mexico, this fruit is used in chili sauces and meat dressings. Plants are occasionally grown as outdoor garden plants in the warmer parts of the United States. Along the west coast, the fruit is sometimes found in domestic markets.

Plants are native to Mexico and naturalized in parts of North America. They are seldom grown as ornamentals because of their weedy appearance. In general, about 454 g of fruit/plant are obtained.

Mixta Squash, Pumpkin, Winter Squash, Squash

Sci. Name(s) : *Cucurbita mixta*
Geog. Reg(s) : America-Central, America-North (U.S. and Canada)
End-Use(s) : Oil, Vegetable
Domesticated : Y
Ref. Code(s) : 132, 170
Summary : Mixta squash (*Cucurbita mixta*) is an annual vine that bears soft- and hard-shelled, edible, nutritious fruit. Its seeds have potential as a protein source but need research. They also yield an edible oil.

Plants require hot temperatures and do not tolerate cool climates. They must be grown at low to medium altitudes in areas with moderate rainfall. Fruits are harvested after 6 months. Mixta squash has several effective sterility barriers and has gained identity as a separate member of the Cucurbita species. Plants are grown in North and Central America.

Moth Bean, Mat Bean

Sci. Name(s) : *Vigna aconitifolia, Phaseolus aconitifolius*
Geog. Reg(s) : Africa, America-North (U.S. and Canada), Asia, Asia-Southeast, Tropical
End-Use(s) : Cereal, Erosion Control, Forage, Pasture & Feed Grain, Vegetable

Domesticated : Y
Ref. Code(s) : 17, 126, 132
Summary : The moth bean is a low-trailing annual vine that provides a matlike cover for the soil and is valuable in southern Asia for food. The seeds are usually fried, boiled, ground to dhal, or made into sprouted bean paste, which is an excellent source of protein and carbohydrates. The young bean pods are eaten as vegetables.

In the hot, dry regions of the United States, the bean plants are important forage. They are well-adapted to poor soils but need good drainage. Plants are grown as a late-season crop under irrigation and are planted in mixtures with sorghum, millet, lablab, and pigeon peas. Crops are successful planted as a 2nd crop after monsoon rains. Moth beans yield up to 2 MT/ha of seed. They provide good weed control and nitrogen for soil improvement. These plants are also grown in tropical Asia and Africa.

Mugwort
(See Spice & Flavoring)

Mung Bean, Green Gram, Golden Gram
Sci. Name(s) : *Vigna radiata*
Geog. Reg(s) : Africa, America-North (U.S. and Canada), America-South, Asia, Asia-Southeast
End-Use(s) : Cereal, Forage, Pasture & Feed Grain, Vegetable
Domesticated : Y
Ref. Code(s) : 17, 126, 132
Summary : The mung bean is widely grown throughout the Orient for its protein-rich seeds, green pods, and young sprouts. Its protein is deficient in 2 amino acids--methionine and cystine--but is well-supplied with lysine and tryptophan, which many other cereals lack. Mung beans are high in carbohydrates, 58%, and are well-supplied with calcium, phosphorus, and vitamins. They provide a useful supplement to cereal grains and other starchy foods. Young sprouts are marketed as the well-known bean sprouts. Beans are eaten boiled or steamed, or the immature pods are eaten fried. Mung beans have potential for freezing or canning. Flour can be made from the seeds. Plants are sometimes grown in the United States as forage.

Mung beans need well-drained soil for optimum productivity. Some varieties are grown under dry conditions. Most plants tolerate alkaline and saline soils. Day length influences maturation. Plants thrive in areas with moderate to hot temperatures and at low to moderate altitudes. Green pod yields are 5-10 MT/ha, with dry seed yields of 2,000 kg/ha. Mung beans are also grown in Southeast Asia, parts of Africa, and South America.

Musk Okra, Musk Mallow, Ambrette
(See Essential Oil)

Native Eggplant
Sci. Name(s) : *Solanum macrocarpon*
Geog. Reg(s) : Africa, Tropical
End-Use(s) : Spice & Flavoring, Vegetable

Domesticated : Y
Ref. Code(s) : 36, 170, 237
Summary : The native eggplant is a perennial crop grown for its edible fruit, collected immature and used as a vegetable or seasoning. The leaves are also edible and used as potherbs. There is limited commercial production of this plant in Africa, as it is primarily grown as a home garden plant. Native eggplants are grown in the tropics.

New Zealand-Spinach

Sci. Name(s) : *Tetragonia tetragonioides*
Geog. Reg(s) : Australia, Pacific Islands
End-Use(s) : Vegetable
Domesticated : Y
Ref. Code(s) : 220
Summary : The New Zealand-spinach is an herb cultivated in New Zealand and Australia for its leaves, eaten as vegetables. There is limited information available concerning the physical characteristics or economic potential for this plant.

Oblique-Seed Jackbean, Oblique Jackbean

Sci. Name(s) : *Canavalia plagiosperma*
Geog. Reg(s) : America-South, Tropical
End-Use(s) : Vegetable
Domesticated : N
Ref. Code(s) : 170
Summary : The oblique-seed jackbean is a bushy herb similar to the jackbean in appearance and structure. This plant was used in tropical South America for its seeds. No other information was found.

Oenocarpus bacaba
(See Oil)

Okra, Lady's Finger, Gumbo

Sci. Name(s) : *Abelmoschus esculentus, Hibiscus esculentus*
Geog. Reg(s) : Africa, Asia-China, Asia-India (subcontinent), Tropical
End-Use(s) : Fiber, Gum & Starch, Medicinal, Oil, Spice & Flavoring, Vegetable
Domesticated : Y
Ref. Code(s) : 148, 154, 170
Summary : Okra is an herb grown for its tender fruit pods, which are eaten as fresh, frozen, or canned vegetables. Ripe okra seeds are 20% edible oil. Medicinally, a mucilaginous preparation of the pods serves as a plasma replacement or a blood volume expander. When dried, preparations are used to treat peptic ulcers. In India, mucilage from the roots and stems has industrial value for clarifying sugarcane juice in gur manufacture. In China, mucilaginous preparations are used for sizing paper. Dried okra powder is used in salad dressings, ice creams, cheese spreads, and confectionery. Okra plant stems produce a fiber of inferior quality.

The herb is native to tropical Africa and grows easily in tropical lowlands in nearly any soil. For best results, okra should be planted in well-manured, sandy loams. Harvesting begins 2-3 months after planting and continues for about 2 months. Green pod yields range from 4,400-5,500 kg/ha.

Onion, Common Onion

Sci. Name(s)	: *Allium cepa*
Geog. Reg(s)	: Africa-North, Asia-India (subcontinent), Europe, Mediterranean, Tropical
End-Use(s)	: Cereal, Spice & Flavoring, Vegetable
Domesticated	: Y
Ref. Code(s)	: 126, 148, 171, 194
Summary	: The onion is an herb grown for its edible, pungent bulbs, which are relatively high in food value, intermediate in protein, and rich in calcium and riboflavin. Onions are cultivated in the Middle East, India, and Europe and throughout the world as a carbohydrate-rich energy food. The bulbs are most often used for flavoring. They can be eaten raw or cooked.

In the tropics, onion crops are most successful in dry, high elevated areas. Plants depend on day length for bulb set. Yields vary, with 7-9 MT/ha possible in the tropics. Crops mature 90-150 days after planting. Commercial production in the United States in 1983 was 2 million MT on a total of just under 50,000 ha. Production was valued at $427.9 million. The United States imported 9,190 MT of fresh, chilled, or frozen onions at a value of about $25 million.

Oysternut, Fluted Cucumber

Sci. Name(s)	: *Telfairia occidentalis*
Geog. Reg(s)	: Africa-West, Tropical
End-Use(s)	: Vegetable
Domesticated	: Y
Ref. Code(s)	: 92, 132, 220
Summary	: The oysternut is a climbing vine cultivated for its edible seeds and leaves. The seeds are cooked and eaten like beans and the plant leaves are eaten as vegetables. Oysternut plants are grown in tropical Africa or other hot, humid climates.

Pak-Choi, Chinese Cabbage, Pe-Tsai

Sci. Name(s)	: *Brassica chinensis, Brassica rapa*
Geog. Reg(s)	: Asia, Asia-China, Asia-India (subcontinent), Asia-Southeast, Tropical
End-Use(s)	: Vegetable
Domesticated	: Y
Ref. Code(s)	: 132, 170
Summary	: Pak-choi is an annual or biennial succulent herb (*Brassica chinensis*) considered a productive vegetable in the tropics. Its leaves are eaten as cooked vegetables or used in salads.

This herb is native to eastern Asia, particularly China and Japan. It is also under cultivation in Malaysia, Indonesia, and western India. Pak-choi is most productive in areas with cool to moderate temperatures and moderate rainfall. Best results are from rich, slightly acidic soils. Harvest occurs after 6-12 weeks. Leaf yields are 23-45 MT/ha.

498

Pamque
(See Dye & Tannin)

Parsnip, Wild Parsnip
(See Tuber)

Pea, Garden Pea, Field Pea

Sci. Name(s)	: *Pisum sativum, Pisum arvense, Pisum sativum* var. *arvense*
Geog. Reg(s)	: Africa-East, Africa-North, America-North (U.S. and Canada), America-South, Asia, Asia-India (subcontinent), Europe, Subtropical, Temperate, Tropical
End-Use(s)	: Cereal, Forage, Pasture & Feed Grain, Vegetable
Domesticated	: Y
Ref. Code(s)	: 126, 170

Summary : The pea is a short-lived, cool-season, annual climbing plant grown primarily as a pulse crop. Peas are high in protein and are valuable supplements to cereals and other starchy foods. They are rich in calcium, phosphorus, iron, and vitamins. Fresh green seeds and pods are cooked and eaten as vegetables. They are canned and frozen and are a leading frozen vegetable in the United States. The ripe dried seeds are a major food source in the tropics and subtropics and are sold whole, split, or as flour. Plants and haulms are used for hay, forage, silage, and green manure. When used for hay, peas are often grown in mixtures with other cereals. They are also planted in home gardens.

Peas are grown during the winter months in the subtropics and at higher altitudes of the tropics. Extensive cultivation occurs in India, Burma, Ethiopia, countries surrounding Lake Victoria in East Africa, Zaire, Morocco, Colombia, Ecuador, and Peru. Yields range from 920 kg/ha of seeds in Ethiopia and Peru to 400 kg/ha of seeds in the other countries. With better cultivation techniques and improved strains, higher yields can be obtained.

Plants originated in southwestern Asia and spread throughout the temperate and tropical zones of the world. They are most successful in deep, fertile, well-drained soils. Peas are a lucrative cash crop in most areas of the world.

In 1983, the United States produced 377,412 MT of green peas for processing, valued at $90.3 million. The United States exported 124,971 MT of dried peas in 1982-83. About 25% of these exports went to Colombia.

Perejil

Sci. Name(s)	: *Peperomia pereskiifolia, Peperomia viridispica*
Geog. Reg(s)	: America-South, Subtropical, Tropical
End-Use(s)	: Spice & Flavoring, Vegetable
Domesticated	: Y
Ref. Code(s)	: 17, 119

Summary : Perejil is a succulent herb of the Peperomia family grown for its young leaves and shoots eaten as vegetables. This herb is found throughout the tropics and subtropics and thrives in warm climates with minimal light. It is however, susceptible to waterlogging. Perejil is native to Venezuela and Colombia.

Pigeon Pea, Red Gram, Congo Pea, No-Eye Pea

Sci. Name(s)	: *Cajanus cajan, Cajanus indicus*
Geog. Reg(s)	: Africa, Asia-India (subcontinent), Subtropical, Tropical, West Indies
End-Use(s)	: Forage, Pasture & Feed Grain, Vegetable
Domesticated	: Y
Ref. Code(s)	: 16, 126, 170
Summary	: The pigeon pea is the 2nd most important pulse crop in India. Plants are grown for grain, green vegetables, and fodder. The protein-rich, dried seeds are widely exported. In Puerto Rico, canning industries are being built to enhance production. The pea is probably native to Africa, where it grows wild. Plants grow throughout the tropics and subtropics. They are annual tropical shrubs 1-3 m in height.

The young green seeds are eaten as vegetables and the ripe dry seeds are eaten as a pulse. In India, ripe dry seeds are split and made into dhal. Dried husks, seeds, and broken dhal are used as cattle feed. Plant tops and fruit provide excellent fodder, hay, and silage.

Pigeon peas are adaptable to most climates and soils. Initial plant growth is slow, but once established, requires little attention. In early varieties, plants begin to pod at 12-14 weeks and mature in 5-6 months. Late varieties mature in 9-12 months. The average green pod yield is 1,100-4,500 kg/ha. In mixed cultivation, 90-360 kg/ha can be obtained.

Polyscias rumphiana

Sci. Name(s)	: *Polyscias rumphiana*
Geog. Reg(s)	: Asia-Southeast
End-Use(s)	: Vegetable
Domesticated	: N
Ref. Code(s)	: 119, 220
Summary	: Polyscias rumphiana is a tropical tree native to Indonesia and the Moluccas and cultivated for its young, edible leaves. There is little information about this particular plant.

Potato, European Potato, Irish Potato, White Potato
(See Tuber)

Princess-Feather

Sci. Name(s)	: *Amaranthus hypochondriacus*
Geog. Reg(s)	: America-Central, America-South, Tropical
End-Use(s)	: Medicinal, Vegetable
Domesticated	: Y
Ref. Code(s)	: 205
Summary	: Princess-feather (*Amaranthus hypochondriacus*) is an herb widely cultivated for medicinal purposes in the Andean regions of South America. The dried flowers are used as an astringent or made into a mouthwash for ulcerated mouths. The young leaves are eaten as vegetables.

The tall, smooth annual is native to the tropics, particularly Central America. Princess-feathers thrive on waste ground and in cultivated fields. The herb is one of many Amaranthus species or varieties to be taken under horticultural cultivation.

Purslane, Pusley, Wild Portulaca, Akulikuli-Kula
(See Ornamental & Lawn Grass)

Radish

Sci. Name(s)	: *Raphanus sativus*
Geog. Reg(s)	: Asia, Worldwide
End-Use(s)	: Miscellaneous, Oil, Vegetable
Domesticated	: Y
Ref. Code(s)	: 111, 116, 170
Summary	: The radish (*Raphanus sativus*) is an annual or biennial herb grown throughout the world for its young, pungent roots and nutritious leaves eaten as salad vegetables. Radish leaves are used as a commercial source of leaf proteins. The seeds are a possible source of nondrying oils used in soap-making and for edible purposes. The expressed oilseed cake is used as fertilizer.

These herbs thrive on rich, friable, sandy loams in areas with cool, humid climates and 87.5-100 cm/year of rain. They grow from sea level to elevations approaching 1,800 m. The roots are usually harvested within 20-40 days. Radish yields from early-maturing varieties are recorded at 7.5 MT/ha. Radishes are native to western Asia and are grown throughout the cooler parts of the world.

Rakkyo, Chiao Tou

Sci. Name(s)	: *Allium chinense*
Geog. Reg(s)	: America-North (U.S. and Canada), Asia, Asia-China
End-Use(s)	: Vegetable
Domesticated	: Y
Ref. Code(s)	: 171, 220
Summary	: Rakkyo is a plant grown for its bulbs, which are pickled. This plant is native to central and eastern China and has spread throughout eastern Asia. Japan and California are centers of commercial rakkyo cultivation.

Rampion

Sci. Name(s)	: *Campanula rapunculus*
Geog. Reg(s)	: Africa-North, Eurasia, Europe
End-Use(s)	: Tuber, Vegetable, Weed
Domesticated	: Y
Ref. Code(s)	: 16, 42
Summary	: Rampion is a biennial or perennial herb reaching heights of 0.6-1 m and cultivated for its radish-shaped roots eaten in salads. This herb has naturalized throughout Europe, Siberia, and North Africa. In Britain, rampion is considered a weed and grows wild in fields, hedge banks, and gravelly soils.

Rice Bean, Ohwi, Ohashi

Sci. Name(s)	: *Vigna umbellata, Phaseolus calcaratus*

Geog. Reg(s) : America-North (U.S. and Canada), Asia, Asia-India (subcontinent), Asia-Southeast, Pacific Islands
End-Use(s) : Forage, Pasture & Feed Grain, Vegetable
Domesticated : Y
Ref. Code(s) : 17, 132
Summary : The rice bean is grown as a pulse crop and green vegetable in southern and Southeast Asia. Mature seeds are eaten boiled or fried and the leaves as spinach. Beans are a good source of calcium, iron, B vitamins, and protein. Plant remains are used as animal feed. Rice beans enrich the soil with nitrogen. Certain cultivars tolerate high temperatures and are moderately drought-resistant. Rice beans are grown as experimental cover crops.

Beans are mainly grown as subsistence crops or as domestic market items in Asia. Limited trade is done through Japan in international markets. Plants are grown in North America, India, and throughout the Pacific islands.

Roquette, Garden Rocket, Rocket Salad, Roka
(See Oil)

Roselle
(See Fiber)

Rutabaga, Swede
Sci. Name(s) : *Brassica napus* var. *napobrassica*
Geog. Reg(s) : America-North (U.S. and Canada), Europe, Temperate
End-Use(s) : Forage, Pasture & Feed Grain, Tuber, Vegetable
Domesticated : Y
Ref. Code(s) : 170, 200
Summary : Rutabaga is a cool-season herb that is a cross between a cabbage and a turnip and is considered one of the best adapted root crops in North America. Its edible roots are grown in Europe for livestock feed. These roots have a longer storage life than turnips and are generally eaten sliced and boiled. Rutabaga is most productive in northern regions and is seldom grown in the tropics.

Saltbush, Fourwing Saltbush, Chamisa
(See Forage, Pasture & Feed Grain)

Scarlet Runner Bean
(See Ornamental & Lawn Grass)

Seakale
Sci. Name(s) : *Crambe maritima*
Geog. Reg(s) : Europe, Mediterranean
End-Use(s) : Vegetable
Domesticated : Y
Ref. Code(s) : 42, 215
Summary : Seakale is a perennial herb occasionally cultivated for its asparaguslike shoots blanched and eaten as vegetables. Plants are common along the Atlantic coast of

Europe and Britain and are less proliferous in the Mediterranean. The shoots are a popular vegetable in England but are not common in United States markets. Plants do well in sandy coastal areas. Seakale is harvested after 3 years.

Seaside Clover

Sci. Name(s) : *Trifolium willdenovii*
Geog. Reg(s) : America-North (U.S. and Canada)
End-Use(s) : Vegetable
Domesticated : Y
Ref. Code(s) : 5, 152, 208, 213
Summary : Seaside clover is a perennial legume that is rarely cultivated. In the United States, plants thrive in southern California on wet ground usually below 2,437 m in elevation. Seaside clover has also been found in coniferous forests and desert areas throughout North America. Its flowers and leaves are edible.

Smooth Loofah, Sponge Gourd, Dishcloth Gourd, Vegetable Sponge, Luffa, Loofah
(See Miscellaneous)

Snakegourd, Club Gourd

Sci. Name(s) : *Trichosanthes anguina*
Geog. Reg(s) : Asia-India (subcontinent), Australia, West Indies
End-Use(s) : Vegetable
Domesticated : N
Ref. Code(s) : 92, 170
Summary : The snakegourd is an annual climbing plant with large edible fruits picked while still immature and cooked as a vegetable. Plants grow wild in India, the West Indies, and Australia.

Sodom Apple

Sci. Name(s) : *Solanum incanum*
Geog. Reg(s) : Africa-West, Asia-India (subcontinent)
End-Use(s) : Medicinal, Miscellaneous, Vegetable, Weed
Domesticated : Y
Ref. Code(s) : 56, 170, 220
Summary : The sodom apple is a shrub reaching 1-3 m in height and grown in Africa for its immature fruit, cooked as a vegetable. It is closely related to the eggplant and the two hybridize easily. In India, the fully ripe fruit has undocumented medicinal uses as an oral contraceptive and home remedy for sore throats and chest pain. The seeds of variety paniya contain 3.8-4.8% of a glyco-alkaloid used in the synthesis of cortisone. This plant is native to tropical India and Africa and is common in the grassy highlands of Kenya. It is unpalatable to cattle and considered a weed.

Sour-Relish Brinjal, Ram-Begun
(See Medicinal)

Soybean
(See Oil)

Spinach

Sci. Name(s)	: *Spinacia oleracea*
Geog. Reg(s)	: Asia-Central, Temperate, Tropical
End-Use(s)	: Vegetable
Domesticated	: Y
Ref. Code(s)	: 17, 170, 237
Summary	: Spinach is an annual temperate herb grown for its edible leaves, which provide an important vegetable often marketed in frozen or canned form. Fresh spinach is also a popular fresh produce item. Plants are cultivated as cool-season crops in areas with long days and temperatures of 16-18 degrees Celsius. They thrive on well-drained sandy or clay loams in areas with high rainfall.

Spinach originated in southwestern Asia. Plants do not fare well in the lowland tropics because of their susceptibility to disease. They are harvested 35-70 days after planting.

Stinging Nettle, European Nettle
(See Weed)

Summer-Cypress, Kochia
(See Cereal)

Swordbean
(See Forage, Pasture & Feed Grain)

Tamarind
(See Fruit)

Tepary Bean

Sci. Name(s)	: *Phaseolus acutifolius* var. *latifolius, Phaseolus acutifolius*
Geog. Reg(s)	: Africa, America-Central, America-North (U.S. and Canada), Asia-India (subcontinent), Asia-Southeast
End-Use(s)	: Cereal, Erosion Control, Forage, Pasture & Feed Grain, Vegetable
Domesticated	: Y
Ref. Code(s)	: 126, 132, 170
Summary	: The tepary bean (*Phaseolus acutifolius* var. *latifolius*) has been introduced to some African territories as a catch-crop and as a rapid food source in low rainfall areas. It thrives in arid regions and succeeds where other bean crops fail, except in the tropics. Plants are susceptible to waterlogging and frost. They grow primarily in Arizona and Mexico. The beans are used as dry shell beans and are ground, fried, or parched. A small portion is processed into a bean meal for commercial use.

Plants are grown as experimental hay and cover crops in the United States. Yields of 5-10 MT/ha of dry hay are possible. Dry farming bean yields are from 400-700 kg/ha with irrigated yields approximating 1,500 kg/ha. The 1st harvest is made in

2-4 months. Tepary beans are being replaced by the common bean (*Phaseolus vulgaris*) and are grown mainly in India and Burma.

Terongan, Turkeyberry, Platebush

Sci. Name(s)	: *Solanum torvum*
Geog. Reg(s)	: America-South, Tropical
End-Use(s)	: Medicinal, Vegetable
Domesticated	: Y
Ref. Code(s)	: 36, 130, 170, 225
Summary	: Terongan is a shrub or tree grown throughout the tropics for its unripe fruits, eaten as vegetables and in curries. The roots are used in some undocumented medicines. Terongan thrives in shady areas and is cultivated in most parts of tropical America.

Tiberato, Indian Nightshade

Sci. Name(s)	: *Solanum indicum*
Geog. Reg(s)	: Asia-China, Asia-Southeast
End-Use(s)	: Medicinal, Spice & Flavoring, Vegetable
Domesticated	: N
Ref. Code(s)	: 36, 98, 170, 225
Summary	: Tiberato is a subshrub 0.3-1.8 m in height that produces edible fruits used as vegetables, seasoning, or in curries. Tiberato roots are used as an undocumented diuretic. Seeds are burned and the smoke inhaled for the treatment of minor toothaches. Tiberato grows at elevations approaching 1,523 m in Malaysia, China, and the Philippines.

Tomato
(See Fruit)

Tsi

Sci. Name(s)	: *Houttuynia cordata*
Geog. Reg(s)	: Asia-China, Asia-India (subcontinent), Asia-Southeast
End-Use(s)	: Medicinal, Vegetable
Domesticated	: Y
Ref. Code(s)	: 16, 220
Summary	: Tsi is a perennial herb about 0.46-1 m high grown in southern Asia, particularly Vietnam, for its leaves, which are eaten as salad and soup vegetables. Tsi leaves are used to treat eye diseases. In China, the whole plant is used to treat bladder and kidney problems and skin diseases. Tsi thrives in moist areas with mild temperatures. In India, plants are successful up to 1,523 m in altitude.

Tuberose
(See Essential Oil)

Turnip

Sci. Name(s)	: *Brassica rapa*
Geog. Reg(s)	: Europe, Tropical, Worldwide

End-Use(s)   : Tuber, Vegetable
Domesticated : Y
Ref. Code(s) : 170, 200
Summary   : The turnip (*Brassica rapa*) is cultivated for its taproots, eaten cooked, and leaf tops, cooked and eaten as vegetables. Turnips grow rapidly and are tolerant of cool climates but susceptible to excessive heat. They are native to central and southern Europe and are cultivated throughout the world with certain cultivars particularly successful in the tropics. Turnip roots are harvested 8-10 weeks after planting.

Tussa Jute, Tossa Jute, Jew's Mallow, Nalta Jute
(See Fiber)

Udo, Spikenard, Oudo
Sci. Name(s) : *Aralia cordata*
Geog. Reg(s) : America-North (U.S. and Canada), Asia
End-Use(s)   : Vegetable
Domesticated : Y
Ref. Code(s) : 15
Summary   : Udo is a shrubby, perennial herb reaching heights of 1-3 m and cultivated for its young shoots, eaten raw in salads or boiled. This herb is cultivated in Japan and is considered a celery alternative. It has been successfully introduced as a home garden vegetable across North America. Large-scale commercial production is nonexistent. In California, the vegetable is grown for local markets but production and price information could not be obtained. This plant thrives in moist, cool regions.

Upland Cress, Wintercress, Yellow Rocket
Sci. Name(s) : *Barbarea vulgaris*
Geog. Reg(s) : Africa-North, America-North (U.S. and Canada), Asia, Australia, Europe, Pacific Islands
End-Use(s)   : Medicinal, Vegetable
Domesticated : Y
Ref. Code(s) : 42, 220
Summary   : Upland cress is a biennial or perennial herb whose leaves are occasionally used in salads. At one time, they were used in a balsam as a healing agent for wounds. Upland cress grows in Europe, North Africa, Asia, North America, New Zealand, and Australia, where it is most successful in damp soils.

Velvetbean
(See Forage, Pasture & Feed Grain)

Water Dropwort
Sci. Name(s) : *Oenanthe javanica, Oenanthe stolonifera*
Geog. Reg(s) : Asia-Southeast
End-Use(s)   : Spice & Flavoring, Vegetable
Domesticated : Y
Ref. Code(s) : 17, 36, 119, 237

Summary   : Water dropwort is a perennial herb whose leaves and young shoots are used as vegetables and flavoring. It is cultivated throughout Southeast Asia. Water dropwort grows wild in damp places from low to medium altitudes.

Water Spinach, Swamp Morning-Glory, Kangkong

Sci. Name(s)   : *Ipomoea aquatica, Ipomoea reptans*
Geog. Reg(s)   : Asia, Subtropical, Tropical
End-Use(s)   : Forage, Pasture & Feed Grain, Vegetable
Domesticated   : Y
Ref. Code(s)   : 132, 170, 220
Summary   : Water spinach is a perennial, succulent trailing vine that produces edible leaves important as vegetables in tropical and subtropical Asia. The leaves are eaten like spinach and are an excellent source of vitamins A and C and good sources of iron, calcium, riboflavin, and protein. Plant stems are tough and used as fodder for pigs and cattle. Water spinach has potential as a domestic market and home garden vegetable in the tropics. Its leaves have a short storage life of a few days.

     Vines grow year-round and have high leaf yields estimated at 53 MT/ha. They are most successful at low to medium altitudes in areas with warm temperatures. They are found wild in swamps, slow-moving streams, and ponds. Water spinach tolerates most soils. In the western hemisphere, it is susceptible to insect damage.

Watercress

Sci. Name(s)   : *Nasturtium officinale, Nasturtium aquaticum*
Geog. Reg(s)   : America-North (U.S. and Canada), Asia, Europe, Temperate
End-Use(s)   : Spice & Flavoring, Vegetable
Domesticated   : Y
Ref. Code(s)   : 132, 170
Summary   : Watercress is a succulent perennial herb whose leafy stem tips are eaten in salads, used as a garnish, or cooked as a vegetable. Its stems are a source of vitamins A and C. This plant grows wild in streams in temperate Europe, Asia, and North America. Stems are harvested after 6 weeks and yields are approximately 50 MT/ha of leafy stems.

Waxgourd, White Gourd

Sci. Name(s)   : *Benincasa hispida*
Geog. Reg(s)   : Asia-Southeast, Tropical
End-Use(s)   : Fat & Wax, Fruit, Nut, Oil, Vegetable
Domesticated   : Y
Ref. Code(s)   : 170
Summary   : The waxgourd is a climbing herb with fruits boiled and eaten as vegetables in many tropical countries. The ripe fruit is candied. The seeds can be fried and eaten, yield an oil, and are covered with a removable wax. The young leaves and flower buds are cooked like spinach.

     The waxgourd is usually grown for domestic consumption but in tropical Asia is cultivated on a small-scale commercial basis. This herb is most successful when grown in semidry, lowland tropics. Waxgourds are harvested after 4-5 months.

Welsh Onion, Japanese Bunching Onion

Sci. Name(s) : *Allium fistulosum*
Geog. Reg(s) : Asia, Asia-China, Tropical
End-Use(s) : Vegetable
Domesticated : Y
Ref. Code(s) : 171
Summary : The Welsh onion is grown for its edible bulbs and leaf bases, eaten raw or cooked similarly to leeks. This plant is agronomically adaptable from the cold regions of Siberia to southern China. Plants are grown at higher altitudes in the tropics.

White Mustard
(See Forage, Pasture & Feed Grain)

Winged Bean, Goa Bean, Asparagus Pea, Four-Angled Bean, Manilla Bean, Princess Pea

Sci. Name(s) : *Psophocarpus tetragonolobus*
Geog. Reg(s) : Asia-Southeast, Subtropical, Tropical
End-Use(s) : Cereal, Erosion Control, Forage, Pasture & Feed Grain, Gum & Starch, Oil, Vegetable
Domesticated : Y
Ref. Code(s) : 56, 126, 132, 170
Summary : The winged bean is an herbaceous annual that is cultivated in the tropics and subtropics for its pods, seeds, tuberous roots, young leaves, shoots, and flowers. The pods are a minor source of protein and oil, while the dried seeds are excellent sources of protein and an oil that is similar to soybean oil and can be used for culinary purposes, illumination, and soap making. Seeds are also a source of an industrial mucilage. The expressed oil cake can be used as livestock feed. In Java, ripe seeds are eaten after parching. In Burma, the tuberous roots are considered a primary vegetable source. Flowers can be fried and eaten and are reported to have a taste similar to mushrooms. Plants have been suggested as a green manure crop, cover crop, and fodder crop. Considerable interest has been paid to goa beans as a potential restorative fallow crop because of their exceptional capacity for forming large and plentiful root nodules.

   Plants are most productive in hot, wet climates, in loamy, well-drained soil with irrigation. They thrive from sea level to elevations approaching 2,000 m. Most varieties require short days for flowering. Plants are slow to bear but will continue cropping for about a year. Tubers are harvested 7-8 months after sowing. Seeds take longer to ripen. A good dry seed yield is 400-600 kg/ha. Edible pod yields are 25 MT/ha. Flowers and tubers are harvested as needed.

Winged Yam, Greater Yam, Water Yam, Ten Months Yam
(See Tuber)

Winter Purslane
Sci. Name(s) : *Montia perfoliata, Claytonia perfoliata*
Geog. Reg(s) : America-North (U.S. and Canada)
End-Use(s) : Medicinal, Vegetable

Domesticated : Y
Ref. Code(s) : 43, 214
Summary : Winter purslane is an herb with leaves rich in vitamin C and eaten as salad vegetables or used as potherbs. These herbs grow in western North America and are introduced to many other countries. The leaves can be used as a medicinal tea.

Wonderberry, Black Nightshade, Poisonberry, Stubbleberry
(See Miscellaneous)

Yam Bean, Erosus Yam Bean, Jicama
(See Tuber)

Yard-Long Bean, Asparagus Bean
Sci. Name(s) : *Vigna unguiculata* subsp. *sesquipedalis*
Geog. Reg(s) : Africa-West, America-South, Asia, Asia-China, Asia-Southeast, Hawaii, Tropical
End-Use(s) : Beverage, Cereal, Spice & Flavoring, Vegetable
Domesticated : Y
Ref. Code(s) : 17, 61, 84, 126, 132, 154
Summary : The yard-long bean plant produces edible bean pods eaten green and prepared like string beans. Beans are high in vitamin A, protein, and carbohydrates, and rich in lysine and tryptophan. They are easily digested and are an extender of animal proteins. Small pods and seeds are eaten as a green boiled vegetable. Seeds are ground into meal or used as a pulse. In West Africa, the seeds are used as a substitute for coffee and as a potherb. This plant is native to southern Asia. It thrives in areas with hot to moderate temperatures and tolerates soil with poor drainage. The yard-long bean is grown in Hawaii. A yield of 1 MT/ha of seed is typical. Plants are grown in South America, China, and Southeast Asia.

Weed

American Licorice, Wild Licorice
(See Spice & Flavoring)

Annual Bluegrass, Low Speargrass, Dwarf Meadow Gold, Six-Weeks Grass, Plains Bluegrass
(See Forage, Pasture & Feed Grain)

Aramina Fiber, Congo Jute, Cadillo, Caesarweed, Aramina
(See Fiber)

Banana Passion Fruit, Curuba, Banana Fruit
(See Fruit)

Barnyardgrass, Barnyard Millet
(See Forage, Pasture & Feed Grain)

Bermudagrass, Star Grass, Bahama Grass, Devilgrass
(See Erosion Control)

Black Medic, None-Such
(See Forage, Pasture & Feed Grain)

Black Oat, Bristle Oat, Sand Oat, Small Oat

Sci. Name(s)	: *Avena strigosa*
Geog. Reg(s)	: Europe
End-Use(s)	: Cereal, Weed
Domesticated	: N
Ref. Code(s)	: 42, 220
Summary	: The black oat is a European cereal grain now considered a weed. It is similar to the wild oat (*Avena fatua*) in appearance.

Bondue, Livid Amaranth, Wild Blite
(See Vegetable)

Burnet, Garden Burnet, Small Burnet

Sci. Name(s)	: *Sanguisorba minor, Poterium sanguisorba*
Geog. Reg(s)	: Africa-North, America-North (U.S. and Canada), Asia, Europe, Mediterranean
End-Use(s)	: Forage, Pasture & Feed Grain, Spice & Flavoring, Vegetable, Weed
Domesticated	: Y
Ref. Code(s)	: 42, 72, 154, 213
Summary	: Burnet is a tufted perennial grass reaching 60 cm in height. The grass has a deep root system, is fairly drought-resistant, and gives good yields. Plants are most common as weeds. Leaves have a taste similar to the cucumber and can be used fresh in salads or dried as spice. They grow in herb gardens.

Plants are found at elevations up to 500 m. They have naturalized on calcareous soils and are especially common in England, Scotland, and other parts of Europe. Burnet also occurs in Persia and Morocco and has become naturalized in North America.

Candytuft, Rocket Candytuft
(See Ornamental & Lawn Grass)

Carpetgrass
(See Forage, Pasture & Feed Grain)

China Jute, Indian Mallow, Velvetleaf, Butterprint
(See Fiber)

Chinese Yam, Cinnamon-Vine
(See Tuber)

Common Bent-Grass, Redtop
(See Forage, Pasture & Feed Grain)

Common Guava, Lemon Guava
(See Fruit)

Crowfoot Grass

Sci. Name(s)	: *Dactyloctenium aegyptium*
Geog. Reg(s)	: Africa, America-North (U.S. and Canada), Asia-India (subcontinent)
End-Use(s)	: Cereal, Erosion Control, Forage, Pasture & Feed Grain, Ornamental & Lawn Grass, Weed
Domesticated	: Y
Ref. Code(s)	: 24, 25, 96, 122, 164, 220
Summary	: Crowfoot grass is an annual tufted grass, with stems 20-40 cm in height, which thrives in warm climates and provides quality grazing. It is also used as a sand binder and lawn grass. This grass grows in fairly dry regions, tolerates saline conditions, and is found from sea level to 2,000 m in altitude.

In the United States, crowfoot grass grows mainly as a weed on cultivated ground. In parts of Africa and India, it is planted along rivers as a grain crop during times of limited food supply but there is a problem with its toxicity.

Crownvetch, Trailing Crownvetch
(See Forage, Pasture & Feed Grain)

Dallisgrass
(See Forage, Pasture & Feed Grain)

Dandelion, Common Dandelion
(See Beverage)

Danicha
(See Fiber)

Edible Canna, Gruya, Queensland Arrowroot
(See Gum & Starch)

Feather Fingergrass

Sci. Name(s)	: *Chloris virgata*
Geog. Reg(s)	: America-North (U.S. and Canada), Temperate, Tropical
End-Use(s)	: Weed
Domesticated	: N
Ref. Code(s)	: 164
Summary	: Feather fingergrass is a tufted annual grass common as a weed in the south-western United States. It is most successful in tropical and warm temperate areas.

Field Horsetail
(See Medicinal)

Giant Reed
(See Miscellaneous)

Goosegrass, Fowl Foot Grass, Yardgrass

Sci. Name(s)	: *Eleusine indica*
Geog. Reg(s)	: Asia-India (subcontinent), Asia-Southeast, Europe, Pacific Islands, Temperate, Tropical
End-Use(s)	: Fiber, Forage, Pasture & Feed Grain, Vegetable, Weed
Domesticated	: Y
Ref. Code(s)	: 25, 35, 122, 171, 220
Summary	: Goosegrass is a tropical grass 30-60 cm tall that was introduced to warm temperate regions. It grows to some extent in the British Isles and New Zealand. In the tropics, goosegrass grows from sea level to 1,500 m.

The grass is hardy and often found in waste places, along roadsides and in nitrogen-rich soils. In some countries, goosegrass is considered valuable fodder, but as it contains hydrocyanic acid, it can be fatal to calves and sheep. In India, the seeds are eaten in times of food scarcity. The seedlings are eaten as vegetables. In Java, grass stems are used to make mats. Goosegrass is often considered a weed.

Grass Pea, Chickling Vetch, Chickling Pea
(See Cereal)

Huisache, Cassie Flower, Kolu, Sweet Acacia, Cassie
(See Essential Oil)

Istle, Mexican Fiber, Lechuguilla, Tula Istle, Tampico Fiber
(See Fiber)

Jaragua Grass
(See Forage, Pasture & Feed Grain)

Jerusalem-Artichoke, Girasole
(See Tuber)

Johnson Grass, Johnsongrass
(See Forage, Pasture & Feed Grain)

Jointed Goatgrass, Goat Grass
Sci. Name(s) : *Aegilops cylindrica, Triticum cylindrica*
Geog. Reg(s) : America-North (U.S. and Canada), Europe
End-Use(s) : Weed
Domesticated : Y
Ref. Code(s) : 164, 218
Summary : Jointed goatgrass is an annual grass with no specific use. This grass was introduced to the midwestern United States in mixtures of Turkey wheat. It is capable of dominating wheat, and its seeds are found as contaminants of wheat seed. Jointed goatgrass is native to southeastern Europe and western Central Europe.

Maidencane
Sci. Name(s) : *Panicum hemitomon*
Geog. Reg(s) : America-North (U.S. and Canada), America-South
End-Use(s) : Weed
Domesticated : Y
Ref. Code(s) : 96
Summary : Maidencane is a grass with culms 50-150 cm in height that is grown in moist soils and is considered a weed. In the United States, maidencane grows on the coastal plains from New Jersey to Florida and Texas and into Tennessee. This grass is found in Brazil.

Nutgrass, Purple Nutsedge, Nutsedge
Sci. Name(s) : *Cyperus rotundus, Cyperus hexastachyus*
Geog. Reg(s) : Asia-India (subcontinent), Temperate, Tropical
End-Use(s) : Miscellaneous, Tuber, Weed
Domesticated : Y
Ref. Code(s) : 220
Summary : Nutgrass is a perennial, grasslike herb that grows in tropical and warm temperate areas. It is a serious weed in the tropics. In India, its dried tubers, called slouchers, are used to perfume hair and clothes.

Pokeweed, Poke, Skoke, Pigeon Berry
Sci. Name(s) : *Phytolacca americana, Phytolacca decandra*
Geog. Reg(s) : America-North (U.S. and Canada), America-South, Hawaii, Tropical
End-Use(s) : Dye & Tannin, Medicinal, Weed
Domesticated : Y
Ref. Code(s) : 17, 86, 205

Summary   : Pokeweed is a perennial herb considered a noxious weed but also used as a potherb. Dried roots are used in internal medicines for throat infections. Root preparations are used to treat chronic rheumatism and upper respiratory tract infections, and roots are applied as ointments for external ailments. Treated berries are used to color wine, confectioneries, and art paints.

This herb is native to tropical America and has become naturalized in the southwestern United States and Hawaii as a weed. Plants thrive in rich, light soils in sunny areas.

Purslane, Pusley, Wild Portulaca, Akulikuli-Kula
(See Ornamental & Lawn Grass)

Quack Grass, Torpedograss

Sci. Name(s)	: *Panicum repens*
Geog. Reg(s)	: Hawaii, Tropical
End-Use(s)	: Erosion Control, Forage, Pasture & Feed Grain, Weed
Domesticated	: Y
Ref. Code(s)	: 24, 36, 125, 154
Summary	: Quack grass is a valuable and nutritious perennial fodder, 0.3-1 m in height,

which grows in grassy waste areas and is highly palatable to horses and cattle. The grass is considered a weed in cultivated fields. It grows well on sandy soils and makes a good sand-binder. In its 1st year, quackgrass produces well at 6.2 MT/ha of dry matter; by the 2nd year, production rises to 20.9 MT/ha; but by the 3rd year, only 8.4 MT/ha are produced. Quackgrass grows well along coastlines and is frequently found in the tropics.

Quackgrass, Twitchgrass, Couchgrass, Couch, Twitch
(See Forage, Pasture & Feed Grain)

Rampion
(See Vegetable)

Rough Pea
(See Forage, Pasture & Feed Grain)

Salsify, Oyster-Plant, Vegetable Oyster
(See Tuber)

Sickle Medic, Yellow-Flowered Alfalfa
(See Forage, Pasture & Feed Grain)

Silver Bluestem

Sci. Name(s)	: *Bothriochloa saccharoides*
Geog. Reg(s)	: America-North (U.S. and Canada)
End-Use(s)	: Weed
Domesticated	: Y
Ref. Code(s)	: 146

Summary   : Silver bluestem is a grass that is considered a weed in the southwestern United States. It is common in vacant fields and usually grows with several other weedy grass species.

Sodom Apple
(See Vegetable)

Stinging Nettle, European Nettle
Sci. Name(s) : *Urtica dioica*
Geog. Reg(s) : Hawaii, Subtropical, Tropical
End-Use(s) : Fiber, Medicinal, Vegetable, Weed
Domesticated : Y
Ref. Code(s) : 83, 86, 170, 220, 233
Summary : Stinging nettle is a shrub with stinging hairs. Its young shoots and leaves are eaten like spinach and the stems provide useful fiber. Parts of the plant have been used in home remedies for rheumatism, gout, skin troubles, and rashes. A decoction of the roots makes an astringent and hair wash, while a decoction of the seeds can be used for washing the hair as well as for a tonic for coughs and bronchitis. The plant is a commercial source of chlorophyll. The stinging hairs are difficult to handle and the base of the plant contains formic acid. This stinging quality is inactivated with cooking. Nettles are thought to be high in vitamin C. They contain about 5% albuminoids and 7% carbohydrates. They are used as a substitute for rennet to coagulate milk and in beer and wine making. Mature stems are retted and spun into a durable cloth.

Plants grow along streams, canyons, and ditches. For the most part, stinging nettles are considered pasture weeds that crowd out desirable forage.

Stinging nettles grow in most tropical and subtropical regions. In Hawaii, they grow on the high plains of Mauna Loa and Mauna Kea at altitudes of 1,523-1,828 m.

Sunolgrass
Sci. Name(s) : *Phalaris coerulescens, Phalaris bulbosa, Phalaris variegata, Phalaris villosula*
Geog. Reg(s) : Mediterranean
End-Use(s) : Weed
Domesticated : Y
Ref. Code(s) : 6, 218
Summary : Sunolgrass is a perennial grass, with culms up to 150 cm in height, which grows mainly in the Mediterranean and is considered a weed of cultivated fields. There is limited information concerning the economic importance of this plant.

Tikus, Tikug
(See Fiber)

Wonderberry, Black Nightshade, Poisonberry, Stubbleberry
(See Miscellaneous)

INDEX A.

ALPHABETICAL LISTING
OF CROPS
BY COMMON NAME

534

To Find a Crop with the Common Name	Look in the End-Use Category	On Page	Under the Crop Name
Gemsbok Bean	Beverage	12	Camel's-Foot
Gentian	Beverage	26	Yellow Gentian
Geocarpa Groundnut	Cereal	33	Geocarpa
Geranium	Essential Oil	101	Rose Geranium
	Spice & Flavoring	387	Costmary
German Chamomile	Essential Oil	110	Wild Chamomile
German Millet	Cereal	32	Foxtail Millet
Ghatti	Gum & Starch	251	Gum Ghatti
Gherkin	Vegetable	489	Gherkin
Ghia	Essential Oil	88	Chia
Ghiabato	Vegetable	483	Chlabato
Giant Alocasia	Gum & Starch	250	Giant Taro
Giant Filbert	Nut	327	Giant Filbert
Giant Granadilla	Fruit	216	Giant Granadilla
Giant Panicum	Forage, Pasture & Feed Grain	144	Blue Panicgrass
Giant Reed	Miscellaneous	308	Giant Reed
Giant Rye	Cereal	38	Polish Wheat
Giant Stargrass	Forage, Pasture & Feed Grain	182	Stargrass
Giant Taro	Gum & Starch	250	Giant Taro
Giant Wildrye	Forage, Pasture & Feed Grain	140	Basin Wildrye
	Forage, Pasture & Feed Grain	157	Giant Wildrye
Gilo	Fruit	216	Gilo
Gingelly	Oil	351	Sesame
Ginger	Spice & Flavoring	391	Ginger
Ginseng	Medicinal	282	Ginseng
Girasole	Tuber	466	Jerusalem-Artichoke
Giri Yam Pea	Tuber	465	Giri Yam Pea
Globe Artichoke	Vegetable	489	Globe Artichoke
Glory	Vegetable	506	Water Spinach
Gmelina	Timber	461	Yemane
Goa Bean	Vegetable	507	Winged Bean
Goat Chili	Spice & Flavoring	392	Goat Chili
Goat Grass or Goatgrass	Weed	512	Jointed Goatgrass
Goat-Nut	Fat & Wax	114	Jojoba
Gold	Forage, Pasture & Feed Grain	136	Annual Bluegrass
Golden Gram	Vegetable	495	Mung Bean
Golden Timothy Grass	Forage, Pasture & Feed Grain	158	Golden Timothy Grass
Golden Wattle	Dye & Tannin	52	Golden Wattle
Golden-Apple	Fruit	195	Ambarella
Gomuti Palm	Gum & Starch	255	Sugar Palm
Goober	Oil	348	Peanut
Gooseberry	Fruit	222	Kiwi
	Fruit	205	Ceylon-Gooseberry
	Fruit	229	Otaheite Gooseberry
	Fruit	204	Cape-Gooseberry
	Fruit	195	American Gooseberry
	Fruit	214	English Gooseberry
Goosefoot	Cereal	32	Fat Hen
	Medicinal	300	Wormseed
Goosegrass	Weed	511	Goosegrass

To Find a Crop with the Common Name	Look in the End-Use Category	On Page	Under the Crop Name
Proso Millet	Cereal	39	Proso Millet
Prune Plum	Fruit	233	Plum
Pulasan	Fruit	234	Pulasan
Pulut-Pulut	Fiber	125	Paroquet Bur
pumilio	Essential Oil	100	Pinus pumilio
Pummelo	Fruit	234	Pummelo
Pumpkin	Vegetable	477	Boston Marrow
	Vegetable	494	Mixta Squash
	Vegetable	485	Crookneck Pumpkin
	Vegetable	493	Marrow
	Oil	334	African Pumpkin
Purging Croton	Medicinal	291	Purging Croton
Purple Bean	Forage, Pasture & Feed Grain	179	Siratro
Purple Granadilla	Beverage	20	Passion Fruit
Purple Guava	Fruit	240	Strawberry Guava
Purple Nutsedge	Weed	512	Nutgrass
Purple Raspberry	Fruit	234	Purple Raspberry
Purple Vetch	Erosion Control	77	Purple Vetch
Purslane	Vegetable	507	Winter Purslane
	Ornamental & Lawn Grass	370	Purslane
Pusley	Ornamental & Lawn Grass	370	Purslane
Putch	Insecticide	260	Tuba Root
Pyrethrum	Insecticide	259	Pyrethrum
Pyrol	Vegetable	476	Black Gram
Quack Grass	Weed	513	Quack Grass
Quackgrass	Forage, Pasture & Feed Grain	171	Quackgrass
Quaker Comfrey	Forage, Pasture & Feed Grain	171	Prickly Comfrey
Quaking Grass	Ornamental & Lawn Grass	371	Quaking Grass
Quarters	Cereal	32	Fat Hen
Quebracho	Dye & Tannin	55	Quebracho
	Dye & Tannin	56	Red Quebracho
Queen Anne's Lace	Vegetable	480	Carrot
Queensland Arrowroot	Gum & Starch	250	Edible Canna
Quetembila	Fruit	205	Ceylon-Gooseberry
Quihuicha	Cereal	34	Inca-Wheat
Quince	Fruit	219	Indian Bael
	Gum & Starch	254	Quince
Quinine Tree	Medicinal	292	Quinine
Quinoa	Cereal	39	Quinoa
	Cereal	34	Inca-Wheat
Rabbit-Eye Blueberry	Fruit	235	Rabbit-Eye Blueberry
Radiata	Timber	441	Monterey Pine
Radish	Vegetable	500	Radish
Ragi	Cereal	32	Finger Millet
Raintree	Timber	449	Raintree
Rakkyo	Vegetable	500	Rakkyo
Ram-Begun	Medicinal	295	Sour-Relish Brinjal
Rambai	Fruit	235	Rambai
Rambutan	Fruit	235	Rambutan
Ramie	Fiber	126	Ramie

To Find a Crop with the Common Name	Look in the End-Use Category	On Page	Under the Crop Name
Samba	Timber	444	Obeche
Sand Bluestem	Erosion Control	78	Sand Bluestem
Sand Dropseed	Erosion Control	78	Sand Dropseed
Sand Lovegrass	Ornamental & Lawn Grass	371	Sand Lovegrass
Sand Oat	Weed	509	Black Oat
Sand Pear	Fruit	237	Sand Pear
Sandalwood	Essential Oil	91	East Indian Sandalwood
	Dye & Tannin	56	Red Sanderswood
Sandberg Bluegrass	Forage, Pasture & Feed Grain	176	Sandberg Bluegrass
Sanderswood	Dye & Tannin	56	Red Sanderswood
Sanduri	Cereal	41	Sanduri
Sann Hemp	Fiber	129	Sunn Hemp
Santonica	Medicinal	293	Russian Wormseed
Sapodilla	Gum & Starch	249	Chicle
Sapote	Fruit	227	Mexican-Apple
	Fruit	202	Black Sapote
	Fruit	237	Sapote
	Fruit	217	Green Sapote
Sappanwood	Dye & Tannin	57	Sappanwood
Sapucaia Nut	Ornamental & Lawn Grass	372	Sapucaja
Sapucaja	Ornamental & Lawn Grass	372	Sapucaja
Sarsaparilla	Medicinal	293	Sarsaparilla
Sarsaparillja	Medicinal	293	Sarsaparilla
Sassafraz	Essential Oil	99	Ocotea
Sativa	Forage, Pasture & Feed Grain	135	Alfalfa
Savanna Grass	Forage, Pasture & Feed Grain	188	Tropical Carpetgrass
Savory	Spice & Flavoring	403	Summer Savory
	Spice & Flavoring	407	Winter Savory
Savoy Cabbage	Vegetable	479	Cabbage
Scarlet Poppy	Medicinal	294	Scarlet Poppy
Scarlet Runner Bean	Ornamental & Lawn Grass	372	Scarlet Runner Bean
Scented	Medicinal	281	Garden Myrrh
Scotch Spearmint	Essential Oil	103	Scotch Spearmint
Sea Island Cotton	Fiber	127	Sea Island Cotton
Sea Rush	Ornamental & Lawn Grass	372	Sea Rush
Sea Wormwood	Medicinal	288	Maritime Wormwood
Seagrape	Fruit	237	Seagrape
Seakale	Vegetable	501	Seakale
Seaside Clover	Vegetable	502	Seaside Clover
Seaving	Oil	351	Seaving
Seed	Vegetable	496	Oblique-Seed Jackbean
	Oil	346	Niger-Seed
Seequa	Fruit	196	Angled Luffa
Seje Ungurahuay	Oil	348	Pataua
Senegal Gum	Gum & Starch	251	Gum Arabic
Senegal Rosewood	Dye & Tannin	58	Senegal Rosewood
Senna	Medicinal	269	Alexandrian Senna
Senna Coffee	Beverage	19	Negro-Coffee
Sere Grass	Essential Oil	109	West Indian Lemongrass
Serendipity-Berry	Sweetener	411	Serendipity-Berry

574

INDEX   B.

ALPHABETICAL LISTING
OF CROPS
BY SCIENTIFIC NAME

To Find a Crop with the Scientific Name	Look in the End-Use Category	On Page	Under the Crop Name
Agropyron trachycaulum	Forage, Pasture & Feed Grain	179	Slender Wheatgrass
Agrostis alba	Forage, Pasture & Feed Grain	153	Creeping Bent-Grass
Agrostis canina	Ornamental & Lawn Grass	377	Velvet Bentgrass
Agrostis gigantea	Forage, Pasture & Feed Grain	151	Common Bent-Grass
Agrostis nigra	Forage, Pasture & Feed Grain	151	Common Bent-Grass
Agrostis palustris	Forage, Pasture & Feed Grain	153	Creeping Bent-Grass
Agrostis stolonifera var. palustris	Forage, Pasture & Feed Grain	153	Creeping Bent-Grass
Agrostis tenuis	Forage, Pasture & Feed Grain	150	Colonial Bentgrass
Agrostis vulgaris	Forage, Pasture & Feed Grain	150	Colonial Bentgrass
Albizia falcata	Timber	418	Batai
Albizia falcataria	Timber	418	Batai
Albizia lebbek	Timber	453	Siris
Albizia moluccana	Timber	418	Batai
Aleurites fordii	Oil	355	Tung-Oil-Tree
Aleurites moluccana	Oil	338	Candlenut
Aleurites montana	Oil	345	Mu-Oil-Tree
Alhagi pseudalhagi	Forage, Pasture & Feed Grain	147	Camelthorn
Allium ampeloprasum	Spice & Flavoring	389	Elephant Garlic
Allium cepa	Vegetable	497	Onion
Allium chinense	Vegetable	500	Rakkyo
Allium fistulosum	Vegetable	507	Welsh Onion
Allium sativum	Spice & Flavoring	391	Garlic
Allium schoenoprasum	Vegetable	483	Chives
Allium tuberosum	Vegetable	482	Chinese Chives
Alocasia macrorrhiza	Gum & Starch	250	Giant Taro
Aloe barbadensis	Medicinal	270	Barbados Aloe
Aloe ferox	Medicinal	273	Cape Aloe
Aloe perryi	Medicinal	294	Socotrine Aloe
Alopecurus arundinaceus	Forage, Pasture & Feed Grain	153	Creeping Foxtail
Alopecurus pratensis	Forage, Pasture & Feed Grain	166	Meadow Foxtail
Alpinia galanga	Spice & Flavoring	392	Greater Galanga
Alpinia officinarum	Spice & Flavoring	394	Lesser Galanga
Alysicarpus nummularifolius	Forage, Pasture & Feed Grain	136	Alyceclover
Alysicarpus vaginalis	Forage, Pasture & Feed Grain	136	Alyceclover
Alyssum gordonii	Oil	343	Gordon Bladderpod
Amaranthus caudatus	Cereal	34	Inca-Wheat
Amaranthus cruentus	Cereal	42	Spanish Greens
Amaranthus hypochondriacus	Vegetable	499	Princess-Feather
Amaranthus lividus	Vegetable	476	Bondue
Amaranthus thunbergii	Vegetable	476	Bondue
Amaranthus tricolor	Vegetable	482	Chinese Amaranth
Ammi copticum	Medicinal	269	Ammi
Ammi visnaga	Medicinal	295	Spanish Carrot
Ammophila arenaria	Erosion Control	73	European Beachgrass
Ammophila breviligulata	Erosion Control	69	American Beachgrass
Amyris balsamifera	Essential Oil	83	Balsam Amyris
Amyris dentata	Fruit	207	Clausena dentata
Anacardium occidentale	Nut	323	Cashew
Ananas comosus	Fruit	232	Pineapple
Andropogon gayanus	Forage, Pasture & Feed Grain	157	Gamba Grass

To Find a Crop with the Scientific Name	Look in the End-Use Category	On Page	Under the Crop Name
Malus pumila	Fruit	196	Apple
Malus sylvestris	Fruit	196	Apple
Mammea americana	Fruit	226	Mammy-Apple
Mandragora officinarum	Medicinal	288	Mandrake
Mangifera indica	Fruit	226	Mango
Manihot dichotoma	Isoprenoid Resin & Rubber	266	Manicoba Rubber
Manihot dulcis	Energy	61	Cassava
Manihot esculenta	Energy	61	Cassava
Manihot glaziovii	Isoprenoid Resin & Rubber	263	Ceara-Rubber
Manihot melanobasis	Energy	61	Cassava
Manihot utilissima	Energy	61	Cassava
Manilkara bidentata	Isoprenoid Resin & Rubber	263	Balata
Manilkara ochras	Gum & Starch	249	Chicle
Manilkara zapota	Gum & Starch	249	Chicle
Manilkara zapotilla	Gum & Starch	249	Chicle
Maranta arundinacea	Gum & Starch	247	Arrowroot
Marjorana hortensis	Spice & Flavoring	404	Sweet Marjoram
Marrubium vulgare	Medicinal	284	Horehound
Matricaria chamomilla	Essential Oil	110	Wild Chamomile
Mauritia flexuosa	Fiber	125	Muriti
Mauritia vinifera	Fiber	125	Muriti
Medicago arabica	Erosion Control	80	Spotted Burclover
Medicago falcata	Forage, Pasture & Feed Grain	178	Sickle Medic
Medicago lupulina	Forage, Pasture & Feed Grain	142	Black Medic
Medicago orbicularis	Forage, Pasture & Feed Grain	146	Buttonclover
Medicago polymorpha	Forage, Pasture & Feed Grain	147	California Burclover
Medicago sativa	Forage, Pasture & Feed Grain	135	Alfalfa
Medicago sativa subsp. sativa	Forage, Pasture & Feed Grain	135	Alfalfa
Medicago scutellata	Forage, Pasture & Feed Grain	181	Snail Medic
Melaleuca cajuputi	Medicinal	273	Cajeput
Melanorrhoea usitata	Natural Resin	315	Burmese Lacquer
Melia azadirachta	Timber	443	Neem
Melicoccus bijugatus	Fruit	239	Spanish Lime
Melilotus alba	Forage, Pasture & Feed Grain	191	White Melilot
Melilotus indica	Forage, Pasture & Feed Grain	161	Indian Melilot
Melilotus officinalis	Forage, Pasture & Feed Grain	193	Yellow Melilot
Melilotus suaveolens	Forage, Pasture & Feed Grain	185	Sweetclover
Melinis minutiflora	Insecticide	258	Molasses Grass
Melissa officinalis	Essential Oil	105	Sweet Balm
Mentha arvensis	Essential Oil	90	Corn Mint
Mentha cardiaca	Essential Oil	103	Scotch Spearmint
Mentha pulegium	Essential Oil	92	European Pennyroyal
Mentha rotundifolia	Spice & Flavoring	379	Applemint
Mentha spicata	Essential Oil	104	Spearmint
Mentha viridis	Essential Oil	104	Spearmint
Mentha X gentilis	Essential Oil	103	Scotch Spearmint
Mentha X piperita	Essential Oil	100	Peppermint
Michelia champaca	Ornamental & Lawn Grass	361	Champac
Micrandra minor	Isoprenoid Resin & Rubber	264	Huemega
Momordica balsamina	Fruit	199	Balsam-Apple

To Find a Crop with the Scientific Name	Look in the End-Use Category	On Page	Under the Crop Name
Olea europaea	Oil	347	Olive
Onobrychis sativa	Forage, Pasture & Feed Grain	175	Sainfoin
Onobrychis viciaefolia	Forage, Pasture & Feed Grain	175	Sainfoin
Onobrychis viciifolia	Forage, Pasture & Feed Grain	175	Sainfoin
Ophiopogon japonicus	Ornamental & Lawn Grass	368	Lily Turf
Opuntia ficus-indica	Fruit	220	Indianfig
Opuntia occidentalis	Fruit	220	Indianfig
Orbignya barbosiana	Oil	335	Babassu
Orbignya cohune	Oil	340	Cohune Nut Palm
Orbignya speciosa	Oil	335	Babassu
Orchidocarpum grandiflorum	Fruit	246	Wooly Pawpaw
Origanum hirtum	Spice & Flavoring	397	Oregano
Origanum majorana	Spice & Flavoring	404	Sweet Marjoram
Origanum onites	Spice & Flavoring	400	Pot Majoram
Origanum virens	Spice & Flavoring	397	Oregano
Origanum vulgare	Spice & Flavoring	397	Oregano
Ornithopus sativus	Forage, Pasture & Feed Grain	177	Serradella
Oryza glaberrima	Cereal	27	African Rice
Oryza sativa	Cereal	40	Rice
Oryzopsis cuspidata	Cereal	35	Indian Rice
Oryzopsis hymenoides	Cereal	35	Indian Rice
Oryzopsis miliacea	Forage, Pasture & Feed Grain	180	Smilograss
Ovina var. duriuscula	Forage, Pasture & Feed Grain	159	Hard Fescue
Oxalis tuberosa	Tuber	467	Oca
Oxycoccus macrocarpon	Fruit	210	Cranberry
Pachyrhizus angulatus	Tuber	471	Yam Bean
Pachyrhizus erosus	Tuber	471	Yam Bean
Pachyrhizus tuberosus	Insecticide	257	Ajipo
Palaquium gutta	Isoprenoid Resin & Rubber	264	Gutta-Percha
Panax quinquefolius	Medicinal	282	Ginseng
Panicum antidotale	Forage, Pasture & Feed Grain	144	Blue Panicgrass
Panicum coloratum	Forage, Pasture & Feed Grain	180	Small Buffalo Grass
Panicum crusgalli	Forage, Pasture & Feed Grain	139	Barnyardgrass
Panicum frumentacea	Cereal	28	Billion Dollar Grass
Panicum hemitomon	Weed	512	Maidencane
Panicum maximum	Forage, Pasture & Feed Grain	158	Guineagrass
Panicum miliaceum	Cereal	39	Proso Millet
Panicum obtusum	Forage, Pasture & Feed Grain	189	Vine Mesquite
Panicum repens	Weed	513	Quack Grass
Panicum virgatum	Forage, Pasture & Feed Grain	185	Switchgrass
Papaver bracteatum	Medicinal	294	Scarlet Poppy
Papaver somniferum	Medicinal	289	Opium Poppy
Parkia filicoidea	Oil	333	African Locust Bean
Parkinsonia aculeata	Timber	432	Horsebean
Parmentiera aculeata	Fruit	210	Cuajilote
Parmentiera cerifera	Forage, Pasture & Feed Grain	148	Candletree
Parthenium argentatum	Isoprenoid Resin & Rubber	264	Guayule
Paspalum dilatatum	Forage, Pasture & Feed Grain	154	Dallisgrass
Paspalum notatum	Forage, Pasture & Feed Grain	138	Bahia Grass

To Find a Crop with the Scientific Name	Look in the End-Use Category	On Page	Under the Crop Name
Portulaca oleracea	Ornamental & Lawn Grass	370	Purslane
Portulacaria afra	Ornamental & Lawn Grass	364	Elephant Bush
Poterium sanguisorba	Weed	509	Burnet
Pouteria campechiana	Fruit	213	Egg-Fruit-Tree
Pouteria sapota	Fruit	237	Sapote
Pouteria viridis	Fruit	217	Green Sapote
Prosopis juliflora	Timber	439	Mesquite
Prosopis tamarugo	Timber	457	Tamarugo
Prunus acuminata	Erosion Control	70	Beach Plum
Prunus amygdalus	Nut	321	Almond
Prunus amygdalus var. amara	Essential Oil	85	Bitter Almond
Prunus armeniaca	Fruit	197	Apricot
Prunus avium	Fruit	241	Sweet Cherry
Prunus cerasifera	Ornamental & Lawn Grass	361	Cherry Plum
Prunus cerasus	Fruit	238	Sour Cherry
Prunus communis	Nut	321	Almond
Prunus domestica	Fruit	233	Plum
Prunus domestica subsp. insititia	Fruit	211	Damson Plum
Prunus dulcis	Nut	321	Almond
Prunus laurocerasus	Erosion Control	72	Cherry Laurel
Prunus maritima	Erosion Control	70	Beach Plum
Prunus myrobalan	Ornamental & Lawn Grass	361	Cherry Plum
Prunus persica	Fruit	231	Peach
Prunus salicina	Fruit	221	Japanese Plum
Prunus subcordata	Fruit	238	Sierra Plum
Prunus triflora	Fruit	221	Japanese Plum
Psathyrostachys juncea	Forage, Pasture & Feed Grain	174	Russian Wildrye
Pseudotsuga menziesii	Timber	427	Douglas-Fir
Psidium cattleianum	Fruit	240	Strawberry Guava
Psidium friedrichsthalianum	Fruit	209	Costa Rican Guava
Psidium guajava	Fruit	208	Common Guava
Psidium guineense	Fruit	203	Brazilian Guava
Psidium littorale	Fruit	240	Strawberry Guava
Psidium montanum	Fruit	246	Wild Guava
Psophocarpus tetragonolobus	Vegetable	507	Winged Bean
Pterocarpus erinaceus	Dye & Tannin	58	Senegal Rosewood
Pterocarpus santalinus	Dye & Tannin	56	Red Sanderswood
Pterocarpus soyauxii	Dye & Tannin	48	Barwood
Pueraria hirsuta	Tuber	466	Kudzu Vine
Pueraria lobata	Tuber	466	Kudzu Vine
Pueraria phaseoloides	Forage, Pasture & Feed Grain	188	Tropical Kudzu
Pueraria thunbergiana	Tuber	466	Kudzu Vine
Pueraria triloba	Tuber	466	Kudzu Vine
Punica granatum	Fruit	233	Pomegranate
Pyrethrum balsamita	Spice & Flavoring	387	Costmary
Pyrethrum cinerariifolium	Insecticide	259	Pyrethrum
Pyrus communis	Fruit	231	Pear
Pyrus pyrifolia	Fruit	237	Sand Pear
Pyrus serotina	Fruit	237	Sand Pear
Quercus suber	Miscellaneous	306	Cork Oak

To Find a Crop with the Scientific Name	Look in the End-Use Category	On Page	Under the Crop Name
Secale cereale	Cereal	40	Rye
Sechium edule	Vegetable	481	Chayote
Semecarpus anacardium	Dye & Tannin	54	Marking-Nut Tree
Sesamum alatum	Oil	354	Tacoutta
Sesamum indicum	Oil	351	Sesame
Sesamum orientale	Oil	351	Sesame
Sesamum radiatum	Oil	356	Wild Sesame
Sesbania aculeata	Fiber	119	Danicha
Sesbania bispinosa	Fiber	119	Danicha
Sesbania exaltata	Forage, Pasture & Feed Grain	177	Sesbania
Setaria italica	Cereal	32	Foxtail Millet
Setaria sphacelata	Forage, Pasture & Feed Grain	158	Golden Timothy Grass
Sicana odorifera	Ornamental & Lawn Grass	363	Curuba
Simarouba glauca	Oil	333	Aceituna
Simmondsia chinensis	Fat & Wax	114	Jojoba
Sinapis alba	Forage, Pasture & Feed Grain	192	White Mustard
Smilax aristolochiifolia	Medicinal	293	Sarsaparilla
Solanum aethiopicum	Vegetable	486	Cut-Egg Plant
Solanum aviculare	Fruit	198	Australian Nightshade
Solanum ferox	Medicinal	295	Sour-Relish Brinjal
Solanum gilo	Fruit	216	Gilo
Solanum hyporhodium	Fruit	207	Cocona
Solanum incanum	Vegetable	502	Sodom Apple
Solanum indicum	Vegetable	504	Tiberato
Solanum khasianum	Medicinal	295	Solanum khasianum
Solanum laciniatum	Fruit	222	Kangaroo Apple
Solanum macrocarpon	Vegetable	495	Native Eggplant
Solanum melongena	Vegetable	486	Eggplant
Solanum muricatum	Fruit	232	Pepino
Solanum nigrum	Miscellaneous	313	Wonderberry
Solanum phureja	Tuber	468	Potato
Solanum quitoense	Fruit	229	Naranjilla
Solanum torvum	Vegetable	504	Terongan
Solanum tuberosum	Tuber	468	Potato
Solenostemon rotundifolius	Tuber	469	Sudan Potato
Sorghastrum avenaceum	Forage, Pasture & Feed Grain	161	Indian Grass
Sorghastrum nutans	Forage, Pasture & Feed Grain	161	Indian Grass
Sorghum bicolor	Cereal	41	Sorghum
Sorghum caffrorum	Cereal	41	Sorghum
Sorghum dochna	Cereal	41	Sorghum
Sorghum drummondii	Cereal	41	Sorghum
Sorghum durra	Cereal	41	Sorghum
Sorghum guineense	Cereal	41	Sorghum
Sorghum halepense	Forage, Pasture & Feed Grain	163	Johnson Grass
Sorghum nervosum	Cereal	41	Sorghum
Sorghum sudanense	Forage, Pasture & Feed Grain	183	Sudangrass
Sorghum vulgare	Cereal	41	Sorghum
Sorghum X almum	Forage, Pasture & Feed Grain	151	Columbus Grass
Spartium junceum	Fiber	129	Spanish Broom
Sphenostylis ornata	Tuber	465	Giri Yam Pea

INDEX C.

INFORMATION SOURCES
BY
REFERENCE CODE

Code	Bibliographic Description

1. *A Handbook of the Principal Trees and Shrubs of the Ancon and Balboa Districts of Panama*, Canal Zone, Panama, 1925.

2. Adams, C.D. *Flowering Plants of Jamaica*. Glasgow: Robert MacLehose and Company, Limited, 1972.

3. Adamson, R.S., and T.M. Salter. *Flora of the Cape Peninsula*. Cape Town: Juta and Co. Ltd., 1950.

4. Allen, Betty. *Malayan Fruits*. Singapore: Donald Moore Press Ltd., 1967.

5. Allen, O.N., and Ethel K. Allen. *The Leguminosae: A Source Book of Characteristics, Uses and Nodulation*. Madison, Wisconsin: The University of Wisconsin Press, 1981.

6. Anderson, Dennis E. "Taxonomy and Distribution of the Genus Phalaris." *Iowa State Journal of Science*. 36(Aug. 15, 1981):1-96.

7. Andes, Louis Edgar. *Vegetable Fats and Oils*. London: Scott Greenwood and Son, 1925.

8. Andrews, F.W. *The Flowering Plants of the Sudan*. Arbroath, Scotland: T. Buncle and Company, Ltd., 1956.

9. Angevine, M.W. "Variations in the Demography of Natural Populations of the Wild Strawberries *Fragaria vesca* and *F. virginiana*." *The Journal of Ecology*. 71(Nov. 1983):959-974.

10. Arber, Agnes. *A Study of Cereal, Bamboo, and Grass*. Cambridge, England: University of Cambridge Press, 1934.

11. Australian Commonwealth of Forestry Bureau. *Forest Trees of Australia*. Department of Interior, 1957.

12. Ayensu, Edward S., and D.G. Coursey. "Yams--The Botany, Ethnobotany, Uses, and Possible Future (Uses) of Yams in West Africa." *Economic Botany*. 26(Oct.-Dec. 1972):301-318.

13. Backer, C.A., and Backhuizer Van Den Brink, Jr. *Flora of Java*. Groningen, Holland: Wolters-Noordhoff N.V., 1968.

14. Bailey, L.H. *Cyclopedia of American Agriculture*. New York: The Macmillan Co., 1907.

15. Bailey, L.H. *Manual of Cultivated Plants*. New York: The Macmillan Co., 1949.

16. Bailey, L.H. *The Standard Cyclopedia of Horticulture*, Vols. I, II, and III. New York: The Macmillan Co., 1928.

17. Bailey, L.H., and Ethel Zoe Bailey. *Hortus Third: A Concise Dictionary of Plants Cultivated in the United States and Canada*. New York: The Macmillan Publishing Co., 1976.

18. Barnard, Carolyn M., and Loren D. Potter. *New Mexico Grasses: A Vegetative Key*. Albuquerque: University of New Mexico Press, 1984.

19. Barnes, Arthur Chapman. *Sugar Cane*. New York: Interscience Publishers Inc., 1964.

20. Beale, William J. *Grasses of North America*, Vols. I and II. New York: Henry Holt and Company, 1896.

21. Bentley, Robert, and Henry Trimen. *Medicinal Plants*. London: J. & A. Churchill, 1880.

22. Berdahl, J.D., and R.E. Barker. "Selection for Improved Seedling Vigor in Russian Wild Ryegrass (*Psathyrostachys juncea*)." *Canadian Journal of Plant Science*. 64(Jan. 1984):131-138.

23. Bews, John W. *The World's Grasses*. New York: Longmans, Green & Co., 1929.

24. Bogdan, A.V. *Tropical Pasture and Fodder Plants*. New York: Longman Group Ltd., 1977.

25. Bor, N.L. *Grasses of Burma, India, Ceylon and Pakistan*. New York: Pergamon Press, 1960.

26. Bor, N.L. *Manual of Indian Forest Botany*. Delhi: J. J. Reprints, 1980.

27. Borg, John. *Descriptive Flora of the Maltese Islands, Including the Ferns and Flowering Plants*. Malta: Government Printing Office, 1927.

28. Boulos, Loutfy. *Medicinal Plants of North Africa*. Algonac, Michigan: Reference Publications, Inc., 1983.

29. Brandis, Dietrich. *Indian Trees*. London: Archibald Constable and Co. Ltd., 1907.

30. Bretaudeau, J. *Trees: A Guide to the Trees of Great Britain and Europe*. London: Paul Hamlyn Ltd., 1966.

31. Brouk, B. *Plants Consumed by Man*. London: Academic Press Inc., Ltd., 1975.

32. Brown, Harry Bates. *Cotton*. New York: McGraw-Hill Book Company, Inc., 1938.

33. Brown, Lauren. *Grasses, An Identification Guide*. Boston: Houghton Mifflin Company, 1979.

34. Bunch, Clarence, and Edd Roberts. "Seeding Native Grasses for Pasture and Erosion Control." Oklahoma Agriculture Extension Service, Circular 646, University of Oklahoma Press, June 1956.

35. Burbidge, Nancy T. *Australian Grasses*, Vol. I. Sydney: Angus and Robertson Ltd., 1966.

36. Burkill, I.H. *A Dictionary of the Economic Products of the Malay Peninsula*. London: Crown Agents for the Colonies, 1935.

618

Code	Bibliographic Description

37. Carter, Jack F., ed. "Sunflower: Science and Technology." *Agronomy*, No. 19. Madison, Wisconsin: American Society of Agronomy, Crop Science Society of America, Soil Science Society of America, Inc., Publishers, 1978.

38. Chittenden, Fred J., ed. *Dictionary of Gardening: A Practical and Scientific Encyclopedia of Horticulture.* Oxford: Clarendon Press, 1956.

39. Chow, Hang-Fan. *The Familiar Trees of Hopei.* Peking: Peking Natural History Bulletin, 1934.

40. Chun, Woon Young. *Chinese Economic Trees.* Shanghai: Commercial Press, Ltd., 1921.

41. Church, A.L. *Food Grains of India.* New Delhi: India Ajay Book Service, 1983.

42. Clapham, A.R., T.G. Tutin, and E.F. Warburg. *Flora of the British Isles.* Cambridge, England: Cambridge University Press, 1952.

43. Clarke, Charlotte Bringle. *Edible and Useful Plants of California.* Berkeley: University of California Press, 1977.

44. Cock, James H. "Cassava: A Basic Energy Source in the Tropics." *Science.* 218(Nov. 19, 1982):755-762.

45. Committee of Revision Pharmacopeia of the United States, 15th revision. Pennsylvania: United States Pharmacopoeial Convention, 1955.

46. Commonwealth Bureau of Horticulture and Plantation Crops. *Horticultural Abstracts.* Farnham Royal, England: Commonwealth Agricultural Bureau. 33(Sept. 1963):598.

47. Cottrell, I.W., and J.K. Baird. "Gums." *Encyclopedia of Chemical Technology*, ed. Kirk-Othmer, Vol. 12, 3rd ed. New York: John Wiley and Sons, Inc., 1980.

48. Coursey, D.G. *Yams: An Account of the Nature, Origins, Cultivation and Utilisation of the Useful Members of the Dioscoreacede.* London: Longmans, Green and Co. Ltd., 1967.

49. Crowe, Andrew. *A Field Guide to the Native Edible Plants of New Zealand, Including Those Plants Eaten by the Maori.* Auckland: William Collins Publishers Ltd., 1981.

50. Dalziel, J.M. *The Useful Plants of West Tropical Africa.* London: The Crown Agents for the Colonies, 1937.

51. Dempsey, James M. *Fiber Crops.* Gainsville, Florida: The University of Florida Press, 1975.

52. Dewey, D.R., and C. Hsiao. "A Cytogenetic Basis for Transferring Russian Wildrye from Elymus to Psathrostachys (*juncea, Elymus junceus*, Interspecific Hybridization)." *Crop Science.* 23(Jan.-Feb. 1983):123-126.

53. Dirr, Michael A. *Manual of Woody Landscape Plants.* Champaign, Illinois: Stipes Publishing Company, 1977.

54. Dore, William G., and J. McNeil. *Grasses of Ontario.* Ottawa: Canadian Government Publishing Centre, 1980.

55. Drury, Heber. *The Useful Plants of India.* Madras: Asylum Press, 1858.

56. Duke, James A. *Handbook of Legumes of World Economic Importance.* New York: Plenum Press, 1981.

57. Duthie, J.F. *Flora of the Upper Gangetic Plain.* Botanical Survey of India, Vols. I and II. Reprinted under the authority of the Government of India, 1960.

58. Duthie, J.F. *The Fodder Grasses of Northern India.* Jodhpur, India: Scientific Publishers, 1978.

59. Elias, Thomas S. *The Complete Trees of North America. Field Guide and Natural History.* New York: Van Nostrand Reinhold Company, 1980.

60. Elliot, Stephen. *A Sketch of the Botany of South Carolina and Georgia*, Vol. II. New York: Hafner Publishing Company, 1971.

61. Everett, Thomas H. *The New York Botanical Garden Illustrated Encyclopedia of Horticulture*, Vol. 6. New York: Garland Publishing Co., 1981.

62. Fernald, Merritt Lyndon, and Alfred Charles Kinsey. *Edible Wild Plants of Eastern North America.* Cornwall-on-Hudson, New York: Idlewild Press, 1943.

63. First International Sago Symposium. "The Equatorial Swamp as a Natural Resource." ed. Koonlin Tan, Kuala Lumpur: University of Malaysia Press, 1977.

64. Foster, Gertrude. *Herbs For Every Garden.* New York: E.P. Dutton and Company, Inc., 1966.

65. Fraser, Samuel. *American Fruits.* New York: Orange Judd Publishing Company, Inc., 1931.

66. Freeman, William George. *The Useful and Ornamental Plants in Trinidad and Tobago.* Trinidad: Government Printery, 1969.

67. Gamble, J.S. *Manual of Indian Timbers.* London: Sampson Low, Marston and Company Limited, 1902.

68. Gamble, J.S. *The Bambuseae of British India.* Calcutta: Bengal Secretariat Press, 1896.

69. Genders, Roy., ed. *Pears Encyclopedia of Gardening Fruit and Vegetables.* London: Pelham Books Ltd., 1973.

Code	Bibliographic Description

70. Gentry, Howard S., and R.W. Miller. "The Search for New Industrial Crop IV. Prospectus of Limnanthes." *Economic Botany.* 19(Jan.-Mar. 1965):25-32.

71. Gildemeister, Edward, and Friedrich Hoffman. *The Volatile Oils,* Vol. III. New York: John Wiley and Sons Inc., 1922.

72. Gill, N.T., and K.C. Vear. *Agricultural Botany 1. Dicotyledonous Crops.* London: Gerald Duckworth and Company, Ltd., 1980.

73. Godin, V.J., and P.C. Spensley. *No. 1 Oils and Oilseeds.* London: Tropical Products Institute, Foreign and Commonwealth (Overseas Development Administration), 1971.

74. Gould, Frank, and Box Thadis. *Grasses of the Texas Coastal Bend.* College Station, Texas: Texas A & M University Press, 1965.

75. Graf, Alfred Byrd. *Tropica: Color Cyclopedia of Exotic Plants and Trees for Tropics and Subtropics.* New Jersey: Roehrs Co., 1978.

76. Grounds, Roger. *Ornamental Grasses.* London: Pelham Books Ltd., 1979.

77. Guenther, Ernest. *Essential Oils,* Volumes I-VI. New York: D. Van Nostrand Company, Inc., 1952.

78. Gustafson, J.P.; E.N. Larter, F.J. Zillinsky, and M. Freuhm. "Carmen Triticale (New Cultivar *X Triticosecale*), Agronomic Performance." *Canadian Journal of Plant Science.* 62(Jan. 1982):221-222.

79. Haarer, A.E. *Modern Coffee Production.* London: Leonard Hill Books Limited, 1958.

80. Hacker, J.B., and J.C. Tothill. *Grasses of Southern Queensland.* St. Lucia, Queensland: University of Queensland Press, 1983.

81. Halpin, Anne Moyer., ed. *Unusual Vegetables: Something New for This Year's Garden.* Emmaus, Pennsylvania: Rodale Press, 1978.

82. Harlan, Jack, C.E. Denman, and W.C. Elder. "Weeping Lovegrass." Forage Crop Leaflets, No. 16, Oklahoma Agricultural Experiment Station, February 1953.

83. Harrington, H.D. *Edible Native Plants of Rocky Mountains.* Albuquerque: University of New Mexico Press, 1967.

84. Harrison, S.G., G.B. Masefield, and Michael Wallis. *The Oxford Book of Food Plants.* London: Oxford University Press, 1969.

85. Hartley, C.W.S. *The Oil Palm,* 2nd ed. New York: Longman Group Ltd., 1977.

86. Haselwood, E.L., and G.G. Motter., eds. *Handbook of Hawaiian Weeds,* 2nd ed. Honolulu: University of Hawaii Press, 1983.

87. Hayes, W.B. *Fruit Growing in India.* Allahabad, India: The Indian Universities Press, 1966.

88. Hedrick, U.P., et al. *The Cherries of New York.* Albany, New York: J.B. Lyon Company & State Printers, 1915.

89. Hedrick, U.P., et al. *The Plums of New York.* Albany, New York: J.B. Lyon Company & State Printers, 1911.

90. Helmut, Muhlberg. *The Complete Guide to Water Plants.* German Democratic Republic: E.P. Publishing Limited, 1982.

91. Heriteau, Jacqueline. *Herbs: How to Grow and Use Them.* New York: Grosset and Dunlap, Inc., 1975.

92. Herklots, G.A.C. *Vegetables in South-East Asia.* Hong Kong: South China Morning Post Ltd., 1972.

93. Higgins, J.J., et al. "Agronomic Evaluation of Prospective New Crop Species II. The American Limnanthes." *Economic Botany.* 25(Jan.-Mar. 1971):44-54.

94. Hill, Albert F. *Economic Botany: A Textbook of Useful Plants and Plant Products.* New York: McGraw-Hill Book Company, Inc., 1937.

95. Hillier, H.G. *Hillier's Manual of Trees and Shrubs.* London: Hillier and Sons, 1981.

96. Hitchcock, A.S. *Manual of the Grasses of the United States.* Washington: United States Government Printing Office, 1950.

97. Holloway, H.L.O. "Seed Propagation of *Dioscoreophyllum cumminsii,* Source of an Intense Natural Sweetener." *Economic Botany.* 31(Jan.-Mar. 1977):47-50.

98. Hooker, J.D. *The Flora of British India.* London: L. Reeve and Co., 1885.

99. Hora, Bayard., ed. *The Oxford Encyclopedia of Trees of the World.* Melbourne: Oxford University Press, 1981.

100. Houghton, A.D. *The Cactus Book.* New York: The Macmillan Company, 1930.

101. Hoyt, Roland Stewart. *Check Lists for Ornamental Plants of Subtropical Regions.* San Diego: Livingston Press, 1958.

102. Humphreys, L.R. *Tropical Pastures and Fodder Crops.* London: Longman Group, 1978.

103. Huxley, Anthony, and Oleg Polunin. *Flowers of the Mediterranean.* London: Chatto and Windus Ltd., 1965.

Code	Bibliographic Description

104. Ingalls, John J. "Handbook of Colorado Native Grasses." Extension Service Bulletin 450-A, Colorado State University, Fort Collins, Extension Service, Nov. 1958.

105. Irvine, Frederick Robert. *West African Crops.* Oxford, England: University Press, 1969.

106. Jain, S.K., and E.H. Abuelgasim. "Some Yield Components and Ideotype Traits in Meadowfoam and New Industrial Oil Crop." *Euphytica.* 30(June 1981):437-443.

107. Jamieson, George S. *Vegetable Fats and Oils.* New York: The Chemical Catalog Co. Inc., 1932.

108. Jaynes, Richard A., ed. *Handbook of North American Nut Trees.* New Haven, Connecticut: Northern Nut Grower's Association, 1969.

109. Judd, Ira. *Handbook of Tropical Forage Grasses.* New York: Garland Publishing, Inc., 1979.

110. Kaul, B.L., and C.K. Atal. "Grow *Solanum khasianum* for Steroids." *Indian Farming.* 27(Mar. 1978):15-16.

111. Kay, D.E. *Root Crops.* Tropical Products Institute, London: Commonwealth Office, 1973.

112. Khedir, K.D., and Fred W. Roeth. "Velvetleaf (*Abutilon theophrasti*) Seed Populations in Six Continuous Corn (*Zea mays*) Fields." *Weed Science.* 29(1981):485.

113. Kirby, R.H. *Vegetable Fibres: Botany, Cultivation and Utilization.* New York: Leonard Hill Publishers, 1963.

114. Kirk, Donald R. *Wild Edible Plants of the Western United States.* Healdsburg, California: Naturegraph Publishers, 1970.

115. Kitagawa, I., H.K. Wang, M. Saito, and M. Yoshikawa. "Saponin and Sapogenol, XXXIII. Chemical Constituents of the Seeds of *Vigna angularis* (Willd.) Ohwi et Ohashi (3). Adzuki Saponins V and VI." *Chemical and Pharmaceutical Bulletin.* 31(Feb. 1983):683- 688.

116. Kochhar, S.L. *Economic Botany in the Tropics.* Delhi: Macmillan India Ltd., 1981.

117. Kohel, R.J., and C.F. Lewis. *Cotton.* Madison, Wisconsin: American Society of Agronomy, Inc., Crop Science Society of America, Inc., Soil Science Society of America, Inc., Publishers, 1984.

118. Kretschmer, A.E., Jr., and G.H. Snyder. "Production and Quality of Limpograss for Use in the Subtropics." *Agronomy Journal.* 71(Jan.-Feb. 1979):37-41.

119. Kunkel, G. *Plants for Human Consumption.* Koenigstein, West Germany: Koeltz Scientific Books, 1984.

120. Kurup, P.N.V., V.N.K. Ramadas, and Shri Joshi Prajapti. *Handbook of Medicinal Plants.* Central Council for Research. New Delhi: Deepak Art Press, 1979.

121. Lassak, E.V., and T. McCarthy. *Australian Medicinal Plants.* North Ryde, Australia: Methuen Australia Pty. Ltd., 1983.

122. Lazarides, M. *The Tropical Grasses of Southeast Asia.* Hirshberg, Germany: Strauss and Cramer, 1980.

123. Lee, Yong No. *Manual of the Korean Grasses.* Seoul: Ehwa Woman's University Press, 1966.

124. Lewkowitsch, J. *Chemical Technology and Analysis of Oils, Fats, and Waxes,* Vol. II, 6th ed. London: Macmillan and Company, Ltd., 1922.

125. Lisboa, J.C. *List of Bombay Grasses and Their Uses.* Delhi: Periodical Experts Book Agency, 1978.

126. Litzenburger, Samuel C., ed. *Guide for Field Crops in the Tropics and the Subtropics.* Agency for International Development, Office of Agriculture, Technical Assistance Bureau, Agency for International Development, Washington D.C.: U.S. Government Printing Office, 1974.

127. Lord, Ernest E., and J.H. Willis. *Shrubs and Trees for Australian Gardens.* Melbourne: Lothian Publishing Company Pty. Ltd., 1982.

128. Lovelock, Yann. *The Vegetable Book: An Unnatural History.* New York: St. Martin's Press, 1972.

129. Macleod, Dawn. *A Book of Herbs.* London: Gerald Duckworth and Co. Ltd., 1968.

130. Maheshwari, J.K. *The Flora of Delhi.* Calcutta: N.K. Gossain and Co. Private Ltd., 1963.

131. Maino, Evelyn, and Frances Howard. *Ornamental Trees: An Illustrated Guide to Their Selection and Care.* Berkeley: University of California Press, 1962.

132. Martin, Franklin W., *Handbook of Tropical Food Crops.* Boca Raton, Florida: CRC Press, Inc., 1984.

133. Martin, Franklin W., and Ruth Ruberte. *Edible Leaves of the Tropics.* Puerto Rico: Antillian College Press, 1975.

134. Martin, William C., and Charles R. Hutchins. *Spring Wildflowers of New Mexico.* Albuquerque: University of New Mexico Press, 1984.

135. Masefield, Geoffrey Bussell. *The Oxford Book of Food Plants.* London: Oxford University Press, 1969.

136. Mauersberger, Herbert, ed. *Textile Fibers.* New York: John Wiley and Sons, 1947.

137. McConnell, D.B., and R.W. Henley. "'Springer' and 'Myers' Ferns." *Foliage Digest.* (Sept. 1983):11.

138. McKee, Roland. "The Vetches." *Forages: The Science of Grassland Agriculture.,* ed. Heath Hughes, Metcalf, Iowa: Iowa State College Press, 1951.